INTERNATIONAL ENE

WORLD ENERGY OUTLOOK 2007

China and India Insights

INTERNATIONAL ENERGY AGENCY

The International Energy Agency (IEA) is an autonomous body which was established in November 1974 within the framework of the Organisation for Economic Co-operation and Development (OECD) to implement an international energy programme.

It carries out a comprehensive programme of energy co-operation among twenty-six of the OECD thirty member countries. The basic aims of the IEA are:

- To maintain and improve systems for coping with oil supply disruptions.
- To promote rational energy policies in a global context through co-operative relations with non-member countries, industry and international organisations.
- To operate a permanent information system on the international oil market.
- To improve the world's energy supply and demand structure by developing alternative energy sources and increasing the efficiency of energy use.
- To promote international collaboration on energy technology.
- To assist in the integration of environmental and energy policies.

The IEA member countries are: Australia, Austria, Belgium, Canada, Czech Republic, Denmark, Finland, France, Germany, Greece, Hungary, Ireland, Italy, Japan, Republic of Korea, Luxembourg, Netherlands, New Zealand, Norway, Portugal, Spain, Sweden, Switzerland, Turkey, United Kingdom and United States. The Slovak Republic and Poland are likely to become member countries in 2007/2008. The European Commission also participates in the work of the IEA.

ORGANISATION FOR ECONOMIC CO-OPERATION AND DEVELOPMENT

The OECD is a unique forum where the governments of thirty democracies work together to address the economic, social and environmental challenges of globalisation. The OECD is also at the forefront of efforts to understand and to help governments respond to new developments and concerns, such as corporate governance, the information economy and the challenges of an ageing population. The Organisation provides a setting where governments can compare policy experiences, seek answers to common problems, identify good practice and work to co-ordinate domestic and international policies.

The OECD member countries are: Australia, Austria, Belgium, Canada, Czech Republic, Denmark, Finland, France, Germany, Greece, Hungary, Iceland, Ireland, Italy, Japan, Republic of Korea, Luxembourg, Mexico, Netherlands, New Zealand, Norway, Poland, Portugal, Slovak Republic, Spain, Sweden, Switzerland, Turkey, United Kingdom and United States.
The European Commission takes part in the work of the OECD.

World leaders have pledged to act to change the future shape of the global energy economy. Since the 2006 edition of the *WEO*, some new policies have been put in place to that effect. Yet in the Reference Scenario in this year's *WEO*, which takes these new policies into account, projected global energy demand in 2030 is higher than before and the supply and emissions trends are worsening. What is going on?

One key answer is sustained high levels of economic growth in the new giants of the world economy. China and India together account for nearly half of the entire growth in world energy demand between 2005 and 2030. China is likely to have overtaken the United States to become the world's largest emitter of energy-related carbon dioxide this year and, by 2015, India will be the third-largest emitter. By around 2010, China will overtake the United States to become the world's largest consumer of energy. In 2030, India will be the third-largest oil importer in the world. Over the period to 2030, China will install more new electricity generating capacity than exists in the United States today.

China and India *need* to sustain a phenomenal rate of economic growth. There are still over 400 million people in India without access to electricity. Access to clean burning fuels for cooking and space heating in rural China is still very limited, despite the near-total success of its rural electrification programme. In both countries, the aspirations of a burgeoning middle class are driving social and economic change. There can be no moral grounds for expecting China and India selectively to curb their economic growth simply because world energy demand is rising unacceptably, with associated risks of supply interruptions, high prices and damage to the environment. These are global problems to be tackled on a global basis.

How those problems might be tackled is illustrated in the Alternative Policy Scenario, which forms an important part of the analysis in this book. Known means exist to cut energy demand and change the fuel mix. Global energy-related CO_2 emissions could be nearly 20% lower by 2030, having levelled off in the 2020s. A volume of oil equal to the entire current oil output of the United States, Canada and Mexico can be removed from world demand by 2030. China and India are increasingly demonstrating their recognition of the need to act – for example, through their commitment to greater energy efficiency, more renewables and cleaner coal technology – with other countries to make the energy future sustainable.

To attain the much more ambitious long-term objective of stabilising the concentration of greenhouse gases in the atmosphere, the measures considered would still not be enough; but the possible implications of that objective, too, are examined in the 450 Stabilisation case, set out in Chapter 5.

On the other hand, growth in the world's economic tigers could be still higher than we have assumed. We set out the consequences of that. They are not all bad. Most countries outside China and India would benefit economically, despite the feedback to higher energy demand and prices.

China and India account for a share of global primary demand which is growing at a phenomenal rate – but it will still be no higher than some 30% in 2030. Through this *Outlook*, the IEA seeks to communicate both parts of this message – the significance of the growth of energy demand in China and India, but also their place in world total demand and their modest use of energy per capita – and then to help realise the global co-operation which, alone, can create a sustainable energy future.

I am immensely proud to have the opportunity to present this latest volume in the acclaimed *WEO* series, a series which has been so carefully nurtured by my predecessor, Claude Mandil. I pay tribute to him, to Fatih Birol, who has again directed with talent his excellent *WEO* team, and to the many others who have contributed to this work. It is particularly gratifying that this edition has been the occasion for close collaboration between the Chinese and Indian authorities and the IEA. This is a relationship which symbolises the interdependence of the global energy community. It is one which I shall do my best to safeguard and develop, hopefully paving the way, with the support of all the governments concerned, to an ultimate objective of their future membership of the International Energy Agency.

Nobuo Tanaka
Executive Director

This study was prepared by the Economic Analysis Division of the International Energy Agency in co-operation with other divisions of the IEA. The Director of the Long-Term Office, Noé van Hulst, provided guidance and encouragement during the project. The study was designed and managed by Fatih Birol, Head of the Economic Analysis Division (EAD). Other members of EAD who were responsible for bringing the study to completion include: Maria Argiri, Raffaella Centurelli, Michael-Xiaobao Chen, Laura Cozzi, Paul Dowling, Hideshi Emoto, Audrey Lee, Lorcan Lyons, Teresa Malyshev, Trevor Morgan, Pawel Olejarnik, Olivier Rech, Fabien Roques and Richard Schimpf. Robert Priddle carried editorial responsibility. Claudia Jones and Lillie Kee provided essential support.

Dagmar Graczyk and Jonathan Sinton (Office of Global Energy Dialogue), Francois Nguyen and Brian Ricketts (Long-Term Office), and Kamel Bennaceur (Energy Technology Office) were part of the *Outlook* team. Viviane Consoli proofread the text.

Our analysis of China and India would not have been possible without the committed support of the Energy Research Institute of the National Development and Reform Commission of China and The Energy and Resources Institute of India. This support was given throughout the project, in relation to all its aspects: data, modelling, policy analysis and evaluation of future prospects. We are deeply grateful.

Centre International de Recherche sur l'Environnement et le Développement, International Institute for Applied Systems Analysis and Technical University of Vienna provided additional modelling support.

The study also benefited from input provided by other IEA colleagues, particularly: Richard Baron, Sankar Bhattacharya, Aad van Bohemen, Toril Bosoni, Richard Bradley, Amos Bromhead, Barbara Buchner, Pierpaolo Cazzola, Ian Cronshaw, Muriel Custodio, Bob Dixon, Lawrence Eagles, Ann Eggington, Jason Elliott, Mark Ellis, Paolo Frankl, David Fyfe, Rebecca Gaghen, Jean-Yves Garnier, Dolf Gielen, Neil Hirst, Didier Houssin, Catherine Hunter, Satoru Koizumi, Jens Laustsen, Eduardo Lopez, David Martin, Isabel Murray, Yo Osumi, Samantha Olz, Antonio Pflüger, Cédric Philibert, Jacek Podkanski, Roberta Quadrelli, Riccardo Quercioli, Loretta Ravera, James Ryder, Bertrand Sadin, Sophie Schlondorff, Maria Sicilia-Salvadores, Dan Simmons, Ralph Sims, Sylvie Stephan, Yuichiro Torikata, Karen Tréanton, Paul Waide and Julius Walker.

The work could not have been achieved without the substantial support and co-operation provided by many government bodies, international organisations, energy companies and research institutions worldwide, notably:

Asian Development Bank, European Commission, Foreign and Commonwealth Office of the United Kingdom, IEA Clean Coal Centre, IHS Energy, International Monetary Fund, Japanese Ministry of Economy, Trade and Industry, Ministry of Economic Development of New Zealand, Norwegian Ministry of Foreign Affairs, Organization of the Petroleum Exporting Countries, Schlumberger, Statoil, The Institute of Energy Economics, Japan, UNDP China, US Department of Energy and World Bank.

Many international experts provided input, commented on the underlying analytical work and reviewed early drafts of each chapter. Their comments and suggestions were of great value. They include:

Part A: Global Energy Prospects – Impact of Developments in China and India

Tera Allas	Department of Trade and Industry, UK
Jun Arima	Ministry of Economy, Trade and Industry, Japan
Paul Bailey	Department of Business, Enterprise and Regulatory Reform, UK
J L Bajaj	The Energy and Resources Institute, India
Paul Baruya	IEA Clean Coal Centre, UK
Jean-Paul Bouttes	EDF, France
Günter Brandes	Federal Ministry of Economics and Technology, Germany
Klaus Brendow	World Energy Council, UK
Nick Butler	Cambridge University, UK
Antonio Canseco	FUELEC, Spain
Graham Chapman	Energy Edge, UK
Xavier Chen	BP, China
Preston Chiaro	Rio Tinto, UK
Paul Conway	OECD, France
Laurent Corbier	AREVA, France
Sylvie Cornot-Gandolphe	ATIC Services, France
Ananda Covindassamy	World Bank, US
Renaud Crassous	AgroParisTech-CIRED, France
William Davie	Schlumberger, France
Yvo De Boer	UNFCCC, Germany

Helen Dickinson	Department of Business Enterprise, and Regulatory Reform, UK
Carmen Difiglio	Department of Energy, US
Simon Dilks	Department of Economy, Trade and Industry, UK
Jane Ellis	OECD, France
Jorgen Elmeskov	OECD, France
Robert Fryklund	IHS Inc., US
Stan Gordelier	OECD Nuclear Energy Agency, France
Nadir Guerer	OPEC, Austria
Céline Guivarch	ENPC-CIRED, France
Reinhard Haas	Vienna University of Technology, Austria
Andrew Hayman	IHS Inc., US
Trevor Houser	China Strategic Advisory, US
Brian Heath	IEA Coal Industry Advisory Board, France
Mikkal Herberg	National Bureau of Asian Research, US
Bert Hofman	World Bank, China
John Houghton	IPCC, Switzerland
Jean-Charles Hourcade	CNRS-EHESS-CIRED, France
Jan-Hein Jesse	Shell, Netherlands
Paul Joskow	Massachusetts Institute of Technology, US
Tor Kartevold	Statoil, Norway
Masami Kojima	World Bank, US
Ken Koyama	The Institute of Energy Economics, Japan
Ranjit J. Lamech	World Bank, US
Alessandro Lanza	Eni SpA, Italy
Stephane Lemoine	ATIC Services, France
Flynt Leverett	New America Foundation, US
Kseniya Lvovsky	World Bank, US
Michael Lynch	Strategic Energy & Economic Research, US
Ritu Mathur	The Energy and Resources Institute, India
Joan McNaughton	Alstom, UK
Bert Metz	IPCC, Switzerland
Kentaro Morita	Ministry of Economy, Trade and Industry, Japan
Nebosa Nakicenovic	Vienna University of Technology, Austria
Pierre Noel	Cambridge University, UK
Peter Nolan	Cambridge University, UK
Francesco Olivieri	ENEL, Italy
Kyran O'Sullivan	World Bank, US

Rajendra Pachauri	The Energy and Resources Institute, India
Keun-Wook Paik	Oxford Institute for Energy Studies, UK
Gustav Resch	Vienna University of Technology, Austria
Derek Riley	The Conference Board, France
Hans-Holger Rogner	International Atomic Energy Agency, Austria
Kenneth Rogoff	Harvard University, US
Jamal Saghir	World Bank, US
Olivier Sassi	ENPC-CIRED, France
Hans-Wilhelm Schiffer	RWE Power, Germany
Robert Schock	World Energy Council, UK
Rajendra Shende	UNEP, France
Rob Sherwin	Foreign Commonwealth Office, UK
Bill Simes	Barlow Jonker, Australia
Jonathan Stern	Oxford Institute for Energy Studies, UK
Anil Terway	Asian Development Bank, Philippines
David Tyrrell	Department of Business, Enterprise and Regulatory Reform, UK
Woochong Um	Asian Development Bank, Philippines
Franck Verrastro	Centre for Strategic and International Studies, US
David Victor	Stanford University, US
Roberto Vigotti	Inergia, Italy
Detlef van Vuuren	RIUM, Netherlands
Henri-David Waisman	ENS Lyon-CIRED, France
Junhui Wu	World Bank, US
Henning Wuester	UNFCCC, Germany
Salman Zaheer	World Bank, US

Part B: China's Energy Prospects

Quan Bai	Energy Research Institute, NDRC, China
Adam S. Chambers	International Institute for Applied Systems, Austria
Xavier Chen	BP, China
Preston Chiaro	Rio Tinto, UK
Yuan-sheng Cui	Institute of Technical Information for Building Materials Industry of China, China
Andrea DeAngelis	UNDP, China
An Feng	Innovation Center for Energy and Transportation, US

Tarhan Feyzioğlu	International Monetary Fund, China
John Fu	Shell, China
Hu Gao	Energy Research Institute, NDRC, China
Wenke Han	Energy Research Institute, NDRC, China
Dave Hawkins	Climate Center, NRDC, US
Ping He	UNDP, China
Thomas Heller	Stanford University, US
Mikkal E. Herberg	National Bureau of Asian Research, US
Bert Hofman	World Bank, China
Trevor Houser	China Strategic Advisory, US
Angang Hu	Tsinghua University, China
Xiulian Hu	Energy Research Institute, NDRC, China
Huiming Gong	Energy Foundation, China
Yuantao Guo	Gao Hua Securities, China
Joseph Jacobelli	Merrill Lynch, China
Jan-Hein Jesse	Shell, Netherlands
Eric Jia	Gao Hua Securities, China
Yi Jiang	Tsinghua University, China
Kejun Jiang	Energy Research Institute, NDRC, China
Junfeng Li	Energy Research Institute, NDRC, China
Boqiang Lin	Xiamen University, China
Michael Lynch	Strategic Energy & Economic Research, US
Cheng Niu	Energy Research Institute, NDRC, China
Peter Nolan	University of Cambridge, UK
Keun-Wook Paik	Oxford Institute for Energy Studies, UK
Zhongyuan Shen	The Institute of Energy Economics, Japan
Pengfei Shi	Chinese Wind Energy Association, China
Jonathan Stern	Oxford Institute for Energy Studies, UK
Paul Suding	Renewable Energy Policy Network for the 21st Century, France
Anil Terway	Asian Development Bank, Philippines
Siqiang Wang	The Office of National Energy Leading Group, China
Xuejun Wang	Peking University, China
Dingming Xu	The Office of National Energy Leading Group, China
Yufeng Yang	Energy Research Institute, NDRC, China
David Yip	Merrill Lynch, China
Kunmin Zhang	Chinese Society for Sustainable Development, China

Jianping Zhao	World Bank, China
Nan Zhou	Lawrence Berkeley National Laboratory, US
Xing Zhuang	Energy Research Institute, NDRC, China

Part C: India's Energy Prospects

P. K. Agarwal	The Energy and Resources Institute, India
J. L. Bajaj	The Energy and Resources Institute, India
R. K. Batra	The Energy and Resources Institute, India
S. C. Bhattacharyya	Centre for Energy, Petroleum and Mineral Law and Policy, University of Dundee, UK
Ranjan Bose	The Energy and Resources Institute, India
Adam S. Chambers	International Institute for Applied Systems Analysis, Austria
Paul Conway	OECD, France
Amit Garg	UNEP Risø Centre, Denmark
Hiroyuki Ishida	The Institute of Energy Economics, Japan
Paul Joskow	Massachusetts Institute of Technology, US
Rekha Krishnan	The Energy and Resources Institute, India
Ritu Mathur	The Energy and Resources Institute, India
Akhilesh Negi	National Engineering and General Industries, India
V. S. Okhde	Indian Oil Corporation, India
Rajendra Pachauri	The Energy and Resources Institute, India
Shonali Pachauri	International Institute for Applied Systems Analysis, Austria
John Paffenbarger	Constellation Energy, US
K. Ramanathan	The Energy and Resources Institute, India
Ashis Rathee	National Engineering and General Industries, India
Lee Schipper	World Resources Institute, US
Surya Sethi	Planning Commission, India
Allen Shaw	Oxford University, UK
Rajendra Shende	UNEP, France
K. Shankar Narayanan	Consultant, India
P. R. Shukla	Indian Institute of Management, India
P. V. Sridharan	The Energy and Resources Institute, India
Leena Srivastava	The Energy and Resources Institute, India
David Victor	Stanford University, US

The individuals and organisations that contributed to this study are not responsible for any opinions or judgments contained in this study. Any errors and omissions are solely the responsibility of the IEA.

**Comments and questions are welcome
and should be addressed to:**

Dr. Fatih Birol
Chief Economist
Head, Economic Analysis Division
International Energy Agency
9, rue de la Fédération
75739 Paris Cedex 15
France

Telephone   (33-1) 4057 6670
Fax   (33-1) 4057 6659
Email: Fatih.Birol@iea.org

TABLE OF CONTENTS

Part A: Global Energy Prospects: Impact of Developments in India and China   **71**

List of Figures

Introduction

Part A: Global Energy Prospects: Impact of Developments in China and India

Chapter 1. Global Energy Trends

Chapter 2. Energy Trends in China and India

Chapter 3. International Trade and the World Economy

Part C: India's Energy Prospects

Chapter 14. Political, Economic and Demographic Context

Chapter 15. Overview of the Energy Sector

Chapter 16. Reference Scenario Demand Projections

List of Tables

List of Boxes

Chapter 17. Reference Scenario Supply Projections

Chapter 18. Alternative Policy Scenario Projections

Chapter 19. High Growth Scenario Projections

Chapter 20. Focus on Energy Poverty

List of Spotlights

Part A: Global Energy Prospects: Impact of Developments in India and China

World Energy Outlook Series

World Energy Outlook 1993
World Energy Outlook 1994
World Energy Outlook 1995
World Energy Outlook 1996
World Energy Outlook 1998
World Energy Outlook: 1999 Insights
    Looking at Energy Subsidies: Getting the Prices Right
World Energy Outlook 2000
World Energy Outlook: 2001 Insights
    Assessing Today's Supplies to Fuel Tomorrow's Growth
World Energy Outlook 2002
World Energy Investment Outlook: 2003 Insights
World Energy Outlook 2004
World Energy Outlook 2005
    Middle East and North Africa Insights
World Energy Outlook 2006
World Energy Outlook 2007
    China and India Insights
World Energy Outlook 2008 (forthcoming)

More information available at www.worldenergyoutlook.org

China and India are the emerging giants of the world economy and international energy markets. Energy developments in China and India are transforming the global energy system by dint of their sheer size and their growing weight in international fossil-fuel trade. Similarly, both countries are increasingly exposed to changes in world energy markets. The staggering pace of Chinese and Indian economic growth in the past few years, outstripping that of all other major countries, has pushed up sharply their energy needs, a growing share of which has to be imported. The momentum of economic development looks set to keep their energy demand growing strongly. As they become richer, the citizens of China and India are using more energy to run their offices and factories, and buying more electrical appliances and cars. These developments are contributing to a big improvement in their quality of life, a legitimate aspiration that needs to be accommodated and supported by the rest of the world.

The consequences for China, India, the OECD and the rest of the world of unfettered growth in global energy demand are, however, alarming. If governments around the world stick with current policies – the underlying premise of our Reference Scenario – the world's energy needs would be well over 50% higher in 2030 than today. China and India together account for 45% of the increase in demand in this scenario. Globally, fossil fuels continue to dominate the fuel mix. These trends lead to continued growth in energy-related emissions of carbon-dioxide (CO_2) and to increased reliance of consuming countries on imports of oil and gas – much of them from the Middle East and Russia. Both developments would heighten concerns about climate change and energy security.

The challenge for all countries is to put in motion a transition to a more secure, lower-carbon energy system, without undermining economic and social development. Nowhere will this challenge be tougher, or of greater importance to the rest of the world, than in China and India. Vigorous, immediate and collective policy action by *all* governments is essential to move the world onto a more sustainable energy path. There has so far been more talk than action in most countries. Were all the policies that governments around the world are considering today to be implemented, as we assume in an Alternative Policy Scenario, the world's energy demand and related emissions would be reduced substantially. Measures to improve energy efficiency stand out as the cheapest and fastest way to curb demand and emissions growth in the near term. But even in this scenario, CO_2 emissions are still one-quarter above current levels in 2030. To achieve a much bigger reduction in emissions would require immediate policy action and technological transformation on an unprecedented scale.

Both the Reference and Alternative Policy Scenario projections are based on what some might consider conservative assumptions about economic growth in the two giants. They envisage a progressive and marked slow-down in the rate of growth of output over the projection period. In a High Growth Scenario, which assumes that China's and India's economies grow on average 1.5 percentage points per year faster than in the Reference Scenario (though more slowly than of late), energy demand is 21% higher in 2030 in China and India combined. The global increase in energy demand amounts to 6%, making it all the more urgent for governments around the world to implement policies, such as those taken into account in the Alternative Policy Scenario, to curb the growth in fossil-energy demand and related emissions.

The World Faces a Fossil Energy Future to 2030

The world's primary energy needs in the Reference Scenario are projected to grow by 55% between 2005 and 2030, at an average annual rate of 1.8% per year. Demand reaches 17.7 billion tonnes of oil equivalent, compared with 11.4 billion toe in 2005. Fossil fuels remain the dominant source of primary energy, accounting for 84% of the overall increase in demand between 2005 and 2030. Oil remains the single largest fuel, though its share in global demand falls from 35% to 32%. Oil demand reaches 116 million barrels per day in 2030 – 32 mb/d, or 37%, up on 2006. In line with the spectacular growth of the past few years, coal sees the biggest increase in demand in absolute terms, jumping by 73% between 2005 and 2030 and pushing its share of total energy demand up from 25% to 28%. Most of the increase in coal use arises in China and India. The share of natural gas increases more modestly, from 21% to 22%. Electricity use doubles, its share of final energy consumption rising from 17% to 22%. Some $22 trillion of investment in supply infrastructure is needed to meet projected global demand. Mobilising all this investment will be challenging.

Developing countries, whose economies and populations are growing fastest, contribute 74% of the increase in global primary energy use in this scenario. China and India alone account for 45% of this increase. OECD countries account for one-fifth and the transition economies the remaining 6%. In aggregate, developing countries make up 47% of the global energy market in 2015 and more than half in 2030, compared with only 41% today. The developing countries' share of global demand expands for all primary energy sources, except non-hydro renewables. About half of the increase in global demand goes to power generation and one-fifth to meeting transport needs – mostly in the form of petroleum-based fuels.

World oil resources are judged to be sufficient to meet the projected growth in demand to 2030, with output becoming more concentrated in OPEC countries – on the assumption that the necessary investment is forthcoming. Their collective output of conventional crude oil, natural gas liquids and non-conventional oil (mainly gas-to-liquids) is projected to climb from 36 mb/d in 2006 to 46 mb/d in 2015 and 61 mb/d in 2030 in the Reference Scenario. As a result, OPEC's share of world oil supply jumps from 42% now to 52% by the end of the projection period. Non-OPEC production rises only slowly to 2030, with most of the increase coming from non-conventional sources – mainly Canadian oil sands – as conventional output levels off at around 47 mb/d by the middle of the 2010s. These projections are based on the assumption that the average IEA crude oil import price falls back from recent highs of over $75 per barrel to around $60 (in year-2006 dollars) by 2015 and then recovers slowly, reaching $62 (or $108 in nominal terms) by 2030. Although new oil-production capacity additions from greenfield projects are expected to increase over the next five years, it is very uncertain whether they will be sufficient to compensate for the decline in output at existing fields and keep pace with the projected increase in demand. A supply-side crunch in the period to 2015, involving an abrupt escalation in oil prices, cannot be ruled out.

The resurgence of coal, driven primarily by booming power-sector demand in China and India, is a marked departure from past *WEOs*. Higher oil and gas prices are making coal more competitive as a fuel for baseload generation. China and India, which already account for 45% of world coal use, drive over four-fifths of the increase to 2030 in the Reference Scenario. In the OECD, coal use grows only very slowly, with most of the increase coming from the United States. In all regions, the outlook for coal use depends largely on relative fuel prices, government policies on fuel diversification, climate change and air pollution, and developments in clean coal technology in power generation. The widespread deployment of more efficient power-generation technology is expected to cut the amount of coal needed to generate a kWh of electricity, but boost the attraction of coal over other fuels, thereby leading to higher demand.

In the Alternative Policy Scenario, global primary energy demand grows by 1.3% per year over 2005-2030 – 0.5 percentage points less than in the Reference Scenario. Global oil demand is 14 mb/d lower in 2030 – equal to the entire current output of the United States, Canada and Mexico combined. Coal use falls most in absolute and percentage terms. Energy-related CO_2 emissions stabilise in the 2020s and, in 2030, are 19% lower than in the

Reference Scenario. In the High Growth Scenario, faster economic growth in China and India, absent any policy changes, boosts their energy demand. The stimulus to demand provided by stronger economic growth more than offsets the dampening effect of the higher international energy prices that accompany stronger demand. Worldwide, the increase in primary energy demand amounts to 6% in 2030, compared with the Reference Scenario. Demand is higher in some regions and lower in others.

China's Share of World Energy Demand will Continue to Expand

That China's energy needs will continue to grow to fuel its economic development is scarcely in doubt. However, the rate of increase and how those needs are met are far from certain, as they depend on just how quickly the economy expands and on the economic and energy-policy landscape worldwide. In the Reference Scenario, China's primary energy demand is projected to more than double from 1 742 million toe in 2005 to 3 819 Mtoe in 2030 – an average annual rate of growth of 3.2%. China, with four times as many people, overtakes the United States to become the world's largest energy consumer soon after 2010. In 2005, US demand was more than one-third larger. In the period to 2015, China's demand grows by 5.1% per year, driven mainly by a continuing boom in heavy industry. In the longer term, demand slows, as the economy matures, the structure of output shifts towards less energy-intensive activities and more energy-efficient technologies are introduced. Oil demand for transport almost quadruples between 2005 and 2030, contributing more than two-thirds of the overall increase in Chinese oil demand. The vehicle fleet expands seven-fold, reaching almost 270 million. New vehicle sales in China exceed those of the United States by around 2015. Fuel economy regulations, adopted in 2006, nonetheless temper oil-demand growth. Rising incomes underpin strong growth in housing, the use of electric appliances and space heating and cooling. Increased fossil-fuel use pushes up emissions of CO_2 and local air pollutants, especially in the early years of the projection period: SO_2 emissions, for example, rise from 26 million tonnes in 2005 to 30 Mt by 2030.

China's energy resources – especially coal – are extensive, but will not meet all the growth in its energy needs. More than 90% of Chinese coal resources are located in inland provinces, but the biggest increase in demand is expected to occur in the coastal region. This adds to the pressure on internal coal transport and makes imports into coastal provinces more competitive. China became a net coal importer in the first half of 2007. In the Reference Scenario, net imports reach 3% of its demand and 7% of global coal trade in 2030. Conventional oil production in China is set to peak at 3.9 mb/d early in the next decade and then start to decline. Consequently, China's oil imports jump from 3.5 mb/d in 2006 to 13.1 mb/d in 2030, while the share of imports in demand rises from 50% to 80%. Natural gas imports also increase quickly, as production growth lags demand over the projection period. China needs to add more than 1 300 GW to its electricity-generating capacity, more than the total current installed capacity in the United States. Coal remains the dominant fuel in power generation. Projected cumulative investment in China's energy-supply infrastructure amounts to $3.7 trillion (in year-2006 dollars) over the period 2006-2030, three-quarters of which goes to the power sector.

China is already making major efforts to address the causes and consequences of burgeoning energy use, but even stronger measures will be needed. China is seeking ways to enhance its energy-policy, regulatory and institutional framework to meet current and future challenges. In the Alternative Policy Scenario, a set of policies the government is currently considering would cut China's primary energy use in 2030 by about 15% relative to the Reference Scenario. Energy-related emissions of CO_2 and local pollutants fall even more. Energy demand, nonetheless, increases by almost 90% between 2005 and 2030 in the Alternative Policy Scenario. Energy-efficiency improvements along the entire energy chain and fuel switching account for 60% of the energy saved. For example, policies that lead to more fuel-efficient vehicles produce big savings in consumption of oil-based fuels. Structural change in the economy accounts for all the other energy savings. Demand for coal and oil is reduced substantially. In contrast, demand for other fuels – natural gas, nuclear and renewables – increases. In this scenario, the government's goal of lowering energy intensity – the amount of energy consumed per unit of GDP – by 20% between 2005 and 2010 is achieved soon after. The majority of the measures analysed have very short payback periods. In addition, each dollar invested in more efficient electrical appliances saves $3.50 of investment on the supply side. And China's efforts to improve the efficiency of vehicles and electrical appliances contribute to improved efficiency in the rest of the world, as the country is a net exporter of these products. Such policies would be all the more critical were China's economy to grow more quickly than assumed in the Reference and Alternative Policy Scenarios. China's primary energy demand is 23% higher in 2030, and coal use alone 21% higher, in the High Growth Scenario than in the Reference Scenario.

India's Energy Use is Similarly Poised for Rapid Growth

Rapid economic expansion will also continue to drive up India's energy demand, boosting the country's share of global energy consumption. In the

Reference Scenario, primary energy demand in India more than doubles by 2030, growing on average by 3.6% per year. Coal remains India's most important fuel, its use nearly tripling between 2005 and 2030. Power generation accounts for much of the increase in primary energy demand, given surging electricity demand in industry and in residential and commercial buildings, with most new generating capacity fuelled by coal. Among end use sectors, transport energy demand sees the fastest rate of growth as the vehicle stock expands rapidly with rising economic activity and household incomes. Residential demand grows much more slowly, largely as a result of switching from traditional biomass, which is used very inefficiently, to modern fuels. The number of Indians relying on biomass for cooking and heating drops from 668 million in 2005 to around 470 million in 2030, while the share of the population with access to electricity rises from 62% to 96%.

Much of India's incremental energy needs to 2030 will have to be imported. It is certain that India will continue to rely on imported coal for reasons of quality in the steel sector and for economic reasons at power plants located a long way from mines but close to ports. In the Reference Scenario, hard coal imports are projected to rise almost seven-fold, their share of total Indian coal demand rising from 12% in 2005 to 28% in 2030. Net oil imports also grow steadily, to 6 mb/d in 2030, as proven reserves of indigenous oil are small. Before 2025, India overtakes Japan to become the world's third-largest net importer of oil, after the United States and China. Yet India's importance as a major exporter of refined oil products will also grow, assuming the necessary investments are forthcoming. Although recent discoveries are expected to boost gas production, it is projected to peak between 2020 and 2030, and then fall back. A growing share of India's gas needs is, therefore, met by imports, entirely in the form of liquefied natural gas. Power-generation capacity, most of it coal-fired, more than triples between 2005 and 2030. Gross capacity additions exceed 400 GW – equal to today's combined capacity of Japan, Korea and Australia. To meet demand in the Reference Scenario, India needs to invest about $1.25 trillion in energy infrastructure – three-quarters in the power sector – in 2006-2030. Attracting electricity investment in a timely manner – a huge challenge for India – will be crucial for sustaining economic growth.

Stronger policies that the Indian government is now considering could yield large energy savings. In the Alternative Policy Scenario, India's primary energy demand is 17% lower than in the Reference Scenario in 2030. Coal savings – mainly in power generation – are the greatest in both absolute and percentage terms, thanks to lower electricity-demand growth, higher power-generation efficiency and fuel-switching in the power sector and in industry. As a result, coal imports in 2030 are little more than half their Reference Scenario level. Oil imports are 1.1 mb/d lower in 2030 than in the Reference Scenario, but oil-import dependence remains high at 90%. Lower fossil-fuel use results in a 27% reduction in CO_2 emissions in 2030, most of which stems from energy-efficiency improvements on the demand and supply sides. Lower energy demand in the power and transport sectors also reduces emissions of local pollutants: SO_2 emissions fall by 27% and NO_x emissions by 23% in 2030, compared with the Reference Scenario. The picture is markedly different in the High Growth Scenario. Primary demand is 16% *higher* than in the Reference Scenario, with coal and oil accounting for most of the difference. Faster economic growth accelerates the alleviation of energy poverty, but results in much higher energy imports, local pollution and CO_2 emissions.

The World Benefits Economically from Growth in China and India

Rapid economic development in China and India will inevitably push up global energy demand, but it will also bring major economic benefits to the rest of the world. Economic expansion in China and India is generating opportunities for other countries to export to them, while increasing other countries' access to a wider range of competitively priced imported products and services. But growing exports from China and India also increase competitive pressures on other countries, leading to structural adjustments, particularly in countries with competing export industries. Rising commodity needs risk driving up international prices for commodities, including energy – especially if supply-side investment is constrained.

Commodity exporters would gain most from even faster economic expansion in China and India than assumed in the Reference Scenario. In the High Growth Scenario, the Middle East, Russia and other energy-exporting countries see a significant net increase in their gross domestic product in 2030, compared with the Reference Scenario. GDP growth in other developing Asian countries, the United States, the European Union and OECD Pacific slows marginally, mainly because of higher commodity import costs. Assuming there are no policy changes in major countries, the average IEA crude oil import price rises to $87 per barrel (in year-2006 dollars) in 2030 – 40% higher than in the Reference Scenario. Overall, world GDP grows by 4.3% per year on average, compared with 3.6% in the Reference Scenario.

Structural changes in China's and India's economies will affect their trade with the rest of the world, including their need to import energy. Light industry and services are expected to play a more important role in driving economic development in both countries in the longer term. The economic policies of all countries will be crucial to sustaining the pace of global economic growth and redressing current imbalances. Rising protectionism could radically change the positive global impact of economic growth in China and India.

By contrast, faster implementation of energy and environmental policies to save energy and reduce emissions worldwide, such as those included in the Alternative Policy Scenario, would boost significantly the net global benefits, by reducing pressures on international commodity markets and lowering fuel-import bills for all. More rapid economic development worldwide may also pave the way for faster development and deployment of emerging, clean energy technologies, such as second-generation biofuels and CO_2 capture and storage, given the right policy environment.

But Threats to the World's Energy Security Must be Tackled

Rising global energy demand poses a real and growing threat to the world's energy security. Oil and gas demand and the reliance of all consuming countries on oil and gas imports increase in all three scenarios presented in this *Outlook*. In the Reference Scenario, China's and India's combined oil imports surge, from 5.4 mb/d in 2006 to 19.1 mb/d in 2030 – more than the combined imports of Japan and the United States today. Ensuring reliable and affordable supply will be a formidable challenge. Inter-regional oil and gas trade grows rapidly over the projection period, with a widening of the gap between indigenous output and demand in every consuming region. The volume of oil trade expands from 41 mb/d in 2006 to 51 mb/d in 2015 and 65 mb/d in 2030. The Middle East, the transition economies, Africa and Latin America export more oil. All other regions – including China and India – have to import more oil. As refining capacity for export increases, a growing share of trade in oil is expected to be in the form of refined products, notably from refineries in the Middle East and India.

The consuming countries' growing reliance on oil and gas imports from a small number of producing countries threatens to exacerbate short-term energy-security risks. Increasing import dependence in any country does not necessarily mean less secure energy supplies, any more than self-sufficiency guarantees uninterrupted supply. Indeed, increased trade could bring mutual economic benefits to all concerned. Yet it could carry a *risk* of heightened short-term energy insecurity for all consuming countries, as geographic supply diversity is reduced and reliance grows on vulnerable supply routes. Much of the additional oil imports are likely to come from the Middle East, the scene of most past supply disruptions, and will transit vulnerable maritime routes to both eastern and western markets. The potential impact on international oil prices of a supply interruption is also likely to increase: oil demand is becoming less sensitive to changes in price as the share of transport demand – which is price-inelastic, relative to other energy services – in overall oil consumption rises worldwide.

Longer-term risks to energy security are also set to grow. With stronger global energy demand, all regions would be faced with higher energy prices in the medium to long term in the absence of concomitant increases in supply-side investment or stronger policy action to curb demand growth in all countries. The increasing concentration of the world's remaining oil reserves in a small group of countries – notably Middle Eastern members of OPEC and Russia – will increase their market dominance and may put at risk the required rate of investment in production capacity. OPEC's global market share increases in all scenarios – most of all in the Reference and High Growth Scenarios. The greater the increase in the call on oil and gas from these regions, the more likely it will be that they will seek to extract a higher rent from their exports and to impose higher prices in the longer term by deferring investment and constraining production. Higher prices would be especially burdensome for developing countries still seeking to protect their consumers through subsidies.

China's and India's growing participation in international trade heightens the importance of their contribution to collective efforts to enhance global energy security. How China and India respond to the rising threats to their energy security will also affect the rest of the world. Both countries are already taking action. The more effective their policies are to avert or handle a supply emergency, the more other consuming countries – including most IEA members – stand to benefit, and vice-versa. In addition, many policies to enhance energy security also directly support policies to address the environmental damage from energy production and use. Diversification of the energy mix, of the sources of imported oil and gas, and of supply routes, together with better emergency preparedness, especially through the establishment of emergency stockpiles and co-ordinated response mechanisms, will be necessary to safeguard their energy security. China and India are increasingly aware that overseas acquisitions of oil assets will do little to help protect them from the effects of supply emergencies. China's and India's oil security – like that of all consuming countries – is increasingly dependent on a well-functioning international oil market.

Unchecked Growth in Fossil Fuel Use will Hasten Climate Change

Rising CO_2 and other greenhouse-gas concentrations in the atmosphere, resulting largely from fossil-energy combustion, are contributing to higher global temperatures and to changes in climate. Growing fossil-fuel use will continue to drive up global energy-related CO_2 emissions over the projection period. In the Reference Scenario, emissions jump by 57% between 2005 and 2030. The United States, China, Russia and India contribute two-thirds of this increase. China is by far the biggest contributor to incremental emissions, overtaking the United States as the world's biggest emitter in 2007. India becomes the third-largest emitter by around 2015. However, China's per-capita emissions in 2030 are only 40% of those of the United States and about two-thirds those of the OECD as a whole in the Reference Scenario. In India, they remain far lower than those of the OECD, even though they grow faster than in almost any other region.

Urgent action is needed if greenhouse-gas concentrations are to be stabilised at a level that would prevent dangerous interference with the climate system. The Alternative Policy Scenario shows that measures currently being considered by governments around the world could lead to a stabilisation of global emissions in the mid-2020s and cut their level in 2030 by 19% relative to the Reference Scenario. OECD emissions peak and begin to decline after 2015. Yet global emissions would still be 27% higher than in 2005. Assuming continued emissions reductions after 2030, the Alternative Policy Scenario projections are consistent with stabilisation of long-term CO_2-equivalent concentration in the atmosphere at about 550 parts per million. According to the best estimates of the Intergovernmental Panel on Climate Change, this concentration would correspond to an increase in average temperature of around 3°C above pre-industrial levels. In order to limit the average increase in global temperatures to a maximum of 2.4°C, the smallest increase in any of the IPCC scenarios, the concentration of greenhouse gases in the atmosphere would need to be stabilised at around 450 ppm. To achieve this, CO_2 emissions would need to peak by 2015 at the latest and to fall between 50% and 85% below 2000 levels by 2050. We estimate that this would require energy-related CO_2 emissions to be cut to around 23 Gt in 2030 – 19 Gt less than in the Reference Scenario and 11 Gt less than in the Alternative Policy Scenario. In a "450 Stabilisation Case", which describes a notional pathway to achieving this outcome, global emissions peak in 2012 at around 30 Gt. Emissions savings come from improved efficiency in fossil-fuel use in industry, buildings and transport, switching to nuclear power and renewables, and the widespread deployment of CO_2 capture and storage (CCS) in power generation and industry. Exceptionally quick and vigorous policy action by all countries, and unprecedented technological advances, entailing substantial costs, would be needed to make this case a reality.

Government action must focus on curbing the rapid growth in CO_2 emissions from coal-fired power stations – the primary cause of the surge in global emissions in the last few years. Energy efficiency and conservation will need to play a central role in curbing soaring electricity demand and reducing inputs to generation. Nuclear power and renewables can also make a major contribution to lowering emissions. Clean coal technology, notably CCS, is one of the most promising routes for mitigating emissions in the longer term – especially in China, India and the United States, where coal use is growing fastest. CCS could reconcile continued coal burning with the need to cut emissions in the longer term – if the technology can be demonstrated on a large scale and if adequate incentives to invest are put in place.

Collective Action is Needed to Address Global Energy Challenges

The emergence of China and India as major players in global energy markets makes it all the more important that *all* countries take decisive and urgent action to curb runaway energy demand. The primary scarcity facing the planet is not of natural resources nor money, but time. Investment now being made in energy-supply infrastructure will lock in technology for decades, especially in power generation. The next ten years will be crucial, as the pace of expansion in energy-supply infrastructure is expected to be particularly rapid. China's and India's energy challenges are the world's energy challenges, which call for collective responses. No major energy consumer can be confident of secure supply if supplies to others are at risk. And there can be no effective long-term solution to the threat of climate change unless all major energy consumers contribute. The adoption and full implementation of policies by IEA countries to address their energy-security and climate-change concerns are essential, but far from sufficient.

Many of the policies available to alleviate energy insecurity can also help to mitigate local pollution and climate change, and vice-versa. As the Alternative Policy Scenario demonstrates, in many cases, those policies bring economic benefits too, by lowering energy costs – a "triple-win" outcome. An integrated approach to policy formulation is, therefore, essential. The right mix of policies to address both energy-security and climate concerns depends on the balance of costs and benefits, which vary among countries. We do not have the luxury of ruling out any of the options for moving the global energy system onto a more sustainable path. The most cost-effective approach will involve market-based instruments, including those that place an explicit financial value on CO_2 emissions. Regulatory measures, such as standards and mandates, will also be needed, together with government support for long-term research, development and demonstration of new technologies. In China and India, the urgent need to tackle local air pollution will undoubtedly continue to provide the primary rationale for further efforts to stem the growth in greenhouse-gas emissions.

There are large potential gains to IEA countries, on the one hand, and to China and India, on the other, from enhanced policy co-operation. IEA countries have long recognised the advantages of co-operation with China and India, reflected in a steady broadening of the range of co-operative activities through the IEA and other multilateral and bilateral agreements. These activities need to be stepped up, with China and India establishing a deeper relationship with the Agency. IEA co-operation with China and India on enhancing oil-emergency preparedness and on developing cleaner and more efficient technologies, especially for coal, remains a priority. Collaboration between IEA countries and developing countries, including China and India, is already accelerating deployment of new technologies – a development that will yield big dividends in the longer term. Mechanisms need to be enhanced to facilitate and encourage the financing of such technologies in China, India and other developing countries. Given the scale of the energy challenge facing the world, a substantial increase is called for in public and private funding for energy technology research, development and demonstration, which remains well below levels reached in the early 1980s. The financial burden of supporting research efforts will continue to fall largely on IEA countries.

Purpose and Scope of the Study

China and India are the world's emerging economic giants and centres of energy use. Phenomenal rates of economic growth in the last two to three decades in China and, more recently, in India have been accompanied by a growing thirst for energy. A rising share of their energy needs has to be met by imports, as demand is outstripping indigenous supply. Increasing fossil-energy consumption has serious implications for the environment, both in terms of local pollution and through rising emissions of greenhouse gases. That these trends will continue is scarcely in doubt. But the future *pace* of growth in energy demand and how it will be met remain very uncertain – as previous *World Energy Outlooks* have pointed out. How rapidly the two countries' energy needs develop and how they are met will have far-reaching consequences for them and the rest of the world.

This *Outlook* provides insights into these very complex and important issues. Within the framework of a comprehensive update of global energy demand and supply projections, it sets out in detail the prospects for energy markets in both China and India, identifying and quantifying the factors that will drive the two countries' energy balances and seeking to answer the question: how will their energy choices affect the world as a whole?

We approach the answer to that question by means of detailed sets of projections of energy markets in both countries, fuel by fuel and sector by sector. These projections reveal how much of each form of energy each country might need in the future, how much could be produced locally, how much will need to be imported and how much might be available for export. In the case of China, in addition to projections for the country as a whole, we have included disaggregated projections for a region comprising the coastal provinces, which has a very different economic and energy profile from the less developed central and western provinces.

The results are intended to provide policy makers and others with a rigorous quantitative framework for analysing future energy developments and energy-policy options in China and India and what they could mean for international energy markets, the world's energy security and the global environment. The analysis builds on a long-standing dialogue with China, India and other major emerging economies on policies to improve energy efficiency, enhance supply security and mitigate climate change. It forms one part of the IEA's response to a call from G8 leaders to expand this dialogue.

It is hard to overstate the growing importance of China and India in global energy developments. After many years of growth, the economies of China and

India are now so big that they are a major force in global energy demand, in trade and in energy-related greenhouse-gas emissions. Their energy demand has soared since the year 2000, with China's energy use expanding as fast as GDP (compared with only one-third as fast over the previous two decades). Together, the two countries accounted for more than half of the estimated global increase in energy use between 2000 and 2006, 45% coming from China alone (Figure 1). Coal accounted for 43% of this increase in global energy demand; 85% of the global increase in coal use arose in China and India. Coal is their primary source of energy and will remain so for decades to come – thanks to abundant low-cost indigenous resources. But more and more of their incremental energy needs are being met by other sources, particularly oil. The emergence of a sizeable middle class, aspiring to modern lifestyles and comfort levels, is leading to a surge in demand for motor vehicles, as well as electrical appliances. Neither country is able to meet all its oil needs from domestic supplies, so imports are rising rapidly. Renewable sources, notably hydro, solar and wind power, and nuclear power are being developed rapidly, but not quickly enough to reduce significantly either country's heavy dependence on fossil fuels.

Figure 1: **Share of China and India in Incremental Energy Demand, Imports and Energy-Related CO_2 Emissions, 2000-2006***

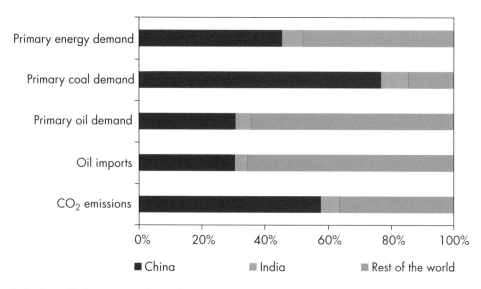

* Based on preliminary estimates for 2006.

Unsurprisingly, the rapid growth in energy demand in China and India – and elsewhere in the world – raises a number of concerns for all energy consumers. Other oil-importing countries question how surging Asian imports will affect their own energy security. Oil prices have risen steeply since the end of the 1990s in response to a tightening of global markets, resulting, at least in part, from stronger Chinese and Indian demand. And rising dependence on imports of oil from the Middle East and elsewhere is increasing the risk of a severe supply disruption. Imports into China and India of natural gas, which has so far played a small role in energy supply in both countries, are taking off. Nonetheless, per-capita consumption of oil and gas in China and India remains far below OECD levels.

Climate change is another big issue. Largely as a result of their heavy reliance on coal, China's and India's energy-related emissions of carbon dioxide are soaring. An astonishing 58% of the global increase in emissions in the six years to 2006 came from China and 6% from India. China is expected to overtake the United States as the world's biggest CO_2 emitter in 2007. However, as for energy demand, per-capita emissions remain about one-third below OECD levels. The recent jump in emissions in China and, to a lesser extent, in India has highlighted the collective need for all countries to act to combat global warming. Both China and India are aware that they are at particular risk from the consequences of climate change and that they need to take action on this issue as well as to address serious problems of local pollution. They recognise that many measures to tackle pollution would bring the added benefit of lower greenhouse-gas emissions. But they also worry that environmental measures might constrain their economic development. Despite impressive rates of economic growth and industrialisation in recent years, both China and India remain poor by OECD standards.

Any assessment of how energy developments in these two countries could affect the rest of the world must take into consideration the overall economic context. Rising energy demand is part and parcel of a broader process of rapid economic and social development, which is lifting millions of people out of poverty. That is cause for cheer. In addition, because China's and India's economic growth is being driven to a large extent by international trade, the rest of the world as a whole benefits too, through opportunities to import cheaper goods and services from those two countries and to export to them. However, the benefits are not shared evenly, with some countries losing out through an inability to adapt to the global structural shifts wrought by the emergence of China and India as major economic powers.

The first part of this study (Part A) assesses the global implications of energy developments in China and India. It provides our updated energy projections, by scenario, for the world and for China and India. It then considers the role of the two giants in international trade and economic growth, the impact of their rising energy use on the world's energy security and the environmental repercussions. A final section in Part A looks at what this means for policy makers. Parts B and C contain a detailed analysis of the prospects for energy demand and supply in China and India under different scenarios. Detailed tables showing the results of the projections can be found in the annexes.

Methodology and Assumptions

As in previous *Outlooks*, a scenario approach has been adopted to examine future energy developments. The projection period runs to 2030. The core projections are derived from a *Reference Scenario*, which assumes that there are no new energy-policy interventions by governments. This scenario is intended to provide a baseline vision of how global energy markets are likely to evolve if governments do nothing more to affect underlying trends in energy demand and supply, thereby allowing us to test alternative assumptions about future government policies.

An *Alternative Policy Scenario* analyses the impact on global energy markets of a package of additional measures to address energy-security and climate-change concerns. The goal is to offer practical guidance to policy makers about the potential impact and cost of the many options they are currently considering. On the basis of this scenario, we assess the implications for energy use of achieving long-term stabilisation of atmospheric greenhouse-gas concentrations at a level that would result in an increase in global temperatures no higher than that which is widely considered to be acceptable: what we call a *450 Stabilisation Case*. These analyses form part of the IEA's response to a request from the G8 leaders at the Gleneagles Summit in 2005 for advice on "alternative energy scenarios and strategies aimed at a clean, clever and competitive energy future".

For this *WEO*, we have also developed a *High Growth Scenario*, which incorporates significantly higher rates of economic growth in China and India than those in the Reference Scenario (though still below current rates). Prospects for economic growth have been systematically underestimated in recent years in both countries, and future rates of growth are extremely uncertain, especially towards the end of the projection period. Were their economies to grow significantly faster than assumed in the Reference Scenario, their energy demand – and that of the world as a whole – could turn out to be much higher by the end of the projection period. This scenario allows us to test the sensitivity of their energy demand to economic growth rates and the implications for global energy trade and energy-related greenhouse-gas emissions.

The projections presented in this study are underpinned by a vast data-collection and modelling effort. Major improvements have been made to the IEA's World Energy Model,[1] including a more detailed representation of end-use sectors in China and India, disaggregated regional models for China and rural-urban models for India (Box 1). Considerable work was also devoted to verifying the accuracy of energy data (see Chapters 8 and 15). The analysis also benefited from major workshops in Beijing and New Delhi in March 2007, organised specifically to provide input for our work, as well as a high-level brainstorming meeting held at the IEA headquarters in Paris in May 2007. This study would not have been possible without the close co-operation of China's National Development and Reform Commission and the Energy Research Institute in Beijing, The Energy and Resources Institute (TERI) in New Delhi and other public and private bodies in China and India. The analysis of macroeconomic linkages and interactions with energy markets also benefited from collaboration with the Centre International de Recherche sur l'Environnement et le Développement (CIRED), a French research institute. The IEA's Coal Industry Advisory Board provided valuable input to the analysis of coal prospects. The International Institute for Applied Systems Analysis (IIASA) assisted with the analysis of local environmental issues. Other international organisations involved included the World Bank, the Asian Development Bank and the International Monetary Fund, all of which made important contributions to the work.

The Reference Scenario

The Reference Scenario is defined in the same way as in previous editions of the *Outlook*. It is designed to show the outcome, on given assumptions about economic growth, population, energy prices and technology, if nothing more is done by governments to change underlying energy trends. It takes account of those government policies and measures that had already been adopted by mid-2007, regardless of whether they have yet been fully implemented – even though the impact on energy demand and supply of the most recent measures does not show up in historical market data.[2] In some cases, policies that were under consideration in 2006 and were included in the Alternative Policy Scenario last year have since been adopted and have, therefore, now been taken into account in the Reference Scenario. These include measures to boost biofuels in the United States, new measures to promote renewables in the European Union and Japan, the national allocation plans for the

---

1. Details of the WEM are available at the *WEO* website at www.worldenergyoutlook.org.
2. Available only up to 2005 for all countries and to 2006 for some fuels and some countries.

second trading period (2008-2012) of the EU Greenhouse Gas Emissions Trading Scheme and the mandatory phase-out of incandescent light bulbs in Australia. In contrast with the Alternative Policy Scenario, the Reference Scenario does *not* take into account possible, potential or even likely future policy actions.

Box 1: **Modelling Improvements for WEO-2007**

The IEA's World Energy Model (WEM) – a large-scale mathematical construct designed to replicate how energy markets function – is the principal tool used to generate detailed sector-by-sector and region-by-region projections for all the scenarios in this *Outlook*. It has been updated using the latest historical data. In addition, more detailed models were developed for China and India to allow more in-depth analysis of energy trends in those countries. For India a rural-urban breakdown in the residential sector was also introduced (see Chapter 16), which includes a new electrification model. For China two separate models were developed – one for the coastal provinces and the other covering the whole country (see Chapter 13).

The other main improvements include a more detailed sectoral representation of end-use sectors, notably the road-transport sector, as well as a field-by-field analysis of oil and gas production prospects in both countries, including the potential impact of enhanced oil recovery. In addition, the WEM has been integrated into a general equilibrium model for the purposes of analysing the interlinkages between energy use and economic activity in China, India and other *WEO* regions (see Box 3.2 in Chapter 3). The results have been used to assess the global impact of structural economic changes in China and India in the High Growth Scenario.

Although the Reference Scenario assumes no change in energy and environmental policies throughout the projection period, it is not always clear exactly how existing policies will be implemented in the future. Inevitably, a degree of judgment is involved in translating stated policies into formal assumptions for modelling purposes. These assumptions vary by fuel and by region. For example, electricity- and gas-market reforms, where approved, are assumed to move ahead, but at varying speeds in different countries and regions. Progress is assumed to be made in liberalising cross-border energy trade and investment, and in reforming energy subsidies, but these policies are expected to be pursued most rigorously in OECD countries.

In all cases, the rates of excise duty and value-added or sales tax applied to different energy sources and carriers are assumed to remain constant. As a result, assumed changes in international prices (see below) have different effects on the retail prices of each fuel and in each region, according to the type of tax applied and the rates currently levied. Consistent with the basic assumption of no policy change in this scenario, nuclear energy is assumed to remain an option for power generation except in those countries that have officially banned it or decided to phase it out.

Demographic Assumptions

Population growth affects the size and composition of energy demand, directly and through its impact on economic growth. Our population assumptions are drawn from the most recent United Nations projections (UNPD, 2007). World population is projected to grow by 1% per year on average, from 6.4 billion in 2005 to almost 8.2 billion in 2030. Population growth slows over the projection period, in line with trends of the last three decades: from 1.1% per year in 2005-2015 to 0.9% in 2015-2030 (Table 1). Population expanded by 1.5% from 1980 to 2005. Projected growth is slightly higher than projected last year, largely because the HIV/AIDS epidemic is expected to be less prevalent and mortality rates lower, thanks to more widely available antiretroviral drugs, in developing countries.

Almost all the increase in world population is expected to arise in developing countries. Their combined population is projected to grow by an average 1.2% per year from 2005 to 2030. This rate is markedly lower than the average rate of 1.9% in the last three decades. Total population in developing countries reaches 6.6 billion in 2030, compared with 4.9 billion in 2005. As a result, the share of the world's population living in developing regions, as they are classified today, will increase from 76% now to 80%. China's population is expected to grow relatively slowly, reaching 1.46 billion in 2030 compared with 1.31 billion in 2005. India's population, which stood at 1.09 billion in 2005, is set to grow much more quickly, catching up that of China by 2030.

Population in the transition economies is expected to decline over the same period. Russia's population drops from 143 million in 2005 to 123 million in 2030, a cumulative fall of around 14%. The OECD's population is expected to grow by an average of only 0.4% per annum, with North America accounting for much of the increase. Most of the population increase in the OECD results from net immigration in North America and Europe.

Table 1: **World Population Growth** (average annual growth rates, %)

	1980-1990	1990-2005	2005-2015	2015-2030	2005-2030
OECD	**0.8**	**0.8**	**0.5**	**0.3**	**0.4**
North America	1.2	1.3	1.0	0.7	0.8
United States	*0.9*	*1.1*	*0.9*	*0.7*	*0.8*
Europe	0.5	0.5	0.3	0.2	0.2
Pacific	0.8	0.5	0.1	−0.2	−0.1
Japan	*0.6*	*0.2*	*−0.1*	*−0.5*	*−0.3*
Transition economies	**0.8**	**−0.2**	**−0.2**	**−0.3**	**−0.2**
Russia	0.6	−0.2	−0.5	−0.6	−0.6
Developing countries	**2.1**	**1.6**	**1.4**	**1.1**	**1.2**
Developing Asia	1.8	1.4	1.1	0.8	0.9
China	*1.5*	*0.9*	*0.6*	*0.3*	*0.4*
India	*2.1*	*1.7*	*1.4*	*1.0*	*1.1*
Middle East	3.6	2.3	2.0	1.5	1.7
Africa	2.9	2.3	2.2	1.9	2.0
Latin America	2.0	1.6	1.2	0.9	1.0
Brazil	*2.1*	*1.5*	*1.2*	*0.8*	*0.9*
World	**1.7**	**1.4**	**1.1**	**0.9**	**1.0**
European Union	*0.3*	*0.3*	*0.1*	*0.0*	*0.0*

Note: These assumptions also apply to the Alternative Policy and High Growth Scenarios.

As a result of declining birth rates and increasing longevity, the populations of a growing number of countries – especially in the OECD – are ageing rapidly. Several European and Pacific countries, notably Germany, Italy, Japan and Korea, face significant population declines and a jump in average age. Population ageing is less advanced in developing countries, but the majority of them are nonetheless expected to enter a period of rapid population ageing. All of the increase in world population will occur in urban areas; rural populations will decline. As a result, access to modern energy services is likely to improve, as it is generally less costly to supply urban communities. Building infrastructure to meet growing urban populations will still be a major challenge.

Macroeconomic Assumptions

Economic growth is by far the most important driver of energy demand. Consequently, our energy projections in every region remain highly sensitive to the underlying assumptions about GDP growth. In most regions, primary demand and gross domestic product have moved broadly in tandem over the last quarter of a century. Final demand for electricity and oil products for transport are particularly closely linked. Between 1991 and 2001, each percentage point increase in world GDP was accompanied by a 0.4% increase in primary demand. Nonetheless, the so-called income elasticity of energy demand – the increase in demand relative to GDP – rose sharply in 2001-2005 to 0.8, mainly due to China.[3]

The world economy expanded briskly in 2006, by 5.5% – up from 4.9% in 2005.[4] Growth was led by developing countries, with China and India the main driving forces. China's real output grew by 11.1% – the highest rate since 1995 – compared with 10.4% the year before. India saw growth of 9.7% in 2006, up from 9% in 2005. Rising investment and surging exports contributed to growth in both countries. High commodity prices and low interest rates continued to support growth in most other developing regions, notably the Middle East. In North America, Europe and Japan, growth picked up with stronger domestic demand. In most OECD countries, robust economic growth boosted tax revenues and trimmed budget deficits.

Inflationary pressures worldwide, which had been building in 2006 with strong household spending and rising oil and other commodity prices, have begun to ease with lower prices in some cases since the second half of 2006. Real long-term bond yields are still below long-term trends. Equity markets, after coming close to reaching all-time highs, fell back sharply in mid-2007 in response to a credit squeeze resulting from worries about the fall-out from a slump in the housing market in the United States. The US dollar has weakened, mainly against the euro and pound sterling. Although it has appreciated a little against the dollar, the Chinese renminbi (yuan) has declined modestly in real terms since 2005 (IMF, 2007). The US current account deficit has continued to rise, reaching $857 billion, or 6.5% of GDP, in 2006, partly as a result of an increase in the oil-import bill. Trade surpluses in China and the oil-exporting countries have increased further.

Globally, the pace of economic growth is expected to ease a little over the next couple of years. Developing countries are likely to continue to set the pace, though some slow-down is expected, associated with generally tighter monetary conditions. Growth in China is expected to ease slightly, to 10.5% in 2008. India's growth is also projected to slow, to 8.4%. Commodity exporters will continue to grow strongly, on the assumption that international prices remain

3. See *WEO 2006* for a detailed analysis of income and price elasticities of demand for different forms of energy.
4. Based on the IMF projections in the July 2007 edition of *World Economic Outlook Updates*, available at www.imf.org/external/pubs/ft/weo/2007/update/01/index.htm.

high (see below). In the longer term, growth rates in all regions are assumed to decline. World GDP is expected to grow on average by 3.6% per year over the projection period (Table 2). That rate is still higher than over the previous two-and-a-half decades, when it averaged 3.2%. Growth is assumed to drop from 4.2% in 2005-2015 to 3.3% in 2015-2030. China, India and other Asian countries are expected to continue to grow faster than all other regions, followed by Africa and the Middle East.[5] In all regions, the share of energy-intensive manufacturing in GDP declines in favour of lighter industries and/or services.

Table 2: **World Real GDP Growth in the Reference Scenario**
(average annual growth rates, %)

	1980-1990	1990-2005	2005-2015	2015-2030	2005-2030
OECD	**3.0**	**2.5**	**2.5**	**1.9**	**2.2**
North America	3.1	3.0	2.6	2.2	2.4
United States	*3.2*	*3.0*	*2.6*	*2.2*	*2.3*
Europe	2.4	2.1	2.4	1.8	2.0
Pacific	4.2	2.2	2.2	1.6	1.8
Japan	*3.9*	*1.3*	*1.6*	*1.3*	*1.4*
Transition economies	**−0.5**	**−0.4**	**4.7**	**2.9**	**3.6**
Russia	n.a.	−0.5	4.3	2.8	3.4
Developing countries	**3.9**	**5.8**	**6.1**	**4.4**	**5.1**
Developing Asia	6.6	7.3	6.9	4.8	5.6
China	*9.1*	*9.9*	*7.7*	*4.9*	*6.0*
India	*5.8*	*6.0*	*7.2*	*5.8*	*6.3*
Middle East	−0.4	4.2	4.9	3.4	4.0
Africa	2.2	3.0	4.5	3.6	3.9
Latin America	1.3	3.0	3.8	2.8	3.2
Brazil	*1.5*	*2.6*	*3.5*	*2.8*	*3.1*
World	**2.9**	**3.4**	**4.2**	**3.3**	**3.6**
European Union	*n.a.*	*2.0*	*2.3*	*1.8*	*2.0*

Note: These assumptions also apply to the Alternative Policy Scenario.

Combining our population and GDP growth assumptions yields an average increase in per-capita income of 2.6% per annum. Per-capita incomes grow most rapidly, by 3.9%, in the developing countries, notably China, where they increase by 5.6% per year on average. Incomes in OECD countries increase much more slowly, by an average of only 1.7% per year.

5. Economic prospects in China and India are discussed in detail in Chapters 7 and 14.

International Energy Price Assumptions

The actual prices paid by energy consumers affect how much of each fuel they wish to consume and how much they are prepared to invest in improving the efficiency of a particular technology used to provide a particular energy service. As in previous editions of the *Outlook*, pre-tax end-use prices for oil, gas and coal in each region are derived from assumed price trends on international markets. Final electricity prices are based on marginal power-generation costs, which are derived from the cost of fossil-fuel inputs to generation, capital costs and non-fuel operating costs. Rates of *ad valorem* taxes and excise duties are assumed to remain constant over the projection period. Energy is often heavily taxed or subsidised and prices regulated. As a result, the final price to the end use changes much less, proportionately, than the change in the international price.

The assumed trajectories of international prices, summarised in Table 3, reflect our judgment of the prices that will be needed to generate sufficient investment in supply to meet projected demand over the *Outlook* period, taking account of market conditions. They should not be interpreted as forecasts. Similarly, although the price paths follow smooth trends, short-term fluctuations in price are inevitable.

Oil prices are expected to remain the main driver of energy prices generally, through inter-fuel competition and price indexation clauses in some long-term gas contracts. International crude oil and refined product prices rose strongly between 2003 and the middle of 2006. They fell back in the second half of 2006, but recovered steadily after January 2007 on stronger demand, OPEC production cuts, supply disruptions in Nigeria and elsewhere and continuing tightness in refining capacity. Persistent geopolitical tensions have helped to keep prices up. In August 2007, the nominal price of Brent crude oil hit a new all-time record of just under $79 per barrel compared with a peak of $75 in 2006 and an average of less than $27 in 2001. Spot prices of oil products – especially gasoline – have generally risen even more than those of crude oil, boosting refining margins. However, a fall in the value of the dollar against most of the main currencies has offset part of the increase in oil prices over the last five years or so.

In this *Outlook*, the IEA crude oil import price – a proxy for international oil prices – is assumed to average around $63 per barrel in 2007 and then fall marginally to around $57 in real year-2006 dollars by 2015.[6] This is based on an assumption that crude oil production and refining capacity will rise marginally faster than demand, as the recent wave of investment in new facilities bears fruit (IEA, 2006; IEA, 2007a). Prices are then assumed to

6. In 2006, the average IEA crude oil import price was $4.45 per barrel lower than first-month West Texas Intermediate (WTI) and $3.34 lower than dated Brent.

Table 3: Fossil-Fuel Price Assumptions in the Reference Scenario
(in year-2006 dollars per unit)

	unit	2000	2006	2010	2015	2030
Real terms						
(year-2006 prices)						
IEA crude oil imports	barrel	32.49	61.72	59.03	57.30	62.00
Natural gas						
United States imports	*MBtu*	*4.49*	*7.22*	*7.36*	*7.36*	*7.88*
European imports	*MBtu*	*3.27*	*7.31*	*6.60*	*6.63*	*7.33*
Japanese LNG imports	*MBtu*	*5.49*	*7.01*	*7.32*	*7.33*	*7.84*
OECD steam coal imports	tonne	39.05	62.87	56.07	56.89	61.17
Nominal terms						
IEA crude oil imports	barrel	28.00	61.72	65.00	70.70	107.59
Natural gas						
United States imports	*MBtu*	*3.87*	*7.22*	*8.11*	*9.08*	*13.67*
European imports	*MBtu*	*2.82*	*7.31*	*7.27*	*8.18*	*12.71*
Japanese LNG imports	*MBtu*	*4.73*	*7.01*	*8.06*	*9.05*	*13.61*
OECD steam coal imports	tonne	33.65	62.87	61.74	70.19	106.14

Note: Prices in the first two columns represent historical data. Gas prices are expressed on a gross calorific-value basis. All prices are for bulk supplies exclusive of tax. Nominal prices assume inflation of 2.3% per year from 2007.

recover slowly, reaching $62 by 2030. In nominal terms, this equates to a price of almost $110. As always, future price trends hinge on the investment and production policies of a small number of countries – mainly Middle Eastern members of the Organization of the Petroleum Exporting Countries (OPEC) – that hold the bulk of the world's remaining oil reserves, as well as on demand prospects. Prices are slightly higher than in last year's *WEO*, mainly because of the continuing tightness of crude oil and product markets and acute supply-side constraints, including growing barriers to upstream investment in several resource-rich countries and refinery bottlenecks.

The near-term outlook too remains very uncertain. New capacity additions in greenfield projects are expected to increase over the next five years, but it is far from clear whether they will be sufficient to compensate for the decline in output at existing fields and to keep pace with the projected increase in demand in the Reference Scenario (see Chapter 1). Any fall in the spare production capacity held by OPEC producers, possibly caused by faster decline rates, stronger demand growth than expected or delays in bringing upstream and downstream projects on stream, would put upward pressure on prices and induce greater price volatility.

Natural gas prices have broadly followed the rise in oil prices since 2003, typically with a lag of up to one year. In most bulk-supply contracts outside North America, Great Britain and Australia, where gas-to-gas competition has developed, gas prices are indexed against oil prices. Even in competitive markets, oil prices still influence gas prices because of competition between gas and oil products. Natural gas markets remain highly regionalised. Yet, averaged over time, regional prices usually move broadly in parallel with one another because of their link to oil prices.

In our Reference Scenario, gas prices are assumed to fall back a little from highs reached in 2007 in all three regions over the next five years or so and then to start to rise steadily early in the next decade in line with oil prices. Rising supply costs contribute to higher gas prices in North America and Europe (IEA, 2007b). Bulk gas import prices nonetheless remain markedly lower than spot crude oil prices on an energy-content basis, reflecting the additional cost of distributing gas to end users and inter-fuel competition in final uses. Increased short-term trading in liquefied natural gas (LNG), which permits arbitrage among regional markets, is expected to lead to some convergence in regional prices over the projection period (Figure 2).

International steam-coal prices have generally risen less rapidly than oil prices since 2002, with the average OECD steam-coal import price jumping from

Figure 2: **Assumed Ratio of Natural Gas and Implied Relation of Coal Prices to Oil Prices in the Reference Scenario**

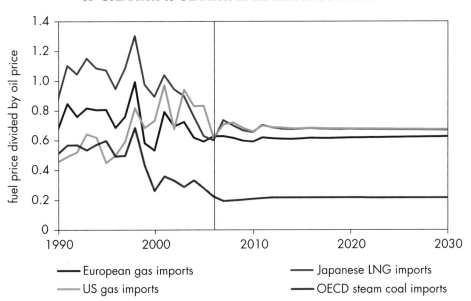

$41 per tonne to $63 in 2006 (in year-2006 dollars). Exceptionally strong demand for coal from power generators and steel manufacturers, especially in China, has boosted overall demand and helped drive up prices – but less in proportionate terms than oil or gas prices. Market fundamentals point to a slight weakening of coal prices in 2007 and 2008, after which they are assumed to remain flat until the middle of the next decade. They are then assumed to rise very slowly, reaching just over $60 per tonne by 2030. Coal prices are broadly constant relative to oil prices from the beginning of the next decade. In the short term, the strength of Chinese and Indian import demand is a particularly important uncertainty for coal prices. In the medium to longer term, the price of coal will remain sensitive to environmental restrictions on coal-burning and developments in clean coal technology, which could allow coal to be used to generate power much more efficiently and in a way that emits less carbon dioxide.

The Alternative Policy Scenario

Since 2000, the *WEO* has presented an Alternative Policy Scenario to assess the potential impact of additional government actions to rein in the growth of energy demand for reasons of energy security or environmental sustainability. The Alternative Policy Scenario takes into account those policies and measures that countries are currently considering and are assumed to adopt and implement, taking account of technological and cost factors, the political context and market barriers. Macroeconomic and population assumptions are the same as in the Reference Scenario. Only policies aimed at enhancing energy security and/or addressing environmental problems, including climate change, are considered. While cost factors are taken into consideration in determining whether or not they are assumed to be implemented, policies are not selected according to their relative economic cost-effectiveness. Rather, they reflect the proposals actually under discussion in the current energy-policy debate.

With each edition of the *Outlook*, the analysis of the Alternative Policy Scenario has been deepened and broadened. *WEO-2006* raised the bar several notches, by compiling and analysing more than 1 400 policies from both OECD and non-OECD countries. This *Outlook* updates that exercise, adding new policies that have been proposed during the year to mid-2007 and moving those policies that have recently been adopted from the Alternative Policy Scenario to the Reference Scenario. Energy and climate policies have remained at the top of the political agenda in many countries, reflecting continuing anxieties about the vulnerability of oil and gas supplies to disruption, rising greenhouse-gas emissions and firmer evidence about the likelihood, extent and long-term economic costs of climate change. The number and strength of policies under consideration continue to grow faster than the number and strength of new policies actually adopted, reflecting a general pattern of growing concern, but more talk than action.

Modelling the impact of the new policies on energy demand and supply involves two main steps. First, the effects of each policy or measure on the main drivers of energy markets are assessed quantitatively, policy by policy. As with the Reference Scenario, a degree of judgment is inevitably involved in translating policies into formal assumptions. Second, these effects are incorporated into the IEA's World Energy Model (WEM) to generate projections of energy demand and supply, related CO_2 emissions and investments. As many of these policies have effects at a micro-level, detailed "bottom-up" sub-models of the energy system have been developed, allowing all policies to be analysed within a coherent and consistent modelling framework. These sub-models explicitly incorporate the energy efficiency of specific technologies, the activities that drive energy demand and the rates of turnover of the physical capital stock of energy-using equipment. The very long life of certain types of energy capital goods limits the rate at which more efficient technology can penetrate and so reduce energy demand. The rebound effect on energy demand of introducing more efficient energy-consuming goods is also modelled. The policies of the Alternative Policy Scenario generally lead to the faster development and deployment of more efficient and cleaner energy technologies, resulting in a more rapid decline in global energy intensity than in the Reference Scenario.

Many of the policies[7] analysed in this scenario were proposed a year or more ago and are still awaiting approval. These include more rigorous action to promote renewables, including biofuels, more stringent energy-efficiency standards and more ambitious plans for nuclear power. A number of new policies have also been proposed, including measures to meet new EU targets to reduce CO_2 emissions and to increase the share of renewables in primary energy supply by 2020, and an EU energy-efficiency action plan, as well as new US Corporate Average Fuel Economy (CAFE) standards for cars (beyond those agreed over the past year) and a long-term target for biofuel use in US road transport. These policies are assumed to be implemented in the Alternative Policy Scenario, though their effectiveness varies, so not all targets are met promptly. High energy prices and concerns about energy security are the principal drivers of some of these moves, but they are increasingly motivated by worries about greenhouse-gas emissions. Other proposed policies aim directly at lowering fossil-energy use and related emissions, such as the bans on the sale of conventional incandescent light bulbs that many countries are now considering.

---

7. The full list of policies and measures analysed for the Alternative Policy Scenario for all countries can be downloaded from the *WEO* website at www.worldenergyoutlook.org.

In line with last year's *Outlook*, we have assumed that international oil prices in the Alternative Policy Scenario are the same as in the Reference Scenario, on the assumption that the investment and production policies of OPEC countries are adjusted to accommodate lower demand for their oil. Gas prices are also assumed to be the same, because of the widespread use of oil-price indexation in long-term gas supply contracts. Coal import prices, however, are assumed to fall – especially towards the end of the projection period – in response to the lower supply-demand equilibrium established in the Alternative Policy Scenario. The average steam coal import price is assumed to fall from $61 per tonne in 2030 in the Reference Scenario to $56. Electricity prices increase in some regions, reflecting changes in the fuel inputs and in the cost of power-generation technologies.

A number of policies in China and India – mainly driven by energy-security and local environmental concerns – have been taken into consideration in the Alternative Policy Scenario. The development of more detailed models for both countries permits a more robust analysis of their impact. China's 11th Five-Year Plan contains a commitment to generate 10% of its electricity from renewables by 2010 and to cut overall energy intensity by 20% between 2005 and 2010 (though recent trends make this a very challenging target). The plan also contains targets for local pollutants. In June 2007, China released a National Climate Change Programme, which includes some new measures to curb the growth in greenhouse-gas emissions. These policies are taken into consideration in the Alternative Policy Scenario. In India, we have analysed 80 different policies and measures, including actions aimed at improving energy efficiency in end-use sectors and promoting the deployment of advanced power-generation technologies.

There is growing support worldwide for radical and urgent action to bring long-term CO_2 emissions down in order to achieve stabilisation of concentrations of the gas at levels compatible with an acceptable increase in global temperatures. At their recent summit in Heiligendamm, G8 leaders "agreed to consider" strategies to halve global emissions by 2050 – an objective in line with long-term stabilisation of the concentration of greenhouse gases in the atmosphere in the range of 445 to 490 parts per million of CO_2 equivalent and a maximum rise in temperature of 2.4°C.[8] We estimate that achieving this goal would require energy-related CO_2 emissions to be reduced to around 23 gigatonnes in 2030 – 11 Gt less than in the Alternative Policy Scenario. We have developed a "450 Stabilisation Case", which describes one possible pathway to achieving this goal, taking the Alternative Policy Scenario as a starting point, in order to illustrate the extent of the challenge of transforming

8. According to IPCC (2007).

the global energy system over the projection period. The results are described in Chapter 5. Achieving this outcome would be possible only with very strong political will worldwide and at substantial economic cost.

The High Growth Scenario

The economies of China and India have continued to grow strongly in recent years, exceeding most forecasts and our own assumptions in previous *Outlooks*. Our projections of energy demand – in all countries and regions – remain highly sensitive to future rates of GDP growth. We assume in the Reference and Alternative Policy Scenarios that GDP growth in China and India slows progressively over the projection period, as their economies mature. But their economic prospects are inevitably uncertain. Past forecasts of GDP have often been revised significantly upwards (see Chapter 3). Were GDP growth to slow less quickly than assumed in the Reference and Alternative Policy Scenarios, energy demand could turn out to be much higher than projected. The cumulative impact of even a marginally higher annual rate of GDP growth means that the level of demand in 2030 could be substantially higher, with far-reaching implications both for China and India and for the rest of the world.

To shed light on the global impact of faster than expected economic growth in China and India, we have developed a High Growth Scenario. The starting point of this analysis is the assumption that GDP growth in both countries is on average 1.5 percentage points per year higher than in the Reference Scenario. This results in an average growth rate to 2030 of 7.5% for China and 7.8% for India. For China, we assume that the main driver of growth in this scenario is sustained high investment and continued rapid productivity gains, as the government pushes ahead with reforms to increase the role of the private sector and to open up the economy to foreign investment. For India, we assume an acceleration and deepening of structural and institutional reforms, combined with faster infrastructure development.

To model the energy-market impact of these higher assumed GDP growth rates, the WEM has been integrated into a general equilibrium model.[9] The resulting hybrid model, WEM-ECO, provides a consistent energy and macroeconomic modelling framework within which energy pathways interact with the macro-economy, in terms of changes in economic structure, productivity and trade that affect the rate, direction and distribution of economic growth and energy demand and supply. It also allows us to quantify the broader economic gains and losses by region. By assuming higher Chinese and Indian growth rates, the integrated model recalculates the global

9. For this purpose we integrated the WEM into the IMACLIM-R framework, developed by CIRED. See Chapter 3 for details.

equilibrium for international trade in energy and non-energy goods and services, energy and other commodity prices and GDP in the rest of the world by major region. The principal channels through which energy markets are affected by faster Chinese and Indian growth are as follows:

- Higher GDP growth in China and India leads to both increased exports and imports of goods and services, and boosts net imports of energy and other commodities, pushing up international prices.

- Other countries benefit from stronger demand from China and India for their exports of goods and services, including commodities. Oil-exporting countries receive a particularly strong boost to both the volume and value of their exports. Exporters of high-tech manufactured goods and services also benefit from stronger Chinese and Indian demand.

- Increased exports of manufactured goods and services *from* China and India offset at least part of the economic stimulus in the rest of the world. Those countries with a pattern of exports similar to that of China and India are most affected by increased Chinese and Indian competition.

- Higher commodity prices, including energy prices, damage the terms of trade and income in net importing countries, offsetting part or all of any boost to income provided by stronger demand from China and India.

- In all countries, higher energy prices push down demand – especially in the long term – offsetting part or all of any stimulus to demand provided by the higher economic growth that results from increased trade with China and India.

The net impact on national income and output, and on energy demand and trade, varies by region according to the structure of the economy, the degree of dependence on energy imports and the potential for raising energy production in response to higher international prices. The detailed results can be found in Chapter 3.

PART A
GLOBAL ENERGY PROSPECTS: IMPACT OF DEVELOPMENTS IN CHINA & INDIA

CHAPTER 1

GLOBAL ENERGY TRENDS

HIGHLIGHTS

- World primary energy demand in the Reference Scenario is projected to grow by more than half between 2005 and 2030, at an average annual rate of 1.8%. Demand reaches 17.7 billion toe, compared with 11.4 billion toe in 2005 – a rise of 55%. Global energy intensity – total energy use per unit of gross domestic product – falls by 1.8% per year over 2005-2030.

- Fossil fuels remain the dominant source of primary energy, accounting for 84% of the overall increase in global demand between 2005 and 2030. Oil remains the single largest fuel, though its share falls from 35% to 32%. Oil demand reaches 116 mb/d in 2030, 32 mb/d, or 37%, up on 2006. Coal sees the biggest increase in demand in absolute terms, jumping by 73% between 2005 and 2030, pushing its share of total energy demand up from 25% to 28%. The share of natural gas increases more modestly, from 21% to 22%. Electricity use almost doubles, its share of final energy consumption rising from 17% to 22%. Some $22 trillion of investment in supply infrastructure is needed to meet projected global demand.

- Developing countries, whose economies and populations are growing fastest, contribute 74% of the increase in global primary energy use. China and India alone account for 45% of the increase. OECD countries account for one-fifth and the transition economies the remaining 6%. China overtakes the United States soon after 2010 to become the world's biggest energy consumer. In 2005, US demand was more than one-third larger.

- Although new oil-production capacity additions from greenfield projects are expected to increase over the next five years, it is very uncertain whether they will be sufficient to compensate for the decline in output at existing fields and keep pace with the projected increase in demand in the Reference Scenario. A supply-side crunch in the period to 2015, involving an abrupt escalation in oil prices, cannot be ruled out.

- In the Alternative Policy Scenario, global primary energy demand grows by 1.3% per year over 2005-2030, 0.5 percentage points less than in the Reference Scenario – resulting in an 11% saving in 2030. Oil demand is 14 mb/d lower in 2030 than in the Reference Scenario – equal to the entire current output of the United States, Canada and Mexico combined. The gap in energy demand between the two scenarios widens progressively over the *Outlook* period, as opportunities grow for installing more energy-efficient equipment.

- In the High Growth Scenario, faster economic growth in China and India boosts their energy demand *vis-à-vis* the Reference Scenario. The stimulus to demand provided by stronger economic growth more than offsets the depressive effect of higher international energy prices. Worldwide, the increase in primary energy demand amounts to 6% in 2030.

Reference Scenario

Global Energy Prospects

World primary energy demand[1] in the Reference Scenario, in which government policies are assumed to remain unchanged from mid-2007, is projected to grow by 55% between 2005 and 2030, an average annual rate of 1.8%. Demand reaches 17.7 billion tonnes of oil equivalent, compared with 11.4 billion toe in 2005 (Table 1.1). The pace of demand growth slackens progressively over the projection period, from 2.3% per year in 2005-2015 to 1.4% per year in 2015-2030. Demand grew by 1.8% per year in 1980-2005.

Table 1.1: **World Primary Energy Demand in the Reference Scenario**
(Mtoe)

	1980	2000	2005	2015	2030	2005-2030*
Coal	1 786	2 292	2 892	3 988	4 994	2.2%
Oil	3 106	3 647	4 000	4 720	5 585	1.3%
Gas	1 237	2 089	2 354	3 044	3 948	2.1%
Nuclear	186	675	721	804	854	0.7%
Hydro	147	226	251	327	416	2.0%
Biomass and waste	753	1 041	1 149	1 334	1 615	1.4%
Other renewables	12	53	61	145	308	6.7%
Total	**7 228**	**10 023**	**11 429**	**14 361**	**17 721**	**1.8%**

* Average annual rate of growth.

The projected level of global demand in 2030 is about 4% higher than in last year's edition of the *Outlook*, almost entirely because of higher demand in China and India. The global fuel mix is little changed, such differences as there are resulting mainly from adjustments to assumed GDP rates and international energy prices.[2] Since *WEO-2006*, some new government policies and measures have been adopted, mostly in OECD countries, and their impact is taken into account in the Reference Scenario. These include tighter fuel-economy

1. World total primary energy demand, which is equivalent to total primary energy supply, includes international marine bunkers, which are excluded from the regional totals. Primary energy refers to energy in its initial form, after production or importation. Some energy is transformed, mainly in refineries, power stations and heat plants. Final consumption refers to consumption in end-use sectors, net of losses in transformation and distribution. In all regions, total primary and final demand includes traditional biomass and waste, such as fuelwood, charcoal, dung and crop residues, some of which are not traded commercially. For details of statistical conventions and conversion factors, please go to www.iea.org.
2. See Introduction for details about the macroeconomic and price assumptions.

standards for vehicles and new measures to boost biofuels in the United States and measures to support renewables generally in the European Union and Japan. However, the overall impact of these actions on total energy demand at the global level is limited. Other measures that have been newly proposed are reflected in the Alternative Policy Scenario.

Box 1.1: **Major Energy Developments since *WEO-2006***

Over the past year, a number of events led to a tightening of global energy markets, helping to drive up prices. Oil supplies from Nigeria were disrupted as a result of a worsening of the civil conflict in the Niger Delta. Several attacks on oil facilities forced companies to halt or slow production, delaying loadings. In mid-2007, a total of 750 thousand barrels per day of Nigerian output was shut in. Civil unrest in Iraq has continued to disrupt oil production, while geopolitical tensions elsewhere in the Middle East have persisted. Technical problems have occurred in the US refining sector, adding to the tightness of global refining capacity. Disagreements between Russia and some of its neighbours over oil and gas pricing and transit fees also created worries in importing countries over the security and continuity of supply. Oil flows through Belarus and gas flows through Ukraine have incurred temporary disruptions since the beginning of 2006. Despite these various constraints and rising demand, the Organization of the Petroleum Exporting Countries (OPEC) announced a production cut of 1.2 mb/d in November 2006 and a further 0.5-mb/d cut in February 2007. In the face of rising prices, OPEC agreed to raise output by 0.5 mb/d in September 2007, but this move was not able to prevent crude oil prices from continuing to rise. The price of Brent crude rose to over $76 per barrel in nominal terms – breaching the all-time highs recorded in 2006.

World coal demand has continued to increase strongly, keeping prices high. Preliminary data show that China continued to account for the bulk of the growth, its demand outstripping production. By early 2007, China had become a net importer of coal. The recent surge in coal use has led to acceleration in the growth of global CO_2 emissions, at a time of growing global attention to the threat of climate change. Major reports have recently been published highlighting the risk of global warming and the potential consequences and costs of inaction. For example, the European Union has adopted a target to reduce CO_2 emissions by 2020 and adopted national allocation plans for the second trading period (2008-2012) of the EU Greenhouse Gas Emissions Trading Scheme. Efforts to reach a global accord on collective action to mitigate emissions beyond the Kyoto Protocol commitment period have been stepped up, though an agreement has not yet been reached.

Fossil fuels – oil, natural gas and coal – remain the dominant sources of primary energy worldwide in the Reference Scenario (Figure 1.1). They account for 84% of the overall increase in energy demand between 2005 and 2030. Their share of world demand rises from 81% in 2005 to 82% in 2030. Oil remains the single largest fuel, though its share falls from 35% to 32%. The share of coal rises from 25% to 28%, and that of natural gas from 21% to 22%. The rise in fossil-energy use drives up related emissions of carbon dioxide by 57% between 2005 and 2030 (see Chapter 5).

Figure 1.1: **World Primary Energy Demand in the Reference Scenario**

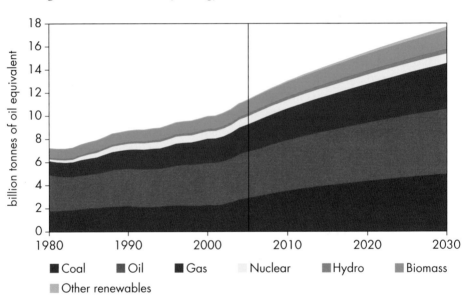

Coal sees the biggest increase in demand among all primary energy sources in absolute terms between 2005 and 2030, closely followed by natural gas and oil (Figure 1.2). Coal demand jumps by 38% between 2005 and 2015 and 73% by 2030 – a faster increase than in previous editions of the *Outlook*. Nuclear power accounts for most of the fall in the share of non-fossil primary fuels, dropping from 6% of total primary energy demand in 2005 to 5% in 2030. There is no change in the share of hydropower, at 2%, and the share of biomass and waste falls slightly, from 10% to 9%. The share of other renewables, a category that includes wind, solar, geothermal, tidal and wave energy, rises from less than 1% to about 2%.

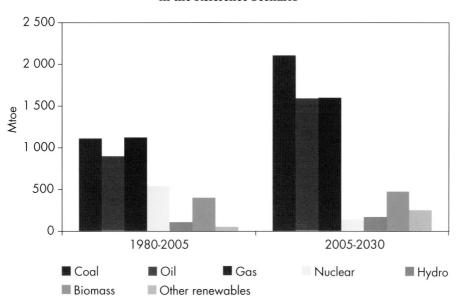

Figure 1.2: **Increase in World Primary Energy Demand by Fuel in the Reference Scenario**

Regional Demand Trends

Developing countries are projected to contribute around 74% of the increase in global primary energy consumption between 2005 and 2030 (Figure 1.3). Their economies and populations grow much faster than those of the industrialised countries, pushing up their energy use. China and India alone account for 45% of the increase in energy use. OECD countries account for one-fifth, the transition economies for 6% and other developing countries for the rest. China is projected to overtake the United States soon after 2010 to become the world's largest energy-consuming country. In 2005, US demand was 34% larger than Chinese demand. In aggregate, developing countries make up 47% of the global energy market in 2015 and more than half in 2030, compared with only 41% today. The OECD's share *falls* from 48% now to 43% in 2015 and to 38% in 2030. The share of the transition economies is flat at 9% through to around 2020 and then drops to 8%.

The developing countries' share of global demand expands for all primary energy sources, except non-hydro renewables. The increase is pronounced for nuclear power, which drops in OECD Europe while expanding in China and other parts of Asia/Pacific. Nuclear output grows marginally in the rest of the world. The developing regions' share of world coal consumption is also projected to increase sharply, mainly because of booming demand in China and India. By 2030, the two countries together account for 60% of total world coal demand, up from 45% in 2005. Over three-quarters, or 25 mb/d, of the

32-mb/d increase in global oil demand between 2006 and 2030 will come from developing regions. China, India and the rest of developing Asia account for most of this increase. Non-OECD countries account for 72% of the growth in world natural gas demand (Figure 1.4).

Figure 1.3: **Primary Energy Demand by Region in the Reference Scenario**

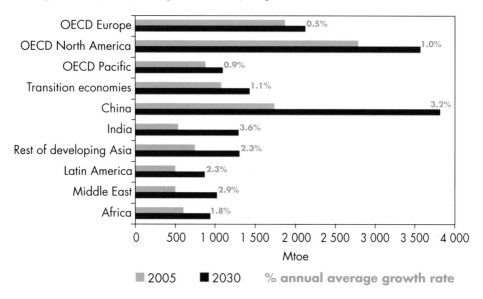

Figure 1.4: **Regional Shares in Incremental Primary Energy Demand by Fuel in the Reference Scenario, 2005-2030**

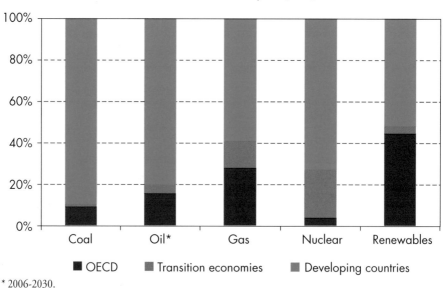

* 2006-2030.

Global primary energy intensity, measured as total energy use per unit of gross domestic product, is projected to fall on average by 1.8% per year over 2005-2030. This compares with a decline of 1.6% over the period 1990-2005 (Figure 1.5). The projected acceleration in the rate of decline is due largely to faster structural economic change away from heavy manufacturing and towards less energy-intensive service activities and lighter industry. For this reason – and due to the strong potential for thermal efficiency gains in power generation – intensity falls most quickly in the non-OECD regions. The transition economies, in particular, become much less energy-intensive as subsidies are lowered, more energy-efficient technologies are introduced and energy waste is reduced.

Figure 1.5: **Primary Energy Intensity in the Reference Scenario**

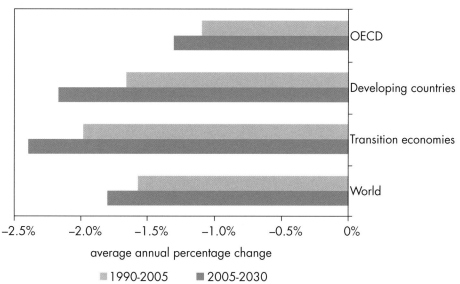

Oil

Oil demand is projected to grow by 1.3% per year, from 83.7 mb/d in 2005 (and 84.7 mb/d in 2006[3]) to 98.5 mb/d in 2015 and 116.3 mb/d in 2030. Some 42% of the increase in 2006-2030 comes from China and India. In absolute terms, their demand grows by 13.3 mb/d (Table 1.2). Indian demand grows fastest, on average by 3.9% per year, while Chinese demand grows at 3.6% per year. China accounts for the biggest increase in oil demand in absolute terms of any country or region.

3. Preliminary data on total oil demand only are available for 2006 by region. Oil does not include biofuels derived from biomass. For this reason, the oil projections in this report are not directly comparable with those published in the IEA's *Oil Market Report.*

Table 1.2: **World Primary Oil Demand in the Reference Scenario**
(million barrels/day)

	1980	2000	2006	2010	2015	2030	2006-2030*
OECD	**41.8**	**46.0**	**47.3**	**49.0**	**50.8**	**52.9**	**0.5%**
North America	20.9	23.4	24.9	26.2	27.7	30.0	0.8%
Europe	14.7	14.2	14.3	14.5	14.7	14.7	0.1%
Pacific	6.3	8.4	8.1	8.3	8.3	8.1	0.0%
Transition economies	**9.4**	**4.2**	**4.5**	**4.7**	**5.1**	**5.6**	**0.9%**
Russia	n.a.	2.6	2.6	2.8	3.0	3.3	0.9%
Developing countries	**11.3**	**23.1**	**28.8**	**33.7**	**38.7**	**53.3**	**2.6%**
China	1.9	4.7	7.1	9.0	11.1	16.5	3.6%
India	0.7	2.3	2.6	3.1	3.7	6.5	3.9%
Other Asia	1.8	4.5	5.5	6.2	6.9	8.9	2.0%
Middle East	2.0	4.6	6.0	7.0	7.9	9.5	1.9%
Africa	1.3	2.3	2.8	3.1	3.4	4.8	2.2%
Latin America	3.5	4.7	4.8	5.2	5.6	7.1	1.6%
Int. marine bunkers and stock changes	2.2	3.6	4.1	3.7	3.9	4.5	n.a.
World	**64.8**	**77.0**	**84.7**	**91.1**	**98.5**	**116.3**	**1.3%**
European Union	*n.a.*	*13.6*	*13.8*	*13.8*	*14.0*	*13.8*	*0.0%*

* Average annual rate of growth.

The transport sector is the principal driver of oil demand in most regions (Figure 1.6). Globally, transport's share of total primary oil use rises from 47% in 2005 to 52% in 2030. Although biofuels take an increasing share of the market for road-transport fuels, oil-based fuels continue to dominate, their share of transport demand falling from 94% to 92% over the projection period. Worldwide, consumption of oil for transport is projected to grow by 1.7% per year over 2005-2030. Demand grows fastest in the developing regions, in line with rising incomes and investment in infrastructure. Today, there are about 900 million vehicles on the world's roads (excluding two-wheelers); by 2030, their number is expected to pass 2.1 billion. Most of the extra vehicles are destined to be used in Asia. The non-OECD vehicle fleet overtakes that of OECD countries in aggregate by around 2025, and is 30% larger by 2030. Major improvements in vehicle fuel economy in all regions slow the growth in demand for gasoline and diesel, but do not reverse it. Industry and the residential and service sectors account for most of the rest of the increase in global oil demand, with most of the growth coming from non-OECD countries. Oil demand for power generation remains small.

Figure 1.6: **Share of Transport in Primary Oil Demand by Region in the Reference Scenario**

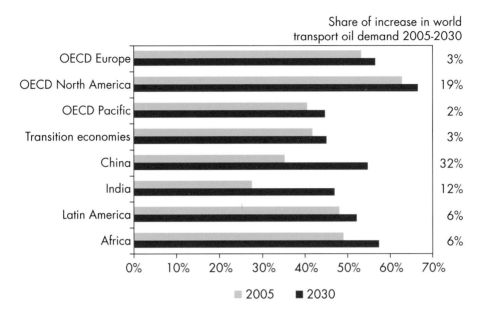

Share of increase in world transport oil demand 2005-2030

Region	%
OECD Europe	3%
OECD North America	19%
OECD Pacific	2%
Transition economies	3%
China	32%
India	12%
Latin America	6%
Africa	6%

■ 2005   ■ 2030

World oil resources are sufficient to meet the projected growth in demand to 2030. Non-conventional supplies such as gas-to-liquids and oil sands are expected to make a growing contribution to output over the projection period. OPEC countries collectively are projected to take a growing share of the world oil market in the Reference Scenario (Table 1.3), as they hold the bulk of remaining proven oil reserves and ultimately recoverable resources, and their development and production costs are generally lower than elsewhere. Their collective output of conventional crude oil, natural gas liquids and non-conventional oil (mainly gas-to-liquids) rises from 36 mb/d in 2006 to 46 mb/d in 2015 and 61 mb/d in 2030, reflecting their share of the global resource base. As a result, OPEC's share of world oil supply jumps from 42% now to 52% by the end of the projection period. These outcomes depend critically on investment and production policies in key OPEC countries.[4]

Non-OPEC production rises slowly to 2030, with most of the increase coming from non-conventional sources – mainly Canadian oil sands – after 2015 as conventional output levels off; only the transition economies, Latin America, Canada and Africa see a continued increase in overall output. Increased output in the transition economies and non-OPEC developing countries taken as a

---

4 The implications of deferred upstream investment in the Middle East and North Africa were analysed in detail in *WEO-2005.*

Table 1.3: **World Oil Production in the Reference Scenario**
(million barrels/day)

	1980	2000	2006	2010	2015	2030	2006-2030*
Non-OPEC	**35.5**	**43.5**	**47.0**	**48.6**	**50.3**	**53.2**	**0.5%**
OECD	17.3	21.8	19.7	18.7	18.3	18.2	–0.3%
North America	*14.2*	*14.2*	*13.9*	*13.8*	*14.1*	*15.2*	*0.4%*
United States	*10.3*	*8.0*	*7.1*	*7.1*	*6.7*	*6.3*	*–0.5%*
Europe	*2.6*	*6.8*	*5.2*	*4.1*	*3.4*	*2.5*	*–3.0%*
Pacific	*0.5*	*0.9*	*0.6*	*0.8*	*0.7*	*0.5*	*–0.6%*
Transition economies	12.1	8.3	12.4	14.0	14.9	17.2	1.4%
Russia	*10.8*	*6.5*	*9.7*	*10.6*	*10.8*	*11.2*	*0.6%*
Developing countries	6.1	13.4	14.9	15.8	17.1	17.8	0.7%
China	*2.1*	*3.2*	*3.7*	*3.9*	*4.0*	*3.4*	*–0.3%*
India	*0.2*	*0.7*	*0.8*	*0.9*	*0.7*	*0.5*	*–1.8%*
Other Asia	*0.6*	*1.6*	*1.9*	*2.0*	*2.0*	*1.6*	*–0.7%*
Latin America	*1.6*	*3.6*	*4.1*	*4.6*	*5.5*	*7.1*	*2.3%*
Brazil	*0.2*	*1.3*	*1.8*	*2.4*	*3.0*	*3.8*	*3.2%*
Africa	*1.1*	*2.1*	*2.6*	*2.8*	*3.3*	*3.6*	*1.3%*
Middle East	*0.6*	*2.1*	*1.7*	*1.6*	*1.6*	*1.6*	*–0.3%*
OPEC**	**28.1**	**31.7**	**35.8**	**40.6**	**46.0**	**60.6**	**2.2%**
Middle East	19.2	21.3	24.1	27.5	31.8	45.0	2.6%
Saudi Arabia	*10.1*	*9.1*	*10.5*	*12.0*	*13.2*	*17.5*	*2.2%*
Non-Middle East	9.0	10.4	11.8	13.1	14.3	15.6	1.2%
OPEC market share	**43%**	**42%**	**42%**	**45%**	**47%**	**52%**	**0.9%**
Processing gains	**1.7**	**1.7**	**1.9**	**2.0**	**2.2**	**2.6**	**1.3%**
World	**65.2**	**76.8**	**84.6**	**91.1**	**98.5**	**116.3**	**1.3%**
Conventional oil***	63.1	73.9	80.9	86.6	92.1	105.2	1.1%
Non-conventional oil****	0.4	1.3	1.8	2.5	4.2	8.5	6.7%
Canada	*0.2*	*0.6*	*1.2*	*1.8*	*2.8*	*4.9*	*6.2%*
OPEC	*0.0*	*0.2*	*0.2*	*0.2*	*0.6*	*1.2*	*8.2%*
Other non-OPEC	*0.2*	*0.5*	*0.5*	*0.5*	*0.8*	*2.5*	*7.1%*

* Average annual rate of growth. ** Includes Angola, which joined OPEC at the beginning of 2007. ***Conventional crude oil and natural gas liquids (NGLs). **** Extra heavy oil, natural bitumen, gas-to-liquids and coal-to-liquids. Biofuels are not included.

whole is insufficient to compensate for a continued decline in conventional OECD production – especially in North America and Europe – after 2015. The increase in non-OPEC supply between 2006 and 2030 comes from

Russia, Central Asia, Latin America and Africa. Output in developing Asia peaks by the beginning of the next decade and then declines.

Inter-regional oil trade grows rapidly over the *Outlook* period, with a widening of the gap between indigenous output and demand in every major *WEO* region (Figure 1.7). The volume of trade expands from 41 mb/d in 2006 to 51 mb/d in 2015 and 65 mb/d in 2030. As refining capacity for export increases, a growing share of trade in oil is expected to be in the form of refined products, notably in the Middle East and India. The biggest increase in net exports occurs in the Middle East. The transition economies, Africa and Latin America also export more oil. All other regions – including China and India – have to import more oil. In fact, the volume increase in imports between now and 2030 in China is larger than that in any other *WEO* region (see Chapter 4 for a detailed discussion of the global implications of these trends).

Figure 1.7: **Net Oil Trade* in the Reference Scenario**

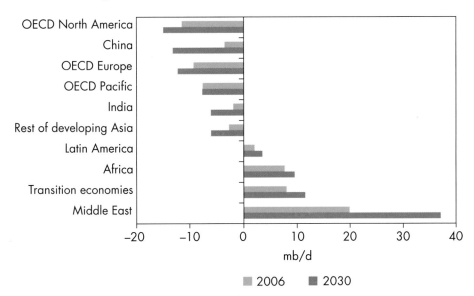

*Trade between *WEO* regions only. Negative figures indicate net imports.

Outlook for Oil Production Capacity to 2015

Although investment and new capacity additions in greenfield projects are expected to increase over the next five years, it is uncertain whether they will be sufficient to offset both the decline in output at existing fields and the projected increase in demand in the Reference Scenario. In addition, there are growing doubts about the willingness and ability of the national oil companies to increase installed capacity once the projects now under construction or sanctioned have been brought on stream. The prospects for oil demand are inevitably uncertain because of uncertainty about future GDP growth: global demand, particularly in China, could grow much more quickly than projected in the Reference Scenario, as the High Growth Scenario shows (see below). In view of these uncertainties, a supply-side crunch in the period to 2015, involving an abrupt run-up in prices, cannot be ruled out.

OPEC countries have launched or plan to launch over 90 major projects that will, in aggregate, add an estimated 11.4 mb/d of gross crude oil and NGLs production capacity to 2006 levels by 2012.[5] We estimate that planned gross capacity additions from new projects in non-OPEC countries (including non-conventional sources, such as oil sands and gas-to-liquids) over the same period will amount to 13.6 mb/d (IEA, 2007a). The bulk of this new capacity will be in Russia, the Caspian region and in deep-water locations, such as the Gulf of Mexico and West Africa. But much of this new capacity – especially in non-OPEC countries – will be needed simply to replace the capacity that will be lost as a result of the depletion and associated decline in production from existing fields. Additional investment at existing fields will undoubtedly occur to combat the natural rate of decline in output. But exactly how much is extremely uncertain, because little information is made available by oil companies on how much they plan to invest in existing fields and what the impact of those investments is expected to be on production rates at each field. In addition, slippage in the completion of projects currently in construction or planned could slow the rate of gross additions to capacity. Slippage on projects completed in the past year has averaged around six months.

The prospects for net installed capacity and, therefore, the oil supply/demand balance are very sensitive to future decline rates, especially in the medium to long term.[6] Worldwide, we estimate that a weighted average *observed* decline rate from fields currently in production of around 3.7% per year would result in a match between global oil-supply capacity and demand in the Reference Scenario to 2012, based on current estimates of new gross capacity additions. At this same decline rate, *12.5 mb/d of gross capacity would need to be added between 2012 and 2015 to meet the increase in demand of 4.2 mb/d and make up for the decline at existing fields of 8.4 mb/d.* In total, 37.5 mb/d of gross capacity (including that needed to compensate for natural declines) needs to be added between 2006 and 2015. But decline rates may, in fact, turn out to be somewhat higher. An increase of a mere 0.5 percentage points in the average observed decline rate would lead to a cumulative shortfall in capacity growth of 2.6 mb/d by 2015 – enough to eat up most of the world's current spare oil production capacity of around 3 mb/d.

5. The 90 projects include only those that had not been brought on line before the start of 2007, according to national upstream investment plans listed on OPEC's website (www.opec.org).
6. The 2008 edition of the *Outlook* will take a detailed look at the issue of decline rates.

It is certainly possible that decline rates will increase in the coming years, as the average age of the world's existing super-giant and giant fields increases and it becomes harder to maintain production levels. Given the very low short-term price elasticities of demand and supply and the modest 3 mb/d of spare capacity available today, any shortfall in net capacity growth could result in a sharp escalation in prices. A small increase in the annual rate of growth of global oil demand projected in the Reference Scenario would have a similar effect (see the results of the High Growth Scenario below). Under-investment in the downstream sector would add to the upward pressures on prices.

Natural Gas

Demand for natural gas grows by 2.1% per year in the Reference Scenario, from 2 854 billion cubic metres in 2005 to 4 779 bcm in 2030 (Table 1.4). As with oil, gas demand increases quickest in developing countries. The biggest regional increase in absolute terms occurs in the Middle East, where gas resources are extensive (Figure 1.8). North America and Europe nonetheless remain the leading gas consumers in 2030, accounting for more than one-third of world consumption, compared with just under half today.

Table 1.4: **World Primary Natural Gas Demand in the Reference Scenario**
(billion cubic metres)

	1980	2000	2005	2015	2030	2005-2030*
OECD	959	1 409	1 465	1 726	2 001	1.3%
North America	659	799	765	887	994	1.1%
Europe	265	477	550	639	771	1.4%
Pacific	35	133	149	201	237	1.9%
Transition economies	439	601	663	789	914	1.3%
Russia	n.a.	395	431	516	586	1.2%
Developing countries	123	528	727	1 174	1 863	3.8%
China	14	28	51	131	238	6.4%
India	1	25	35	58	112	4.8%
Other Asia	22	131	177	262	360	2.9%
Middle East	36	182	261	394	639	3.6%
Africa	14	62	85	136	211	3.7%
Latin America	36	100	118	193	302	3.8%
World	1 521	2 539	2 854	3 689	4 779	2.1%
European Union	*n.a.*	*482*	*541*	*621*	*744*	*1.3%*

* Average annual rate of growth.

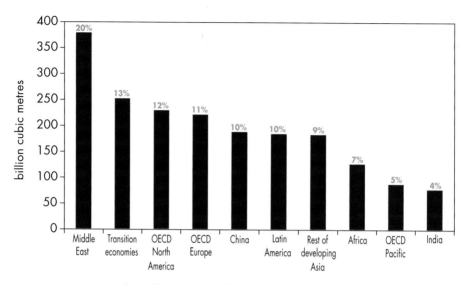

Figure 1.8: **Incremental Primary Natural Gas Demand by Region in the Reference Scenario, 2005-2030**

% share of increase in world primary natural gas demand 2005-2030

New power stations, mostly using combined-cycle gas turbine technology, are projected to absorb over half of the increase in gas demand over the projection period. In many parts of the world, gas remains the preferred generating fuel for economic and environmental reasons. Gas-fired generating plants are very efficient at converting primary energy into electricity and are cheap to build, compared with coal-based and nuclear power technologies. Gas is also favoured over coal and oil for its lower emissions, especially of carbon dioxide. However, the choice of fuel and technology for new power plants will hinge on the price of gas relative to other generating options: higher gas prices in recent years have tempered investment in new gas-fired plants and are projected to do so again from the middle of the next decade.

Worldwide, gas resources are more than sufficient to meet projected demand to 2030. Gas production is projected to increase in all major *WEO* regions except OECD Europe, where output from the North Sea is expected to decline gradually over the projection period. As with demand, the Middle East sees the biggest increase in production in volume terms between 2005 and 2030, its output trebling to 940 bcm (Table 1.5). Output also increases markedly in Africa and Latin America. Natural gas supplies will continue to come mainly from conventional sources, though coal-bed methane and other non-conventional supplies are expected to play a growing role in some regions, notably North America. As with oil, projected gas-production trends

Table 1.5: **World Primary Natural Gas Production in the Reference Scenario**
(billion cubic metres)

	1980	2000	2005	2015	2030	2005-2030*
OECD	879	1 114	1 106	1 199	1 219	0.4%
North America	650	769	743	820	839	0.5%
Europe	217	304	315	292	251	−0.9%
Pacific	12	42	48	87	129	4.0%
Transition economies	480	732	814	947	1 155	1.4%
Russia	n.a.	576	639	702	823	1.0%
Developing countries	155	691	944	1 543	2 405	3.8%
China	14	28	51	103	111	3.2%
India	1	25	29	45	51	2.3%
Other Asia	43	190	240	310	431	2.4%
Middle East	38	212	304	589	940	4.6%
Africa	23	131	186	279	501	4.0%
Latin America	35	104	134	217	372	4.2%
World	1 514	2 538	2 864	3 689	4 779	2.1%

* Average annual rate of growth.

generally reflect the relative size of reserves and, given the high costs of transporting gas over long distances, their proximity to the main consuming markets. How major resource holders will respond to increasing demand is a matter of considerable uncertainty.

Although most regions continue to be supplied mainly with indigenously produced gas, the share of gas supply that is traded between regions grows sharply – from 13% in 2005 to 22% in 2030. All the regions that already import gas on a net basis become more import-dependent by 2030, both in terms of volume and, with the exception of OECD Pacific, the share of total consumption (Figure 1.9). Imports in OECD Europe increase most in absolute terms, from 234 bcm to 520 bcm. North America, which only recently started importing liquefied natural gas (LNG) in significant quantities, becomes a major importer. About two-thirds of the increase in global inter-regional exports over the projection period comes from the Middle East and Africa. Most of these additional exports go to Europe, OECD Asia and North America. Some 13% comes from Russia and other transition economies, most of which is destined for Europe. China, Korea and Japan emerge as new importers of gas from Russia and Central Asia, though the volumes are expected to remain modest.

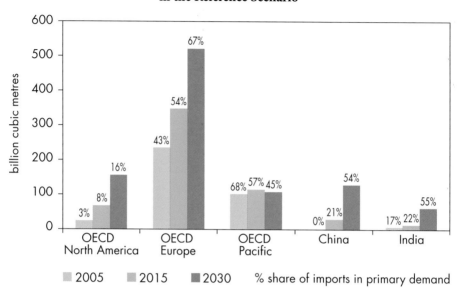

Figure 1.9: **Net Imports of Natural Gas by Major Region in the Reference Scenario**

■ 2005　■ 2015　■ 2030　% share of imports in primary demand

LNG accounts for about 84% of the increase in total inter-regional trade[7], with exports growing from 189 bcm in 2005 (Cedigaz, 2007) to 393 bcm in 2015 and 758 bcm in 2030. LNG is generally the cheapest form of gas transportation for distances in excess of about 4 000 kilometres, even where it is technically feasible to build a pipeline. LNG is also more flexible than piped gas, both logistically and politically. A wave of construction of LNG supply chains is currently under way, which is expected to result in a near doubling of liquefaction and shipping capacity between now and the beginning of the next decade (IEA, 2007b).

Coal

Demand for coal is projected to rise from 4 154 million tonnes of coal equivalent in 2005 to 7 173 Mtce in 2030 – an average annual rate of increase of 2.2% (Table 1.6).[8] China and India, which already account for 45% of world coal use, account for over three-quarters of the increase to 2030. In the OECD, coal use grows only very slowly, with most of the increase coming from North America; consumption in the European Union drops by more than 10%, mainly as a result of policies (such as the EU Greenhouse Gas Emissions Trading Scheme) to promote less carbon-intensive power-generation technologies. In all regions, the outlook for coal use depends largely on

---

7. Using the regional definitions and country groupings of this *Outlook* (see Annex B).
8. A tonne of coal equivalent is defined as 7 million kilocalories (1 tce = 0.7 toe).

developments in clean coal technology in power generation (see Chapter 5), government policies on fuel diversification, climate change and air pollution, and relative fuel prices. In most regions the power sector is the main driver of coal use. Worldwide, the sector absorbs close to three-quarters of incremental coal demand. The deployment of clean coal technology is expected to drive up the average thermal efficiency of coal-fired power stations (see Chapter 5). The past few years have shown that coal is the fall-back fuel for power generation when oil and gas prices rise.

Table 1.6: **World Primary Coal* Demand in the Reference Scenario**
(million tonnes of coal equivalent)

	1980	2000	2005	2015	2030	2005-2030**
OECD	**1 373**	**1 561**	**1 615**	**1 751**	**1 883**	**0.6%**
North America	571	828	846	954	1 083	1.0%
Europe	657	468	457	442	448	−0.1%
Pacific	145	266	311	354	352	0.5%
Transition economies	**515**	**292**	**292**	**341**	**328**	**0.5%**
Russia	n.a.	158	148	179	187	1.0%
Developing countries	**663**	**1 421**	**2 225**	**3 604**	**4 923**	**3.2%**
China	446	899	1 563	2 669	3 426	3.2%
India	75	235	297	472	886	4.5%
Other Asia	51	117	173	241	337	2.7%
Middle East	2	11	13	20	28	3.2%
Africa	74	128	146	162	188	1.0%
Latin America	16	30	33	40	59	2.4%
World***	**2 570**	**3 176**	**4 154**	**5 723**	**7 173**	**2.2%**
European Union	*n.a.*	*459*	*453*	*416*	*393*	*−0.6%*

* Includes hard coal (steam and coking), brown coal (lignite) and peat. ** Average annual rate of growth.
*** Includes statistical differences and stock changes.

China alone is projected to account for 56% of the increase in total world coal production in the Reference Scenario. Its share rises from about 39% now to 46% in 2030. Even so, China became a net importer of coal in 2007 and is projected to increase its imports over the projection period. However, given that the difference between China's demand and output is proportionately small, even by 2030, it is certainly possible that China could once again become a net *exporter* over the projection period (see Part B). India's output also grows, but similarly not fast enough to meet rising demand, partly because the quality of indigenous resources does not match consumers' needs. Production increases in all other regions, except OECD Europe (Table 1.7). Rising output of brown coal in Europe is insufficient to compensate for the continued decline in European hard coal production as remaining subsidies are eliminated and mines are closed in several countries. The global output of steam coal grows faster than that of both coking coal and brown coal (Figure 1.10). By 2030, steam coal accounts for 82% of total production in energy terms, compared with 78% in 2005.

Table 1.7: **World Coal Production in the Reference Scenario**
(million tonnes of coal equivalent)

	1980	2000	2005	2015	2030	2005-2030*
OECD	**1 378**	**1 384**	**1 433**	**1 612**	**1 843**	**1.0%**
North America	672	835	859	1 010	1 172	1.3%
Europe	603	306	276	244	218	–0.9%
Pacific	104	243	299	358	452	1.7%
Transition economies	**515**	**306**	**343**	**421**	**455**	**1.1%**
Russia	n.a.	167	199	286	340	2.2%
Developing countries	**677**	**1 487**	**2 378**	**3 689**	**4 876**	**2.9%**
China	444	928	1 636	2 604	3 334	2.9%
India	77	209	266	358	644	3.6%
Other Asia	47	115	206	369	476	3.4%
Middle East	1	1	1	1	1	0.5%
Africa	100	186	202	247	285	1.4%
Latin America	9	48	67	111	136	2.9%
World	**2 570**	**3 176**	**4 154**	**5 723**	**7 173**	**2.2%**
European Union	*n.a.*	*306*	*280*	*231*	*183*	*–1.7%*

* Average annual rate of growth.

Global inter-regional trade in hard coal is projected to grow at a rate of 3% per year, more than doubling from 648 Mtce (721 million tonnes) in 2005 to 1 354 Mtce (1 523 Mt) in 2030. Between 2000 and 2006, hard-coal trade grew by 5% per year. The share of total world hard-coal output that is traded between *WEO* regions rises from 17% in 2005 to 20% in 2030. Steam coal accounts for most of the growth in hard-coal trade. OECD Asia remains the largest net importer of coal, while India catches up with OECD Europe. European imports grow very slowly. China's net imports grow substantially in volume terms, reaching 92 Mtce (133 Mt) in 2030. They represent only 3% of the country's total coal needs, but 9% of inter-regional coal trade. Indian imports grow seven-fold and reach 18% of internationally traded coal in 2030. US exports, mainly in the form of coking coal, grow steadily, from 46 Mtce in 2005 to 63 Mtce in 2030. Australia, with some of the lowest production costs in the world (Figure 1.11), remains the biggest exporter of both steam coal and coking coal.

Figure 1.10: **World Coal Production by Type in the Reference Scenario**

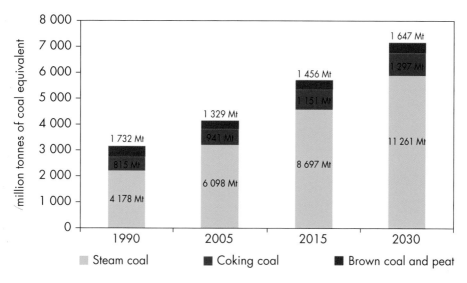

Figure 1.11: **FOB* Cash Costs and Prices of World Steam Coal from Major Exporters, 2005**

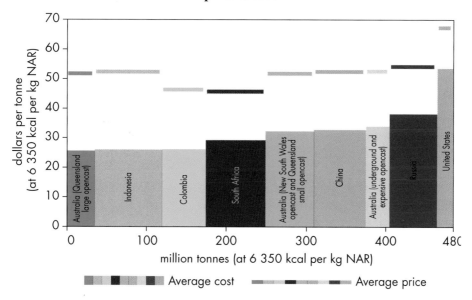

* Free on board (free at ship-side for the United States).
Source: IEA-CCC (2007).

Non-Fossil Energy Sources

Nuclear power supply worldwide is projected to grow slowly over the projection period in the Reference Scenario, from 2 771 terawatt-hours in 2005 to 3 275 TWh in 2030. This is an average annual growth rate of 0.7%, compared with 2.7% per year for total electricity generation. Installed capacity increases from 368 gigawatts to 415 GW. The most significant increases occur in China, Japan, India, Russia, the United States and Korea, with 83% coming from China and India alone. Because of the adopted assumption that existing policies continue unchanged, nuclear capacity in OECD Europe falls from 131 GW to 74 GW – in large part due to phase-out policies in Germany, Sweden and Belgium, which result in the closure of all nuclear power plants in these three countries before 2030.

World hydropower production is projected to grow by an average 2% per year over the *Outlook* period, its share of primary demand remaining broadly constant at about 2%, while its share of electricity generation drops from 16% to 14%. Developing countries account for over three-quarters of the increase in hydropower production. There are few low-cost hydroelectric resources left to exploit in OECD countries, but there are still opportunities for adding capacity in the developing world.

Consumption of biomass and waste continues to grow in absolute terms between 2005 and 2030, though, at an average rate of 1.4% per year, less rapidly than demand for energy as a whole. Their use remains highly concentrated in poor households in developing countries, where modern fuels are too expensive or, in the case of some rural areas, simply not available. The share of biomass in household energy use gradually declines as, with rising incomes, they are replaced by modern commercial fuels. In contrast, the use of biomass for the production of biofuels grows strongly from 19 Mtoe in 2005 to 102 Mtoe in 2030, an average annual rate of 7%. The use of biomass for power and heat generation increases, particularly in OECD countries.

Other renewables – a group that includes geothermal, solar, wind, tidal and wave energy – grow faster than any other energy source, at an average rate of 6.7% per year over the *Outlook* period. But they still make only a small contribution to meeting global energy demand in 2030, because they start from a very low base. Most of the increase in the use of this category of renewables is in the power sector. The increase is much bigger in OECD countries, many of which have adopted strong measures aimed at encouraging the take-up of modern renewable-energy technologies.

Electricity Demand and Generation

Global electricity demand in the Reference Scenario is projected to almost double over the next 25 years, from 15 016 TWh in 2005 to 29 737 TWh in 2030. On average, demand grows by 2.8% per year worldwide. In developing countries, it grows three times as fast as in the OECD, tripling by 2030

(Table 1.8). India and China experience the fastest rates of demand growth. The share of electricity in total final energy consumption increases in all regions. Globally, it rises from 17% in 2005 to 22% in 2030.

Table 1.8: **World Electricity Demand in the Reference Scenario**
(TWh)

	1980	2000	2005	2015	2030	2005-2030*
OECD	4 738	8 226	8 948	10 667	12 828	1.5%
North America	2 385	4 140	4 406	5 227	6 390	1.5%
Europe	1 709	2 700	2 957	3 467	4 182	1.4%
Pacific	645	1 386	1 585	1 973	2 257	1.4%
Transition economies	1 098	1 015	1 099	1 381	1 729	1.8%
Russia	n.a.	607	647	792	968	1.6%
Developing countries	958	3 368	4 969	9 230	15 180	4.6%
China	259	1 081	2 033	4 409	7 100	5.1%
India	90	369	478	950	2 104	6.1%
Other Asia	129	575	766	1 306	1 927	3.8%
Middle East	75	371	501	779	1 228	3.6%
Africa	158	346	457	669	1 122	3.7%
Latin America	248	626	734	1 116	1 700	3.4%
World	6 794	12 609	15 016	21 278	29 737	2.8%
European Union	*n.a.*	*2 524*	*2 755*	*3 179*	*3 786*	*1.3%*

* Average annual rate of growth.

Total power generation is projected to grow from 18 197 TWh in 2005 to 35 384 TWh in 2030.[9] The share of coal-fired power stations in total generation increases from 40% now to 45% in 2030, while the share of gas-fired generation grows from 20% to 23%. The share of non-hydro renewable energy sources continues to increase, from 2% now to about 7% by the end of the projection period. Oil use in power generation continues to decline, from 7% to 3%, while hydropower's share also edges lower from 16% to 14%. Nuclear power suffers the largest fall in market share, dropping from 15% in 2005 to 9% in 2030. Coal continues to dominate the fuel mix in most regions, though its share increases in non-OECD regions and falls in the OECD (Figure 1.12).

9. Electricity generation is equal to final demand plus network losses and own use of electricity at power plants.

Figure 1.12: **Fuel Mix in Power Generation in the Reference Scenario**

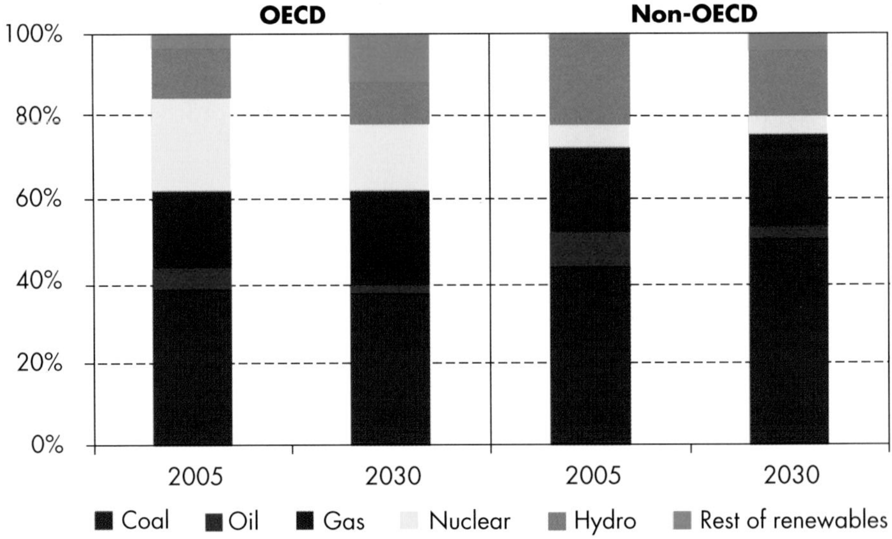

Energy Investment

The Reference Scenario projections in this *Outlook* call for cumulative investment in energy-supply infrastructure of around $22 trillion (in year-2006 dollars) for the period 2006-2030 (Table 1.9).[10] This projection is about $1.7 trillion more than last year's. Although the period is one year shorter, higher units costs – particularly in the upstream oil and gas industry – have pushed up overall capital needs. Projected investment will be needed to expand supply capacity and to replace existing and future supply facilities that will be closed during the projection period as they become obsolete or resources are exhausted.

The power sector requires $11.6 trillion of capital expenditure over the *Outlook* period, accounting for more than half of total energy-supply investments. The share is closer to about two-thirds if investment in the

10. The projections of investment in each of the three scenarios presented in this *WEO* for the period 2006-2030 are derived from the projections of energy supply for each fuel and each region. The methodology used involved estimating new-build capacity needs for production, transportation and (where appropriate) transformation, and unit capital costs for each component in the supply. Incremental capacity needs were multiplied by unit costs to yield the amount of investment needed. Capital spending is attributed to the year in which the plant in question becomes operational. It does not include spending that is usually classified as operating costs.

Table 1.9: **Cumulative Investment in Energy-Supply Infrastructure in the Reference Scenario, 2006-2030**
($ billion in year-2006 dollars)

	Coal	Oil	Gas	Power	Total
OECD	**146**	**1 377**	**1 774**	**4 661**	**8 082**
North America	78	1 023	1 291	2 246	4 669
Europe	35	247	315	1 728	2 417
Pacific	33	107	168	687	997
Transition economies	**40**	**769**	**657**	**681**	**2 148**
Russia	27	568	492	292	1 379
Developing countries	**369**	**2 968**	**1 716**	**6 220**	**11 338**
China	251	547	168	2 764	3 740
India	57	169	63	956	1 249
Other Asia	33	251	303	846	1 441
Middle East	0	1 074	430	406	1 911
Africa	19	494	460	484	1 461
Latin America	10	432	292	762	1 536
Inter-regional transport	41	246	82	0	369
World	**597**	**5 360**	**4 229**	**11 562**	**21 936**

Note: Regional totals include biofuels. Coal includes mining, processing, international ports and shipping.

supply chain to meet the fuel needs for power generation is taken into account. More than half of the investment in the electricity industry is needed for transmission and distribution networks and the rest for power stations. Investment in the oil sector, mostly for upstream developments and mainly to replace capacity that will become obsolete over the projection period, amounts to $5.4 trillion, equal to one-quarter of total energy investment. Investment totals $4.2 trillion in the gas sector and $600 billion in the coal industry. Investment in bio-refineries is projected to total $188 billion, most of which will occur in OECD Europe, Latin America and OECD North America.

About half of global energy investment goes to developing countries, where demand and production increases most (Figure 1.13). China alone needs to invest about $3.7 trillion – 17% of the world total and more than all other developing Asian countries put together. India's investment needs are more than $1.2 trillion, most of it – as in developing countries generally – in the power sector. The Middle East will require about $1.9 trillion, of which about 60% is for upstream oil and gas projects. OECD countries will account for almost 40% of global investment and Russia and other transition economies for the remaining 10%, much of it to replace ageing infrastructure.

Figure 1.13: **Cumulative Investment in Energy Infrastructure in the Reference Scenario by Fuel and Region, 2006-2030**

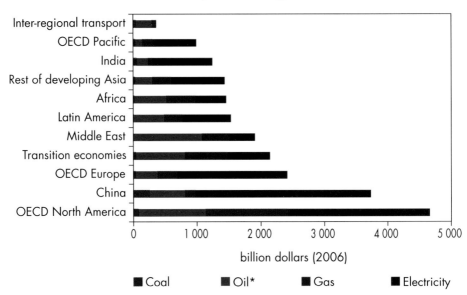

*Oil includes investment in biofuels.

Alternative Policy Scenario

Global Energy Prospects

The Alternative Policy Scenario analyses the impact of the adoption of a set of policies and measures that governments around the world are currently considering to address energy-security and climate-change concerns (see Introduction for a detailed explanation of the methodology and assumptions). In this scenario, global primary energy demand in 2030 reaches 15 783 Mtoe – 1 937 Mtoe, or 11%, less than in the Reference Scenario. That saving is roughly equal to the entire current energy consumption of China. Demand grows at a rate of 1.3% per year over 2005-2030, compared with 1.8% in the Reference Scenario (Table 1.10). The gap in demand between the two scenarios widens progressively over the projection period, as opportunities grow for retiring and replacing capital equipment using more efficient technologies. Yet the energy savings in 2015, at about 4%, are far from negligible. The timing of policy implementation is critical: delaying the implementation of these policies and measures by ten years would reduce the savings in 2030 by two-thirds. Global energy-related CO_2 emissions are 19% lower in 2030 than in the Reference Scenario, but are still 27% higher than in 2005 (see Chapter 5).

Table 1.10: **World Primary Energy Demand in the Alternative Policy Scenario**
(Mtoe)

	2005	2015	2030	2005-2030*	Difference from the Reference Scenario in 2030	
					Mtoe	%
Coal	2 892	3 643	3 700	1.0%	−1 294	−26
Oil	4 000	4 512	4 911	0.8%	−675	−12
Gas	2 354	2 938	3 447	1.5%	−501	−13
Nuclear	721	850	1 080	1.6%	226	27
Hydro	251	352	465	2.5%	48	12
Biomass and waste	1 149	1 359	1 738	1.7%	122	8
Other renewables	61	165	444	8.2%	136	44
Total	**11 429**	**13 818**	**15 783**	**1.3%**	**−1 937**	**−11**

* Average annual rate of growth.

Lower fossil-energy consumption, resulting from the introduction of more efficient technologies, accounts for most of the energy savings in the Alternative Policy Scenario.[11] Nonetheless, demand for each of the three fossil fuels continues to grow. They make up 76% of primary energy demand in 2030, compared with 82% in the Reference Scenario. The biggest savings in both absolute and percentage terms are in coal use (Figure 1.14). OECD countries contribute 26% of the global energy savings in 2030, developing countries 66% and the transition economies 7%. The savings in China – at 29% of the world total – are bigger than those in any other *WEO* region.

Primary energy intensity falls at an average rate of 2.3% per year in the Alternative Policy Scenario – 0.5 percentage points faster than in the Reference Scenario. The difference is bigger for the developing countries and the

11. Carbon capture and storage is assumed not to be deployed in either the Reference or Alternative Policy Scenarios, because of doubts about whether technical and cost challenges can be overcome before 2030 (see Introduction).

What is Stopping Governments from Implementing New Policies?

We all stand to gain from national and regional efforts to address the energy-security and environmental challenges posed by rising energy use. But, in practice, there are formidable hurdles to the adoption and implementation of the policies and measures in the Alternative Policy Scenario, largely caused by strong resistance from industry and consumer interests.

Improving energy efficiency is often the cheapest, fastest and most environment-friendly way to save fossil energy. Cost-reflective, market-based mechanisms such as carbon penalties are, in principle, the most economically efficient approach to encouraging more energy-efficient and cleaner technologies in power generation and industrial applications. But the public and industry are, unsurprisingly, very reluctant to pay higher prices for their energy services, without clear evidence of the needs and long-term benefits, making politicians correspondingly reluctant to push up taxes and prices. Even where it is politically feasible to use market-based instruments, market barriers or the low price elasticity of demand (for example, for transport) can inhibit their effectiveness.

Where there is a readiness to act, regulatory approaches may sometimes be preferred to market mechanisms. For example, measures to regulate appliance efficiency and vehicle fuel economy are among the most cost-effective ways of curbing energy-demand and emission growth. Yet they, too, can be politically difficult. For example, some car makers may oppose increases in mandatory fuel-efficiency standards on the grounds that a switch to smaller and more efficient vehicles will increase costs and lower sales and margins, while car-industry workers may worry about the impact on their jobs.

Financial incentives can be a powerful instrument for change, but they too have limitations. They can be costly, either to governments and taxpayers (through increased public spending) or to consumers (through higher taxes or prices). Even where they encourage new initiatives, local opposition to some types of renewables projects, such as wind farms and hydropower, can outgun the wider community's readiness to accept them. In many parts of the world, barriers to the adoption of policies encouraging the construction of nuclear reactors are particularly high.

Overcoming these barriers to policy action and pushing through the kinds of policies described above takes considerable political will and courage – even when the public is familiar with the energy-security and environmental advantages of action to encourage more efficient energy use and reduce fossil-fuel use. Governments must give a clear lead, in order to generate a collective sense of responsibility. The prospective benefits to the economy and to society as a whole must be clear. Since private capital will finance much energy-related investment, governments remain responsible for creating the appropriate investment environment.

transition economies, because there is more potential in these regions than in OECD for improving energy efficiency in power generation and end uses (Figure 1.15). This is because much of the current energy-related capital stock

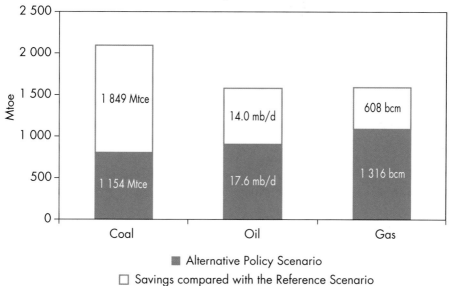

Figure 1.14: **Incremental World Primary Fossil-Energy Demand in the Alternative Policy Scenario, 2005-2030**

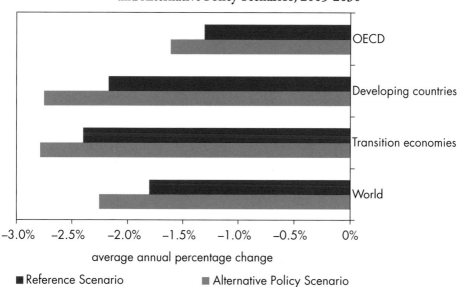

Figure 1.15: **Change in Primary Energy Intensity in the Reference and Alternative Policy Scenarios, 2005-2030**

in these regions is relatively inefficient and because more new capacity will be added over the projection period, creating an opportunity for deploying the most efficient technologies commercially available. Per-capita energy use worldwide in the Alternative Policy Scenario rises to about 1.9 toe around 2020 and then levels off; in the Reference Scenario, it keeps rising through to 2030.

Oil

Global oil savings reach 14 mb/d, or about 12% of total oil demand in 2030. Demand grows by 0.8% per year between 2005 and 2030 – 0.5 percentage points less than in the Reference Scenario. Oil's share of total primary energy demand falls from 35% now to 31% in 2030 in the Alternative Policy Scenario, compared with 32% in the Reference Scenario. By 2015, oil savings reach 4.3 mb/d, or more than 4%. Nearly two-thirds of the savings come from the transport sector, thanks to increased fuel efficiency in new conventional vehicles and the faster introduction of alternative fuels and vehicles (Figure 1.16). Most of the rest comes from more efficient oil use in industry and in residential and commercial buildings. Oil savings are most pronounced for developing countries, at

Figure 1.16: **Oil Demand and Savings by Sector in the Alternative Policy Scenario**

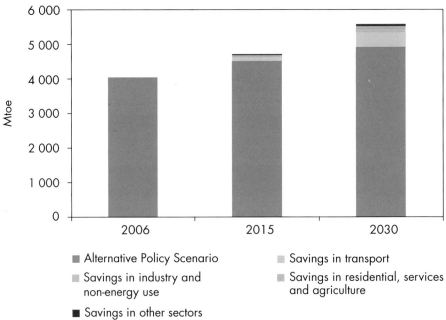

15.1% in 2030, compared with 11.7% for the transition economies and 9.5% for the OECD.

In line with last year's *Outlook*, we have assumed that international oil prices in the Alternative Policy Scenario are the same as in the Reference Scenario. As global oil demand is lower, the call on oil from OPEC members and other exporting countries is reduced. However, the investment and production policies of OPEC member countries are assumed to be adjusted accordingly, resulting in a rate of utilisation of installed production capacities in the longer term similar to that in the Reference Scenario.

OPEC production grows only half as fast as in the Reference Scenario, reaching 42 mb/d in 2015 and 47 mb/d in 2030 (Table 1.11). Its market share rises from 42% now to 46% in 2030, but this is six percentage points lower than in the Reference Scenario. The risk of an oil-supply crunch within ten years, as discussed in the previous section, would clearly be reduced in the Alternative Policy Scenario, as demand grows more slowly.

Table 1.11: **Oil Production in the Alternative Policy Scenario**
(mb/d)

	2006	2015	2030	2006-2030*	Difference from the Reference Scenario in 2030	
					mb/d	%
OPEC	35.8	41.7	46.6	1.1%	−13.9	−23.0
Non-OPEC	47.0	50.4	53.4	0.5%	0.3	0.5
World**	**84.6**	**94.1**	**102.3**	**0.8%**	**−14.0**	**−12.0**

* Average annual rate of growth. ** World total includes processing gains.

In the Alternative Policy Scenario, all the major net oil-importing regions except OECD Pacific – OECD North America and Europe, China, India and the rest of developing Asia – see their oil imports rise over the projection period, but markedly less than in the Reference Scenario. OECD imports reach a peak of 31 mb/d around 2015 and then begin to fall, though they are still higher in 2030 than in 2006. In the Reference Scenario, they rise continuously. By contrast, oil imports into developing countries continue to increase over the whole period, but more slowly. The biggest reduction in imports occurs in developing Asia (Figure 1.17).

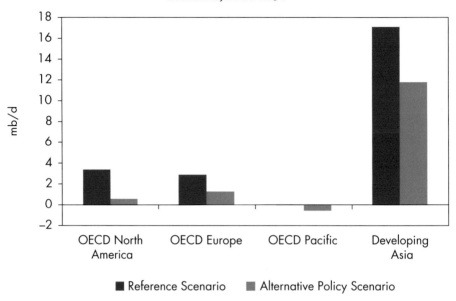

Figure 1.17: **Increase in Net Oil Imports in the Reference and Alternative Policy Scenarios, 2006-2030**

■ Reference Scenario   ■ Alternative Policy Scenario

Natural Gas

Natural gas demand grows by 1.5% per year over the *Outlook* period, 0.6 percentage points lower than in the Reference Scenario. As a result, demand is 3% lower in 2015 and 13% lower in 2030 (Table 1.12). The saving in gas use in 2030 reaches 608 bcm, an amount equal to the current consumption of the United States – the world's biggest consumer. Reduced gas use for power generation, resulting from less demand for electricity and fuel switching to non-carbon fuel, is the main contributor to global gas savings. Global demand is still 46% higher than now, but this is less than the increase of 67% in the Reference Scenario. Gas demand continues to rise in all regions through to 2030, except in the United States and Japan, where it begins to decline late in the projection period. Gas consumption actually increases in China, because of measures to encourage switching away from coal for environmental reasons.

With lower global demand, gas production is lower in all exporting regions compared with the Reference Scenario. The biggest falls in volumetric and proportionate terms occur in the Middle East, Russia and Africa. Their combined production grows from 1 129 bcm in 2005 to 1 721 bcm in 2030 – 542 bcm, or one-quarter less than in the Reference Scenario. Gas production in OECD countries is assumed to be the same as in the Reference Scenario. Inter-regional gas trade consequently grows more slowly in the Alternative Policy Scenario. It totals 872 bcm in 2030, or 21% of world production, against 1 053 bcm (22%) in the Reference Scenario. All the major net

Table 1.12: **World Primary Natural Gas Demand in the Alternative Policy Scenario** (billion cubic metres)

	2005	2015	2030	2005-2030*	Difference from the Reference Scenario in 2030	
					bcm	%
OECD	**1 465**	**1 683**	**1 792**	**0.8%**	**–210**	**–10**
North America	765	869	919	0.7%	–75	–8
Europe	550	617	667	0.8%	–103	–13
Pacific	149	197	205	1.3%	–31	–13
Transition economies	**663**	**761**	**788**	**0.7%**	**–127**	**–14**
Russia	431	488	509	0.7%	–77	–13
Developing countries	**727**	**1 116**	**1 591**	**3.2%**	**–272**	**–15**
China	51	150	268	6.9%	30	13
India	35	57	107	4.6%	–5	–4
Other Asia	177	238	298	2.1%	–62	–17
Middle East	261	357	492	2.6%	–147	–23
Africa	85	130	179	3.0%	–33	–15
Latin America	118	184	247	3.0%	–55	–18
World	**2 854**	**3 560**	**4 170**	**1.5%**	**–608**	**–13**
European Union	*541*	*601*	*645*	*0.7%*	*–99*	*–13*

* Average annual rate of growth.

importing regions need to import more gas in 2030 than now, but – with the exception of China – less than in the Reference Scenario.

Coal

Relative to the Reference Scenario, demand for coal falls more than demand for any other primary energy source in absolute and percentage terms, by 9% in 2015 and 26% in 2030. The rate of growth of coal demand, at 1% per year, over the period 2005-2030, is less than half that in the Reference Scenario. Demand continues to grow through to 2020, but then levels off. More than three-quarters of the coal saved is in the power sector, largely thanks to fuel switching and lower electricity demand. China alone accounts for 43% of the global savings in coal consumption (Figure 1.18). A further 38% comes from the European Union, the United States and India.

Coal production in each region adjusts to the lower demand levels. However, the lower international coal prices that are assumed to result from policies that reduce demand affect most those producers with the highest marginal production costs. In the United States, the decline in domestic production is more marked than the

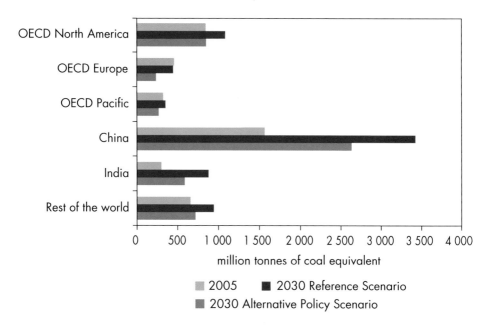

Figure 1.18: **Coal Demand in the Reference and Alternative Policy Scenarios**

decline in domestic demand as high domestic production costs cause export demand to fall more steeply. Globally, coal trade grows markedly more slowly, such that by 2030, exports are about 48% lower than in the Reference Scenario.

Non-Fossil Energy Sources

Demand for energy from all non-fossil fuel primary sources combined is 17% higher in 2030 than in the Reference Scenario (Figure 1.19). Nuclear power accounts for 42% of the additional demand for non-fossil fuel energy, hydropower for 9%, biomass for 23% and other renewables for 26%. Nuclear energy grows over the projection period more than twice as fast as in the Reference Scenario, and is 27% higher in 2030. The largest increases in net capacity are in OECD Europe (36 GW), where the implementation of policies to phase out nuclear energy is assumed to be delayed, in China (24 GW) and the United States (16 GW).

Global consumption of biomass is 7.6% higher in the Alternative Policy Scenario in 2030 than in the Reference Scenario. This results from several factors. Switching away from traditional biomass for cooking and heating in developing countries and, to a lesser extent, improvements in efficiency in industrial processes, drive demand down. But these changes are outweighed by the increased use of biomass in combined heat and power production, in

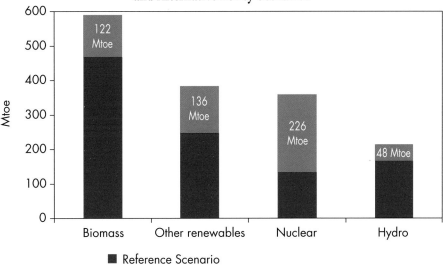

Figure 1.19: **Incremental Non-Fossil Energy Demand in the Reference and Alternative Policy Scenarios**

Reference Scenario

Additional demand in the Alternative Policy Scenario

electricity-only power plants and in liquid biofuels for transport. Most of the increase in biofuels production over and above the Reference Scenario occurs in Europe and the United States.

Electricity Use and Generation

The policies on energy efficiency and conservation taken into consideration in the Alternative Policy Scenario reduce global electricity consumption by 12% in 2030 vis-à-vis the Reference Scenario. More efficient appliances in the residential and services sectors account for most of these savings and more efficient motors in industry for most of the rest. Electricity intensity – consumption per unit of GDP – falls much faster over 2005-2030 as a result: by 1.7% per year, compared with 1.2% in the Reference Scenario.

Lower electricity demand and measures to boost the thermal efficiency of power stations reduce demand for fossil fuels as inputs to generation. As a result of these measures and other policies to boost non-fossil fuels and technologies, the fuel mix in power generation changes markedly (Figure 1.20). The share of coal in 2030 drops most relative to the Reference Scenario, from 45% to 34%. The absolute amount of coal burned in power stations in 2030 is around two-thirds that in the Reference Scenario. The share of gas in 2030 also falls, from 23% to 21%. In contrast, the shares of nuclear power and renewables increase significantly.

Energy Investment

The policies and measures analysed in the Alternative Policy Scenario lead to a major shift in the pattern of energy investment. Consumers – households and

Box 1.2: **Renewable Energy in the Alternative Policy Scenario**

Renewable energy plays a much greater role in the Alternative Policy Scenario, reflecting increased government support worldwide. Excluding traditional biomass, renewable energy increases from 713 Mtoe in 2005 to 1 976 Mtoe in 2030 – 27% more than in the Reference Scenario.

Most of the policies currently under consideration (and therefore taken into account in the Alternative Policy Scenario) focus on the power sector and on transportation. The most significant increases come from the power sector, where renewable energy is projected to account for 29% of global electricity generation in 2030, compared with 18% now. Renewables overtake gas to become the second-largest source of electricity after coal. Renewables account for 43% of incremental electricity generation between 2005 and 2030, with most of the increase coming from hydropower, wind power and biomass. In transport, global biofuel consumption increases from 19 Mtoe to 164 Mtoe, displacing around 3.4 mb/d of diesel and gasoline in 2030 (compared with 2.1 mb/d in the Reference Scenario). Excluding traditional biomass and hydropower, other renewables' share of primary energy demand increases five-fold to 1 512 Mtoe in 2030, 33% higher than in the Reference Scenario.

The European Union already has ambitious policies in place to promote renewable energy. Implementation of these policies results in a 15% share of renewables in primary energy demand in 2030 in the Reference Scenario.

The target set by EU governments is to reach 20% in 2020. To meet it, additional measures going beyond those already announced must be put in place. In the Alternative Policy Scenario, renewables account for 19% of primary energy demand and 38% of electricity generation in 2030. Nonetheless, the indicative EU target, to reach a 34% share in 2020, is projected to be met only in 2024. Renewables for electricity represent about 60% of capacity additions between now and 2030. This would call for investment of $603 billion (in year-2006 dollars), or almost two-thirds of total investment in power generation. The proposed target for biofuels, to meet at least 10% of road transport fuel needs by 2020, is reached in 2022 in this scenario.

firms – invest more in energy-efficient appliances and equipment, while energy suppliers generally invest less in new energy-production and transport infrastructure, in response to lower demand, compared with the Reference Scenario. Overall, the net investment required by the energy sector – ranging from end-use appliances to production and distribution of energy – is $386 billion less over 2006-2030 (in year-2006 dollars) in the Alternative

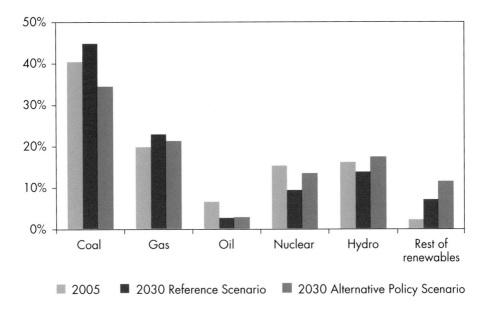

Figure 1.20: **Fuel Mix in World Power Generation in the Reference and Alternative Policy Scenarios**

■ 2005 ■ 2030 Reference Scenario ■ 2030 Alternative Policy Scenario

Policy Scenario than in the Reference Scenario. Consumers spend $2.3 trillion more, helping to reduce supply-side capital needs by $2.7 trillion, or 12%. The biggest reduction in supply-side investment in dollar terms is in the power sector (Table 1.13). The payback period on demand-side investments is typically very short, especially in developing countries and for those policies introduced before 2015.

Table 1.13: **Change in Cumulative Investment in Energy-Supply Infrastructure in the Alternative Policy Scenario,* 2006-2030**
($ billion in year-2006 dollars)

	Coal	Oil	Gas	Power	Total
OECD	−36	−97	−36	−944	−1 082
Transition economies	−11	−16	−95	−100	−222
Developing countries	−90	−426	−118	−780	−1 372
World	**−163**	**−565**	**−253**	**−1 824**	**−2 732**

* Relative to the Reference Scenario.
Note: Totals includes biofuels and intra-regional transport.

The High Growth Scenario

Global Energy Prospects

The High Growth Scenario assumes higher rates of GDP growth in China and India. These higher rates – which bring major benefits in quality of life – result in faster growth in energy demand in both countries. But it also boosts international trade between each of the two countries and the rest of the world. Higher growth in energy demand, in turn, coupled with supply constraints, drives up international energy prices (see Chapter 4 for details). The impact of faster growth in China and India on global economic activity and international energy (and other commodity) prices was modelled using a general equilibrium model. The results were fed into the World Energy Model to analyse their overall impact on energy demand and supply for each *WEO* region, including China and India. A more detailed description of the methodological approach can be found in the Introduction to this *Outlook*.[12]

Higher GDP growth in China and India boosts energy demand in those countries vis-à-vis the Reference Scenario (Table 1.14). In effect, the stimulus to demand provided by stronger economic growth more than offsets the depressive effect of higher prices. In 2030, total primary energy demand is 23% higher in China and 16% higher in India. Worldwide, the increase in demand amounts to 6% in 2030. The impact on energy demand in other regions varies, depending on the extent to which trade and GDP are affected by faster growth in China and India and on the sensitivity of demand to higher prices. Demand increases in some regions and falls in others. The Middle East sees the biggest increases in demand, reaching 11% in 2030, because their economies grow more strongly – thanks to stronger demand for its oil and gas exports (mainly from China and India) and to higher prices. Demand in all three OECD regions, other developing Asian countries and Latin America falls slightly, due to slower GDP growth resulting from higher commodity-import costs. Global energy-related CO_2 emissions are 7% higher in 2030 than in the Reference Scenario and 32% higher than in the Alternative Policy Scenario (see Chapter 5).

Globally, coal sees the biggest increase in demand in volume terms (Table 1.15). This is mainly because incremental coal use is concentrated in China and India. The share of coal in global primary energy demand reaches 30% in 2030, compared with 28% in the Reference Scenario. Oil and gas demand also increases in China and India, as well as in energy-exporting regions, though this growth is partially offset by lower demand in the rest of the world. Excluding biomass, renewables grow most rapidly in percentage terms, though much less than in the Alternative Policy Scenario.

---

12. The global economic effects of this scenario are described in Chapter 3 while the impact on energy security and the environment is described in Chapters 4 and 5. The results for China and India are summarised in the next chapter and are described in more detail in Chapters 12 and 19.

Table 1.14: **World Primary Energy Demand by Region in the High Growth Scenario** (Mtoe)

	2005	2015	2030	2005-2030*	Difference from the Reference Scenario in 2030 Mtoe	Difference from the Reference Scenario in 2030 %
OECD	**5 542**	**6 135**	**6 663**	**0.7%**	**–136**	**–2.0**
North America	2 786	3 139	3 501	0.9%	–72	–2.0
Europe	1 874	2 011	2 118	0.5%	–9	–0.4
Pacific	882	986	1 045	0.7%	–55	–5.0
Transition economies	**1 080**	**1 266**	**1 422**	**1.1%**	**–12**	**–0.8**
Russia	645	767	873	1.2%	3	0.3
Developing countries	**4 635**	**7 045**	**10 433**	**3.3%**	**1 163**	**12.5**
China	1 742	3 135	4 691	4.0%	872	22.8
India	537	804	1 508	4.2%	209	16.1
Other Asia	749	986	1 272	2.1%	–36	–2.8
Middle East	503	748	1 138	3.3%	112	10.9
Africa	606	729	954	1.8%	11	1.2
Latin America	500	643	869	2.2%	–4	–0.5
World**	**11 429**	**14 636**	**18 739**	**2.0%**	**1 018**	**5.7**
European Union	*1 814*	*1 923*	*2 002*	*0.4%*	*–4*	*–0.2*

* Average annual rate of growth.
**Includes international marine bunkers.

Table 1.15: **World Primary Energy Demand by Fuel in the High Growth Scenario** (Mtoe)

	2005	2015	2030	2005-2030*	Difference from the Reference Scenario in 2030 Mtoe	Difference from the Reference Scenario in 2030 %
Coal	2 892	4 164	5 571	2.7%	576	12
Oil	4 000	4 765	5 771	1.5%	186	3
Gas	2 354	3 066	4 105	2.3%	157	4
Nuclear	721	810	881	0.8%	27	3
Hydro	251	333	437	2.2%	21	5
Biomass and waste	1 149	1 351	1 650	1.5%	34	2
Other renewables	61	147	324	6.9%	16	5
Total	**11 429**	**14 636**	**18 739**	**2.0%**	**1 018**	**6**

* Average annual rate of growth.

Fossil Fuels

In the High Growth Scenario, global demand for oil grows by 1.5% per year over the projection period – 0.1 percentage points faster than in the Reference Scenario. Demand reaches 120 mb/d in 2030 – 3.6 mb/d more than in the Reference Scenario. Demand is 6.7 mb/d higher in China and India combined and 0.4 mb/d higher in the Middle East. This offsets a 2.4-mb/d drop in the OECD, concentrated in North America, and an overall 1 mb/d decline in other developing Asian countries, Latin America and Africa (Figure 1.21). The fall in demand is biggest in North America, where GDP growth is reduced most and where demand is relatively sensitive to international prices because of low excise taxes.

Figure 1.21: **Change in World Primary Oil Demand by Region in 2030 in the High Growth Scenario Relative to the Reference Scenario**

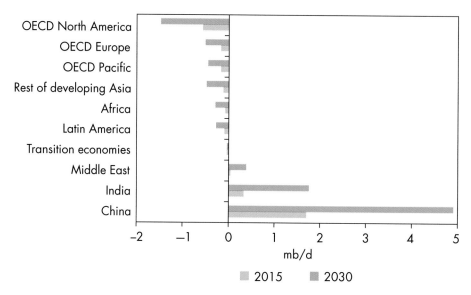

World oil production adjusts upwards to meet the higher level of oil demand in the High Growth Scenario. Most of the additional output comes from non-OPEC countries, where higher prices bring forth more investment and capacity.[13] OPEC countries are assumed to fill the difference between total world demand and non-OPEC supply, accounting for about one-quarter of the total increase in production. Non-conventional oil in China and North America (Oil sands in Canada and CTL in the United States) and conventional oil in the transition economies account for the bulk of the increase in non-OPEC output (Figure 1.22).

13. In the High Growth Scenario, additional oil output in China and India is projected by assessing the impact of enhanced recovery techniques on a field-by-field basis, whereas additional output from the other non-OPEC producers is based on a top-down approach.

The regional effect on oil imports and exports of changes in oil demand and production differs. China and India see their net imports increase sharply, as the increase in demand outstrips the growth in indigenous output. By contrast, all other importing regions – notably North America – need to import less oil, as their demand falls and their indigenous output rises in response to higher prices (Table 1.16). Overall, OPEC and other exporting countries generally increase their exports. Total inter-regional oil trade reaches 67 mb/d in 2030, compared with 65 mb/d in the Reference Scenario. Trade as a share of total consumption remains stable at 56% in both the Reference and High Growth Scenarios.

Figure 1.22: **Incremental Oil Production by Region in 2030 in the High Growth Scenario Relative to the Reference Scenario**

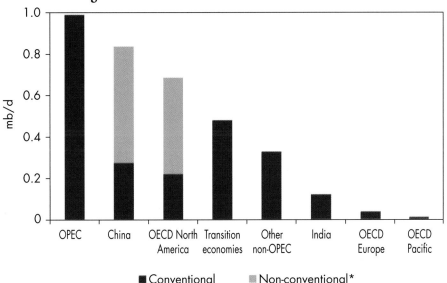

* Oil sands, shale oil, extra-heavy oil, gas-to-liquids and coal-to-liquids.

Table 1.16: **Net Oil Imports by Major Importing Region in the High Growth Scenario** (mb/d)

	2015	2030	Difference from the Reference Scenario in 2030	
			mb/d	%
OECD North America	13.0	12.7	–2.2	–15
OECD Europe	11.1	11.6	–0.5	–4
OECD Pacific	7.5	7.2	–0.5	–6
Developing Asia	15.2	30.3	5.2	21
China	*8.3*	*17.2*	*4.1*	*31*
India	*3.2*	*7.7*	*1.6*	*27*

Global natural gas demand is marginally higher in 2030 in the High Growth Scenario than in the Reference Scenario. Higher oil demand pushes up gas prices significantly, as they are linked to oil prices through long-term contracts and inter-fuel competition. Higher gas prices, in turn, depress demand for gas in some regions, especially where the impact on GDP of higher economic growth in China and India is negative – notably in OECD Europe and Pacific. Nonetheless, the increase in gas demand in China, India and energy-exporting regions that results from faster GDP growth is big enough to outweigh the *drop* in gas demand in other regions. Higher gas prices also stimulate indigenous production marginally in the net importing regions, which, together with lower demand, reduce their gas-import needs, especially by the end of the projection period (Figure 1.23). Total inter-regional gas trade is higher, by around 70 bcm or 7%, in 2030 relative to the Reference Scenario.

Figure 1.23: **Change in Gas Imports in the High Growth Scenario Relative to the Reference Scenario**

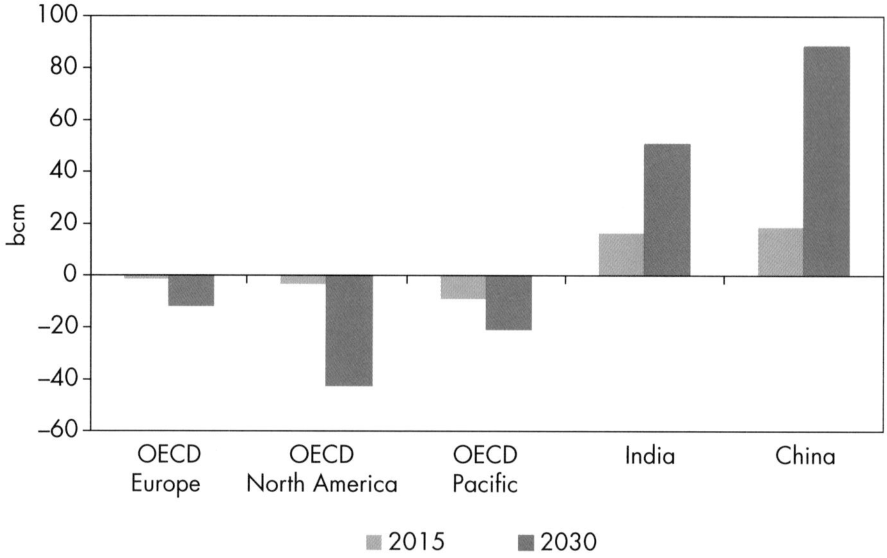

World coal demand increases sharply, reaching 7 958 Mtce in 2030 in the High Growth Scenario – 12% more than in the Reference Scenario. Most of the increase, unsurprisingly, comes from China and India, where stronger economic growth and lower coal prices relative to gas and oil stimulate demand – especially in power generation. Elsewhere, coal demand is boosted by switching away from natural gas and, to a lesser extent, from oil. Higher prices lift production levels in all regions, though by less than the increase in demand in some cases. Consequently, imports in the major consuming regions – including China and India – rise, boosting overall inter-regional coal trade in 2030 from 1 354 Mtce in the Reference Scenario to 1 481 Mtce (Table 1.17). Russia and Australia are the main sources of the additional coal needs in China and India. The increase in China's coal imports is roughly equal to 50% of Australia's current coal exports.

Table 1.17: **Net Inter-Regional Hard Coal Trade* in Selected Regions in the High Growth Scenario** (Mtce)

	2005	2015	2030	Difference from the Reference Scenario in 2030	
				Mtce	%
OECD North America	−19	−63	−106	−20	23
OECD Europe	186	235	249	9	4
OECD Pacific	11	−70	−237	−146	160
Asia	*228*	*260*	*245*	*−37*	*−13*
Oceania	*−216*	*−330*	*−482*	*−108*	*29*
Russia	−48	−103	−126	−52	69
China	−48	112	199	106	115
India	36	125	282	39	16
Indonesia	−98	−180	−200	−16	9

* Negative figures denote exports; positive figures imports.

Non-Fossil Energy Sources

Nuclear power is marginally higher than in the Reference Scenario, with all of the difference assumed to come from China and India. Their combined capacity is 11 GW, or 23%, higher in 2030. The use of modern renewable energy sources is boosted by higher fossil-fuel prices, increasing by 5% relative to the Reference Scenario in 2030.

Electricity Use and Generation

Higher GDP growth boosts final demand for electricity globally and, therefore, fuel inputs to power generation – the major driver of higher energy demand in this scenario. Power generation costs rise in line with higher fossil-fuel prices, pushing up final prices to consumers. Yet this has little impact on demand.[14] In most of the regions that see higher GDP growth, the positive effect of that growth on demand more than outweighs the negative effect of higher prices. Coal-fired generation – mostly in China and India – accounts for most of the increase in power needs in absolute terms (Figure 1.24). Natural gas accounts for most of the rest of the global increase in generation.

14. See *WEO-2006* (Chapter 11) for a discussion of the price sensitivity of demand for electricity and other fuels.

Figure 1.24: **Change in Power Generation by Fuel in the High Growth Scenario Relative to the Reference Scenario**

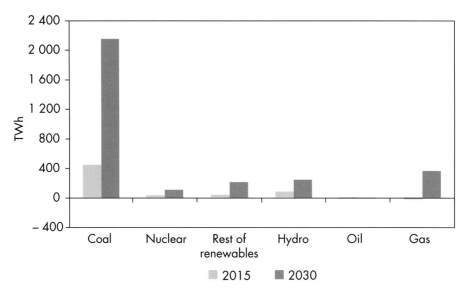

2015   2030

Energy Investment

The higher level of energy demand in the High Growth Scenario boosts investment needs for energy-supply infrastructure over 2006-2030 by almost $2 trillion (in year-2006 dollars). The biggest increase in investment needs in dollar terms occurs in the developing countries, mostly in China and India (Table 1.18). In the OECD, most of the increase in investment is accounted for by oil, where upstream development costs are high. In non-OECD countries, the power sector, where there is a shift towards capital-intensive coal-fired plants, accounts for the overwhelming bulk of the increase in investment. This is the main reason why the increase in global investment needs vis-à-vis the Reference Scenario is larger than the increase in energy demand.

Table 1.18: **Change in Cumulative Investment in Energy-Supply Infrastructure in the High Growth Scenario*, 2006-2030**
(\$ billion in year-2006 dollars)

	Coal	Oil	Gas	Power	Total
OECD	10	59	−2	−106	−43
Transition economies	−2	19	28	6	51
Developing countries	71	309	56	1 416	1 866
World	**85**	**392**	**83**	**1 316**	**1 886**

* Relative to the Reference Scenario.
Note: Totals include biofuels and intra-regional transport.

The High Growth Scenario projections, were they to materialise, would make it all the more urgent for governments around the world to implement policies – such as those taken into consideration in the Alternative Policy Scenario – to curb the growth in fossil-energy demand and related emissions. We have not attempted to build a formal hybrid High Growth/Alternative Policy Scenario to avoid overwhelming the reader with numbers. It is, nonetheless, evident from the above analysis that new policies could go a considerable way to offsetting the adverse energy-security and environmental effects of faster economic development in China and India.

CHAPTER 2

ENERGY TRENDS IN CHINA AND INDIA

HIGHLIGHTS

- China's and India's importance in world energy will continue to grow steadily over the coming decades, reflecting rapid economic development, industrialisation, urbanisation and improved quality of life. In the Reference Scenario, primary energy needs expand at an average annual rate of 3.2% in China and 3.6% in India – much faster than in the rest of the world. Together, they account for 45% of the increase in world energy demand through to 2030.

- All primary fuels except biomass see continuing growth in demand in both countries over the projection period. Their economies remain heavily dependent on coal, mostly produced indigenously. By the end of the projection period, coal – used mainly in power stations – makes up 59% of the two countries' combined energy use, up from 57% in 2005. Oil demand also grows swiftly in both countries. Their combined oil use increases from 9.3 mb/d in 2005 to 23.1 mb/d in 2030 – growth of 3.7% per year and 42% of the global increase in oil demand in 2005-2030.

- Chinese and Indian output of coal expands over the *Outlook* period, but not quickly enough to keep pace with demand. Coal imports rise markedly in India, while China emerged as a net importer in 2007, and its imports are projected to reach 92 Mtce in 2030. Oil production falls between now and 2030 in both China and India. Consequently, net imports surge, from 3.5 mb/d in 2006 to 13.1 mb/d in 2030 in China and from 1.9 mb/d to 6 mb/d in India.

- In the Alternative Policy Scenario, energy demand in China and India grows more slowly as existing government policies to curb demand growth are enforced more strictly and new policies now being discussed are introduced. Primary demand expands by 0.7 percentage points less per year in China and 0.8 points less in India than in the Reference Scenario. In both countries, coal demand falls most thanks to more efficient coal-burning technology, especially in power stations, and switching to less carbon-intensive fuels and zero-carbon technologies, including nuclear power and renewables.

- In the High Growth Scenario, by contrast, faster economic development drives energy demand higher. In China, primary energy demand in 2030 is nearly a quarter higher than in the Reference Scenario. In India, the increase is 16%. Coal and oil account for most of the increase in both countries' primary energy demand. Together, China and India account for 54% of the increase in world primary energy demand between 2005 and 2030 in the High Growth Scenario.

Reference Scenario

Energy Demand

The importance of China and India in the world's energy outlook is set to continue to grow steadily over the coming decades. Rapid economic development, industrialisation, urbanisation and improved lifestyles will undoubtedly drive energy demand yet higher, though at a less rapid rate than in the recent past. In our Reference Scenario, which illustrates the outcome were there to be no new government policies, China's primary energy needs expand from 1 742 million tonnes of oil equivalent in 2005 to 3 819 Mtoe in 2030, an average annual rate of increase of 3.2%. India's needs grow even faster, by 3.6% per year, from 537 Mtoe to 1 299 Mtoe (Table 2.1).[1] Their energy needs grow much faster than in the rest of the world. China and India account for 45% of the total increase in world energy demand over the projection period, and 82% of the increase in coal demand (Figure 2.1). Today, the two countries together account for 20% of the world's primary energy use. By 2030, this share increases to 29%. Their share of

Figure 2.1: **Shares of China and India in the Increase in World Primary Energy Demand by Fuel in the Reference Scenario, 2005-2030**

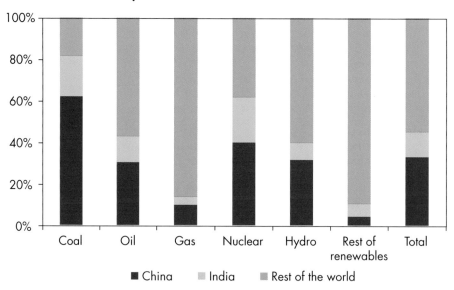

1. There are some important differences between IEA energy statistics for China and India and official national data, because of methodological differences and statistical discrepancies. IEA data for total primary energy use are markedly higher than official estimates for China, largely because the latter do not include traditional biomass. The differences are smaller for India, but there are significant differences at the sectoral level. See Box 8.1 in Chapter 8 and Box 15.1 in Chapter 15 for further details.

Table 2.1: **Primary Energy Demand in China and India in the Reference Scenario**
(Mtoe)

	1990	2000	2005	2015	2030	2005-2030*
China	**874**	**1 121**	**1 742**	**2 851**	**3 819**	**3.2%**
Coal	534	629	1 094	1 869	2 399	3.2%
Oil	116	230	327	543	808	3.7%
Gas	13	23	42	109	199	6.4%
Nuclear	0	4	14	32	67	6.5%
Hydro	11	19	34	62	86	3.8%
Biomass and waste	200	214	227	225	227	0.0%
Other renewables	0	0	3	12	33	9.9%
India	**320**	**459**	**537**	**770**	**1 299**	**3.6%**
Coal	106	164	208	330	620	4.5%
Oil	63	114	129	188	328	3.8%
Gas	10	21	29	48	93	4.8%
Nuclear	2	4	5	16	33	8.3%
Hydro	6	6	9	13	22	3.9%
Biomass and waste	133	149	158	171	194	0.8%
Other renewables	0	0	1	4	9	11.7%
Total	**1 194**	**1 580**	**2 279**	**3 622**	**5 119**	**3.3%**
Coal	640	794	1 302	2 199	3 018	3.4%
Oil	178	345	456	730	1 136	3.7%
Gas	23	44	71	157	292	5.8%
Nuclear	2	9	18	48	100	7.0%
Hydro	17	26	43	75	109	3.8%
Biomass and waste	334	363	385	396	422	0.4%
Other renewables	0	0	4	16	41	10.2%

* Average annual rate of growth.

energy-related emissions of carbon dioxide also increases sharply.[2] Global demand in these projections grows significantly faster than in *WEO-2006*, largely because of faster growth in China and India, reflecting more detailed sectoral analysis and an upward revision to our GDP growth assumptions.

In both countries, the pace of growth in energy demand slows progressively over the projection period in line with economic growth. Primary energy intensity – the amount of energy needed to produce a unit of GDP – continues to fall in India and,

2. The implications of China's and India's energy trends for global climate change and regional pollution are discussed in Chapter 5.

having rebounded in the early part of this decade, resumes its steady downward path in China. Intensity is projected to fall more quickly in the Alternative Policy Scenario (see below). Per-capita demand grows sharply: from 1.3 toe in 2005 to 2.6 toe in 2030 in China, and from 0.5 toe to 0.9 toe in India. By comparison, it rises from 4.8 toe to 5.2 toe in the OECD. Wide disparities among provinces in China and between urban and rural areas in India remain.

All primary fuels except biomass and waste see continuing growth in demand in both countries over the projection period. In the absence of new government policy action, both China and India will remain heavily dependent on coal, mostly produced indigenously, to energise their economies. China's coal demand in the Reference Scenario grows on average as fast as total primary energy demand, so the share of coal remains broadly constant. In India, the share of coal actually increases. By the end of the projection period, coal makes up 59% of the two countries' combined energy needs: 63% in China and 48% in India. In both countries, power generators remain the main consumers of coal, accounting for almost 68% of China's incremental coal needs between 2005 and 2030 and 70% of India's. In China, coal-to-liquids (CTL) emerges as a significant new market for coal.

China's and India's combined oil consumption increases from 9.3 mb/d in 2005 to 23.1 mb/d in 2030 – growth of 3.7% per year.[3] Demand is already 60% higher, at 14.8 mb/d, in 2015. The two countries together account for 43% of the global increase in oil demand between 2005 and 2030. Almost two-thirds of the increase in oil use between 2005 and 2030 comes from the transport sector (Figure 2.2). As a result, the share of transport in total oil demand rises sharply, from 33% in 2005 to 52% in 2030. Oil continues to play a more important role in meeting energy needs in India than in China. It accounts for 25% of India's primary fuel demand in 2030, up from 24% in 2005, while in China it rises from 19% to 21%. This is because of the relatively greater importance of energy-intensive industry, which depends heavily on coal, in China. Per-capita oil demand nonetheless remains higher in China than India, largely because incomes are higher.

Natural gas use grows rapidly over the *Outlook* period from current very low levels in both countries, boosting its share in the overall primary energy mix. The governments of both China and India are keen to see gas play a bigger role, in order to reduce reliance on dirtier coal. But gas remains a marginal fuel – largely consumed in the power generation, industry and residential sectors – as it struggles to compete with coal, which is more competitively priced. Its share of primary energy demand rises from 2% in 2005 to 5% in 2030 in China, and from 5% to 7% in India. The shares of nuclear and

3. The analysis of China's and India's oil demand in this *Outlook* benefited from discussions held at the 5th OPEC-IEA Workshop in Bali in May 2007.

Figure 2.2: **Primary Oil Demand in China and India by Sector in the Reference Scenario**

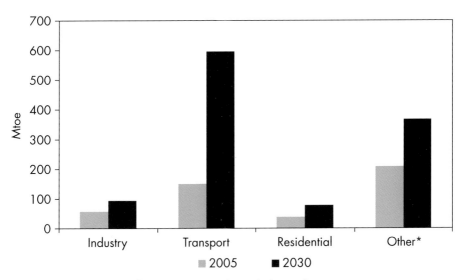

* Includes power generation, other energy sector, services, agriculture and non-energy use.

hydropower also rise, while that of biomass drops. In absolute terms, biomass use is broadly flat in China, as growing numbers of households – mainly in rural areas – switch to modern fuels for cooking and heating as they become richer and the availability of those fuels improves. Biomass consumption increases by almost a quarter in India, though its share in the residential energy mix falls with rising use of modern fuels.

The power sector alone accounts for 53% of the increase in primary energy demand in China over the *Outlook* period and for just over half in India. Its share of primary demand reaches 46% in 2030 in China and 45% in India. The growth in power-sector energy demand would be even faster were it not for the expected improvement in the thermal efficiency of power stations. Coal continues to be the dominant fuel input for generation, though it falls from 89% to 84% in China and from 81% to 76% in India (Figure 2.3). In both countries, power stations remain the main source of air pollution and of energy-related carbon-dioxide emissions.

Among final sectors, transport sees the fastest growth in energy demand, though industry is the single biggest contributor to the growth in final energy demand over the projection period and remains the single largest consumer in both countries. Road transport – freight and passenger cars – accounts for the bulk of the increase in transport fuel use (Figure 2.4). As people get richer, their demand for mobility takes off, especially once average per-capita GDP (in purchasing power parity terms) reaches a level of between \$3 000 and \$10 000,

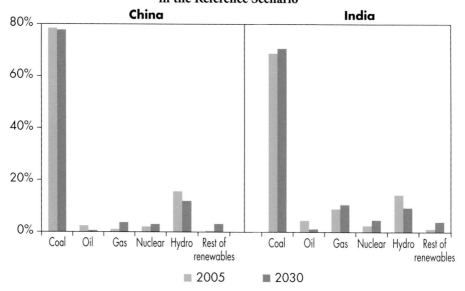

Figure 2.3: **Fuel Mix in Power Generation in China and India in the Reference Scenario**

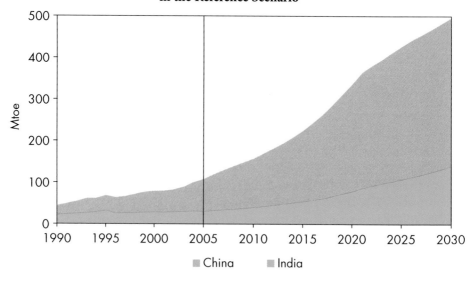

Figure 2.4: **Road Transport Fuel Consumption in China and India in the Reference Scenario**

the point at which a large portion of the population can afford to own a motor vehicle (ADB, 2006). Vehicle sales are already booming in China and India, and the total number of light-duty vehicles on the road is projected to soar from about 22 million in 2005 to more than 200 million in 2030 in China, and from 11 million to 115 million in India. The two countries, China followed by India, have the biggest markets for new cars worldwide well before the end of the *Outlook* period. The share of road transport in total primary oil demand rises from 24% in 2005 to 43% in 2030 in China and from 23% to 41% in India.

At 42%, industry accounts for a significantly larger share of final energy use in China than in the rest of the world on average, because of the importance to the economy of heavy industry, which is highly energy-intensive, and because residential and transport demand is, as yet, low, as a result of relatively low household incomes. Industry's share in India, at 28%, is slightly below the world average of 32%. The share of industry in final energy consumption rises in both countries over the projection period in the Reference Scenario (Table 2.2), mainly because residential energy use is moderated by households continuing to switch away from traditional biomass to modern fuels, which are used more efficiently. Coal continues to dominate industrial energy demand, but the use of electricity grows more quickly in percentage terms.

Table 2.2: **Sectoral Shares in Final Energy Consumption in China and India in the Reference Scenario** (%)

	China			India		
	2005	2015	2030	2005	2015	2030
Industry	42	46	44	28	32	34
Transport	11	13	19	10	14	20
Residential	30	22	19	44	37	29
Services	4	5	6	3	3	4
Other*	13	14	12	14	14	13

*Includes agriculture and non-energy use.

Energy Supply
Oil and Gas

Neither China nor India currently produces enough crude oil or natural gas to meet its needs. These shortfalls grow substantially in the Reference Scenario. China's oil production totalled 3.7 mb/d in 2006, of which about 90% was onshore. Output has grown by about 500 kb/d since the beginning of the decade, but is expected to level off at 4 mb/d early in the next decade and then decline to about 3.4 mb/d in 2030 (Table 2.3). Most of China's fields already in production have reached or passed their peak and discovered fields awaiting development do not have large enough reserves to make good the decline. Undiscovered resources are not thought to be large enough to maintain output levels in the longer term. The projected fall

Are China and India Following the Same Energy Development Path?

There are notable similarities in the current energy systems of China and India, but some important differences too. Both countries' economies rely heavily on indigenous coal resources, especially for power generation. But the share of coal in primary energy use is much higher in China – at 63% in 2005, compared with only 39% in India. The main reason is the much bigger share in China's economy of the industrial sector – which uses large amounts of coal in both countries. Biomass and waste also play a much bigger role in India than in China. In India, they are the second-largest source of primary energy, accounting for 29% of the country's needs. In China, the share is only 13%. A striking similarity is the contribution of nuclear power to electricity generation – around 2% in both countries. China's per-capita energy use is significantly higher than that of India, mainly because China is at a later stage of economic development. Climate and geography also affect energy use.

China and India certainly face similar energy challenges, some shared with other major energy-consuming countries. Rising fossil-energy use is causing air quality to worsen in most major cities and putting pressure on the authorities to require the installation of pollution-control equipment and to relocate power stations and industrial facilities. And rising energy-related emissions of greenhouse-gases will increase the threat of climate change, which could prove very costly to both countries in the long term. China is seeking to address these problems in large part by rebalancing the economy away from energy-intensive heavy industry and towards lighter manufacturing and services. India's economy is already much more geared towards services. India faces the additional challenge of raising finance for much-needed investment in energy infrastructure, especially in the power sector. Generating capacity and household access to electricity remain far below Chinese levels. Market reforms, aimed at expanding the role of the private sector, establishing cost-reflective prices and improving the financial health of electricity companies, are the key to mobilising investment.

in China's crude oil production is offset to a large degree by increased production from CTL plants, such that overall oil output falls by just 0.2 mb/d between 2006 and 2030. Nonetheless, the rapid increase in oil demand means that net imports rise sharply, from 3.5 mb/d in 2006 to 7.1 mb/d in 2015 and 13.1 mb/d in 2030 (Figure 2.5). Most of this oil is in the form of crude oil, as China's refining capacity is expected to grow broadly in line with domestic demand for oil products.

India is also facing the prospect of increasing dependence on oil imports. Despite some major discoveries since the late 1990s, India is a mature oil-producing

Table 2.3: **Oil Production in China and India in the Reference Scenario**
(million barrels per day)

	China			India		
	2006	2015	2030	2006	2015	2030
Crude oil	3.67	3.84	2.70	0.69	0.62	0.39
Natural gas liquids	–	–	–	0.11	0.11	0.12
Coal-to-liquids	–	0.18	0.75	–	–	–
Total	**3.67**	**4.02**	**3.45**	**0.79**	**0.73**	**0.52**

country. Most major oilfields in production today were discovered in the 1970s and 1980s and have passed their production peak. In the Reference Scenario, India's oil production is projected to increase from just over 790 kb/d in 2006 to 870 kb/d in 2010 and then to fall back to 730 kb/d in 2015 and 520 kb/d in 2030. Higher NGLs production tempers the decline in oil output through to 2020. There are no plans to develop CTL production, as it is not commercially viable. As in China, demand for oil outpaces output over the projection period. Net oil imports increase steadily from 1.9 mb/d in 2006 to 2.3 mb/d in 2010, 3 mb/d in 2015 and 6 mb/d in 2030 (Figure 2.5). Gross oil imports – crude oil and other types of refinery feedstock – are projected to be even higher in order to supply India's export-oriented refineries as well as its domestic needs, reaching 7.6 mb/d in 2030. Net product exports reach close to 1.6 mb/d by 2015 and then stabilise. India's overall dependence on imports net of exports rises from 70% today to 92% by the end of the projection period.

Figure 2.5: **Net Oil Imports in China and India in the Reference Scenario**

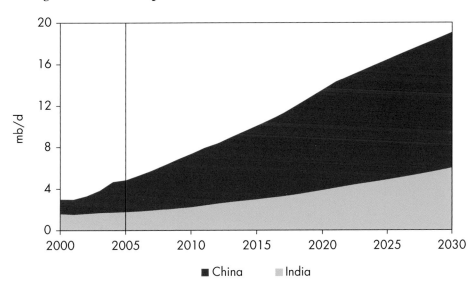

For both China and India, the story is similar for natural gas. China's gas production is projected to rise from 57 billion cubic metres in 2006 to 103 bcm in 2015 and 118 bcm in 2020, falling back to 111 bcm by 2030. Demand grows much faster. As a result, imports of gas, which began in 2006 with the commissioning of the Shenzhen liquefied natural gas (LNG) terminal, increase sharply, to 12 bcm in 2010, 28 bcm in 2015 and 128 bcm in 2030. Gas comes from LNG projects and, later in the projection period, via pipeline from neighbouring countries. In India, demand also outstrips gas production, which rises from 29 bcm today to 51 bcm in 2030. Most of the increase comes from recently discovered fields in the Krishna-Godavari basin. Gas imports – entirely as LNG – jump by about 30% between now and 2010, reaching about 12 bcm. Imports remain stable over the next five years or so before quadrupling between 2020 and 2030, reaching 61 bcm at the end of the projection period. In both China and India, however, the prospects for imports are highly uncertain, as they depend critically on the balance of production and demand, which in turn is very sensitive to the relative prices of coal and gas, and the affordability and availability of imported gas.

Coal

China's and India's heavy dependence on coal stems from their abundant indigenous resources. Until very recently, China met all its domestic coal needs from domestic production. But the figures for 2007, when available, are expected to show that surging demand for steam coal – largely for power generation – has turned China into a net coal importer. As recently as 2005, China was a significant coal exporter. In the Reference Scenario, overall mining and transport capacity is projected to continue to lag behind demand, particularly as coal has to be transported over ever-longer distances from the areas identified for future exploitation, which raises costs. As a result, China emerges as a sizeable net importer of coal, with volumes reaching almost 66 million tonnes of coal equivalent (93 million tonnes) in 2015 and 92 Mtce (133 Mt) in 2030 (Figure 2.6).[4] Most imports are of steam coal. In 2005, China *exported* a small quantity of steam coal on a net basis. China has less need to import coking coal, as it has ample high-quality resources. Although imports cover only a small proportion of the country's total coal needs by 2030, they make up a major part of international coal trade – especially in the short to medium term.

---

4. Historical data on coal trade are normally reported in tonnes and are converted by the IEA Secretariat to tonnes of coal equivalent (tce), based on an energy content of 7 000 kcal (corresponding to 0.7 toe). Differences between tce and tonnes in the trade and production data and projections reported here are, therefore, explained by the differences in the quality and types of coal. Coking coal typically has a much higher calorific value than steam coal.

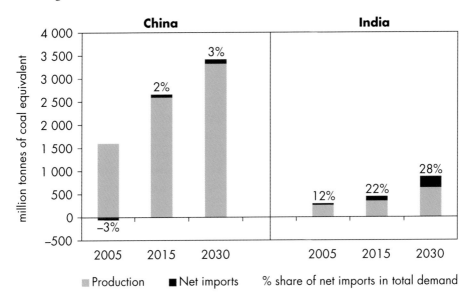

Figure 2.6: **Coal Balance in China and India in the Reference Scenario**

■ Production ■ Net imports   % share of net imports in total demand

In the Reference Scenario, coal production in China is projected to increase from 1 611 Mtce in 2005 to 2 604 Mtce in 2015 and 3 334 Mtce in 2030. Output of steam coal, which currently accounts for 85% of production in energy terms, increases faster than that of coking coal. This expansion of coal output hinges on the continued restructuring and modernisation of the coal-mining industry and massive investment in transport infrastructure to move coal to market. Despite some progress, there remains considerable scope for improving mine productivity through the closure of small, inefficient mines and the more widespread use of modern techniques such as long-wall mining. The government's policy of closing small mines will reduce coal output from the thousands of town and village mines in China. To compensate, the government aims to increase production by consolidating smaller state-owned mines into larger, more efficient companies.

India's coal production is already insufficient in quantity and quality to meet its domestic needs, despite moves to open up mining to private investors. The country produced 252 Mtce (403 Mt) of hard coal and 10 Mtce (30 Mt) of lignite in 2005. Total output is projected to rise to 354 Mtce (580 Mt) in 2015 and to 637 Mtce (1 059 Mt) in 2030. India imported a total of 36 Mtce (39 Mt) of coal in 2005, covering about 12% of demand. Steam coal imports were 18 Mtce (20 Mt). Coal imports, which have grown strongly over the past decade, continue to grow with buoyant demand in the Reference Scenario, for both technical reasons – the quality of indigenous coking coal is often not high enough for steel producers – and for reasons of cost. Steam coal imports are projected to rise further, to 52 Mtce (54 Mt) in 2015 and 139 Mtce (151 Mt) in 2030; coking coal imports are projected to rise more slowly.

Coal-import needs in both China and India are inevitably very uncertain. Because the volume of imports projected is small relative to demand (especially in China), marginally faster or slower demand or output growth rates would have a big impact on the volume and direction of trade. For example, a slow-down in production growth of just 1 percentage point per year than projected here – possibly resulting from slower reform of the mining industry – would increase China's imports in 2030 eight-fold and India's by 56%. Faster demand growth would have a similar effect (see High Growth Scenario below). The increase in imports would be very large relative to total world hard coal trade (see Chapter 4).

Non-Fossil Energy Sources

The Chinese and Indian governments plan to expand significantly the role of nuclear power and modern renewable energy technologies. In the Reference Scenario, the share of nuclear in electricity generation in China is projected to rise from 2.1% in 2005 to 3% in 2030, with capacity jumping from under 7 GW to 31 GW. This capacity nonetheless falls 9 GW short of the government's target, which is ambitious in view of the long construction times and current global bottlenecks in nuclear component manufacturing. All new nuclear power plants are expected to be built in coastal areas. Hydropower capacity also increases, though its share of primary demand and electricity generation declines. Total biomass consumption remains broadly unchanged through to 2030, but its utilisation pattern changes considerably. Traditional biomass consumption, mainly for household cooking and heating, declines, but biomass to fuel power plants and to make biofuels for transport increases. The supply of energy from other renewable sources increases rapidly, but from a very low base. Wind-power capacity is projected to climb from a little over 1 GW in 2005 to 49 GW in 2030, accounting for 1.6% of China's total electricity supply. Solar thermal and photovoltaic energy supply is also projected to grow strongly.

In India, nuclear power capacity is projected to surge from 3 GW in 2005 to 17 GW by 2030, with the share of nuclear power in electricity generation rising from 2.5% in 2005 to 4.6% in 2030. Nonetheless, this is well below the rate of increase targeted by the government, as difficulties in building nuclear power plants, including high construction costs and problems in gaining access to technology and materials are expected to persist into the future. Energy from renewable sources in total expands slowly, with traditional biomass continuing to dominate consumption. Hydropower output more than doubles, yet its share of power generation falls from 14% in 2005 to 9% in 2030. Demand for other renewable energy sources grows more quickly, but still accounts for only a very small share of primary energy demand in 2030. Among power-generation technologies, wind power sees the fastest growth, with capacity rising from just over 6 GW in 2006 – the fourth-largest in the world – to 27 GW by 2030, so that its share of total electricity generation rises from just under 1% to 2.5%. Biomass use for biofuels, which only recently started in India, is expected to grow to almost 2 Mtoe in 2030, though this represents little more than 1% of road-transport fuel demand.

Investment Needs

The projected energy supply in China and India in the Reference Scenario calls for cumulative infrastructure investment of $5 trillion (in year-2006 dollars) over the period 2006-2030, or $200 billion per year (Figure 2.7). This investment is needed both to expand supply capacity and to replace existing and future supply facilities that are retired during the projection period. China's overall investment needs are three times those of India. China accounts for 17% of projected world energy investment and India 6%.

In both countries, the electricity-supply industry – covering power generation, transmission and distribution – takes the lion's share of energy investment. China needs to invest $2.8 trillion in electricity facilities, or close to three-quarters of the country's total energy investment. India's electricity sector

Figure 2.7: **Cumulative Investment in Energy Supply in China and India by Fuel in the Reference Scenario, 2006-2030**

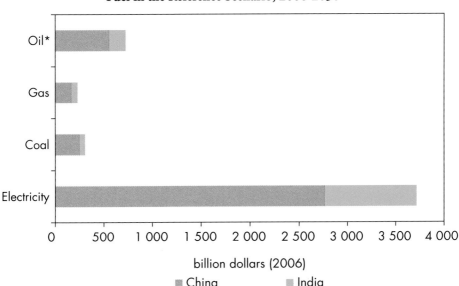

*Oil includes investment in biofuels.

requires $956 billion of capital spending, just over three-quarters of the total for the energy industry. More than half of each country's electricity investment goes to reinforcing and extending networks. Achieving this rate of investment is most uncertain in India, where the poor financial health of the public utilities has held back the development of electricity infrastructure in the past. If the power sector is not reformed as planned, electricity sector investments will continue to place an unsustainable burden on the budgets of the central and state governments. Price reform, better management and reduced losses will be crucial to improving the propensity and ability of the public utilities to invest in the future and to attract private investors (see Chapter 17 for more details).

The oil sector requires $547 billion of investment in China, equal to 15% of total energy-investment needs, and $169 billion in India (14%). In China, the upstream accounts for 47% of total oil investment. Of downstream investment, CTL accounts for about $41 billion. At 77%, the downstream share is higher in India, because of the more rapid expansion of refining capacity relative to demand (to supply export markets). The gas-supply projections call for cumulative investment of $168 billion in China and $63 billion in India. Around 55% of this investment is needed upstream in both countries. The rest goes to LNG terminals, transmission and distribution networks, and storage facilities.

Investment in the coal-mining industry is relatively modest, at $251 billion (less than 7% of total energy investment) in China and $57 billion (5%) in India.[5] The share rises to 40% in China and 39% in India if coal-fired power stations are included. Coal mining is much less capital-intensive than other energy sectors. Together, the two countries account for the overwhelming bulk of projected global coal investment: China for 42% and India 10%.

Alternative Policy Scenario

The results of the Alternative Policy Scenario demonstrate that China and India can both move to a more economically and environmentally sustainable development path by enforcing existing government policies more strictly and introducing new policies that are now being discussed.[6] These actions result in a significant reduction in energy demand and switching to less polluting, low- and zero-carbon fuels and technologies. Importantly, these outcomes produce a net financial benefit for energy consumers and lower costs to the economy as a whole – even without putting a monetary value on the energy-security and environmental benefits.

---

5. The coal-investment projections presented here do not include investment in coal-transportation infrastructure.
6. See Introduction and Chapters 11 and 18 for details about the methodology used and assumptions made.

Primary energy demand grows markedly less quickly in this scenario, by 2.5% per year on average in China (0.7 percentage points less than in the Reference Scenario) and 2.8% in India (0.8 points less). In both countries, coal demand falls most – by 794 Mtce, or 23%, in 2030 in China and 293 Mtce, or 34%, in India (Figure 2.8 and Table 2.4). This results from the use of more efficient coal-burning technology, especially in power stations, and switching to less carbon-intensive fuels. Demand for oil is 3.2 mb/d, or 19%, lower in China and 1.1 mb/d, or 17%, lower in India in 2030. Most of the oil savings – 68% in China and 69% in India – come from the transport sector, with the introduction of more fuel-efficient vehicles and the expanded use of alternative fuels, notably biofuels. Consumption of natural gas is reduced slightly in India. But China's gas use is higher than in the Reference Scenario, because of switching from coal in power generation. Nuclear power output and renewables supply increase significantly in both countries, as a result of policies to curb pollution and lower greenhouse-gas emissions.

Figure 2.8: **Incremental Primary Fossil Fuel Demand in China and India in the Alternative Policy Scenario, 2005-2030**

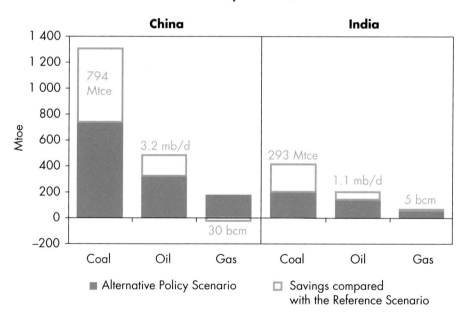

In both countries, the biggest reductions in energy demand in absolute terms occur in the power sector, thanks mainly to lower demand for electricity, higher thermal efficiency of coal-fired power stations and reduced network losses. Electricity demand is reduced by government policies that encourage the

Table 2.4: Primary Energy Demand in China and India in the Alternative Policy Scenario (Mtoe)

	2005	2015	2030	2005-2030*	Difference from the Reference Scenario in 2030 Mtoe	Difference from the Reference Scenario in 2030 %
China	**1 742**	**2 743**	**3 256**	**2.5%**	**–563**	**–14.7**
Coal	1 094	1 743	1 842	2.1%	–556	–23.2
Oil	327	518	653	2.8%	–155	–19.2
Gas	42	126	225	6.9%	25	12.6
Nuclear	14	44	120	9.0%	53	79.4
Hydro	34	75	109	4.8%	23	26.4
Biomass and waste	227	223	255	0.5%	28	12.4
Other renewables	3	14	52	11.9%	19	57.4
India	**537**	**719**	**1 082**	**2.8%**	**–217**	**–16.7**
Coal	208	289	411	2.8%	–209	–33.7
Oil	129	173	272	3.0%	–56	–17.1
Gas	29	47	89	4.6%	–4	–4.3
Nuclear	5	19	47	9.9%	14	41.9
Hydro	9	17	32	5.3%	9	42.3
Biomass and waste	158	168	211	1.2%	17	8.5
Other renewables	1	6	21	15.8%	12	145.5
Total	**2 279**	**3 462**	**4 339**	**2.6%**	**–780**	**–15.2**
Coal	1 302	2 032	2 253	2.2%	–765	–25.3
Oil	456	692	925	2.9%	–211	–18.6
Gas	71	173	313	6.1%	21	7.3
Nuclear	18	62	167	9.2%	67	66.9
Hydro	43	92	141	4.9%	32	29.6
Biomass and waste	385	392	466	0.8%	45	10.6
Other renewables	4	20	73	12.7%	31	75.5

* Average annual rate of growth.

deployment of more efficient appliances and equipment. Transport demand is reduced most in percentage terms, with new policies to promote public transport and the faster introduction of more fuel-efficient vehicles. In China, a large part of the overall reduction in energy use comes about as a result of rebalancing the economy away from heavy industry and towards services. In both countries, the energy savings rise progressively over the projection period, but are already significant in 2015, amounting to 3.8% in China and 6.7% in India.

Most of the measures analysed in the Alternative Policy Scenario have very short payback periods. The higher initial cost to energy end users of improved motors in industry and more efficient appliances and cars is paid back quickly in China. Payback periods for more efficient appliances in India are significantly longer for households, because of subsidised electricity prices: introducing economically efficient electricity pricing would result in much shorter paybacks and much faster gains in efficiency improvements.

Indigenous output of conventional oil and gas is no different in the Alternative Policy Scenario, as oil and gas prices are assumed to be unchanged. Output of biofuels, however, increases as a result of government policies to boost their role in meeting transport demand. In addition, output of CTL is significantly higher in China. With lower demand, both countries see a big reduction in their oil and gas imports. China remains largely self-sufficient in coal over the *Outlook* period (mainly because coking-coal exports increase), whereas it becomes a major net importer of coal in the Reference Scenario. Although India has to import increasing amounts of coal, the level in 2030 is 60% of that in the Reference Scenario level, mostly because of lower demand for steam coal for power generation.

High Growth Scenario

The High Growth Scenario assumes higher rates of GDP growth – 1.5 percentage points higher than in the Reference Scenario – in both China and India.[7] These higher rates, unsurprisingly, result in faster growth in energy demand in both countries. In China, total primary energy demand in 2030 is 23% higher than in the Reference Scenario. In India, the increase is 16% (Figure 2.9). The difference is explained by the higher income elasticity of energy demand in China, resulting from the bigger share of industry (which is relatively energy-intensive) in GDP. Together, China and India account for 54% of the increase in world primary energy demand between 2005 and 2030 in the High Growth Scenario, compared with 45% in the Reference Scenario.

Coal and oil account for most of the increase in both countries' primary energy demand. Coal consumption is 21% higher in 2030 than in the Reference Scenario in China and 13% higher in India. Most of the additional use is in power generation, where demand for fuel inputs is pushed up by much

---

7. The energy-policy assumptions in the High Growth Scenario match those of the Reference Scenario, in that no new government policies and measures beyond those already enacted by mid-2007 are taken into account.

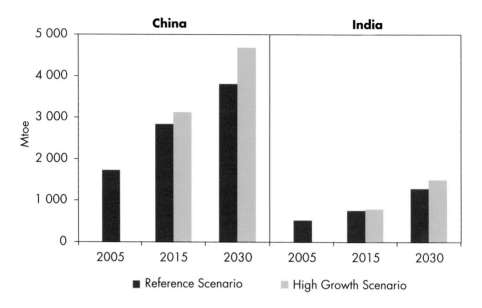

Figure 2.9: **Primary Energy Demand in China and India in the Reference and High Growth Scenarios**

stronger final demand for electricity. Electricity consumption in 2030 is 28% higher in China and 25% higher in India. Oil demand reaches 21.4 mb/d in 2030 in China – 4.9 mb/d, or 30%, more than in the Reference Scenario. In India, it is 1.8 mb/d, or 27%, higher, at 8.3 mb/d. About two-thirds of this increase comes from the transport sector in China and three-quarters in India. Gas demand also grows faster, mainly driven by the power and residential sectors.

China and India rely much more on imported fuels in the High Growth Scenario, as higher prices stimulate only marginal increases in indigenous production. Though coal production is 19% higher in 2030 in China and 13% higher in India, with higher coal prices *vis-à-vis* the Reference Scenario, the development of mining and inland transportation capacity trails the projected surge in demand. China's dependence on coal imports reaches 5% in 2030, compared with 3% in the Reference Scenario. In India, the share of imports in total coal supply in 2030, at 28%, is the same as in the Reference Scenario. Although China's oil imports are 31% higher in 2030, import dependence increases by only one percentage point to 80% in 2030, due to the rise in domestic production. For the same reason, India's dependence is roughly stable at 92%, even though imports are 27% higher than in the Reference Scenario. Imports of natural gas rise strongly to meet higher demand, because indigenous output is barely affected by higher prices.

CHAPTER 3

INTERNATIONAL TRADE AND THE WORLD ECONOMY

HIGHLIGHTS

- China and India are the emerging giants of the world economy. Growth in both countries has accelerated in recent years, boosting their shares of world GDP. Economic developments in China and India will increasingly affect the rest of the world by dint of their sheer size and their growing weight in international trade and cross-border financial flows.

- How China's and India's international trade develops in the future depends on structural changes in their domestic economies. Investment in infrastructure and in labour-intensive manufacturing for export will continue to drive economic development in China for some time to come. There is potential for both countries to boost their share of trade in higher-value manufacturing and services in line with policy objectives.

- Economic expansion in China and India is generating opportunities for other countries to export to them, while increasing the other countries' access to a wider range of cheaper imported products and services. But growing exports from China and India also increase competitive pressures on other countries, leading to structural adjustments, particularly in countries with competing export industries. Rising commodity needs risk driving up international prices for commodities, including energy.

- Continuing rapid economic growth in China and India would boost growth in most other regions, especially those that are net commodity exporters. In the High Growth Scenario, in which China's and India's GDP is assumed to grow by around 1.5 percentage points faster in 2005-2030 than in the Reference Scenario, the Middle East, Russia and other energy-exporting countries see a significant net increase in their GDP in 2030. GDP in the United States, the European Union, OECD Pacific and other developing Asian countries falls marginally, mainly because of higher commodity import costs. Assuming no energy-policy changes in major countries, the average IEA crude oil import price rises to $87 per barrel (in year-2006 dollars) in 2030 – 40% higher than in the Reference Scenario. Overall, world GDP grows by 4.3% per year on average compared with 3.6% in the Reference Scenario.

- Economic policies will play a critical role in sustaining the pace of global economic growth and redressing current imbalances. Rising protectionism could radically change the positive global impact of economic growth in China and India. By contrast, faster implementation of energy and environmental policies to save energy and reduce emissions, such as those included in the Alternative Policy Scenario, would boost significantly the net global benefits by reducing pressures on international commodity markets and lowering fuel-import bills.

China and India in the Global Economy

China and India are the emerging giants of the world economy. Growth in both countries has accelerated in recent years, boosting their shares of world gross domestic product. In 2006, China accounted for 5.5% of global GDP at market exchange rates (15% in purchasing power parity terms) and India 1.8% (6.3%). High growth rates in the 1980s made little difference to the world economy, because both countries' economies were relatively small. Today, their size means that continuing high growth makes a much bigger difference to the world economy. For example, 10% growth in China is equivalent to almost 2% of US growth at market exchange rates. Large populations, which are fuelling the labour pool, and still low levels of income compared with the industrialised countries are expected to maintain the momentum of economic development. Over 37% of all the people in the world are either Chinese or Indian. Their expanding skills base, combined with high rates of investment (especially in China), points to enormous potential for raising productive capacity.

Development in China and, to a lesser extent, in India has been driven by massive domestic and inward investment, high saving rates, and a concomitant expansion of exports of manufactured goods and services as well as capital. Thus, development is proceeding hand in hand with their integration into the global economy. China and India are at the heart of the current wave of economic globalisation, involving rising international trade and capital flows, and integration of financial markets. As a result, developments in these two countries are increasingly affecting the economic health and the structural evolution of the economy of the rest of the world, with inevitable consequences for global energy markets.

A common question is: how does the rise of China and India affect my country economically? This chapter tries to answer that question. The next two sections review recent macroeconomic trends in China and India, and the role of trade between them and the rest of the world. The following sections look at the implications of growing economic interdependence – between China and India on the one hand and the rest of the world on the other – and assess quantitatively which regions gain or lose from economic development in China and India, and why.

Explaining China's and India's Economic Growth

Measuring GDP using market exchange rates – the most appropriate basis when assessing the impact of one economy on another (Box 3.1) – China had the world's fourth-largest economy and India the 13th-largest in 2006.[1] The economies of both countries have been growing very rapidly, with China

1. GDP measured in US dollars at current price and market exchange rates (IMF, 2007).

averaging growth in real gross domestic product of 9.8% per year since 1980 and India 5.9%. The world as a whole grew by only 2.8% per year over the same period. India saw growth of 9.7% in 2006, up from 9% in 2005, while China's growth reached 11.1% in 2006, up from 10.4% in 2005.

No other large country has grown as fast as China since 1980. Such a high rate of growth is not unprecedented – double-digit rates have been recorded in some countries over other periods – but no large country has sustained such a rate for such a long period (Figure 3.1). The rate of expansion of China's share of the world economy has been much larger than that of any other country yet recorded, jumping from barely 1% in 1980 to over 5% in 2006. India's economic take-off is more recent, as growth began to accelerate in the 1990s, and is now approaching that of China. GDP growth in India has averaged 7% per year since 2000, compared with 5.7% per year in the 1980s and 1990s. More detail about China's and India's economic development can be found in Chapters 7 and 14.

Box 3.1: **Measuring and Comparing Gross Domestic Product**

For energy-modelling purposes, the gross domestic product of different countries is converted into constant US dollars using purchasing power parities (PPPs) rather than market exchange rates. PPPs compare the costs in different currencies of a fixed, wide-ranging basket of goods and services, including items that are traded and not traded on international markets. Adjusting GDP for PPP provides a more reliable measure of the physical economy, including the amount of infrastructure and industrial activity, and standards of living. This is a better explanatory variable for energy demand than simple measures of income. It also aids comparisons of energy intensity and energy-use patterns among countries. However, the use of PPPs is not without problems, one being that people and business consume different baskets of goods and services in different countries. For assessing the impact of economic developments in one country on another, actual market exchange rates provide a better basis. This is because international effects result from the international exchange of goods, services and assets – the prices of which tend not to vary much across countries.

There are some similarities and some important differences between the characteristics of economic development in China and India. China has followed a similar development path to that of other East Asian countries, involving the recycling of export revenues and domestic savings into fixed investment. China is often characterised as the world's workshop, with growth driven largely by production and exports of manufactured goods. In fact, China's growth has been remarkably broad-based across agriculture, industry and services, though industry's share of GDP has risen steadily. India's growth has been driven in large part by service-related activities, both export and domestically oriented, which accounted for 54% of GDP in 2006 (compared with 39% in China). This has entailed lower investment and exports relative to GDP than in China. Even so, the rate of growth in services in China since 1990 has exceeded that of India.

Figure 3.1: **Real Output in China, India, Other Asian and Newly Industrialised Economies**

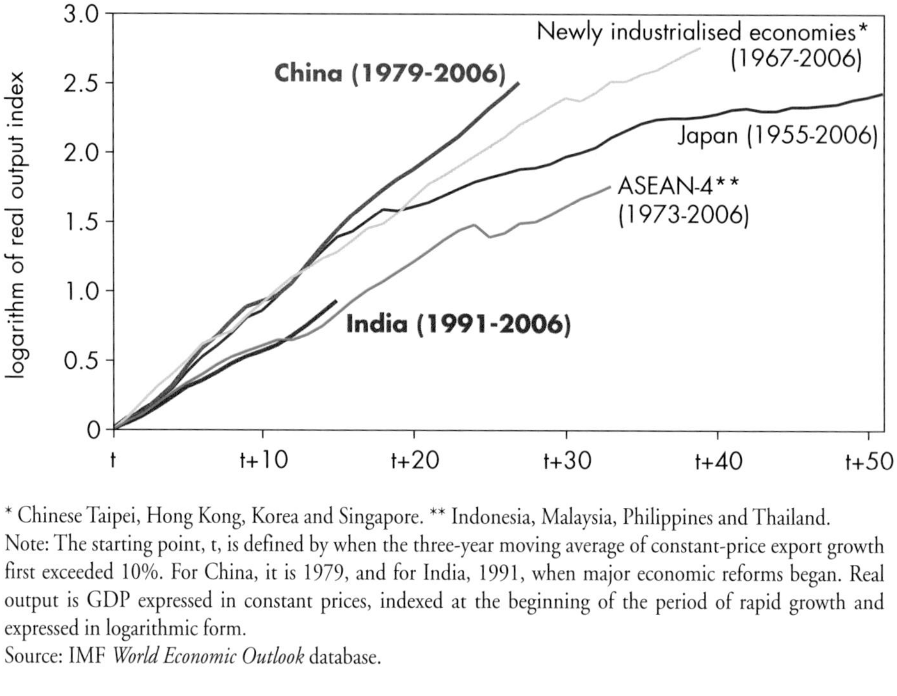

* Chinese Taipei, Hong Kong, Korea and Singapore. ** Indonesia, Malaysia, Philippines and Thailand.
Note: The starting point, t, is defined by when the three-year moving average of constant-price export growth first exceeded 10%. For China, it is 1979, and for India, 1991, when major economic reforms began. Real output is GDP expressed in constant prices, indexed at the beginning of the period of rapid growth and expressed in logarithmic form.
Source: IMF *World Economic Outlook* database.

China's economic expansion has been largely based on capital formation – underpinned by the country's extremely high savings rate – and rising total factor productivity; increased labour input has made only a marginal contribution to GDP growth. Productivity has grown at a similar rate in India, but labour inputs have grown slightly faster than in China, mainly due to a higher rate of population increase. India's capital formation has been much slower than China's and is the main reason why its overall GDP growth has lagged that of China since the 1980s (Bosworth and Collins, 2007). The GDP-weighted average rate of gross capital accumulation in 1990-2003 was 42% in China and 24% in India (World Bank, 2007). In both countries, much of the increase in productivity stems from the reallocation of labour from the farming and state sectors to private industry and services, associated with migration towards cities and industrial districts.

Sustaining productivity growth will be a major challenge for both countries. A main uncertainty facing China's economic prospects is the extent to which heavy industry can continue to drive growth, in view of rising raw-material costs, resource constraints and environmental effects. The political leadership decided in 2004 to adjust the structure of the economy in the medium to long term from investment- and export-led growth to more consumption-led growth, with services and lighter industrial activities playing a bigger role. Private domestic consumption is expected to account for a growing share of GDP, reducing the reliance on investment and exports. Household consumption, at barely 40% of GDP at present, is low by international standards (in most other Asian countries with high savings, the share is between 50% and 70%). High dependence on investment and exports makes China vulnerable to the global economic cycle. Industrial development is also placing strains on the availability of natural resources and on the environment (see Chapter 5). Though savings and investment rates are expected to fall from their current high levels, this could be compensated by higher productivity resulting from institutional and trade reform and from the continued migration of labour from agriculture to services and industry.

The Chinese government has adopted several measures in pursuit of structural adjustment, including raising minimum wages, reducing income taxes and increasing public spending, as well as taking the steps needed to contain rapid growth in investment and to promote consumption. The government has lifted interest rates, imposed duties on some exports and instructed state banks to rein back lending to overheated sectors. But these efforts will take time to take effect. Both investment and savings have continued to grow strongly in recent years, pushing the trade balance into massive surplus (see below). Yet there are signs that production is starting to shift towards less resource-intensive goods and higher-value industrial products and services that generate better wages.

In contrast to China, India is faced with a need to increase the share of investment in GDP and to relieve infrastructure constraints to sustain growth, including inadequate roads and electricity networks. Low real interest rates in recent years have driven consumption up more than investment. Another major challenge is to develop human capital and provide job opportunities for a large pool of underemployed and undereducated workers. The continued movement of labour from the farming sector to urban industry and services could underpin further advances in labour productivity. Faster and deeper labour-market and product-market reforms, improved management of government finances and more effective public sector administration could boost the long-term rate of economic growth, but the pace of such progress remains uncertain.

International Trade and Financial Flows

Economic growth in China and India affects growth in the rest of the world largely through international transactions. The global impact of growing trade with the two giants – especially China – has already been huge and, on current trends, will increase further in the future. China has one of the world's most open economies: exports account for a remarkable 37% of the country's GDP (Figure 3.2). In India, the share is about 14% but has been growing rapidly since the early 1990s. China and, to a smaller degree, India have been major contributors to the massive expansion of global trade.

Figure 3.2: **Share of Exports in GDP in Selected Countries, 2006**

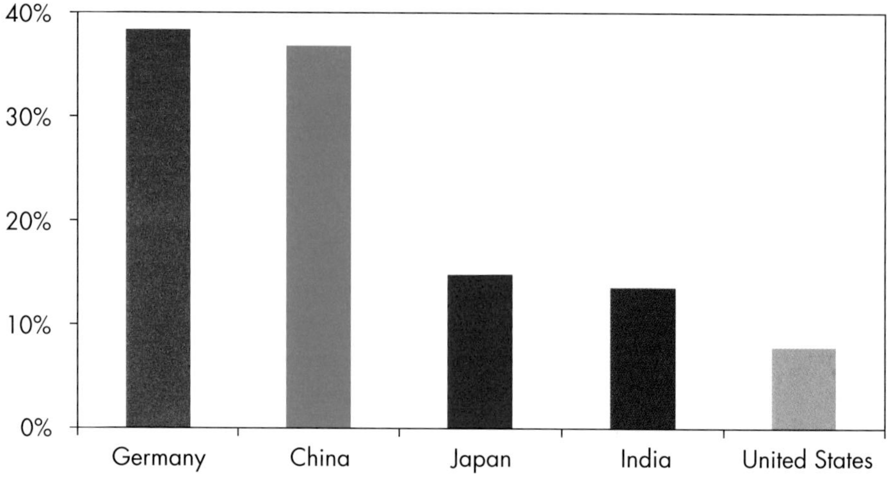

Sources: WTO and IMF databases.

The expansion of China's trade with the rest of the world has been one of the most striking global economic phenomena of the last quarter of a century. China accounts for close to 8% of world exports and about 6% of world imports (Table 3.1) – well above the country's contribution to world GDP. China is the world's third-largest exporter, having recently overtaken Japan (Figure 3.3). Chinese exports grew by 25% per year between 2000 and 2006 compared with 15% per year in the previous ten years. The importance of trade to China's economy reflects the high degree of integration of Chinese industry into international production chains, particularly within Asia. Up to one-third of the value of gross exports is estimated to come from imported inputs – mainly parts and components for assembly into finished products and capital equipment; most exports are finished goods (Winters and Yusuf, 2007). Between 2000 and 2005, China accounted for 13% of the increase in world exports of goods and services, and 10% of the increase in imports in 2000-

2006. India's contribution to international trade remains much smaller, at about 1% of global trade, but has been growing in recent years. In the three years to 2005, trade grew, on average, by 25% per year.

3

Table 3.1: **Share of China, India and the United States in World Trade in Goods and Services**

	China	India	United States
Exports (2006)	8%	1%	9%
Imports (2006)	6%	1%	16%
Share of Increase in world exports (2000-2006)	13%	1%	5%
Share of Increase in world imports (2000-2006)	10%	2%	12%

Source: WTO database.

Figure 3.3: **China's and India's Share in World Trade* in Goods and Services Compared with Other Countries**

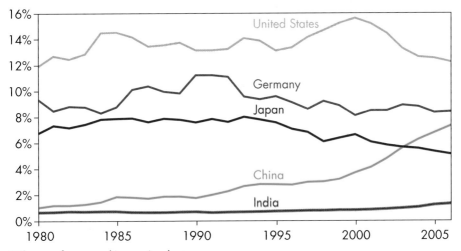

* The sum of exports and imports in value terms.
Source: WTO database.

For many commodities, China is the world's leading consumer and importer. It ranks first as consumer of all the main metals, including aluminium, copper, iron ore[2] and lead. It is also by far the largest consumer of coal, accounting for 41% of global consumption of hard coal (steam and coking coal) and 38% of

2. China is also the largest producer of steel and aluminium, accounting for more than one-third of world steel production (IISI, 2007), and more than 30% of aluminium (ABARE, 2007).

total coal use in 2005, and the second-largest consumer of oil (Figure 3.4). China consumes about 30% of the world's output of minerals and other raw materials and was responsible for two-thirds of the increase in world metals consumption between 1999 and 2005 (Streifel, 2006). China is also the largest consumer of several types of agricultural products including wheat, rice, palm oil, cotton and rubber. India is also among the largest consumers of agricultural goods and is the world's biggest consumer of sugar. It is the third-biggest market for coal, fifth for electricity and sixth for oil. China and India continue to meet the bulk of their food and energy needs from domestic production, but imports of some items – notably oil – have been growing rapidly. Imports continue to meet most of their demand for metals.

Figure 3.4: **Share of China, India and the United States in World Primary Commodity Consumption in 2005**

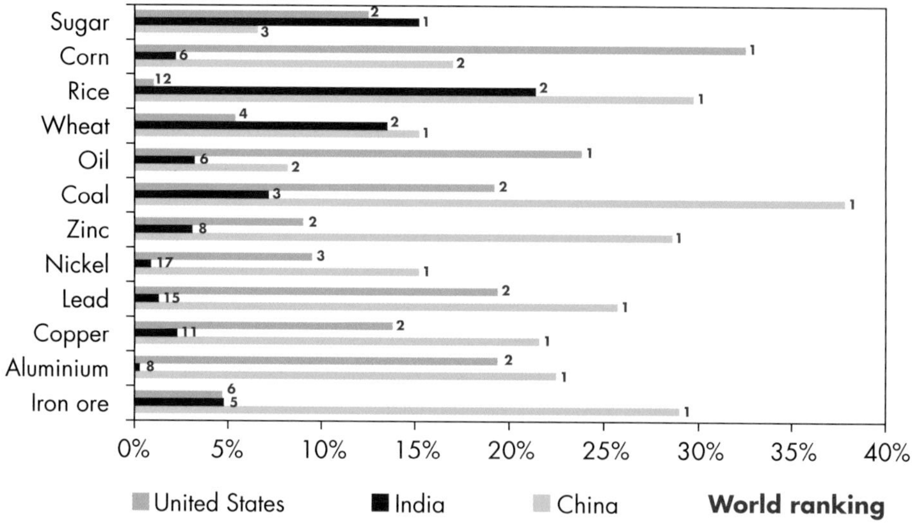

Sources: IEA databases; Streifel (2006).

Rising demand in China, India and elsewhere has contributed to the higher prices of most commodities in recent years, though supply-side constraints and dollar depreciation have played a part. The prices of copper, iron ore, lead, nickel and zinc, of which China is the leading consumer, have increased most. China is a net importer of all these metals. By contrast, China is a modest net exporter of aluminium, the price of which has increased much less in percentage terms. For example, the price of copper increased almost five-fold between the beginning of 2002 and the middle of 2006, while aluminium prices rose by 80%.

There are major differences in the breakdown of exports from China and India. Although clothing and textiles account for a similar share of both countries' exports, other types of manufacturing account for a significantly higher share in China than in India. Exports of services, made up largely of information technology software services and IT-enabled business process services (such as call centres and software application, design and maintenance), are much larger in percentage terms in India. The total *value* of such exports is nonetheless higher in China: $62 billion in 2004 compared with $40 billion in India (Winters and Yusuf, 2007). India benefits from an abundant supply of cheap, qualified English-speaking labour. Nonetheless, both countries still account for a small share of the total value of IT-related services (1.8% for India and 2.8% for China). There is considerable scope for China to boost the share of services in its exports.

The main sources of increased commodity imports into China and India in recent years have been Latin America, Africa and the Middle East. Most of the two countries' purchases from Latin America are agricultural products and metals (Brazil is China's third-largest supplier of iron ore), while they buy mainly oil and metals from Africa. Commodity exporters have benefited from the rise of China and India, both through stronger demand and higher prices for their commodities and through cheaper imports of manufactured products.

The increasing integration of China and India into the global trading system has been accompanied in recent years by liberalisation of both inward and outward capital flows, though some restrictions remain. But, in both cases, there is a marked asymmetry in the composition of their gross liabilities and assets. The liabilities of both countries are mainly foreign direct investment (FDI), debt and portfolio equity, which usually yield a significantly higher rate of return than domestic assets. FDI has played a bigger role in China, which was in 2005 the world's second-largest recipient of FDI. It amounted to $108 billion, or 12% of the world total (Figure 3.5). Even so, FDI still represents a small proportion of total investment in China, the overwhelming bulk of which is financed domestically. Although FDI amounted to only $6.6 billion in India in the year to March 2007, this represented a three-fold increase over March 2006 (OECD, 2007). Portfolio investment is the main type of capital flow into India.

In contrast, the bulk of China's and India's assets are held in low-return liquid foreign reserves, such as US treasury bills. Both countries have accumulated large amounts of such assets over the past decade or so. China's foreign exchange reserves totalled $1.3 trillion at the end of June 2007. However, the imbalance between assets and liabilities is starting to change, with growing overseas direct investment (ODI) by Chinese and Indian firms. The bulk of China's ODI[3] is going to other Asian countries, but a significant share is going to energy and other natural resource projects in Africa and Latin America. The total stock of Chinese ODI amounted to $73 billion in 2006. India's ODI assets remain small, at about $5 billion. However, ODI flows by Indian companies are developing rapidly, reaching nearly $1.4 billion in 2005 (see Chapter 14).

A reduction in the net liabilities of both countries has been a striking feature of the integration of China and India into the global financial system. Growing surpluses in the capital account have been accompanied by surpluses in trade of goods and services. China, in particular, has been running a trade surplus since the early 1990s, while India's trade balance went into deficit in 2003. China's surplus has grown considerably larger in the last four years, mainly due to increased bilateral imbalances with the United States and Europe. These surpluses, which have caused the surge in foreign exchange reserves, complicate monetary policy operations and have put upward pressure on the yuan. The Chinese authorities have been reluctant to revalue the currency too quickly, for fear of undermining exports and slowing job creation.[4] Excessive liquidity has fuelled a boom in property and stock markets, which endured considerable turbulence in mid-2007. China's large foreign-exchange reserves and its large trade surplus, particularly with the United States, make China's monetary policy and its economic restructuring policies critical factors in maintaining the stability of the world's economic system (Roach, 2007).

How China's and India's international trade develops in the future depends on how their domestic economies evolve structurally.[5] Investment in infrastructure and in labour-intensive manufacturing for export will continue to drive China's economic development for some time to come. Investment in higher value-added goods is also set to grow. Investment in less-skilled sectors could shift inland from coastal areas to take advantage of the vast reserve of low-cost, underemployed farm labour. There may also be some shift in investment to other countries, including India. Services could account for a growing share of China's exports, depending on the success of structural adjustment policies, aided by the growing tradability of many types of business-related service activities. Similarly, India looks set to retain its competitive advantage in textiles, clothing and other relatively low value-added, labour-intensive sectors. But there is potential for India to boost its share of trade in other, higher-value manufactures, including pharmaceuticals and specialised engineering, in which it already has a significant presence.

3. Between 2002 and 2005, the overseas investment of Chinese multinationals grew on average by 66% per year (UNCTAD, 2006).
4. McKinsey Global Institute (2007) investigates the implications of different scenarios for reducing the Chinese-US trade imbalance.
5. See Chapter 7 and Chapter 14 for a more detailed discussion of the future evolution of China's and India's economies.

Figure 3.5: **Foreign and Overseas Direct Investment, 2005**

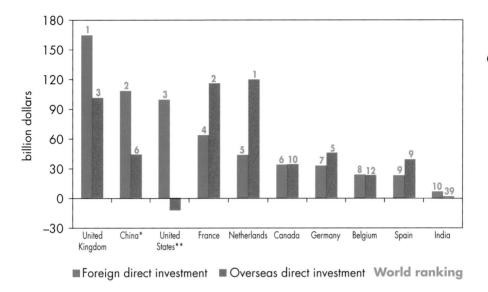

■ Foreign direct investment ■ Overseas direct investment **World ranking**

*China includes Mainland China and Hong Kong.
**In 2005, US overseas direct investment fell sharply and turned negative, largely as the temporary effect of a change in tax rules, which make it very profitable for US companies to repatriate earnings from abroad (OECD, 2006).
Source: UNCTAD database.

Global Economic and Energy Market Linkages

Both China and India will continue to play an increasingly important role in global trade in view of their potential for boosting exports of both manufactures and, increasingly, services. This will pose challenges for the rest of the world, but will also bring opportunities. The two countries' growing weight in international trade will undoubtedly intensify competition in export markets. In industrialised countries, the share of China and India (and other developing countries) in their imports will continue to rise, putting further downward pressure on margins and wage levels in the sectors that are least able to compete – especially labour-intensive, low value-added production of goods and services that can be traded easily, such as textiles. These countries will continue to focus on higher value-added, more specialised manufacturing and services. But they worry that China and India can readily acquire and master the newest technologies, such that their exports displace the rich countries' sales in their own domestic and export markets. Growing reliance on China and India for business services threatens further to undermine profitable opportunities for investment and jobs in the industrialised world. Similar pressures are building in higher-income developing countries, many of whom fear that Chinese and Indian exports might swamp their domestic markets and prevent them from entering new export markets.

These factors explain the apprehension of many countries – rich and poor – about the pace of economic expansion in the two emerging giants and their growing importance in international trade and financial flows. Rapid economic growth, fuelled by high saving rates, will also drive China's and India's acquisition of overseas assets. China's growing trade surplus is adding to these concerns. But there is a positive side to the story. These developments will, on balance, bring net economic benefits to the rest of the world in the medium to long term. Growth in China and India opens opportunities for other developing and industrialised countries to increase exports. In order to be able to increase their exports, China and India will need to increase imports of intermediate inputs, raw materials, energy resources and products, technology and investment goods. The surge in exports from Africa and Latin America over the past decade has been largely driven by demand from China and India. Their demand for high-technology goods will also continue to rise, boosting imports from industrialised countries. In addition, rising incomes and living standards in China and India, together with a probable increase in those countries' exchange rates, will create opportunities for low-income countries to move into low-skill activities abandoned by producers in the giants. Wages have been rising much faster in China than in many other developing countries (World Bank, 2007). Increasing capital flows from China and India could also boost investment and growth in the rest of the world.

Increasing demand for mineral resources to fuel China's and India's economic expansion, including metals, clays and aggregates for the construction industry, is expected to put upward pressure on prices in the long term.[6] This will have major consequences for the competitiveness of processing industries, the terms of trade and economic growth. Resource-poor countries will inevitably be hit hardest. Prices have already increased sharply in recent years (Figure 3.6), driven partly by strong Chinese demand. Prices of some resources, notably aluminium, lead and tin, have started to fall back from recent peaks as new production capacity has come on stream. But increasing extraction of these finite resources could lead to higher marginal production costs in many cases, pushing up prices over the coming decades. This is expected to occur for oil and gas, the prices of which are assumed to rise over most of the projection

---

6. Bloch *et al.* (2007) finds that commodity prices rise on average by 1.5% for every 1% increase in world industrial output, with a maximum lag of one quarter. The barter terms of trade of commodities to finished good also rise when world industrial growth exceeds 4% per year. Higher US interest rates and a stronger dollar have a generally negative impact on commodity prices.

Figure 3.6: **International Prices of Major Commodities**

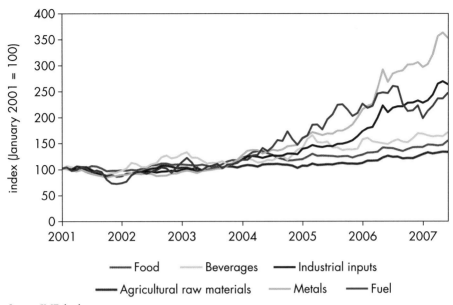

Source: IMF database.

Which Countries Gain Most from Economic Growth in China and India?

Economic developments in China and India will increasingly affect the rest of the world, by dint of their sheer size and their growing weight in international trade and cross-border financial flows. The effects are both positive and negative: economic expansion in China and India generates opportunities for other countries to export to them, while also increasing the other countries' access to a wider range of cheaper imported products. Both factors can boost productivity growth and raise average living standards in the countries trading with China and India. But growing exports from China and India also increase competitive pressures on other countries, particularly those with competing export industries leading to structural adjustments. Demand from the two giants could also contribute to higher international prices for commodities – including energy – potentially tempering incomes and economic growth in all countries that rely heavily on imports. Although most countries are likely to benefit overall, some may lose out. Even in countries that stand to gain overall from expanding trade with the emerging giants, some sections of society – particularly unskilled workers – may be worse off.

.../...

In general, exporters of natural resources and other primary commodities, that have other export industries that do not compete head-on with Chinese and Indian goods and services, could expect to benefit most from China's and India's economic expansion (Goldstein *et al.*, 2006; Stevens and Kennan, 2006; and Winters and Yusuf, 2007). On the other hand, countries that export largely goods and services that match those of China and India are likely to suffer economic losses, at least in the short term. The economic impact on other countries, with different mixes of exports and imports, depends on the precise mix of exports and imports and the extent to which Chinese and Indian growth pushes up demand for commodities and international prices (Figure 3.7). In the long term, structural economic adjustments resulting from higher growth in China and India may make good any economic losses suffered by other countries. This depends on how flexible and efficient their economies are in responding to world market changes triggered by external economic growth, as well as on the magnitude of the shocks, the steps taken to mitigate them and the implementation of transition agreements to allow for adaptation. Even countries that stand to gain eventually may endure painful adjustment effects, including job losses and lower wage pressures in some sectors. Some workers may lose out, even if the majority are ultimately better off.

The policy response of other countries to the challenges and opportunities created by the rise of China and India will affect the extent to which they benefit or not. Policies that embrace rather than resist global integration will lay the foundations for future growth and job creation (World Bank, 2007). A slow-down in the pace of globalisation triggered by revived protectionism is one of the major downside risks to world economic growth prospects. Openness to trade and inward investment – and a generally attractive investment climate – will be particularly important for the poorest countries, which most need the technologies and know-how that will allow them to profit from rising demand from China and India.

period (see Introduction). The conclusion of long-term contractual arrangements may also restrict access by some metal producers to raw material supplies.[7] Mineral exporters will clearly benefit from any increase in international prices, while importers will suffer.

---

7. Access to raw materials was discussed by G8 leaders at their summit meeting in Heiligendamm, Germany in June 2007. They signed a joint Declaration on Responsibility for Raw Materials: Transparency and Sustainable Growth, which addresses priorities for ensuring a sustainable and transparent approach to this question.

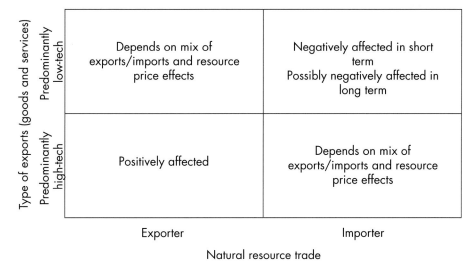

Figure 3.7: **Potential Economic Impact from High GDP Growth in China and India**

	Exporter	Importer
Predominantly low-tech	Depends on mix of exports/imports and resource price effects	Negatively affected in short term Possibly negatively affected in long term
Predominantly high-tech	Positively affected	Depends on mix of exports/imports and resource price effects

Type of exports (goods and services)

Natural resource trade

Simulating the Impact of Faster Growth in China and India

To determine how the economy in each region may be affected by the rise of China and India in the future, we made use of a general equilibrium model of the world economy to simulate the effect of faster growth in those two countries on the rest of the world. That model was integrated into the IEA World Energy Model (WEM) – the primary tool for generating global energy projections in all the scenarios in the *Outlook* – to capture energy-market effects and interactions (Box 3.2) The integrated model, WEM-ECO, ensures internal macroeconomic consistency with energy trends and allows constraints on the supply of major resources – including energy – to be taken into account explicitly.

In the Reference Scenario, GDP is assumed to grow by an average of 6% per year in China and 6.3% in India over the projection period (see Introduction). But the future rate of GDP growth and the evolution of the structure of the economies of China and India are inevitably uncertain. This is reflected in the wide range of projections made by various organisations, and their frequent revision. Recent projections have often significantly underestimated growth rates. For example, the IMF's short- and medium-term projections of China's and India's GDP have been revised upwards significantly over recent years. For China, the projected growth rate for the year ahead was 7.5% in 2003; the actual rate of growth in 2004 turned out to be 10.1%. In its 11th Five-Year Plan, the Chinese government currently targets 7.5% GDP growth in 2006-2010. Yet China's National Bureau of Statistics recently revised the estimated growth rate in 2006

Box 3.2: **Modelling Economic and Energy Interlinkages: the WEM-ECO Model[8]**

The IEA World Energy Model (WEM) is a partial equilibrium model with a rich technology representation of all energy sectors. The bottom-up structure of the model makes use of the extensive IEA statistical databases. Regional GDP growth rates and international fuel prices are exogenous. For the purposes of the High Growth Scenario, we developed in collaboration with the Centre International de Recherche sur l'Environnement et le Développement (CIRED) – a French research institute – a hybrid model, WEM-ECO, by integrating WEM into CIRED's IMACLIM-R model, a general equilibrium model (GEM) of the world economy.[9] GEMs allow the effects on the rest of the economy of changes in a particular sector, such as energy, to be simulated quantitatively. WEM-ECO thus provides a consistent energy and macroeconomic modelling framework, integrating the macroeconomic feedback effects of different energy pathways, including changes in the economic structure, productivity and trade – all of which affect the rate, direction and distribution of economic growth.

WEM-ECO has twelve regions, including both China and India as individual regions. Economic activity is portrayed both in monetary and physical terms. Short-term rigidities and long-term flexibilities in the production function in each sector are explicitly represented, using an input-output (Leontief) approach. Economic growth is driven by exogenous assumptions of population growth, savings behaviours and labour productivity growth in each region. Oil and other commodity prices are determined by the model, based on assumptions about oil resources, supply-side constraints and the dynamics of oil demand.

The energy projections of WEM were incorporated into the IMACLIM-R structure in an iterative manner. IMACLIM-R was first calibrated to the Reference Scenario, using the GDP and energy prices assumptions in the WEM and adjusting the labour productivity gains to obtain the same growth rate and the Middle Eastern oil producers' production behaviour to reproduce the same oil prices. The energy-related technical coefficients in the WEM were also incorporated into IMACLIMR-R, to ensure consistency between two models. To generate the High Growth Scenario projections, WEM was first run to obtain the change in energy demand induced by the assumed higher economic growth. These results were then integrated into IMACLIM-R, to yield the change in energy prices and other commodity prices and the overall impact of increased trade, as well as new prices, on other countries' economic growth. Those results were then used to re-run WEM. These steps were repeated until the two models converged on a consistent energy and macroeconomic trajectory.

8. More details about WEM-ECO are available at www.worldenergyoutlook.org.
9. See Hourcade *et al.* (2006) for a discussion of hybrid modelling.

up to 11.1% – an all-time high (NBS, 2007). Revisions have also been large for India: the IMF's projection of GDP growth for the year ahead was 6.7% in 2004, while the actual rate of growth in 2005 turned out to be 9%.

It is certainly possible that both economies could grow much faster than assumed in our Reference Scenario. These uncertainties led us to develop a High Growth Scenario, in which GDP is assumed to grow by around 1.5 percentage points per year faster in both countries. That corresponds to an average rate of GDP growth to 2030 of 7.5% per year for China and 7.8% per year for India. Faster growth in the two countries would have major implications for international trade in goods and services, including energy.

The primary goal of the High Growth Scenario is to test the sensitivity of the energy demand and supply projections for China, India and the rest of the world to faster growth in the emerging giants. But, thanks to WEM-ECO, that scenario also allows us to quantify the extent to which economic growth in China and India affects other regions' economies. The results of this economic analysis, which feed into the energy projections, are presented below.[10] They should be interpreted with care, given the large uncertainties surrounding the pattern of growth and underlying capital flows.

Energy and Other Commodity Prices and Expenditures

Higher growth in China and India affects the economies of the rest of the world through its impact on international commodity prices and on overall trade in all types of goods and services. Energy is the most important commodity category, as it is an indispensable input to all productive activities. The higher GDP growth rates assumed in the High Growth Scenario result in faster growth of energy demand in both countries, as described in Chapter 2. Higher growth in energy demand, combined with supply-side constraints (limited investment response by major oil and gas producers), drives up international energy prices (see Chapter 1 for a description of oil and gas supply-side constraints). WEM-ECO recalculates the global equilibrium for international trade in energy and non-energy goods and services, and for energy and other commodity prices in the rest of the world by major region. The average IEA crude oil import price – a proxy for international oil prices – reaches $87 per barrel in year-2006 dollars ($150 in nominal terms) in 2030, 40% higher than in the Reference Scenario. Natural gas prices rise in the same proportion. Increased coal demand drives the price up to about $73 per tonne in 2030, 19% higher than in the Reference Scenario (Table 3.2).[11] Other commodity prices also increase significantly.

10. The energy supply-side assumptions in the Reference and High Growth Scenarios are described in Chapter 1, while the Introduction details the methodology and main assumptions in the different scenarios.
11. Chapter 4 describes in more detail the implications of higher fossil-fuel demand from China and India on energy markets and security of supply.

Table 3.2: **Fossil-Fuel Prices in the High Growth Scenario**
(in year-2006 dollars)

	unit	2006	2010	2015	2030
Real terms (year-2006 prices)					
IEA crude oil imports	barrel	61.7	64.4	66.8	87.0
Natural gas					
US imports	*MBtu*	*7.2*	*8.0*	*8.6*	*11.1*
European imports	*MBtu*	*7.3*	*7.2*	*7.7*	*10.3*
Japanese LNG imports	*MBtu*	*7.0*	*8.0*	*8.6*	*11.0*
OECD steam coal imports	tonne	62.9	57.6	60.9	72.7
Increase over the Reference Scenario					
IEA crude oil imports	%	0	9	17	40
Natural gas	%	0	9	17	40
OECD steam coal imports	%	0	3	7	19

Note: 2006 prices represent historical data. Gas prices are expressed on a gross calorific-value basis. All prices are for bulk supplies exclusive of tax.

An increase in energy and other raw-material costs alters the relative profitability and competitiveness of the production of goods and services. In the long term, it leads to a shift towards less energy- and more capital-intensive productive capacities. Because of inertia, sectors using equipment with a long life are particularly vulnerable to a loss of competitiveness. The magnitude of this effect is related to the rate of growth of the sector as the penetration of new efficient equipment is fastest during a period of rapid economic growth. It also depends on the availability of cheap finance that could facilitate early scrapping of inefficient capacities. In industry, the share of energy in total production costs increases in all regions in the High Growth Scenario, but to differing degrees. The increase, compared with the Reference Scenario, averages 16% in the OECD, 12% in China and 22% in India. The differences are explained mainly by differences in the rate of change in energy intensity.

International Trade

In the High Growth Scenario, worldwide trade in goods and services expands much faster than in the Reference Scenario, as the relative competitiveness of China's and India's exports improves (Figure 3.8). Global inter-regional trade is 12.5% higher in 2030. China's share of international trade in 2030 increases from 9.2% in the Reference Scenario to 10.2% in the High Growth Scenario. India's share grows from 1.7% to 2.1 %.

Higher commodity costs also affect trade balances and currency flows. Commodity exports are an important source of income for some regions, while imports account for a large share of total expenditure in others. The

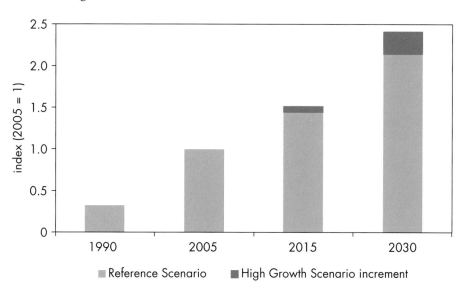

Figure 3.8: **Value of Worldwide Trade in Goods and Services**

Reference Scenario   ■ High Growth Scenario increment

importance of energy in total international trade has been growing in recent years in both China and India. This reflects their increasing dependence on imports, as demand has outstripped indigenous output, and rising international energy prices (see Chapter 4). Energy – mostly crude oil and refined products – made up 10% of China's total import bill in 2005, up from 9% in 2000. In India, the share increased from 35% to 37%. Similarly, the share of China and India in total *exports* of oil from the main exporting countries has risen sharply.

Worldwide, the share of energy in international trade has also been growing with higher prices, reaching 13.5% in 2005, up from 7.5% in 1995. In the High Growth Scenario, this trend continues through to 2015, when the share reaches 16%, but then falls back a little to 14.7% in 2030, as the growth in international trade generally more than outweighs the rising prices and volume of traded energy (Figure 3.9). In the Reference Scenario, the share barely changes over the *Outlook* period. The difference between the High Growth Scenario and the Reference Scenario reaches 3.5% in 2015, dropping to less than 2% in 2030.

The impact of faster GDP growth in China and India on inter-regional trade varies markedly by region. All regions are affected by stronger demand for imports into China and India, by their higher exports and by the rise in international commodity prices. In the OECD, total commercial goods exports – to China, India and other regions that enjoy higher GDP growth – and imports, which are assumed to rise in line with exports, increase by between 5% and 8% in the High Growth Scenario. But this is more than offset by the increase in commodity imports. The energy-import bill alone in 2030 increases by between 28% and 31% compared with the Reference Scenario (Figure 3.10) – mostly due to more costly oil products. Brazil is an interesting case, as rising biofuels production allows it to remain self-sufficient in oil throughout the projection period, shielding its commercial trade balance from

Figure 3.9: **Share of Energy in World International Trade Value**

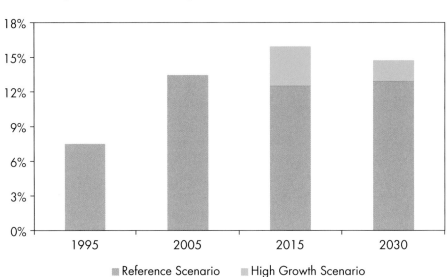

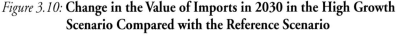

Figure 3.10: **Change in the Value of Imports in 2030 in the High Growth Scenario Compared with the Reference Scenario**

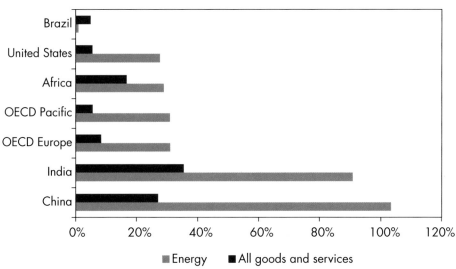

higher oil prices. The total volume of Brazil's imports of goods increases at more or less the same rate as in the OECD – by around 5% in 2030 compared with the Reference Scenario, while its energy bill increases by only 1%. China and India also face an increase of their domestic needs in terms of both energy and goods to sustain their accelerated growth. Total import expenditures increase by 27% in China and 35 % in India in the High Growth Scenario, while China's energy import bill more than doubles and India's rises by 91%, compared with the Reference Scenario.

Economic Growth and Structure in the Rest of the World

Higher economic growth in China and India affects the world economy as a whole through the intertwined channels described earlier in this section. On the one hand, higher demand for energy and raw materials, combined with supply-side constraints, leads to a tightening of commodities markets. This adversely affects economic growth in commodity-importing countries, but boosts growth in exporting countries. On the other hand, larger volumes of trade associated with higher demand in China and India (and to a lesser degree in energy-exporting countries) draw in additional imports from the rest of the world and, therefore, stimulate economic activity in other countries. The latter effect is partly offset by losses in market share for tradable goods produced in other regions, because of increased exports by China and India.[12]

The net effect varies. Most regions – Canada, Brazil, the rest of Latin America, the Middle East, Russia and other transition economies – enjoy a net increase in GDP in 2030. But the United States, the European Union, OECD Pacific and other developing Asian countries as a whole see marginal reductions in their GDP (Table 3.3 and Figure 3.11).[13] Although OECD countries export more advanced technology equipment and high value-added goods to China and India, their competitiveness is diminished by higher energy prices, as they import large amounts of oil and gas. The cumulative reduction in GDP is, nonetheless, very small when seen against the large uncertainties surrounding economic prospects over the *Outlook* period. GDP gains in Russia, Canada, the Middle East, Africa and Latin America are mainly associated with their higher energy exports and commodity prices. Brazil logically falls into an intermediate position, as it does not directly compete with China and India for international trade in goods and its dependence on oil is reduced thanks to the development of biofuels. Overall, world GDP grows faster, by 4.3% per year on average compared with 3.6% in the Reference Scenario, as the "demand-pull" effect offsets the depressive impact of higher commodity prices.

12. We assumed that only part of the additional GDP growth in the High Growth Scenario in China and India was based on an increase in domestic demand, the rest coming from higher exports.
13. PPP – based rates are used to determine the weights in the regional and global aggregations of GDP.

Table 3.3: **World Real GDP Growth in the High Growth Scenario**

	Average annual growth rate, 2005-2030	Difference from Reference Scenario	
		Average annual growth rate, 2005-2030	Level of GDP in 2030
OECD	**2.1%**	**−0.06%**	**−1.4%**
North America	2.4%	−0.02%	−0.4%
United States	*2.3%*	*−0.04%*	*−1.0%*
Europe	1.9%	−0.10%	−2.4%
Pacific	1.8%	−0.07%	−1.8%
Japan	*1.3%*	*−0.07%*	*−1.7%*
Transition economies	**3.6%**	**0.02%**	**0.4%**
Russia	3.5%	0.03%	0.6%
Developing countries	**6.2%**	**1.06%**	**30.2%**
Developing Asia	6.9%	1.28%	37.3%
China	*7.5%*	*1.50%*	*45.2%*
India	*7.8%*	*1.50%*	*45.1%*
Middle East	4.4%	0.41%	10.9%
Africa	4.0%	0.05%	1.4%
Latin America	3.3%	0.06%	1.4%
Brazil	*3.1%*	*−0.00%*	*−0.1%*
World	**4.3%**	**0.61%**	**16.3%**
European Union	*1.9%*	*−0.10%*	*−2.4%*

Regional shares in global GDP change markedly in the High Growth Scenario. The shares of China and India, unsurprisingly, increase even more substantially than in the Reference Scenario (Figure 3.11). By 2030, their combined share of world GDP at market exchange rates reaches 21% (marginally below that of the European Union) compared with about 16% in the Reference Scenario (Figure 3.12).

Higher growth rates in China and India are assumed not to induce dramatic changes in the structure of their economies. At the aggregate level, the share of industry remains the same as in the Reference Scenario, the negative effect of higher energy prices on energy-intensive industrial production offsetting their higher shares on world industrial markets associated with higher internal demand and with their relative gains in competitiveness. There is nonetheless some shift from heavy industry towards lighter manufacturing. There are bigger changes in other regions. Resource-rich countries such as Russia and Middle Eastern oil exporters, suffer losses in competitiveness in non-resource sectors, leading to lower production growth. This is caused by the so-called "natural resource curse" or "Dutch Disease": windfall revenues from natural resources give rise to real exchange rate appreciation, which in turn reduces the competitiveness of the manufacturing sector (Sachs and Warner, 2001).

Figure 3.11: **Gross Domestic Product by Region**

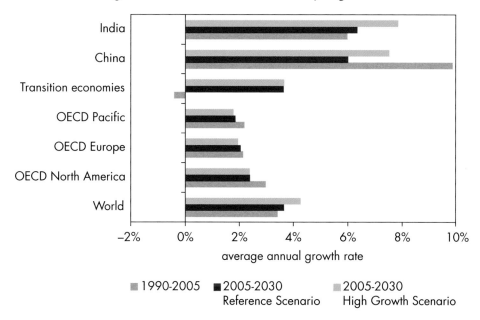

■ 1990-2005   ■ 2005-2030   ▨ 2005-2030
              Reference Scenario   High Growth Scenario

Figure 3.12: **Regional Shares of World GDP at Market Exchange Rates**

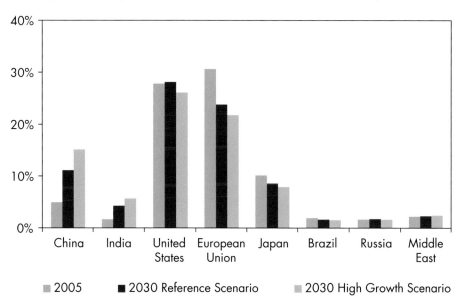

■ 2005   ■ 2030 Reference Scenario   ▨ 2030 High Growth Scenario

The results presented here should be treated with caution. They are very sensitive to a range of modelling assumptions, especially those concerning capital flows and the speed with which trade and capital imbalances are corrected. The sensitivity of energy prices to supply-side constraints and to assumed rigidities in productive sectors is also important. Economic policies will also play a critical role in sustaining the pace of global economic growth: rising protectionism could radically change the global impact of economic growth in China and India. With respect to energy, faster implementation of policies to use energy more efficiently and reduce emissions, as described in the Alternative Policies Scenario, would increase the net benefits to China and India, as well as to the rest of the world, by reducing pressures on international commodity markets and lowering fuel-import costs.

CHAPTER 4

THE WORLD'S ENERGY SECURITY

HIGHLIGHTS

■ Rising global energy demand, to which China and India contribute, has important implications for the world's energy security. The reliance of all consuming countries on oil and gas imports will grow markedly in the absence of new policies in major countries to curb demand. Ensuring reliable and affordable supply will be a formidable challenge.

■ Increased trade could bring mutual economic benefits to all concerned, but it carries heightened risks, for consuming countries generally, of short-term supply interruptions, as geographic supply diversity is reduced and reliance grows on a few supply routes. Much of the additional oil imports required by China, India and other countries will come from the Middle East, the scene of most past supply disruptions, and will transit vulnerable maritime routes to both eastern and western markets. Supply disruptions drive up prices to *all* consuming countries, regardless of where they obtain their oil.

■ The increasing concentration of the world's remaining oil reserves in a small group of countries – notably a few Middle Eastern producers and Russia – will increase their market dominance and their ability to impose higher prices in the longer term. Their share of gas supply is also likely to rise. In the Reference Scenario, almost three-quarters of the growth in world oil production and 43% of gas production come from the Middle East and Russia.

■ With stronger global energy demand, all regions would be faced with higher energy prices in the medium to long term in the absence of concomitant increases in supply-side investment or stronger policy action to curb demand growth in consuming countries. The faster the increase in the call on oil and gas from the leading exporters, the more likely it will be that they will seek to extract a higher rent from their exports in the future.

■ China's and India's growing participation in international trade amplifies the importance of their contribution to collective efforts to enhance global energy security. The more effective are their policies to avoid or handle a supply emergency, the more other consuming countries stand to benefit. Equally, efforts by other consuming countries bring important benefits to China and India. Most energy-security policies also bring environmental benefits.

Energy Security in a Global Market

In a global energy market, changes in the supply/demand balance and fuel mix in one country inevitably affect all other market participants. Energy security is one aspect of the change. Rising energy needs in China and India – on top of rising demand in all other regions – will call for increased investment in developing indigenous and external resources. Mobilising this investment so as to expand capacity will be crucial to the world's long-term energy security. Energy trends in China and India (summarised in Chapter 2), like those in the OECD, mean rising dependence on imports of hydrocarbons in the coming decades, as demand outstrips indigenous production – unless radical new policies are introduced that go well beyond those of the Alternative Policy Scenario or unless there are major technological breakthroughs. These trends carry increased threats to supply security, both in terms of the long-term adequacy and price of supply and the risk of short-term supply disruptions.

The consequences are potentially serious for China, India and all other countries that depend on imports. How China and India respond to these threats will also affect the rest of the world. The more effective are their policies to avert or handle a supply emergency, the more other consuming countries[1] – including most IEA members – stand to benefit, and *vice-versa*. In addition, many policies to enhance energy security also directly support policies to address the environmental damage from energy production and use.

Oil will remain the main focus of attention for the Chinese and Indian governments in their efforts to address growing worries about energy security. This reflects both the prospect of a sharp increase in their import needs and the limited scope for switching away from oil products – especially in the transport sector. Thus, as for other consuming countries, the implications of rising oil use and rising oil-import dependence in China and India are of primary concern. Both China's and India's natural gas imports are also expected to rise, but, in all three scenarios described in this *Outlook*, the volumes remain small relative to the size of their energy markets and to global international trade in gas. Their coal imports are expected to grow as a share of international coal trade. However, the security of coal supplies is attracting less scrutiny, as most of the world's needs are met by indigenous resources and what coal is traded comes from a variety of different sources. For these reasons, the main focus of the discussion in this chapter is oil.

Defining Energy Security

Energy security, broadly defined, means adequate, affordable and reliable supplies of energy. It matters because energy is essential to economic growth

1. The term "consuming countries" is used to describe countries that rely on imports to meet at least part of their energy needs net of exports. "Producing countries" refers to net exporters.

and human development. Yet no energy system can be entirely secure in the short term, because disruptions or shortages can arise unexpectedly, whether through sabotage, political intervention, strikes, technical failures, accidents or natural disasters. In the longer term, under-investment in crude oil production, refining or transportation capacity, or other market failures can lead to shortages and consequently unacceptably high prices. So energy security, in practice, is best seen as a problem of risk management, that is reducing to an acceptable level the risks and consequences of disruptions and adverse long-term market trends. Secure energy supply is a public good, as the benefit derived from it by one consumer does not reduce the benefit to everyone else. Markets alone do not reflect the cost to society of a supply failure because it is beyond the power of an individual supplier or consumer to guarantee security. Put another way, all market players benefit from action to safeguard energy security, whether or not they have contributed to it. For these reasons, governments must take ultimate responsibility for ensuring an adequate degree of security within the framework of open, competitive markets. This applies to producers, as well as to consumers: they benefit from more secure energy supplies if the demand for their resources is not reduced by the adverse macroeconomic effects of higher prices or logistical supply problems that might result from a supply disruption.

Short-term threats to security concern unexpected disruptions, whether of a political, technical, accidental or malevolent nature. Long-term threats relate to a lack of deliverability caused by deliberate or unintentional under-investment in capacity. Both short-term disruptions and under-investment result in higher prices, causing hardship to consumers and harming economic prospects. The two are linked: under-investment also renders the energy system more vulnerable to sudden supply disruptions, accentuating their impact on prices, while experience of short-term disruptions shakes market confidence in supply, increasing the risk of under-investment in production.

Concerns about energy security have evolved over time with changes in the global energy system and perceptions about the risks and potential costs of supply disruptions. In the 1970s and 1980s, the focus was on oil and the dangers associated with over-dependence on oil imports. Today, worries about energy security extend to natural gas, which is increasingly traded internationally, and the reliability of electricity supply. Increasing attention is being given to the adequacy of investment in all types of energy infrastructure. There are growing concerns about whether competitive markets for electricity and gas, as they currently operate, provide sufficient incentive for building capacity.

Most governments have developed policies to protect against failures in the energy supply system that arise from weakness in market mechanisms or that cannot be handled by the market alone. Long-term policies aim to encourage:

n Adequate investment in production, processing, transportation and storage capacity to meet projected needs.

n More efficient energy use, to reduce the risk of demand running ahead of deliverability.

n More diversity in the fuel mix, geographic sources and types of supply, transportation routes and market participants.

n More market transparency, to help suppliers and consumers make economically efficient investment and trading decisions, and governments to take informed policy decisions.

Policies and measures to respond to short-term disruptions include co-ordinated use of emergency energy stocks, redirected supply flows and demand-side management (IEA, 2005). Their purpose is to alleviate rapidly the effects of any loss of physical supply, by making good all or part of the shortfall or by reining back demand. These measures can help to minimise the economic and social cost of a supply disruption by facilitating the movement by the market of scarce supplies to where they are most needed. Emergency stocks and co-ordinated responses to a supply disruption form a central pillar of the energy-security policies of IEA countries (Box 4.1). Governments also adopt measures aimed at preventing supply disruptions, such as protection of pipelines, maritime ports and sea lanes, enforcement of health and safety regulations to prevent accidents, and early-warning systems for severe weather.

Box 4.1: **IEA Emergency Response Mechanisms**

The IEA's emergency response mechanisms were set up under the 1974 Agreement on an International Energy Program (IEP). The Agreement requires IEA countries to hold oil stocks (now standing at the equivalent of at least 90 days of net oil imports) and, in the event of a major oil supply disruption, to release stocks, restrain demand, switch to other fuels or increase domestic production in a co-ordinated manner.

Implementation of IEP measures was initially designed for oil supply disruptions involving a loss of 7% or more of normal oil supply, either for the IEA as a whole or for any individual member country. To supplement the mechanisms defined in the IEP, the IEA has elaborated more flexible arrangements known as the Coordinated Emergency Response Measures (CERM). They provide a rapid and flexible system of response to actual or imminent oil supply disruptions of any size. The response may include stock release, demand-restraint measures and/or use of surge-production capacity. The last time CERM was deployed was in September 2005, when IEA countries agreed to make available to the market 60 million barrels of oil to help offset the loss of 1.5 million barrels per day of crude oil

.../...

production and 2 mb/d of refining capacity caused by hurricanes Katrina and Rita in the United States.

To ensure the potential of IEA countries to respond rapidly and effectively to oil emergencies in changing oil market conditions, the IEA Standing Group on Emergency Questions (SEQ) conducts a regular cycle of Emergency Response Reviews of IEA member countries. These peer reviews cover procedures and institutional arrangements, and result in recommendations for improvements. In addition, the IEA carries out a series of workshops and emergency-response exercises every two years to train personnel and test policies and procedures.

4

How Supply Disruptions Affect Consuming Countries

The consequences of a disruption in energy supplies for a consuming country or region depend on several factors, including the type of fuel, the nature and size of the disruption or shortage, expectations about how long the disruption will last and the fuel-import intensity of the economy. In practice, economic vulnerability depends not just on the nature and duration of a disruption, but also on the flexibility and resilience of the economy to respond to and withstand the physical loss of supply and the higher prices that result. Experience has shown that the sudden loss of even a modest volume of oil can lead to sharp increases in prices, particularly when global spare capacity is tight or when geopolitical tensions are high.

A well-functioning, competitive market will reallocate supplies according to ability to pay, though macroeconomic damage may result from the increase in price. In this case, particularly where supplementary emergency measures are available, a supply disruption should not, in principle, cause a physical shortage, as price adjusts upwards to bring demand back into balance with the new, lower level of supply. Similarly, where prices are driven higher by a lack of supply capacity as demand outstrips capacity additions, more investment would normally be forthcoming, eventually driving prices back down. But there may be important time lags. Where prices are not free to adjust because of price controls or infrastructure constraints on deliverability, physical shortages can occur at local or national levels. In the case of oil, all OECD countries and many other non-member states have liberalised their oil markets, so prices are free to rise in response to a supply disruption. In these countries, the risk of physical unavailability is largely reduced to extreme events – such as weather-related catastrophes, strikes or terrorism.

The effects of a disruption in oil supplies, regardless of where it takes place and which buyers are directly affected, mainly depend on the extent of the global price response – not on whether the consuming country obtains its oil physically from the country from which supply is disrupted. Crude oil and refined products are global commodities. The prices of all crude oils are linked, via explicit formulas in term contracts or through direct competition in the spot market, to the futures or spot prices of a small number of benchmark grades. The international spot prices of refined products are closely correlated with crude oil prices. Thus, a shortfall in oil supply to one country, by driving up the price of all grades and types of oil, affects *all* consuming countries, regardless of whether their supplies are directly affected or not. Even in the theoretical case of a country isolated from the market by self-sufficient supply lines, the value of the oil would rise in response to an external supply disruption and that additional value would be forgone if domestic prices did not increase accordingly.

In the event of a disruption in the supply of natural gas, physical shortages can occur because of inflexible infrastructure, price controls or rigidities in supply contracts to end users (even in competitive markets). In such circumstances, gas use must be reduced by administrative means. In many OECD countries, gas distribution companies are tasked by the authorities with determining how to allocate scarce supplies during a supply emergency. With oil-price indexation, gas-consuming countries are also exposed to the price risk of the oil market. Though some contracts, for example in Japan and Korea, have price arrangements that moderate this risk, this feature also delays the market response to the new circumstances.

The impact of a disruption in the supply of gas, in contrast with oil, *does* depend to a large extent on the source of the gas. Gas-pipeline infrastructure is inflexible, so that a loss of supply through a particular pipeline system cannot always be made good by supplies from other sources. LNG supply is more flexible, as it may be possible to replace the loss of supply from one source by output from another, as has happened in several recent cases. The share of LNG in world gas trade is set to rise strongly over the projection period, particularly to supply OECD countries, which will contribute to more flexibility in gas supply. But, in practice, there may not be enough spare liquefaction and shipping capacity immediately available to compensate for a large supply disruption. In addition, most LNG is at a present sold under long-term contracts, with rigid clauses covering delivery, though many new projects to come on stream in the next few years have more flexible terms.

Measuring Energy Security

There is no single universally recognised way of measuring a country's level of energy security. Such assessments are normally a matter of expert judgment, as the perceived risk of a serious disruption or shortfall in investment for any given country or at any given time depends on a large array of different factors.

Some of these factors, such as political stability, are inherently difficult to measure. Nonetheless, most discussions centre on the following variables, or indicators:

- Diversity of the primary fuel mix.
- Import dependence and fuel substitutability.
- Market concentration (the dominance of a small number of producing countries in total trade of any one fuel).
- Share of politically unstable regions in imports.

For a given consuming country, what matters is both its own situation with respect to these indicators and that of all consuming countries. A given country may have a geographically highly diversified mix of imports from what are considered politically stable and reliable producing countries, but it still faces the risk of a price shock from a disruption to supplies from less stable producing countries to other consuming countries. For this reason, a reduction in a given country's imports does not necessarily enhance *its* own overall energy security, if the *world's* reliance on supplies from politically unstable countries is increasing. Likewise, rising import dependence does not necessarily mean less secure energy supplies: a flourishing international market can respond flexibly to unexpected events. Increased fuel diversity can contribute to lower import dependence for particular fuels.[2]

The Role of China and India in International Energy Trade

Oil

In all three scenarios described in this *Outlook*, the shares of China and India in world oil demand grow significantly (Figure 4.1). This reflects rapid rates of economic growth in the two countries, which drive up demand for mobility and for stationary energy-related services. Oil-based fuels continue to dominate transport energy demand – even in the Alternative Policy Scenario – and to meet a significant share of rising energy needs for space and water heating and cooking in the residential and commercial sectors, as well as process energy needs in industry. The combined share of China and India in global oil use increases markedly: from 12% in 2006 to 20% in 2030 in the Reference Scenario and to 19% in the Alternative Policy Scenario. Their share rises even more in the High Growth Scenario, to around 25%. In all three scenarios, both countries account for a bigger share of the increase in world oil demand between 2006 and 2030 than any other *WEO* country or region (Table 4.1).

2. The IEA has developed composite indicators that attempt to measure the degree of security for a given country (IEA, 2007a).

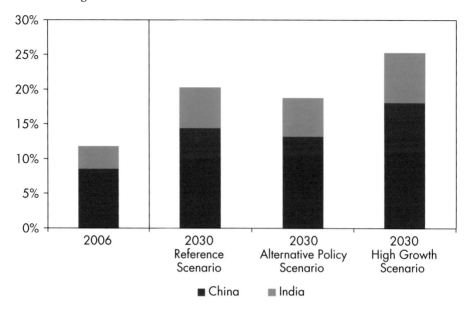

Figure 4.1: **Share of China and India in World Oil Demand**

■ China ■ India

Table 4.1: **Increase in World Primary Oil Demand by Region and Scenario, 2006-2030**

	Reference Scenario		Alternative Policy Scenario		High Growth Scenario	
	mb/d	% of world	mb/d	% of world	mb/d	% of world
China	9.4	30	6.2	36	14.3	41
India	3.9	12	2.8	16	5.6	16
OECD	4.9	16	−0.1	−1	2.5	7
Rest of world	13.2	42	8.5	49	12.5	36
World	**31.4**	**100**	**17.4**	**100**	**35.0**	**100**

Oil demand outstrips indigenous production in both China and India in all three scenarios, pushing up net oil-import needs. The extent of the increase in imports varies significantly by scenario. In the case of China, for example, net imports rise from 3.5 mb/d in 2006 to 9.7 mb/d in 2030 in the Alternative Policy Scenario and 17.2 mb/d in the High Growth Scenario. In the latter scenario, the combined imports of the two countries, 24.8 mb/d, approach those of the entire OECD, 31.5 mb/d (Figure 4.2). The increase in OECD imports between 2006 and 2030, at 3.3 mb/d, is much less marked.

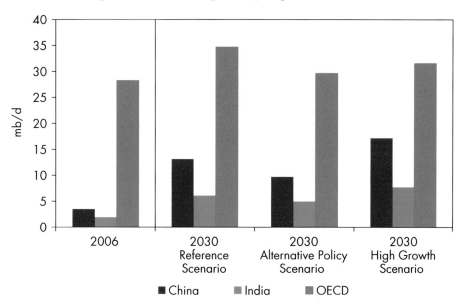

Figure 4.2: **Net Oil Imports* by Region and Scenario**

*Based on trade between *WEO* regions only.

As a result of their much faster growth in import requirements, the shares of China and India in inter-regional oil trade increase sharply between 2006 and 2030. Their combined share reaches 29% in 2030 in the Reference Scenario – up from 13% now. In the High Growth Scenario, it hits 37% (Figure 4.3). Imports represent about half of China's total oil consumption at present. This share reaches 80% in 2030 in both the Reference and High Growth Scenarios (Table 4.2). Import dependence grows slightly less rapidly in India, though it remains proportionately more dependent than China in 2030 in all three scenarios.

Most of the additional oil that China will need between now and 2030 is likely to come from two major sources: the Middle East and, to a much lesser extent, the former Soviet Union. India is expected to be supplied mainly from the Middle East. These exporting regions have the resources to meet a significant share of the increase in global demand and are geographically well placed to supply the Chinese and Indian markets. About two-thirds of India's oil imports currently come from the Middle East. Middle Eastern producers supply about 45% of China's imports, with the rest coming from Russia, Africa and other developing countries.

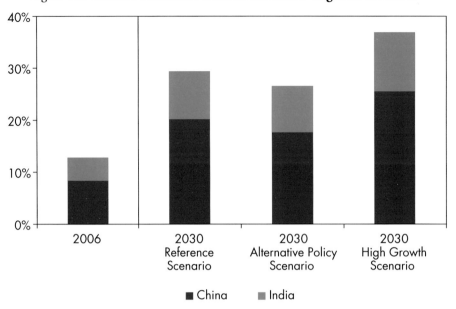

Figure 4.3: **Share of China and India in Total Inter-Regional Oil Trade***

■ China   ■ India

*Trade between *WEO* regions only.

Table 4.2: **Oil Net Imports in China and India**

		Reference Scenario		Alternative Policy Scenario		High Growth Scenario	
	2006	2015	2030	2015	2030	2015	2030
China							
mb/d	3.5	7.1	13.1	6.5	9.7	8.3	17.2
% of primary oil demand	52	64	79	61	72	65	80
% of primary energy demand	10	12	17	12	15	12	17
India							
mb/d	1.9	3.0	6.0	2.7	4.9	3.2	7.7
% of primary oil demand	70	80	92	79	90	79	92
% of primary energy demand	17	20	23	19	23	19	23

The bulk of China's and India's crude oil and refined product imports is shipped by sea. China imports oil through a pipeline from Kazakhstan, the first leg of which was commissioned in 2006, and by rail from Russia. The Chinese and Kazakh governments recently agreed to double the capacity of the pipeline to around 400 kb/d. China is seeking to boost imports of Russian oil by pipeline, but Russia has not yet taken a decision on whether to proceed with the construction of a spur line to Daqing in northern China. This line would link with the 600 kb/d line from Taishet (in East Siberia) to Nakhodka (on the coast of the Sea of Japan), the first leg of which is under construction. China has also proposed building a pipeline running across Myanmar to the Chinese border carrying Middle Eastern oil, which would circumvent the Straits of Malacca.

Rising Chinese and Indian imports of oil from the Middle East will push up both countries' reliance on two critical shipping channels. In the medium term, at least, most of the Middle Eastern oil shipped to both countries will continue to transit the Straits of Hormuz at the mouth of the Persian Gulf – the world's busiest oil-shipping lane. The straits comprise two 3-km-wide inbound and outbound lanes. Only a small proportion of the oil could be transported along alternative routes. In 2006, approximately 13.4 mb/d, or 16% of the world's total oil supply, passed along this route. The Middle Eastern oil destined for China is subsequently shipped through the Malacca Straits between Indonesia, Malaysia and Singapore – another busy and narrow route. In 2006, volumes shipped through this channel (including a small amount from West Africa) reached about 12 mb/d, of which approximately 2.5 mb/d went to China. However, alternative, slightly longer routes exist, such as the Straits of Lombok and Sunda in Indonesia. The volumes shipped through the Straits of Hormuz and the Straits of Malacca are projected to increase significantly in all three scenarios (Figure 4.4).

Although the Straits of Hormuz have never been closed to shipping (though oil shipping was attacked during the Iran-Iraq war of 1980-1988), growing tensions over Iran's nuclear policy have highlighted the risk of disruptions to shipping in the event of a major regional conflict. In response to growing concerns about this risk among Persian Gulf oil exporters, a trans-Gulf pipeline has been proposed. The line would start in Kuwait, cross Saudi Arabia and the United Arab Emirates and end in Oman, Yemen or Fujairah outside the straits, picking up oil along the way. It is uncertain whether the project will receive political and financial backing. A smaller line from Abu Dhabi to Fujairah has already been given the green light.

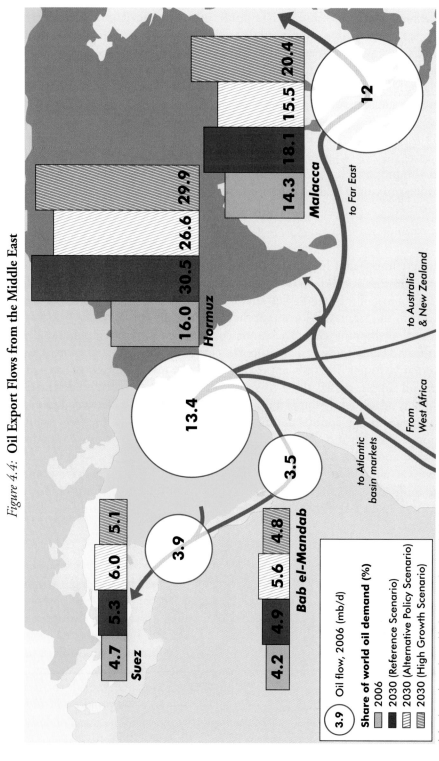

Figure 4.4: Oil Export Flows from the Middle East

The boundaries and names shown and the designations used on maps included in this publication do not imply official endorsement or acceptance by the IEA.

Natural Gas

The share of natural gas in primary energy use is expected to grow over the projection period in China and India in all scenarios, though it remains relatively low by 2030 (Figure 4.5). Both countries have only modest proven reserves of gas and the potential for raising production substantially is limited. As a result, in the absence of large discoveries, they will become increasingly reliant on imports. For now, China and India import only small volumes of gas, entirely in the form of LNG. India started importing LNG in 2004 and China in 2006. Volumes are set to grow substantially, especially in the second half of the projection period. In the High Growth Scenario, imports as a share of total gas consumption reach as much as 65% in 2030 in China and 68% in India (Table 4.3). Nonetheless, the share of imported gas in both countries' total primary energy mix remains small in 2030, regardless of the scenario.

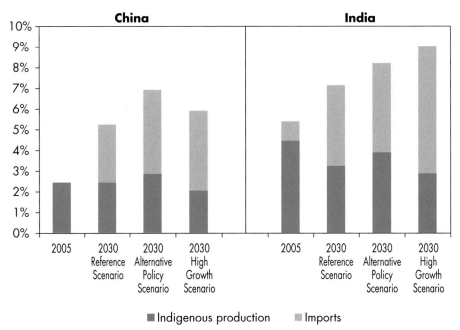

Figure 4.5: **Share of Natural Gas in Total Primary Energy Demand in China and India**

Although their current import needs are small in volume terms, the importance of China and India in global gas trade will increase – particularly towards the end of the *Outlook* period. In the High Growth Scenario, their combined share of world inter-regional trade reaches 29% in 2030 – 11 percentage points more than in the Reference Scenario and up from less than 2% in 2005 (Figure 4.6). China accounts for most of the increase in all three scenarios.

Table 4.3: Net Natural Gas Imports in China and India

	2005	Reference Scenario		Alternative Policy Scenario		High Growth Scenario	
		2015	2030	2015	2030	2015	2030
China							
bcm	0	28	128	48	158	47	216
% of primary gas demand	0	21	54	32	59	31	65
% of primary energy demand	0	1	3	1	4	1	4
India							
bcm	6	13	61	12	56	29	112
% of primary gas demand	17	22	55	21	53	39	68
% of primary energy demand	1	1	4	1	4	3	6

Figure 4.6: **Share of China and India in Total Inter-Regional Natural Gas Trade***

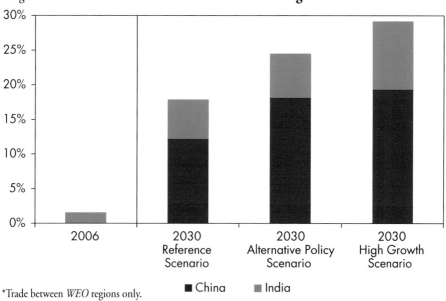

*Trade between *WEO* regions only.

For India, almost all gas imports come from the Middle East in all scenarios (Table 4.4). For China, gas is imported from several suppliers, with Australia accounting for between 25% and 40% of imports. The Middle East's share of China's imports in 2030 is 19% in the High Growth Scenario, higher than in the other two scenarios. China is the most important market for Australian LNG in all scenarios, accounting for 60% of the latter's output in 2030 in the High Growth Scenario (Figure 4.7). As for oil, most LNG shipped from the Middle East to both India and China will have to pass through the Straits of Hormuz. In the case of China, Middle East and Australian LNG will have to transit the Indonesian archipelago.

Table 4.4: **Natural Gas Imports into China and India by Source** (bcm)

	2005	Reference Scenario		Alternative Policy Scenario		High Growth Scenario	
		2015	2030	2015	2030	2015	2030
China	**0**	**28**	**128**	**48**	**158**	**47**	**216**
Transition economies	0	10	38	10	38	10	38
Middle East	0	0	8	0	0	12	40
Australia	0	11	50	15	53	14	55
Other developing Asia	0	7	31	23	66	10	54
Africa	0	0	0	0	0	0	30
India	**6**	**13**	**61**	**12**	**56**	**29**	**112**
Middle East	6	12	60	10	55	28	111
Africa	0	1	1	1	1	1	1
Australia	0	0	0	0	0	0	0

Figure 4.7: **Share of China and India in Natural Gas Exports by Source**

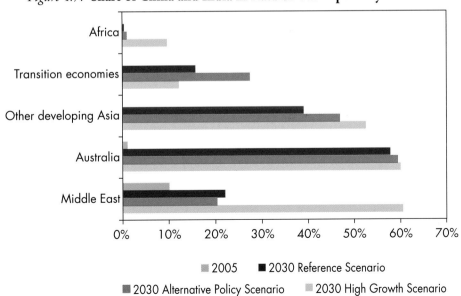

Coal

Coal dominates energy use in China and India. Today, the two countries account for a combined 45% of world coal demand. This share is projected to grow in all three scenarios. China has historically been self-sufficient in coal, but is thought to have become a net importer in 2007. India imported 12% of its total coal needs in 2005. In the Reference Scenario, China imports increasing quantities of coal. India's imports also continue to grow. In the Alternative Policy Scenario, China's coal production remains broadly in balance with demand, but India's imports are lower because of slower demand growth. In the High Growth Scenario, China's coal imports are twice as high in 2030 as in the Reference Scenario while India's imports are about one-fifth higher (Table 4.5).

Table 4.5: **Net Hard Coal Trade in China and India**

		Reference Scenario		Alternative Policy Scenario		High Growth Scenario	
	2005	2015	2030	2015	2030	2015	2030
China							
Mtce	−49	66	92	24	4	112	199
% of production	−3	3	3	1	0	4	5
% of world trade*	−7	6	7	3	1	10	13
India							
Mtce	36	97	244	87	147	125	282
% of production	14	27	38	27	33	35	39
% of world trade*	6	9	18	11	21	11	19

* Inter-regional trade between main *WEO* regions.
Note: Negative figures denote exports; positive figures imports.

Because coal is relatively expensive to move, and because many countries are endowed with resources that meet part or all of their coal needs, only 17% of world hard-coal consumption is currently traded between *WEO* regions. Total hard-coal trade is projected to grow from 648 million tonnes of coal equivalent in 2005 to 1 354 Mtce in 2030 in the Reference Scenario, to 709 Mtce in the Alternative Policy Scenario and 1 481 Mtce in the High Growth Scenario. China and India play an increasingly important role, albeit a relatively small one, in world coal trade over the projection period. The projections of coal trade for China are particularly sensitive to the projections of demand and production, as trade represents only a very small share of the country's total coal market. The faster development and deployment of clean coal technology in power generation could boost coal demand significantly, though this might be partially offset by higher thermal efficiencies in the case of advanced combustion technologies (see Chapters 5 and 9).[3] The results would be a proportionately much larger increase in coal imports.

The Energy Security Policies of China and India

China and India have differing perceptions and concerns related to the security of their energy supplies. There are nonetheless important similarities between the two countries' energy-security policies. The policy focus in both countries is on oil, reflecting increasing imports in recent years and the expectation that this trend will continue in the medium term. Chinese and Indian leaders are worried that growing dependence on imported oil will bring foreign-policy and economic pressures that might threaten national security and social and political stability. Those concerns have grown since the events of 11 September 2001 and the US-led military intervention in Iraq in 2003. Both countries have stepped up their military and naval capabilities, but recognise that they will continue to rely to a large degree on the United States for protection of international sea lanes in the Middle and Far East for many years to come.

Consequently, China and India are pursuing policies to alleviate the increase in import dependence, diversify the sources and routes of imported oil and prepare for supply disruptions.[4] Those policies are intended to minimise the vulnerability of oil supply to external events and influences, and to limit the economic damage wrought by a supply disruption and subsequent price shock. The projections of oil import requirements described above take account of current policies and measures in the case of the Reference and High Growth Scenarios; new policies to curb oil import dependence (and address environmental concerns) are taken into consideration in the Alternative Policy Scenario. The principal policies and measures in place or planned are summarised in Table 4.6. The development of indigenous resources, particularly coal, has always been the primary thrust of both countries' policies to minimise the need to import energy. Increasing emphasis is now being given to energy efficiency and conservation.

3. The deployment of carbon capture and storage, which lowers efficiency, could boost coal demand.
4. The IEA collaborates actively with China and India on a range of policy issues relating to energy security, including emergency preparedness (see Chapter 6).

Table 4.6: **Main Policy Responses in China and India to Rising Energy Insecurity**

	China	India
Energy savings	Energy intensity target set for all provinces. Fuel economy standards (2005) and taxes based on weight of the cars enacted (2006). Increased export taxes for energy-intensive products (2006). Closure of inefficient heavy industry and power plants. Domestic oil prices lifted to better reflect international prices (though subsidies remain) and transport fuel tax proposed.	Establishment of the Bureau of Energy Efficiency in 2001 to co-ordinate policies and programmes. Renovation and modernisation of power stations and electricity networks leading to increased efficiency and reduced losses. Introduction of mandatory appliance labelling and energy efficiency standards for buildings.
Diversification of fuels	Development of natural gas market, including recent completion of West-East pipeline and an LNG terminal (two new terminals are under construction). Support to renewables, including hydropower, biofuels, solar and wind, and nuclear power (10 existing reactors, two under construction and others planned). Research, development and deployment of clean coal, coal-to-liquids and biofuel technologies.	Policy to increase use of hydropower, nuclear, wind, solar and biomass in power generation. Promotion of biofuels (ethanol and biodiesel) and mandatory use of CNG for public vehicles in several cities. Iran-Pakistan-India gas pipeline, bypassing the Straits of Hormuz, proposed. Solar water heating systems and solar air heating/steam generating systems for community cooking promoted. Coal-bed methane production started in 2007.

Table 4.6: **Main Policy Responses in China and India to Rising Energy Insecurity** *(continued)*

	China	India
Diversification of oil supply sources/routes	Increased purchases of oil from Africa, Russia and Central Asia, reducing share of Middle East. Construction of Kazakhstan-China Oil Pipeline (eastern leg completed in 2005). Russia-China pipeline proposed. Opening-up of onshore production to foreign investment.	New Exploration Licensing Policy attracting growing interest from foreign companies, with 7th round planned for 2007. Possible shift to an open acreage licensing policy in the coming years.
Equity oil overseas acquisitions	Government facilitates foreign investment by state companies through financial support and energy diplomacy (sometimes involving development aid) under "going-out" policy. Equity oil production is about 600 kb/d in 2006.	Hydrocarbon Vision 2025 encourages overseas equity investment, backed by government's energy diplomacy. Production is currently about 100 kb/d. Rapid development of ONGC Videsh's assets portfolio.
Strategic oil reserves	Four sites with a combined storage capacity of close to 100 million barrels (about 27 days of current net imports) under construction, to be completed by 2008. Two additional phases, involving 200 Mb each, are planned.	Construction of a Strategic Petroleum Reserve with capacity of 36 Mb (19 days of current net imports) due to start in 2007, with completion expected in 2012. In two additional phases, stocks are due to rise to 110 Mb.

In China, oil security has emerged as a central policy issue and is increasingly affecting domestic economic and foreign policy. The government's response to rising imports in the 1990s was focused on the supply side, characterised by efforts to diversify the geographic sources of oil and physical supply routes, aimed at reducing the heavy reliance on maritime shipping through the Indonesian archipelago (Downs, 2006). A particular concern was – and remains – the threat of an oil blockade in the event of a military conflict over Chinese Taipei. The share of China's oil supplied by sea from Middle Eastern countries has been reduced in recent years, thanks to increased purchases from Africa, Central Asia and Russia, though this trend is set to reverse in the coming years. Proportionately more oil is now supplied overland, by pipeline from Kazakhstan and by rail and road from Russia, helping to alleviate the risk of disruptions to seaborne transportation. Pipelines from Russia and across Myanmar have also been proposed. China is stepping up its military and naval protection of maritime routes in Asia, including expanding bases, ports and patrols. China is also building strategic oil-storage facilities and, in a first phase, has begun to fill a stockpile of up to 100 million barrels of oil by 2008; 400 Mb is due to be added in two later phases.

Another facet of China's official energy-security policy is the acquisition of equity stakes in exploration and production assets overseas. This "going-out" policy was initiated in the early 1990s. It was motivated both by the perceived need to secure oil supplies to meet growing import needs and by the ambition of the state companies to increase their reserves, diversify their activities and increase profits, with the ultimate aim of creating internationally competitive world-scale businesses (see Chapter 10 for more detail). Today, China's national companies control about 600 kb/d of oil production overseas and India's about 100 kb/d. China's overseas equity oil output could reach 1.1 mb/d in 2015 (Paik *et al.*, 2007). Neither the Chinese government nor the companies have drawn up a comprehensive national plan for acquiring overseas assets: the companies – often in competition with one another – take decisions about acquisitions and then obtain state approval. It is doubtful whether Chinese equity oil investments contribute materially to improving the country's energy security or whether this objective still drives continuing overseas expansion by Chinese oil firms (see Spotlight on next page). The volume of overseas equity oil is small relative to the country's oil demand and is, in any case, mostly sold on the international market rather than physically shipped to China.

In recent years, China has placed more emphasis on demand-side measures to curb the growth in oil imports. Stringent vehicle fuel economy standards came into effect in 2005 and a new car-tax regime, that penalises large cars, was introduced in 2006. A new road-fuel tax, which could significantly lower fuel demand in the longer term, is still under consideration. Other policies, including the development of the natural gas market and nuclear power capacity, are aimed at diversifying the fuel mix in buildings and in power generation.

Do China's and India's Equity Oil Acquisitions Improve Energy Security?

For both China and India, acquisitions of equity oil overseas by state companies have formed a central plank of energy-security policy for several years. Yet it is far from obvious that the availability of equity oil would, in practice, enhance either country's physical oil supply or protect them from the effects of higher prices in the event of a supply crisis.

Though the emphasis by Chinese and Indian policy makers is waning, they have long argued that equity oil enhances security because it cannot be taken for granted that the international market would make sufficient oil available in a supply crisis, as other countries might intervene to divert the physical flow of oil. US resistance to the attempt by the Chinese company, CNOOC, to buy the American company Unocal in 2005 reinforced that perception. Chinese and Indian concerns about securing supply in an emergency are understandable. But today, oil – including most of China's and India's equity oil – is traded openly on the international market and any disruption to physical supplies quickly leads to an increase in international prices and adjustments to regional price differentials, which have the effect of redirecting supplies. The scope for governments to intervene in trade flows is extremely limited. In addition, if equity oil were shipped to domestic markets, it would face the same transportation risks as oil bought by Chinese and Indian oil importers on the spot market. For example, equity oil from the Middle East (assuming it were available) would be of no use in alleviating any blockage in the flow of oil to either country through the Straits of Hormuz. Transportation costs would also be higher than if the equity oil were sold onto the world market. In addition, the amount of equity oil available remains small relative to both countries' needs (though it is set to grow) and reliance on a single source, as history shows, can deny countries the flexibility of the international market.

Another argument is that equity oil provides protection against a price hike. The government could intervene to cap the prices of equity oil, while obliging the national companies to divert that oil entirely to the domestic market – assuming there is no transportation constraint. Such a policy would, to some degree, insulate the domestic market from international price fluctuations. But equity oil could only cover part of each country's needs. More importantly, holding prices below market levels would remove incentives to use oil more efficiently. It would deprive the companies of profits and undermine their ability and incentive to invest in other upstream projects in the longer term.

Experience in IEA countries and elsewhere has shown that open, well-functioning, transparent international markets in oil coupled with effective emergency measures, such as the use of emergency stocks and short-term demand-side responses, provide more effective and efficient protection against a supply disruption than directing equity oil to the domestic market. Working with markets rather than against them has been shown to be the most effective approach to enhancing supply security.

India has adopted similar measures to reduce its vulnerability to oil supply disruptions (Madan, 2006). Supply-side measures include the introduction in 1998 of a new exploration licensing policy aimed at encouraging investment in the upstream oil and gas sectors. There have been six licensing rounds under the new rules, though interest from the major international oil companies has been limited. The government has decided to build a strategic petroleum reserve with a capacity of 15 million tonnes (around 110 million barrels), with a first phase of 5 million tonnes (36 million barrels) or around 19 days of net imports at current rates. Construction was due to start in 2007. The Indian government has also encouraged state-owned companies to acquire oil assets overseas, though to a lesser degree than the Chinese government. The government is promoting the development of natural gas, clean coal technology, nuclear power and renewables to diversify energy use away from oil in both non-transport and transport uses (see Chapter 16). Several measures have been introduced to promote more efficient energy use and reduce waste, including the phasing-out of state subsidies on all petroleum products, except kerosene and liquefied petroleum gas, and higher taxes on transport fuels.

For both China and India, energy diplomacy – involving the development of a broad network of bilateral relationships with producer countries – is considered an important element of energy-security strategy. Diplomatic efforts have been focused on the Middle East and Africa. Policy makers in both countries believe that, in an oil or gas crisis, relationships with producers will count for more than just ownership of assets or ability to pay. Energy diplomacy is intended to help improve security by assisting domestic companies to win deals involving equity oil, ensuring privileged treatment in the event of a supply disruption and attracting inward investment and technology. In particular, encouraging investment from producer countries in China's and India's downstream sector is seen as a way of ensuring that the producers have a mutual interest in maintaining the flow of hydrocarbons.

High-level diplomacy is also considered necessary to help national companies counter the dominant position of the major international oil companies in securing access to resources, even if equity investments do not necessarily contribute to energy security. The Chinese and Indian governments support their national companies through summit meetings in oil-producing states, direct involvement in project negotiations and energy co-operation agreements. The Chinese government also provides direct and indirect support to its national oil companies through loans, sometimes at below-market interest rates.

Implications for Consuming Countries

The outlook for energy use and supply in China and India has important and complex implications for the energy security of other consuming countries – all of which could see a significant increase in their reliance on both oil and gas imports. How China and India respond to the prospects of growing reliance on imported energy will affect not just their own energy security, but also that of other consuming countries.

Impact of Rising Energy Demand

Increasing import dependence in any country does not *necessarily* mean less secure energy supplies, any more than self-sufficiency guarantees uninterrupted supply. Yet rising trade does carry a *risk* of heightened short-term energy insecurity for all consuming countries, to the extent that geographic supply diversity is reduced and reliance on specific supply routes is increased. The degree of risk at any given time hinges on myriad technical, climatic, geopolitical and economic factors. In the long run, the prospect of increased collective dependence on the part of consuming countries on supplies of oil and gas from a small number of producers carries the risk of the latter seeking to impose higher prices through investment or production constraints or other cartel action as their market power grows. China's and India's growing participation in international trade heightens the importance of their contribution to collective efforts to enhance global energy security.

The susceptibility of the global oil system to a supply disruption and a resulting price shock is likely to grow as consuming countries, as a group, become more dependent on imports. Most of the additional imports are expected to come from the Middle East, along vulnerable maritime routes to both eastern and western markets. Any disruption to supplies from that region or any other major source would drive up international prices and the import bills of all consuming countries. The potential impact on international oil prices of a supply emergency is also likely to increase: oil demand is becoming less sensitive to changes in price as the share of transport demand – which is price-inelastic relative to other energy services – in overall oil consumption rises worldwide. Rising incomes in China, India and other parts of the world are driving up demand for mobility and, therefore, oil-based transport fuels. As a result, oil demand will become less and less responsive to movements in international crude oil prices.[5]

Longer-term risks to energy security are also set to grow. The growing concentration of remaining oil reserves in a small group of countries – notably Middle East members of the Organization of the Petroleum Exporting Countries (OPEC) and Russia – will increase their influence over the market and prices. OPEC's global market share increases in all scenarios – most of all in the Reference and High Growth Scenarios (Figure 4.8). An increasing share of gas demand is also expected to be met by imports from these countries, via pipeline or in the form of liquefied natural gas. OPEC countries can be expected to seek to avoid raising prices so fast as to depress global demand and to encourage investment in production of higher-cost oil in other regions and in alternative sources of energy. Nonetheless, the greater the increase in the call

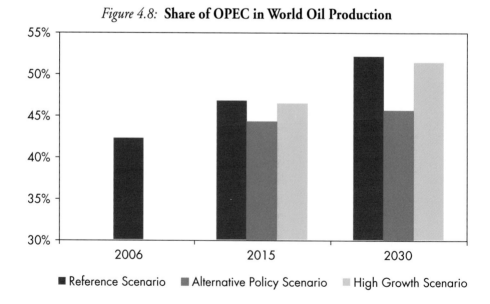

Figure 4.8: **Share of OPEC in World Oil Production**

■ Reference Scenario   ■ Alternative Policy Scenario   ■ High Growth Scenario

---

5. The removal of price subsidies, on the other hand, would make demand more responsive.

on oil and gas from these regions, the more likely it will be that they will seek to extract a higher rent from their exports.

The results of the High Growth Scenario provide an indication of the extent to which the rate of expansion of energy demand in China and India affects energy prices and, therefore, the affordability of energy – a critical component of supply security – for the whole of the world. It also shows how sensitive international energy trade is to economic growth in the two countries, through its impact on energy demand. In this scenario, GDP is assumed to grow on average by around 1.5 percentage points per year more than in the Reference Scenario in both China and India. As a result, their international trade in goods and services with the rest of the world grows more quickly, boosting global GDP. But higher economic growth in China and India also drives up their demand for energy and other raw materials, crowding out demand from other regions and pushing up international energy prices.[6] In some regions, the overall impact on GDP is negative. The impact on energy demand of increased economic output in China and India, and of higher international prices, differs markedly by region, increasing in some and falling in others.

The average IEA crude oil import price over 2007-2030 is $72 a barrel (in year-2006 dollars), or 21% higher, in the High Growth Scenario than in the Reference Scenario; the gap reaches $25 a barrel, or 40%, in 2030 (Figure 4.9).[7] Gas prices rise as much as oil prices in percentage terms, as the prices are linked through indexation clauses in long-term contracts and, in competitive gas markets, through inter-fuel competition, though to a lesser extent. Coal prices increase only modestly in response to stronger world demand. The price elasticity of supply is relatively high – that is, investment in new coal capacity is more sensitive to price than for oil and gas – because the marginal cost of coal production rises slowly with higher global production, thanks to ample reserves worldwide.

In China and India combined, primary energy demand in 2005-2030 grows on average by 4.1% per year in the High Growth Scenario – 0.8 percentage points more than in the Reference Scenario. As a result, demand is 9% higher in 2015 and 21% higher in 2030. The increase in demand relative to the Reference Scenario is of the same magnitude for coal, which is 8% higher in

6. See the Introduction for a detailed explanation of the methodology used to simulate the impact of higher economic growth on global energy demand and prices. The results of the High Growth Scenario for energy markets are described in summary form in Chapters 1 and 2 and in more detail in Chapter 12 (China) and Chapter 19 (India). The impact on world economic growth and trade is discussed in Chapter 3.

7. The oil price is an assumption in the Reference Scenario. In the High Growth Scenario, the change in the price *vis-à-vis* the Reference Scenario is projected using a hybrid model, WEM-ECO (see Chapter 3 for details).

2015 and 20% higher in 2030. The rise in coal demand occurs because, among final forms of energy, the income elasticity of demand is highest for electricity and strong growth in electricity demand drives up the use of coal for power generation. Oil demand increases by 29% in 2030. Higher oil prices stimulate increased investment in oil exploration and development, boosting output – but not by enough to meet all of the increase in demand. As a result, oil imports rise sharply in both countries, by a combined 1.4 mb/d, or 14%, in 2015 and 5.7 mb/d in 2030, a 30% increase compared with the Reference Scenario.

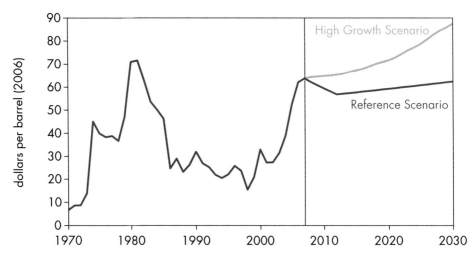

Figure 4.9: **Average IEA Crude Oil Import Price in the Reference and High Growth Scenarios**

In the rest of the world as a whole, total primary energy demand actually falls slightly in the High Growth Scenario compared with the Reference Scenario. This is because the negative impact on demand of the increase in energy prices outweighs the demand effect of the higher GDP that results from more trade with China and India. Higher oil prices boost oil output in the net oil-importing (consuming) countries, as well as in exporting countries. As a result, all consuming regions other than China and India need to import less oil. Overall, OPEC and other exporting countries see their exports rise – because the increase in imports in China and India exceeds the drop in imports in other regions. Total inter-regional oil trade reaches 67 mb/d in 2030, compared with 65 mb/d in the Reference Scenario. Most of the increase in exports is expected to come from the Middle East.

Chinese and Indian imports of gas are also higher in the High Growth Scenario, though this effect is partially offset by lower imports in the rest of the world (because of higher gas prices). Global inter-regional gas trade reaches 1 122 bcm in 2030, compared with 1 053 bcm in the Reference Scenario – an increase of 7%. Unlike oil, additional gas exports come mostly from the transition economies and Africa. Exports from the Middle East drop by 20% compared to the Reference Scenario, as a result of the rise in that region's domestic consumption.

This analysis suggests that – all other things being equal – the higher the rate of growth in energy demand in China and India, the greater concerns will be about the security of energy supplies to consuming countries generally. China's and India's oil and gas imports would be higher with faster economic growth. As a result, the *risk* of a disruption may be higher, as more of the oil and gas consumed worldwide is traded internationally and more of that oil is supplied by a small number of countries – especially in the Middle East. Dependence on Russian and Central Asian oil and gas would also grow. To the extent that stronger demand reduces spare crude oil production capacity in OPEC countries, the impact of an oil-supply disruption would also be more severe.

Most of the major oil-supply disruptions in recent decades have occurred in the Middle East (Figure 4.10). Since 1967, close to 90% of all the oil lost in major supply disruptions was caused by events in that region. One of the most recent – in Iraq following the US-led invasion in 2003 – caused an initial loss of 2.3 mb/d of production, equal to 2.9% of world demand. Although much of that capacity has now been restored, persistent conflict in the country continues to deter investment, making the short- and long-term outlook for exports extremely uncertain. Other parts of the region remain politically and socially unstable, casting doubt about investment in oil and gas infrastructure and the reliability of exports, now and in the future. Geopolitical tensions related to Iran's nuclear programme are a major source of uncertainty and have recently contributed to higher and more volatile prices. The Israel-Palestine dispute and conflict between Israel and Lebanon continue to cloud the political and economic climate in the region and relations with the rest of the world.

Although protection of oil facilities in the region has been stepped up in recent years, various threats to the integrity of production and transportation facilities remain. In April 2007, a major al-Qaida plot to attack a number of oil installations in Saudi Arabia was uncovered before it could be put into effect. In 2006, a terrorist attack on the Abqaiq oil-processing plant, which handles more than 60% of Saudi production, was also thwarted. But in Iraq, persistent insurgent attacks on oil wells and pipelines continue severely to curtail exports. Kuwaiti oil wells were sabotaged during the Iraqi occupation of 1990-1991.

Figure 4.10: **Major World Oil Supply Disruptions***

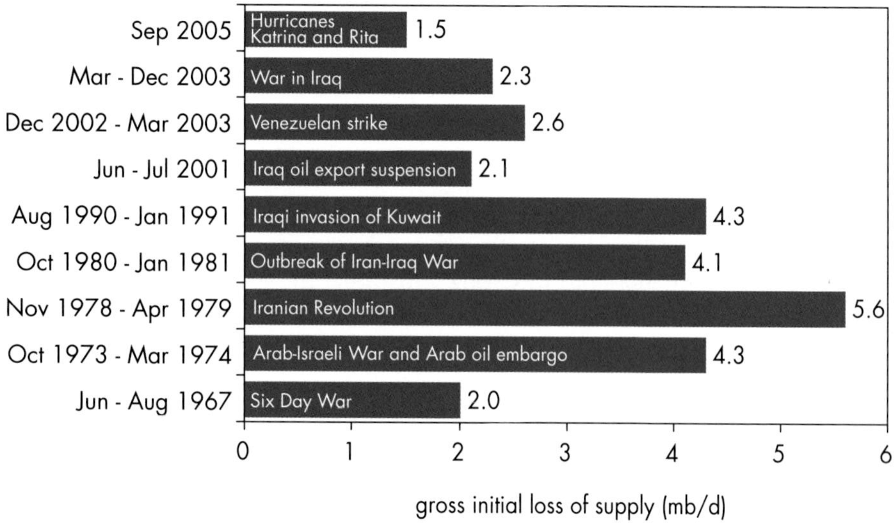

Period	Event	gross initial loss of supply (mb/d)
Sep 2005	Hurricanes Katrina and Rita	1.5
Mar - Dec 2003	War in Iraq	2.3
Dec 2002 - Mar 2003	Venezuelan strike	2.6
Jun - Jul 2001	Iraq oil export suspension	2.1
Aug 1990 - Jan 1991	Iraqi invasion of Kuwait	4.3
Oct 1980 - Jan 1981	Outbreak of Iran-Iraq War	4.1
Nov 1978 - Apr 1979	Iranian Revolution	5.6
Oct 1973 - Mar 1974	Arab-Israeli War and Arab oil embargo	4.3
Jun - Aug 1967	Six Day War	2.0

* Include all disruptions involving a gross supply loss of at least 1.5 mb/d.
Sources: US Department of Energy and IEA Secretariat.

The sudden loss of a significant part of the region's production and export capacity through terrorism or deliberate political acts by producing countries cannot be ruled out.

The increased dependence of China – together with that of Europe – on Russian gas gives rise to another issue of energy security. Gas imports from Russia would contribute to China's energy-supply diversity and Russia has been a reliable supplier of gas to Europe for several decades. Yet the temporary cut-off of Russian supplies to central and western European customers in January 2006 that resulted from a dispute with Ukraine has drawn attention to the risks associated with political control of strategic pipeline routes. Moreover, there are doubts about the adequacy of investment in Russia's gas industry to meet rising domestic and export demand, exacerbated by the lack of transparency over future capacity plans. There are also concerns about the possibility of Russia formally co-ordinating its investment and production plans with other gas-exporting countries in order to support prices in a similar way to OPEC. Russia has indicated that it is interested in pursuing the idea of more closely co-ordinating export pricing and even creating a formal cartel. It signed a memorandum of understanding on upstream co-operation with Algeria in 2006. Although a meeting of the Gas Exporting Countries Forum in April 2007, which brought together government representatives of most gas exporters, failed to reach agreement on such a move, concerns remain about gas pricing in the future (Box 4.1). Faster growth in Chinese and Indian gas demand could stimulate further concentration of the development of gas reserves in Russia and the Middle East, making it easier for gas exporters to exert more control over gas pricing. Concerns such as this underline the common interest of all consuming countries in sharing experience on moderating demand and in co-operating on emergency preparedness.

Box 4.2: **Gas Exporting Countries Forum**

The Gas Exporting Countries Forum (GECF) was set up in 2001 to provide a means for gas producers to discuss issues of common interest. It is currently made up of Algeria, Bolivia, Brunei, Egypt, Indonesia, Iran, Libya, Malaysia, Nigeria, Norway (as an observer), Oman, Qatar, Russia, Trinidad and Tobago, the United Arab Emirates and Venezuela. Collectively, these countries account for 73% of the world's gas reserves and 42% of production. The group meets once a year, most recently in Qatar in April 2007. It has no headquarters, budget or staff.

It has been suggested that the GECF could evolve into a group like OPEC. However, several factors limit the likelihood of such a development, not least the prevalence of long-term contracts, the regionalised nature of gas markets and the growing number of competing suppliers and energy sources (IEA, 2007b). The price of gas in export contracts is mainly indexed to oil prices, in order to ensure that gas remains competitive. Gas can be substituted by other fuels more easily than oil. And uncertainty over future gas pricing could drive consumers away from gas. For these reasons, it has proved difficult for the diverse membership of the GECF to find common ground on co-ordinating pricing.

Nonetheless, the eventual concentration of gas exports could pave the way for formal co-ordination of production and pricing policies in the future, to the detriment of consuming countries. If undertaken, such a move, by undermining the development of gas markets, would be unlikely to be in the interests of gas exporters in the long run.

Impact of China's and India's Energy Security Policies

China's and India's policies on energy security will have important implications for the security of other consuming countries. Any improvement to China's and India's security would generally yield benefits for all other consuming countries. Equally, efforts by other consuming countries could bring important benefits to China and India. The success of all countries' efforts to save energy, diversify away from oil and, thus, curb their need to import hydrocarbons will clearly affect the energy security of others. The extent to which the policies that China and India are currently considering to enhance their energy security *and* tackle environmental problems could reduce their imports is demonstrated by the Alternative Policy Scenario. China's oil imports are cut by 26% in 2030, compared with the Reference Scenario, while India's imports are reduced by 19%. Most of the imports savings come after 2015, as the effects of new policies build up.

Establishing emergency stocks could play a particularly important role in enhancing short-term security. On current plans, net import coverage from these stocks is expected to be around 20 days in China and 16 days in India by the beginning of the next decade. In the longer term, it is likely that both countries will seek to increase this coverage: India plans to increase it to 90 days — the coverage of total stocks that IEA member countries are required to maintain. However, a date for achieving that goal has not been set, mainly because of the high cost of building and maintaining storage facilities and the cost of buying the oil itself. Uncertainties about future import needs can lead to reluctance to act. The IEA is collaborating with China and India on enhancing their collective emergency response capabilities (see Chapter 6).

By way of an example, at current import levels, to achieve forward net import coverage of even 45 days on a similar basis to IEA countries (*i.e.* including commercial stocks), China would need to store a total of 156 million barrels (Table 4.7). For the same level of coverage, China's stocks would need to rise massively in 2030, to 589 Mb, in the Reference Scenario and 772 Mb in the High Growth Scenario. In India, stocks would need to rise to 84 Mb to achieve 45 days of net import coverage at 2006 levels. The required level of stocks would need to rise to 271 Mb in 2030 in the Reference Scenario and 345 Mb in the High Growth Scenario. Achieving these levels of stocks would involve heavy financial commitments. Assuming that all the oil stored is crude, the total cumulative cost of building stocks to cover 45 days of net imports would reach $38 to $74 billion in China and $19 to $33 billion in India by 2030 (Figure 4.11).[8] In the Reference Scenario, the cost would be equivalent to close to 10% of total oil investment in China and 15% in India. The use of emergency stocks in the event of a supply disruption would be much more effective if co-ordinated with the use of stocks under the IEA emergency response system.

Much has been written about the impact on consuming countries of moves by China and India to exert control over hydrocarbon resources in the producing countries through equity oil and direct government-to-government deals. To the extent that Chinese and Indian companies over-bid for resources, costs and

---

8. Based on estimated capital costs of storage facilities of $16.50 per barrel. Total costs include buying crude oil, operation and maintenance, and capital.

Table 4.7: **Volume of Oil Stocks to Ensure 45 Days of Net Imports in China and India** (million barrels)

	2006	2015	2030
China			
Reference Scenario	156	318	589
Alternative Policy Scenario	156	291	435
High Growth Scenario	156	372	772
India			
Reference Scenario	84	135	271
Alternative Policy Scenario	84	123	221
High Growth Scenario	84	144	345
Total			
Reference Scenario	240	454	860
Alternative Policy Scenario	240	414	655
High Growth Scenario	240	516	1 117

4

Figure 4.11: **Cumulative Cost of Maintaining Oil Stocks to Ensure 45 Days of Net Imports in China and India, 2006-2030**

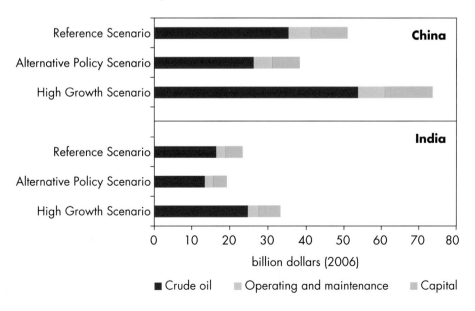

prices to all consuming countries might be driven higher (USCESRC, 2006). However, it may be that China's and India's overseas investments will actually improve global energy security by bringing to market oil that might not otherwise have been developed and thereby augmenting global oil supplies (Andrews-Speed, 2006; Douglas *et al.*, 2006; Rosen and Houser, 2007). At a time of growing worries about the adequacy of investment in oil-production capacity, it is clearly of benefit to other consuming countries that China and India are investing in bringing more oil to market – so long as it does not crowd out investment elsewhere.

Whatever the overall impact on the world's energy security of China's and India's increasingly assertive policies in the area of energy diplomacy, partly aimed at securing equity oil, they will undoubtedly have broad ramifications for international relations. The real risk is not so much commercial competition for scarce hydrocarbon resources, as that upstream developments get caught up in broader foreign-policy issues. This makes it all the more important for all consuming countries, including China and India, to work together to enhance their collective energy security in a mutually beneficial way (see Chapter 6).

CHAPTER 5

GLOBAL ENVIRONMENTAL REPERCUSSIONS

HIGHLIGHTS

- Rising global fossil fuel use will continue to drive up energy-related CO_2 emissions over the projection period. In the Reference Scenario, emissions jump by 57% between 2005 and 2030, from 26.6 to 41.9 Gt. The United States, China, Russia and India contribute two-thirds of this increase.

- China is by far the biggest contributor to incremental emissions, overtaking the United States as the world's biggest emitter in 2007. India becomes the third-largest emitter around 2015. But these figures need to be looked at in a historical context. From 1900 to 2005, the United States and the EU countries combined accounted for just over half of cumulative global emissions. China accounted for only 8% and India 2%. In the Reference Scenario, China's share of emissions from 1900 to 2030 rises to 16%, approaching that of the United States (25%) and the European Union (18%). India's cumulative emissions (4%) approach those of Japan (4%).

- Rising CO_2 and other greenhouse-gas concentrations in the atmosphere resulting mainly from fossil-energy combustion and other human activities are contributing to rising global temperatures and to changes in climate. There is growing support worldwide for urgent action to stabilise greenhouse-gas concentrations at a level that would prevent dangerous anthropogenic interference with the climate system, as agreed by G8 leaders at their recent summit in Heiligendamm.

- In the most ambitious of the IPCC's scenarios, in which CO_2-equivalent concentrations are stabilised at around 450 ppm, global CO_2 emissions would need to peak by 2015 at the latest and to fall by between 50% and 85% below 2000 levels by 2050. Energy-related CO_2 emissions do not peak before 2020 in any of the scenarios in this *Outlook*, though emissions stabilise in the mid-2020s in the Alternative Policy Scenario and are 19% lower in 2030 than in the Reference Scenario.

- In our "450 Stabilisation Case", energy-related CO_2 emissions would need to peak in 2012 at around 30 Gt and then decline to 23 Gt in 2030 – 19 Gt less than in the Reference Scenario and 11 Gt less than in the Alternative Policy Scenario. Emissions savings come from improved efficiency in fossil-fuel use in industry, buildings and transport, switching to nuclear power and renewables, and the widespread deployment of CO_2 capture and storage in power generation and industry. Exceptionally strong and immediate policy action would be essential for this to happen and the associated costs would be very high.

Energy-Related CO$_2$ Emissions

Global Trends

Rising global fossil energy use will continue to drive up energy-related CO$_2$ emissions over the projection period (Figure 5.1). A range of government policies, including those intended to address climate change, air pollution and energy security, have helped to slow the rate of growth in emissions in some countries in recent years, but have not stopped it. In the Reference Scenario, which examines the implications of governments adopting no new policies, world emissions jump by 57% between 2005 and 2030 to 41.9 gigatonnes, an average rate of growth of 1.8% per year. The increase is 27% in the Alternative Policy Scenario (1.0% per year) and 68% (2.1% per year) in the High Growth Scenario. By comparison, emissions grew by 1.7% per year over 1990-2005. Emissions in 2030 in the Reference Scenario are 1.5 Gt higher than in last year's *Outlook*, mainly because of higher coal use in China and India, while emissions in the Alternative Policy Scenario are lower as more policies are under consideration. Although emissions grow in all three scenarios, the path between 2005 and 2030 differs markedly, with important consequences for the prospects for reducing emissions and stabilising concentrations beyond 2030.

Figure 5.1: **Energy-Related CO$_2$ Emissions by Scenario**

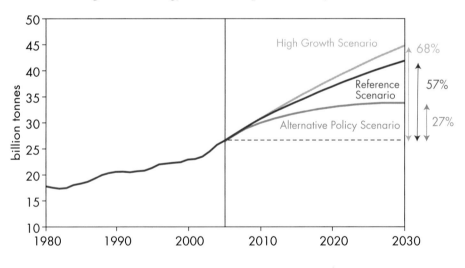

Emissions continue to rise in all regions through to 2030 in the Reference Scenario, but peak and begin to decline in the OECD in 2015 in the Alternative Policy Scenario causing global emissions to stabilise by around 2025. In the OECD Europe and Pacific regions, emissions in 2030 are lower than current levels (Figure 5.2). In OECD North America, emissions level off soon after 2015 and then decline to 2030 to 6% above that of 2005.

Regardless of the scenario, coal remains the biggest contributor to global emissions throughout the projection period (Figure 5.3). Coal overtook oil as the leading source of emissions in 2004. With the exception of the Alternative Policy Scenario, coal's share of emissions increases over time.

Figure 5.2: **Incremental Energy-Related CO$_2$ Emissions by Scenario, 2005-2030**

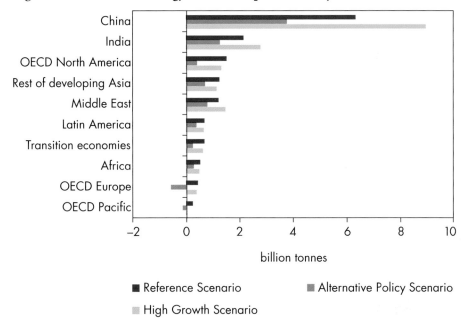

Figure 5.3: **Energy-Related CO$_2$ Emissions by Fuel and Scenario**

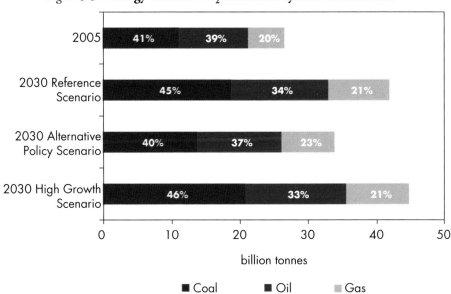

In the past two-and-a-half decades, global emissions rose more slowly than primary energy demand, mainly because the shares of carbon-neutral nuclear power and natural gas – the least carbon-intensive fossil fuel – expanded. In addition, the share of renewables in the energy mix increased. Emissions grew by 1.61% per year, while energy use rose by 1.85%. This trend is projected to reverse in the Reference and High Growth Scenarios, as the share of nuclear power declines while that of coal rises. By contrast, in the Alternative Policy Scenario, the carbon intensity of energy use falls (Figure 5.4). Average carbon-dioxide content per toe of energy is projected to rise from 2.33 tonnes in 2005 to 2.36 tonnes in 2030 in the Reference Scenario and 2.39 in the High Growth Scenario. It declines to 2.15 in the Alternative Policy Scenario. Non-OECD regions in aggregate account for all of the increase in carbon intensity in every scenario: rapid growth in renewables reduces intensity in the OECD in all three scenarios.

Figure 5.4: **Average Annual Growth in World Energy-Related CO$_2$ Emissions and Primary Energy Demand by Scenario**

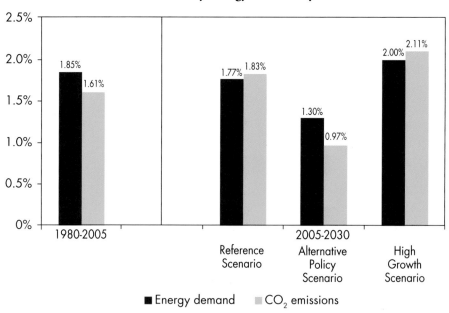

Power generation contributes around half the increase in global emissions from 2005 to 2030 in the Reference Scenario. This share is significantly lower in the Alternative Policy Scenario, at 38%, and a fraction higher in the High Growth Scenario. The share of the power sector in total emissions continues to rise, from 41% in 2005 to around 45% in 2030 in the Reference Scenario and 46% in the High Growth Scenario, driven by the sector's increasing share in primary energy use and the growing dependence on fossil energy (Figure 5.5). But, in the Alternative Policy Scenario, the sector's share of emissions falls markedly, reversing past trends, as the share of nuclear power and renewables in the generation fuel mix increases significantly.

Figure 5.5: **Share of Power Generation in World Energy-Related CO$_2$ Emissions and in Primary Energy Demand, 1980-2030**

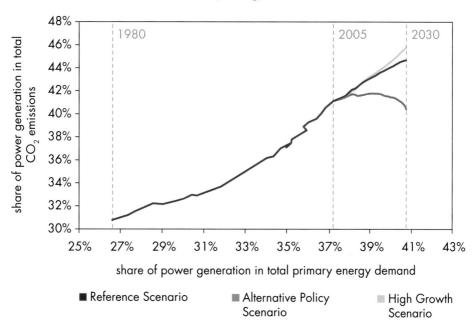

In all three scenarios, most of the increase in emissions from power stations comes from developing countries, mainly because their electricity production increases faster than that of the OECD and the transition economies. In addition, their reliance on coal will remain much higher. For the world as a whole, average global emissions per kWh of electricity produced fall slightly in the Reference and High Growth Scenarios as a result of continuing improvements in the thermal efficiency of power plants. In the Alternative Policy Scenario, emissions intensity falls much more sharply, again thanks to faster efficiency gains and faster growth in nuclear power and renewables (Figure 5.6).

Transport contributes roughly a fifth of the increase in global emissions to 2030 in all three scenarios, consolidating its position as the second-largest sector for CO_2 emissions worldwide. Most of the increase in transport emissions comes from developing countries, where car ownership and freight transport are expected to grow rapidly. Other final uses – mainly industry and the residential sector – account for the rest.

Figure 5.6: **Carbon Dioxide Intensity of Electricity Generation by Scenario**

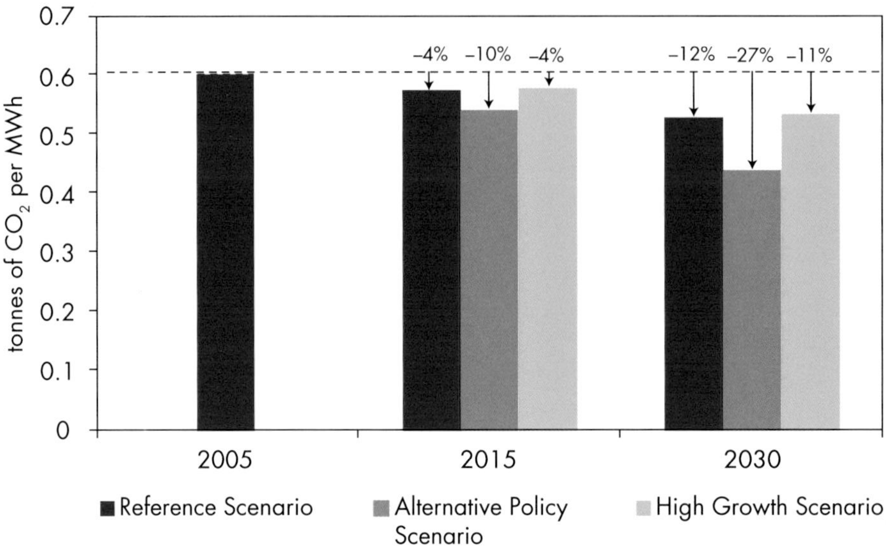

Contribution of China and India to Global Emissions

In each scenario described in this Outlook, most of the increase in energy-related CO_2 emissions comes from China, India and other developing countries, though local pollution will remain the primary environmental concern for these countries (Box 5.1). China and India together account for 56% of the increase in emissions between 2005 and 2030 in the Reference Scenario, 69% in the Alternative Policy Scenario and 65% in the High Growth Scenario (Figure 5.7 and Table 5.1). China is by far the biggest single contributor to incremental emissions between 2005 and 2030 in all three scenarios. It is expected to overtake the United States in 2007 as the world's biggest emitter, though its per-capita emissions are far lower (Box 5.2). One reason for the strong increase in China's emissions is the significant quantity of fossil energy and, therefore, carbon embodied in the goods that China produces for export, which far outweighs the carbon embodied in its imports (see Spotlight in Chapter 9).

Rising energy consumption and the continuing heavy reliance on coal is contributing not just to higher CO_2 emissions but also to worsening air pollution in China and India. Fossil-energy use gives rise to various toxic and noxious emissions, notably SO_x, NO_x, carbon monoxide and particulate matter (soot). These emissions contribute directly to health problems, ground-level and atmospheric ozone and acid rain. Many of these problems are of a local nature. Despite some improvements in recent years, air pollution remains a major public health issue in all large Chinese and Indian cities.

Most of the effects of rising air pollution are felt at the local level, close to the sources of emissions. But the effects are being increasingly felt in neighbouring countries too. All Asian countries suffer to some degree from pollution from China and India. Most transboundary energy-related pollution takes the form of acid rain and soot deposition. Both are caused mainly by burning coal and oil products. The seriousness of these problems to other countries depends on their proximity to the sources of pollution and on prevailing winds. Japan, Korea and, to a lesser extent, the United States and Canada suffer from pollution emanating from China, while Bangladesh suffers from pollution from India.

Acid rain or precipitation occurs when SO_2 and NO_x emitted into the atmosphere undergo chemical transformations to form acidic compounds, which are then absorbed by water droplets in clouds. The droplets fall to earth as rain or snow, increasing the acidity of the soil, lakes and rivers. This, in turn, upsets ecosystems and can render land infertile and damage forests. Acid rain also contributes to public health problems and damages buildings. Most of the acidic compounds are deposited close to the sources of pollution, but they can be transported over hundreds or thousands of miles. Acid rain has become a major problem across Asia. Pollution from China, India and other Asia-Pacific countries is also contributing to acid rain in North America, counteracting part of the considerable progress that has been made there in reducing emissions in the last two decades.[1]

In the absence of new policies in China and India to constrain emissions of SO_2 and NO_x, they will continue to rise steadily. In the Reference Scenario, for example, China's SO_2 emissions are projected to increase from 26 million tonnes in 2005 to 31 Mt in 2015, before levelling off to

.../...

1. See www.epa.gov/airmarkets/progress/arp05.html for more details about trends in acid rain in the United States.

30 Mt by 2030. Emissions of NO_x rise even faster. Emissions of these and other pollutants also rise steadily in India; SO_2 emissions more than double from 7 Mt in 2005 to 16.5 Mt in 2030 in the Reference Scenario.[2] Technologies already exist to address local and regional pollution from fossil-energy use. Most industrialised countries have made considerable progress in improving air quality, despite growing energy consumption. The health benefits generally far outweigh the financial costs associated with the more stringent environmental standards on energy equipment and fuel quality. Some of the gains have come from improving energy efficiency, which reduces the need to burn fossil fuels. Integrating air pollution abatement and climate change mitigation policies offers potentially large cost reductions compared to treating those policies in isolation. The near-term health benefits from reduced air pollution as a result of actions to reduce greenhouse-gas emissions can be substantial and may offset a substantial fraction of mitigation costs (IPCC, 2007). Other benefits, such as enhanced energy security, increased agricultural production and reduced pressure on natural ecosystems from lower ozone concentrations would also add to the potential cost savings.

Figure 5.7: **Incremental Energy-Related CO_2 Emissions by Region and Scenario, 2005-2030**

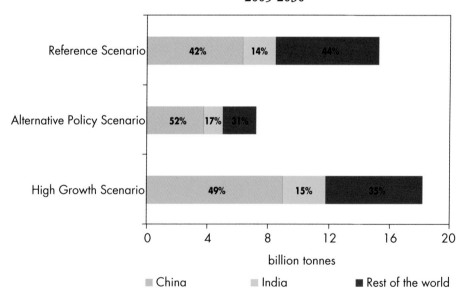

2. See Chapters 9 and 16 for details of the outlook for Chinese and Indian air pollution.

Table 5.1: **Energy-Related CO$_2$ Emissions by Region and Scenario** (billion tonnes)

	2005	Reference Scenario		Alternative Policy Scenario		High Growth Scenario	
		2015	2030	2015	2030	2015	2030
OECD	**12.8**	**14.1**	**15.1**	**13.2**	**12.5**	**13.9**	**14.6**
North America	6.7	7.5	8.3	7.2	7.1	7.5	8.1
United States	*5.8*	*6.4*	*6.9*	*6.2*	*6.0*	*6.3*	*6.7*
Europe	4.0	4.2	4.5	3.8	3.5	4.3	4.4
Pacific	2.1	2.3	2.3	2.2	1.9	2.2	2.1
Transition economies	**2.5**	**3.0**	**3.2**	**2.9**	**2.8**	**3.0**	**3.2**
Russia	1.5	1.8	2.0	1.7	1.7	1.8	2.0
Developing countries	**10.7**	**16.4**	**22.9**	**15.2**	**17.9**	**17.4**	**26.3**
China	5.1	8.6	11.4	8.1	8.9	9.5	14.1
India	1.1	1.8	3.3	1.6	2.4	1.9	3.9
Other Asia	1.4	2.0	2.7	1.8	2.1	2.0	2.6
Middle East	1.2	1.8	2.5	1.7	2.0	1.8	2.7
Africa	0.8	1.0	1.4	0.9	1.1	1.0	1.3
Latin America	0.9	1.2	1.6	1.1	1.3	1.2	1.6
World*	**26.6**	**34.1**	**41.9**	**31.9**	**33.9**	**34.9**	**44.8**
European Union	*3.9*	*4.0*	*4.2*	*3.6*	*3.2*	*4.1*	*4.2*

* Includes emissions from international marine bunkers.

Despite the strong increase in emissions in both China and India over the past few years, their historical share in cumulative emissions, measured over the period 1900 to 2005,[3] amounted to only 8% for China and 2% for India. By comparison, the United States and the EU countries combined accounted for just over half of all cumulative emissions (Figure 5.8). This pattern changes radically over the *Outlook* period. In the Reference Scenario China's share of cumulative emissions from 1900 to 2030 rises to 16%, approaching that of the United States (25%) and the European Union (18%). India's cumulative emissions (4%) approach those of Japan (4%) (Figure 5.9). In the High Growth Scenario, China's cumulative emissions are the same as those of the European Union by 2030, while India's exceed those of Japan.

3. Cumulative emissions over a long period provide an indication of a country's total contribution to greenhouse-gas concentrations in the atmosphere. The time frame shown here reflects the availability of data of reasonable accuracy.

Box 5.2: **Which Countries Emit the Most CO$_2$?**

The world's top five CO$_2$ emitting countries – the United States, China, Russia, Japan and India – currently account for 55% of global energy-related CO$_2$ emissions. By 2030, that share rises to 59% in the Reference and Alternative Policy Scenarios and 62% in the High Growth Scenario. Those countries remain the top five emitters, but their relative position changes (Table 5.2) – the same way in each scenario. According to preliminary fuel consumption data released by the US Energy Information Administration and by the Chinese National Bureau of Statistics, US emissions fell slightly to 5.7 gigatonnes (billion tonnes) in 2006, while Chinese emissions jumped by 9.4% to 5.6 Gt. On these trends, China will overtake the United States in 2007. The gap between the emissions of China and the United States widens progressively over the *Outlook* period in all three scenarios. China's emissions are 35% larger than those of the United States in 2015 and 66% bigger in 2030 in the Reference Scenario. India rises from fifth- to third-largest emitter by 2015, overtaking Japan and Russia. For comparison purposes, emissions by all EU countries combined rise from 3.9 Gt in 2005 to 4.2 Gt in 2030, still bigger than India's yet smaller than those of China and the United States.

Table 5.2: **Top Five Countries for Energy-Related CO$_2$ Emissions in the Reference Scenario**

	2005		2015		2030	
	Gt	rank	Gt	rank	Gt	rank
US	5.8	1	6.4	2 ↓	6.9	2 =
China	5.1	2	8.6	1 ↑	11.4	1 =
Russia	1.5	3	1.8	4 ↓	2.0	4 =
Japan	1.2	4	1.3	5 ↓	1.2	5 =
India	1.1	5	1.8	3 ↑	3.3	3 =

China's per-capita emissions are projected to approach those of OECD Europe by the end of the projection period in the Reference Scenario. But China's per-capita emissions are less than half those of the United States and about two-thirds those of the OECD as a whole in the Reference Scenario. In India, they remain far lower than those of both OECD countries and the transition economies in 2030, even though they grow faster than in almost any other

Figure 5.8: **Energy-Related CO$_2$ Emissions by Region, 1900-2005***

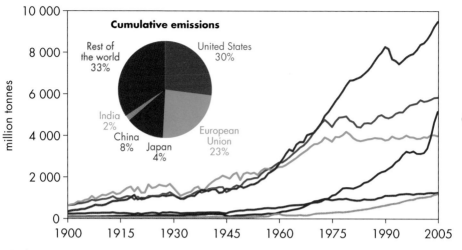

* See footnote 3.

Figure 5.9: **Cumulative Energy-Related CO$_2$ Emissions in Selected Countries/Regions in the Reference Scenario**

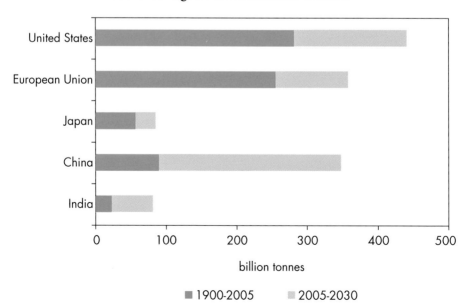

region (Table 5.3 and Figure 5.10). Per-capita emissions are markedly lower in all regions in the Alternative Policy Scenario than in the Reference Scenario. In the High Growth Scenario, both China's and India's per-capita emissions are about one-fifth higher than in the Reference Scenario. The change is less marked in the rest of the world.

Table 5.3: **Per-Capita Energy-Related CO_2 Emissions by Region and Scenario**
(tonnes)

	2005	Reference Scenario		Alternative Policy Scenario		High Growth Scenario	
		2015	2030	2015	2030	2015	2030
OECD	**11.0**	**11.4**	**11.6**	**10.7**	**9.7**	**11.3**	**11.3**
North America	15.5	15.8	15.6	15.2	13.5	15.6	15.2
United States	*19.5*	*19.6*	*19.0*	*18.9*	*16.5*	*19.4*	*18.5*
Europe	7.5	7.6	7.9	6.8	6.1	7.7	7.8
Pacific	10.3	11.4	11.8	10.9	9.8	10.8	10.7
Transition economies	**7.5**	**8.9**	**10.1**	**8.5**	**8.7**	**8.9**	**9.9**
Russia	10.7	13.3	16.0	12.6	14.1	13.3	16.0
Developing countries	**2.2**	**2.9**	**3.5**	**2.7**	**2.7**	**3.1**	**4.0**
China	3.9	6.2	7.9	5.8	6.1	6.8	9.7
India	1.0	1.4	2.3	1.3	1.7	1.5	2.7
Other Asia	1.5	1.8	2.0	1.6	1.6	1.7	1.9
Middle East	6.7	8.0	8.7	7.3	7.2	8.1	9.6
Africa	0.9	0.9	0.9	0.9	0.8	0.9	0.9
Latin America	2.1	2.3	2.8	2.2	2.3	2.3	2.8
World	**4.1**	**4.7**	**5.1**	**4.4**	**4.1**	**4.8**	**5.5**
European Union	*8.0*	*8.0*	*8.4*	*7.2*	*6.5*	*8.2*	*8.3*

In contrast to per-capita emissions, carbon intensity, measured as emissions per unit of GDP, falls sharply in both China and India in all scenarios over the *Outlook* period (Figure 5.11). This is because the contribution to GDP of energy-intensive manufacturing industry, which relies heavily on coal, falls over the projection period, with faster growth in services. The reduction in the share of coal in the country's primary energy mix also drives down carbon intensity. Carbon intensity falls less rapidly in other developing countries and in the OECD.

Figure 5.10: **Per-Capita Energy-Related CO$_2$ Emissions and Population by Region in the Reference Scenario**

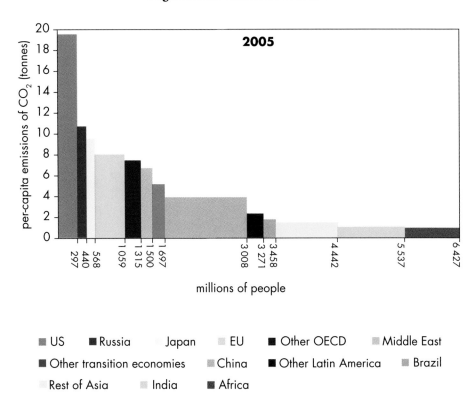

- ■ US ■ Russia ▫ Japan ▫ EU ■ Other OECD ▫ Middle East
- ■ Other transition economies ▫ China ■ Other Latin America ▫ Brazil
- ▫ Rest of Asia ▫ India ■ Africa

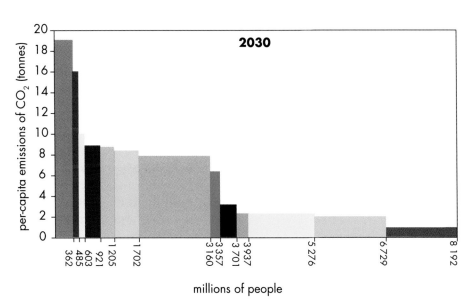

Figure 5.11: **Change in Carbon Intensity by Region and Scenario, 2005-2030**

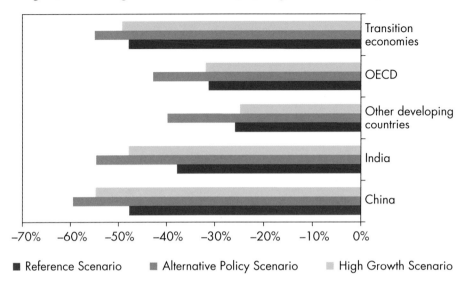

■ Reference Scenario ■ Alternative Policy Scenario ■ High Growth Scenario

Implications for Climate Change

There now exists a major body of scientific evidence that rising greenhouse-gas concentrations in the atmosphere resulting from fossil-energy combustion and other human activities are contributing to rising global temperatures and to changes in climate patterns. Today, the link between greenhouse-gas emissions and climate change is accepted by the overwhelming majority of scientists, even if the magnitude and nature of the changes that will follow from a given rise in emissions are still very uncertain. The primary source of the increased atmospheric concentration of greenhouse gases since the pre-industrial period has been the burning of fossil fuels, with land-use change providing another significant, but smaller, contribution.

The latest assessment of the Intergovernmental Panel on Climate Change (IPCC) concludes that most of the observed increase in global average temperatures since the mid-20th century is very likely to be due to the observed increase in anthropogenic greenhouse-gas concentrations, notably CO_2 (Box 5.3). Continued greenhouse-gas emissions at or above current rates would drive atmospheric concentrations even higher, causing further warming and inducing many changes in the global climate system during the 21st century that would very likely be larger than those observed during the 20th century (IPCC, 2007). These conclusions were somewhat stronger than those of the previous assessment, reflecting important advances in the knowledge and understanding of climate change. Even if concentrations are stabilised, some anthropogenic warming and rises in sea levels are expected to continue for centuries, due to the time-scales associated with climate processes and feedbacks.

Box 5.3: **IPCC Fourth Assessment Report**

The World Meteorological Organisation (WMO) and the United Nations Environment Programme (UNEP) established the Intergovernmental Panel on Climate Change (IPCC) in 1988 to assess scientific, technical and socio-economic information relevant for the understanding of climate change, its potential impacts and options for adaptation and mitigation. The IPCC has three working groups (WGs) which assess knowledge of the climate system (WG I), the impact of the adaptation to climate change (WG II) and the mitigation of climate change (WG III), plus a Task Force on national greenhouse-gas inventories. The assessments are performed by scientists nominated by governments and scientific organisations. They are managed by an elected bureau and are supported by the IPCC Secretariat and the Technical Support Units of the working groups. The IPCC meets in plenary session about once a year to approve the work programme and the assessment reports,.

A main activity of the IPCC is to provide at regular intervals an assessment of the state of knowledge about climate change. It recently finalised and released the three working group reports that will make up its Fourth Assessment Report, providing a comprehensive and up-to-date appraisal of the current state of knowledge on climate change. A synthesis report is due to be released in November 2007.

The global atmospheric concentration of CO_2 has increased from a pre-industrial level of about 280 parts per million to 379 ppm in 2005. This level exceeds by far the natural range over the last 650 000 years (180 to 300 ppm) as determined from ice cores. It is estimated that, were CO_2 concentrations to rise to 400 to 440 ppm and stabilise at that level, the eventual rise in global average temperature would amount to around 2.4° to 2.8°C (IPCC, 2007). In order to stabilise the concentration of CO_2 in the atmosphere, emissions would need to peak and decline thereafter. The lower the target stabilisation level, the more quickly this peak and decline would need to occur. For this reason, mitigation efforts over the next two to three decades will affect our ability to achieve lower stabilisation levels. Based on current understanding of climate-carbon cycle feedback, the IPCC concludes that, in order to stabilise CO_2-equivalent[4] concentrations at 445-490 ppm, CO_2 emissions would need to peak by 2015 at the latest and fall to between 50% and 85% below 2000 levels by 2050. A later peak and less sharp reductions in emissions would lead to higher concentrations and bigger increases in temperature (Table 5.4).

4. All greenhouse gases expressed in CO_2-equivalent terms (adjusted for differences in radiative forcing).

Table 5.4: **CO$_2$ Concentrations and Emissions**

CO$_2$ concentration (ppm)	CO$_2$-equivalent concentration (ppm)	Global mean temperature increase above pre-industrial level at equilibrium*(°C)	Peaking year for CO$_2$ emissions	Global change in CO$_2$ emissions in 2050 (% of 2000 emissions)
350 – 400	445 – 490	2.0 – 2.4	2000 – 2015	−50 to −85
400 – 440	490 – 535	2.4 – 2.8	2000 – 2020	−30 to −60
440 – 485	535 – 590	2.8 – 3.2	2010 – 2030	+5 to −30
485 – 570	590 – 710	3.2 – 4.0	2020 – 2060	+10 to +60
570 – 660	710 – 855	4.0 – 4.9	2050 – 2080	+25 to +85
660 – 790	855 – 1 130	4.9 – 6.1	2060 – 2090	+90 to +140

* Based on the "best estimate" of climate sensitivity.
Source: IPCC (2007).

Assessing the impact of the projections of energy-related CO$_2$ emissions in this *Outlook* on global concentrations of carbon dioxide and long-term global changes in temperature is extremely difficult. Climate-carbon cycle coupling – the inter-relationship between changes in climate and natural carbon emissions and absorption processes – is expected to add CO$_2$ to the atmosphere as the climate system warms, but the magnitude of this feedback is uncertain. This increases the uncertainty about the trajectory of emissions required to achieve stabilisation of the atmospheric CO$_2$ concentration at a particular level. Our projections run only to 2030, though the trend in emissions over that time frame will clearly influence strongly the longer-term trajectory. In addition, emissions of other greenhouse gases – not modelled in this *Outlook* – will affect the overall concentration of these gases in CO$_2$-equivalent terms.

To determine the likely CO$_2$-equivalent long-term concentration of greenhouse gases corresponding to each *WEO* scenario, we took into account projected emissions of other greenhouse gases to 2030, using IIASA's integrated assessment scenarios,[5] and coupling each *WEO* scenario with the closest IIASA scenario. We then compared overall greenhouse-gas emissions trends with the IPCC's assessment of the resulting eventual change in concentration and the associated increase in global temperature.[6] The Reference and High Growth

5. The data underlying IIASA scenarios are available at www.iiasa.ac.at/research/GGI.
6. A detailed methodology and analysis can be found in "World Energy Outlook 2007: CO$_2$ Emissions Pathways Compared to Long-Term CO$_2$ Stabilisation Scenarios in the Literature and IPCC AR4", by N. Nakicenovic, available at www.worldenergyoutlook.org.

Scenarios energy projections are both consistent with stabilisation of a CO_2-equivalent concentration at levels of 855 to 1 130 ppm (or CO_2 of 660 to 790 ppm). By contrast, assuming continued emissions reduction after 2030, energy-related CO_2 emissions in the Alternative Policy Scenario are consistent with a CO_2-equivalent concentration of about 550 ppm – a level that corresponds to an increase in average temperature of around 3°C above pre-industrial levels.[7]

The 450 Stabilisation Case[8]

There is growing support worldwide for early action to tackle climate change in order to stabilise the concentration of greenhouse gases at a level that would prevent dangerous interference with the climate system. At their summit in Heiligendamm in 2007, G8 leaders, meeting with the leaders of several major developing countries and heads of international organisations, including the IEA, committed to "taking strong and early action to tackle climate change in order to stabilise greenhouse-gas concentrations at a level that would prevent dangerous anthropogenic interference with the climate system".[9] In their declaration, they also pledged to "consider seriously the decisions made by the European Union, Canada and Japan which include at least a halving of global emissions by 2050".

We estimate that stabilising the greenhouse-gas concentration in the range of 445-490 ppm of CO_2-equivalent – the most ambitious of the IPCC's scenarios – would require energy-related CO_2 emissions to be reduced to around 23 Gt in 2030 – some 19 Gt less than in the Reference Scenario and some 11 Gt less than in the Alternative Policy Scenario.[10] This level is 13% lower than 2005 emissions and 12% higher than 1990 emissions.

In principle, there are many ways in which energy-related CO_2 emissions could be reduced to 23 Gt in 2030. In response to requests from policy makers, we describe here one possible pathway – which we have called the *450 Stabilisation Case* – to achieving this very ambitious target in order to

7. Taking account of all greenhouse gases, emissions in the Alternative Policy Scenario are lower than in IIASA's B1 590 ppm scenario and higher than in the B1 520 ppm scenario.
8. The 2008 edition of the *World Energy Outlook* will explore in detail a range of climate-change scenarios and their implications for global energy markets.
9. Heiligendamm G8 Summit Declaration. Available at: www.iea.org/G8/docs/declaration_2007.pdf.
10. The range of emissions identified by the IPCC in a specific year is large, since higher emissions in earlier years can be compensated by stronger cuts in later years. In 2030, the estimated range of CO_2 emissions compatible with stabilisation of CO_2-equivalent at 445-490 ppm is 10 to 29 Gt. We decided to use 23 Gt as an illustrative target, allowing for up to 6 Gt of CO_2 from non-energy-related sources, notably land use, land-use changes and forestry.

illustrate the magnitude and urgency of the challenge of transforming the global energy system over the projection period. We have not used the same modelling tools as those used to prepare the Reference, Alternative Policy and High Growth Scenario projections. Rather, a backcasting methodology has been used, which involved identifying a combination of technological changes that would allow the target to be met, based on the expected availability of end-use and power-generation technology options and estimates of potential efficiency gains by sector. In the 450 Stabilisation Case, cleaner and more advanced technologies are deployed more quickly than in the Alternative Policy Scenario. In addition, technologies that are not yet financially viable, including CO_2 capture and storage and second-generation biofuels technologies, are assumed to be widely deployed. This case requires that existing energy-using capital would be prematurely retired, at substantial cost.

Energy and CO_2 Emission Trends

In the 450 Stabilisation Case, global energy-related CO_2 emissions peak in 2012 at around 30 Gt and then decline, reaching the goal of 23 Gt in 2030 (Figure 5.12). Improved efficiency in fossil-fuel use in industry and buildings accounts for more than a quarter of total avoided CO_2 emissions in 2030, compared with the Alternative Policy Scenario. Lower electricity demand, resulting from more efficient electricity use in buildings, represents 13% of the savings. Switching to second-generation biofuels in transport accounts for 4% and renewables in the power sector for 19%. Increased reliance on nuclear generation is responsible for 16%. CO_2 capture and storage (CCS) in power generation and industry accounts for the remaining 21%. In practice, rapid deployment of CCS and expansion of nuclear power face major policy and regulatory hurdles that may take considerable time to resolve (the prospects for CCS and other types of clean coal technology are discussed below). Clearly, exceptionally vigorous policy action – entailing substantial costs – would be needed to make the 450 Stabilisation Case a reality. Such action would need to start immediately: each year of delay would reduce substantially the likelihood of achieving the target.

Primary demand reaches 14 031 Mtoe in 2030 – a reduction of about 11% relative to the Alternative Policy Scenario and 21% relative to the Reference Scenario (Table 5.5). The saving compared with the Alternative Policy Scenario is comparable to the current energy demand of OECD Europe. The reduction in the use of fossil fuels is more marked than the reduction in primary energy demand, even though fossil fuels still account for two-thirds of primary energy demand by 2030 (compared with 82% in the Reference Scenario and 76% in the Alternative Policy Scenario). Coal demand peaks around 2015 and declines thereafter, reaching in 2030 a level close to that of 2003. Oil demand increases slightly through to 2015, but then falls. Use of nuclear power is significantly higher. Biomass use increases sharply in combined heat and power production and electricity-only power plants, as well as for making biofuels for transport. Reliance on hydropower and other renewables – wind, geothermal, and solar power – is also significantly higher.

Figure 5.12: **CO$_2$ Emissions in the 450 Stabilisation Case**

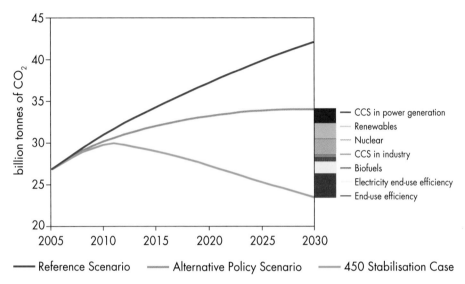

Table 5.5: **World Energy Demand in the 450 Stabilisation Case** (Mtoe)

	2005	2015	2030	2005-2030*	Difference from the Reference Scenario in 2030	Difference from the Alternative Policy Scenario in 2030
Coal	2 892	3 213	2 559	–0.5	–2 435	–1 140
Oil	4 000	4 278	4 114	0.1	–1 471	– 797
Gas	2 354	2 736	2 644	0.5	–1 304	– 802
Nuclear	721	1 037	1 709	3.5	855	629
Hydro	251	393	568	3.3	152	104
Biomass	1 149	1 484	1 966	2.2	350	228
Other renewables	61	223	471	8.5	163	28
Total	**11 429**	**13 364**	**14 031**	**0.8**	**–3 689**	**–1 752**

* Average annual rate of growth.

Energy Demand by Sector

CO_2 emissions from direct combustion of fossil fuels in end uses and other transformation other than power generation are reduced to 16.2 Gt in 2030 – 3.3 Gt, or 17%, lower than in the Alternative Policy Scenario. Energy savings by sector in 2030, over and above those achieved in the Alternative Policy Scenario, assume that best-practice commercial technologies available are quickly and widely deployed where the potential costs less than $50 per tonne of CO_2 (IPCC, 2007). Taking into account the carbon content of crude oil, this is equivalent to an increase in oil prices of $18 per barrel.

The use of fossil fuels in *industry* is reduced by 18% in 2030 compared with the Alternative Policy Scenario, yielding 1.3 Gt of CO_2 savings. The biggest savings are in the iron and steel and cement industries. Widespread adoption of best-practice technology, which is already commercial, or will become so, would allow this potential to be harvested (IEA, 2007a). In practice, financial incentives or regulations would be required to ensure that less efficient equipment is retired early. Electricity savings in less energy-intensive industries and improved motor efficiency are already fully exploited in the Alternative Policy Scenario. Therefore, we do not assume any additional electricity savings in the industrial sector above those in the Alternative Policy Scenario. Equipping some refineries, ammonia, cement and iron and steel plants with CO_2 capture and storage (CCS) brings about an additional reduction of 0.5 Gt. Strong policies, such as regulations or subsidies, would be required for this to happen.

In the *residential and services* sector there is only limited remaining potential to reduce coal, oil and natural gas direct use beyond the level achieved in the Alternative Policy Scenario. Additional fossil-fuel savings amount to 7%, yielding savings of 0.3 Gt of CO_2. Increased electricity savings are also 8% more than in the Alternative Policy Scenario. Indirectly avoided CO_2 emissions, through the reduced need to generate power, amount to 0.7 Gt. Widespread use of minimum efficiency standards in a wide range of appliances and equipments could help capture this potential.

Additional savings are achieved in the *transport* sector, mainly through improved efficiency of light-duty vehicles, increased use of biofuels and more efficient aircraft. Together, these outcomes would cut global oil use in 2030 by more than 10 mb/d, saving 1.4 Gt of CO_2. The fuel efficiency of light-duty vehicles in 2030 is 14% better than in the Alternative Policy Scenario. To achieve this, the average car sold in 2030 would need to consume 60% less fuel than the average car sold in 2005. With current technologies, only plug-in hybrids are capable of this. In addition, such cars reduce the need for oil-based fuels even more, because they use electricity from the grid. As power generation becomes less carbon-intensive, emissions are reduced by even more than energy demand. Improvements of up to 50% in the efficiency of gasoline and diesel internal combustion engines, and even of full-hybrid vehicles, would also be needed. Policies to promote hybrid technology could include vehicle-purchase subsidies, regulatory standards and higher taxes on the least efficient vehicles.

Biofuels use in 2030 is twice as big as in the Alternative Policy Scenario, at 330 Mtoe. Given the constraints on land, water and biomass availability, this level of production of biofuels could be achieved only through the large-scale introduction of second-generation biofuels, based mainly on ligno-cellulosic feedstock (IEA, 2006a). Energy use in non-power sector transformation – including refineries and oil and gas extraction – is reduced by 16% over and above the Alternative Policy Scenario, thanks to the reduced need to supply hydrocarbons to end users.

Power Generation

To reach the overall 23 Gt target for 2030, emissions from the power sector would need to be limited to 6.3 Gt, compared with 13.7 Gt in the Alternative Policy Scenario and 18.7 Gt in the Reference Scenario.[11] Given the long lead times in bringing new capacity on line in the power sector and the current policy framework, we assume that the installed power generating capacity follows the Alternative Policy Scenario trend until 2012. Even if no new power plants were built after 2012 and taking retirements into account, emissions in 2030 would still be around 10 Gt. This is well in excess of the level compatible with the 450 ppm of CO_2-equivalent stabilisation target. Therefore, some of the power plants in operation in 2012 would need to be retired before the end of their economic lifetime and any new capacity added would need to be zero-carbon. We calculate that some 15% of the fossil-fuel generating capacity would need to be retired early between 2012 and 2030 on the assumption that all new generating capacity is either nuclear power, renewables-based or, after 2015, fossil-based with CO_2 capture and storage (Figure 5.13). If retrofitting of fossil-based generation with CCS was considered, the need for early retirements would be lower (CCS is discussed later in this chapter).

Electricity demand in the 450 Stabilisation Case grows from 18 200 TWh in 2005 to 29 300 TWh in 2030, by 1.9% per year. Electricity generated by the power plants in use in 2012 declines from 22 930 TWh in 2012 to some 15 100 TWh in 2030. The balance comes from zero- or low-carbon power plants – renewables, and nuclear power. After 2015 we assume the gradual introduction of coal- and gas-fired power plants equipped with CCS (Figure 5.14). As there is an infinite number of combinations of capacity

---

11. Implementation of energy-efficiency measures in end-use sectors at a higher cost than assumed in this case ($50 per tonne of CO_2) would increase energy and CO_2-emissions savings in final consumption, requiring less reductions in the power sector.

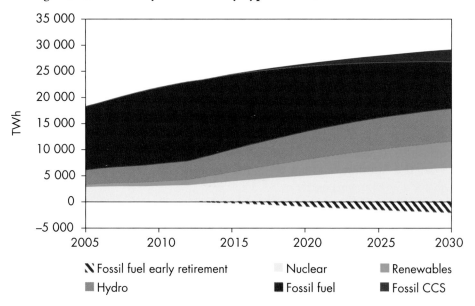

Figure 5.13: **Electricity Generation by Type in the 450 Stabilisation Case**

Legend: ✕ Fossil fuel early retirement ▪ Hydro ▫ Nuclear ▪ Fossil fuel ▪ Renewables ▪ Fossil CCS

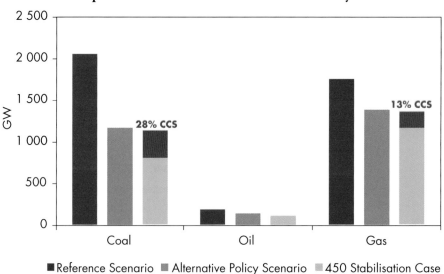

Figure 5.14: **Fossil-Energy Generating Capacity in 2030 in the 450 Stabilisation Case Compared with the Reference and Alternative Policy Scenarios**

Legend: ▪ Reference Scenario ▪ Alternative Policy Scenario ▪ 450 Stabilisation Case

that could meet the gap, we have applied a simple rule, whereby new generating needs by 2030 are met equally by nuclear and hydropower combined, other renewables and CCS. As a result, the total share of renewables-based power generation increases to 40% in 2030 (Table 5.6).

Hydropower generation reaches 6 610 TWh and other renewables reach 5 170 TWh. In this scenario, 80% of the economic potential for renewables in 2030 would be used. Most of the increase in renewables-based generation comes from hydropower and biomass. Intermittent renewables would account for 10% of generation in 2030, below the maximum level of 15% to 30% considered safe for the grid (Implementing Agreement on Wind Energy, 2006). However, backup generating capacity would be needed at additional cost. Exceptionally rigorous new government measures, involving strong financial incentives and/or regulations, would be needed to effect such a rapid expansion of renewables-based generation.

5

Table 5.6: **Renewables-Based Electricity Generation**

	2005		2030					
			450 Stabilisation Case		Reference Scenario		Alternative Policy Scenario	
	TWh	%	TWh	%	TWh	%	TWh	%
Hydro	2 922	16.1	6 608	22.5	4 842	13.7	5 403	17.3
Biomass	231	1.3	2 056	7.0	840	2.4	1 166	3.7
Wind	111	0.6	2 464	8.4	1 287	3.6	1 800	5.8
Geothermal	52	0.3	219	0.7	173	0.5	190	0.6
Solar	3	0.0	406	1.4	161	0.5	352	1.1
Tidal/wave	1	0.0	28	0.1	12	0.0	24	0.1
Total	**3 321**	**18.2**	**11 781**	**40.2**	**7 315**	**20.7**	**8 935**	**28.6**

Installed nuclear generating capacity reaches 833 GW in 2030, compared with 525 GW in the Alternative Policy Scenario and 415 GW in the Reference Scenario. Electricity generation from nuclear more than doubles, compared with 2005 levels, from 2 770 TWh to 6 560 TWh in 2030. About 4 600 TWh of electricity in 2030 is still produced by coal-fired plants that are installed before 2012. An increasing proportion of new coal plants built after 2012 are assumed to be equipped with CCS. In 2030, some 1 750 TWh of electricity is generated from coal plants equipped with CCS, equivalent to capacity of about 310 GW. Gas-fired plants produce 4 370 TWh, of which 13% is from plants with CCS. Oil will by then have become a marginal source of electricity, accounting for only 1% of electricity generation. CCS will have to be particularly widely deployed in the United States, China and India.

The capital costs involved in stabilising CO_2-equivalent concentrations at around 450 ppm would be very large. Unlike the Alternative Policy Scenario, in which investment needs are lower than in the Reference Scenario, the 450 Stabilisation Case implies much higher investment in the power-generation sector compared with the Reference Scenario. Cumulative investments in this case are $7.5 trillion, compared with $5.7 trillion in the Reference Scenario (an increase of 31%) and $5.5 trillion in the Alternative Policy Scenario (36% more). Early retirement of fossil-fuel generating capacity will comprise almost $1 trillion of the additional investment.

The average capital cost of new capacity is 56% higher than in the Reference Scenario. Generating-capacity needs are lower compared with the Reference and Alternative Policy Scenarios because of the increased efficiency of electricity use. But this is outweighed by the much higher capital cost of zero- and low-carbon technologies. The implication is substantially higher electricity prices for consumers. CCS accounts for a fifth of cumulative power-generation investment needs in 2006-2030 (Figure 5.15).

Figure 5.15: **Share of Cumulative Power-Generation Investment by Technology, 2006-2030**

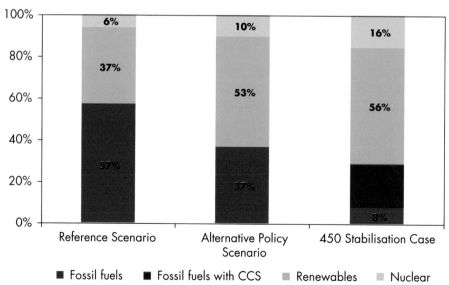

Can China and India Ever Mirror Western Lifestyles?

In principle, the continued economic and social development of China and India need not be incompatible with protecting the local or global environment. But major shifts in resource use, policies and technologies – as well as public attitudes and expectations – will be needed worldwide. Quite simply, the resource-intensive economic model currently being pursued throughout the world cannot be sustained indefinitely. A level of per-capita income in China and India comparable with that of the industrialised countries would, on today's model, require a level of energy use beyond the world's energy resource endowment and the absorptive capacity of the planet's ecosystem.

A couple of simple calculations illustrate this very clearly: if per-capita oil use in China and India were to rise to the current level in the United States, their oil demand would increase by a combined 160 mb/d – almost twice the current level of world oil demand (not allowing for future increases in population). Without major changes elsewhere, total world demand of close to 240 mb/d would deplete remaining proven reserves fully in just 15 years, and estimated ultimately recoverable oil and natural gas liquid resources (including proven reserves, reserves growth and undiscovered resources) in 26 years.[12] Similarly, if per-capita CO_2 emissions in China and India reached current US levels, again assuming no major departures from trends elsewhere, world emissions would be three times higher than today. The implications for climate change of such an increase could be catastrophic. Even sustained global fossil-energy consumption at current levels risks causing a substantial increase in CO_2 concentrations and global temperatures.

Up to now, China and India have focused on economic growth on traditional lines on the path to national goals – including reducing poverty, modernising lifestyles and raising comfort levels. But there is a growing recognition in both countries of the need to seek out a radically different development path to that adopted in the west, leapfrogging to new technologies and involving different lifestyles. There are some signs of this happening. The sheer size of the two economies and the pace of their economic growth makes it essential that all countries – China, India, the industrialised countries and the rest of the global community – co-operate on moving quickly towards a genuinely sustainable lifestyle.

12. Based on reserves estimates from the *Oil and Gas Journal* (18 December 2006) and the US Geological Survey's mean estimates for reserves growth and undiscovered resources (USGS, 2000).

Focus on Prospects for Clean Coal Technology for Power Generation

Clean coal technologies[13] in power generation could play an important role in minimising the environmental impact of coal use by reducing emissions of dust, sulphur dioxide, oxides of nitrogen and CO_2, in part through improved thermal efficiency. Emerging CCS technologies hold out the prospect of generating power from coal with very low CO_2 emissions. The share of coal in global emissions is set to rise significantly over the projection period in the Reference Scenario (Table 5.7). The potential impact of accelerating the deployment of clean coal technology is greatest in China and India, where most of the rise in global demand for coal will come from and where there is more scope to move to the most advanced technologies currently available. Indeed, we calculate that were both China and India to reach the OECD level of efficiency for new coal power plants by the year 2012, the cumulative saving in emissions through to 2030 would be of the order of 6.8 Gt in the Reference Scenario. In 2030, the emission saving is 650 Mt CO_2 – equal to about 2% of global emissions. Environmental concerns have come more to the fore in China and India in recent years, but they remain subordinate to the demands of economic development and poverty alleviation. There remains considerable scope in both countries and elsewhere to adopt more advanced coal technologies and, thereby, to reduce significantly the environmental damage caused by coal-based generation.

Table 5.7: **Share of Coal in CO_2 Emissions in the Reference Scenario** (%)

	1990	2000	2005	2015	2030
China	85	80	82	82	78
India	69	65	67	68	69
OECD	37	34	34	34	33
World	40	38	41	44	45

CO₂ Capture and Storage

CO_2 capture and storage (CCS) is one of the most promising options for mitigating emissions from coal-fired power plants and other industrial facilities. It plays a major role in stabilising CO_2 concentrations in the 450 Stabilisation Case described above. CCS is a three-step process involving the capture of CO_2 emitted by large-scale stationary sources and the

13. There is no definitively adopted definition of the term "clean coal technology". Some prefer "cleaner coal" because it is impossible to mine and use coal without environmental consequences.

compression of the gas and its transportation (usually via pipelines) to a storage site, such as a deep saline formation, depleted oil/gas field or unmineable coal seam (Figure 5.16). The CO_2 may also be used for enhanced oil or gas recovery. CCS processes can currently capture more than 85% of the CO_2 that would otherwise be emitted by a power plant, but they reduce the plant's thermal efficiency by about 8 to 12 percentage points and, thus, increase fossil-fuel inputs, because of the additional energy consumed in capturing the gas. Initially, CCS is expected to be deployed primarily in coal-fired power stations, because the CO_2 emissions to be captured are proportionately larger than in oil- or natural-gas-fired plants, reducing the per-tonne cost.

Figure 5.16: **CO_2 Capture, Transport and Storage Infrastructure**

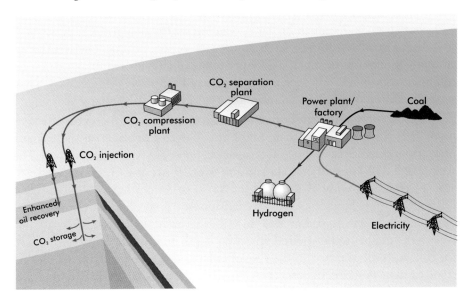

The process of capturing CO_2 generally represents the largest component of CCS costs. There are three main processes currently available:

- *Pre-combustion capture:* This process can be used in plants using coal or natural gas. The fuel is reacted with air or oxygen, generating carbon monoxide and hydrogen, which is further processed in a shift reactor to produce a mixture of hydrogen and CO_2. The gases are separated, with the hydrogen used to generate electricity and heat in a combined-cycle gas turbine.

- *Post-combustion capture:* This involves removing the CO_2 from flue gas from power stations or industrial plants. The gas contains between 3% and 4% of CO_2 by volume in a gas-fired plant and around 15% in a coal-fired plant.

Technologies involving the absorption of CO_2 in solvents and subsequent solvent regeneration, sometimes in combination with membrane separation, are the most prevalent.

■ *The oxy-combustion process:* This involves the combustion of fossil fuel in a mixture of near-pure oxygen and recycled flue gas, producing a secondary flue gas stream consisting essentially of CO_2 and water, which can then be separated.

CO_2 capture from combustion processes is highly energy-intensive and expensive. Separation of CO_2 from natural gas at the well head is necessary to meet quality standards and offers a cheap source of CO_2 for storage (Box 5.4). CCS in power generation is cheapest for large, highly efficient coal-fired plants. The capital cost of a demonstration power plant with CCS is estimated to range from $0.5 to $1 billion, 50% of which is for the capture and compression equipment alone. The typical cost of CCS in power plants ranges from $30 to $90 per tonne of CO_2, but costs can be much higher depending on technology, CO_2 purity and site. The cost of retrofitting CCS equipment on an existing power plant is currently thought to be much higher per tonne of CO_2 abated than the cost of equipping a greenfield power plant. No large coal-fired power plant with CCS has yet been built, though a large number are at an advanced planning stage.

CO_2 is most commonly transported by pipeline, as this is the most cost-effective mode over distances of less than 1 000 km. The cost depends on the terrain, pressure requirements, distance and capacity. There are very large economies of scale. For a 250-km onshore pipeline and over 10 Mt of CO_2 transported per year, costs are currently of the order of $1.50 to $4 per tonne of CO_2 (UK-DBERR, 2007). Storage costs vary enormously, according to infrastructure requirements (new injection and monitoring wells or retrofitting existing facilities), volumes to be injected, injection depth and whether the CO_2 is used for enhanced oil recovery. In the case of enhanced oil recovery, the net cost of storage can be negative. In other cases, costs may be up to $10/tonne for onshore aquifers and $40/tonne for depleted offshore oil and gas fields (UK-DBERR, 2007). Offshore storage is generally two to three times more costly than onshore disposal. Monitoring costs depend on the risk of leakage; they are estimated to be generally less than $1 per tonne of CO_2 injected.

Using cost-effective technologies and favourable siting, the lowest costs achievable for CCS at greenfield coal-fired plants are currently estimated to be of the order of $50/tonne (IEA, 2006b). This includes capture costs of $20 to $40/tonne, large-scale transportation by pipeline costing $1 to $5/tonne per 100 km and storage costs of $2 to $5/tonne. Short-distance transport and storage together might cost less than $10/tonne, if monitoring costs are small. Assuming reasonable rates of technology learning, the total cost of CCS might be expected to drop to below $25/tonne of CO_2 by 2030 (IEA, 2006c).

Box 5.4: **Major CO$_2$ Capture and Storage Projects**

There are at present three large-scale CCS projects in operation around the world, each involving around 1 Mt of CO$_2$ per year: Sleipner in Norway, Weyburn in Canada and the United States, and In Salah in Algeria. A fourth project, at the Snøhvit gasfield in Norway, is due to begin operation at the end of 2007. In addition to these projects in the oil and gas sector, around 20 other major projects in the power sector have been announced.

In the off-shore ***Sleipner*** project, which began operation in 1996, CO$_2$ is separated from produced natural gas and injected into a saline aquifer. Over 1 Mt per year has so far been stored and a total of 20 Mt is expected to be stored during the life of the project. Extensive monitoring, to track the dispersion of CO$_2$ in the aquifer, has been carried out, including the use of 4-D seismic techniques.

The ***Weyburn*** project involves the capture of over 1.7 Mt per year of CO$_2$ in a coal-gasification plant in North Dakota in the United States. The gas is compressed and transported via a 330-km pipeline to EnCana's Weyburn field in Saskatchewan in Canada, where it is used for enhanced oil recovery. Injection, which started in 2000, is expected to boost cumulative oil output by over 120 million barrels.

Like Sleipner, the gas produced at the ***In Salah*** (and neighbouring) fields has a CO$_2$ content of between 4% and 9 %, which exceeds the permitted amount in sales contracts. A processing plant at Krechba uses a chemical solvent to separate out the CO$_2$ from the gas produced. Four compression stages are then used to pressurise CO$_2$ and inject it into a 20 metre-thick reservoir, which lies under the gas-producing zone. Storage capacity is 1 Mt of CO$_2$ per year. A total of 17 Mt is expected to be stored over the life of the project, at a cost of $6 per tonne of CO$_2$ (Wright, 2006).

At present, there are limited financial incentives for operators of power stations or large industrial facilities to install CCS. But this may change in the future. For this reason, and because power plants have very long lives (typically over 40 years), the IEA is investigating the possibility of making power plants "CCS-ready" as part of the Agency's G8 Gleneagles Plan of Action (IEA-GHG, 2007). The aim is to lower the cost of retrofitting existing plants. At present, the total cost of CCS is estimated at between $66 and $122/tonne of CO$_2$ for a retrofit of a coal-fired pulverised coal plant (WEC, 2006). Retrofit costs are expected to fall as operational experience grows and technology improves, and with the introduction of CCS-ready plants.

There are a number of barriers to the widespread and rapid deployment of CCS, which will need to be addressed if it is to make a major contribution to mitigating energy-related CO_2 emissions, as in the 450 Stabilisation Case:

- *Commercial and financial issues:* CO_2 must be given a value, either through carbon taxation or the trading of emission credits.

- *Legal and regulatory issues:* There is a need to establish legal guidelines with respect to the injection of CO_2, to define regulatory frameworks, to allocate long-term liabilities and to develop risk-management procedures, including monitoring and remediation (IEA, 2007b).

- *Technical issues:* Capture technology needs to be improved in order to improve reliability and lower costs. Potential leakage routes need to be identified in different types of reservoir and long-term isolation procedures established.

- *Public awareness:* Key messages on CCS need to be effectively communicated.

In addition to several national programmes, several international initiatives have been launched, by both the public and private sectors, to study, develop and promote CCS technologies. Given the magnitude of the challenges, including the cost of research, development and demonstration, greater international co-operation and sharing of best practices are required to accelerate the pace of technology development and deployment (see Chapter 6).

Power-Generation Technologies

The combustion of pulverised or powdered coal to raise steam in boilers has been the mainstay of coal-based power generation worldwide for almost a hundred years. The efficiency of the current generation of pulverised coal units has steadily improved and today ranges between 30% and 45% (on a lower heating-value basis) depending on the quality of coal used, ambient conditions and the back-end cooling employed. A number of advanced power-generation technologies have been or are being developed to improve thermal efficiency and to reduce other emissions, notably nitrogen oxides (NO_x) and sulphur dioxide (SO_2). These technologies hold out the prospect of significantly raising the efficiency of the new coal-fired plants that will be built in the coming decades and reducing their emissions (Figure 15.17).

The most important of current technologies and others in development are:

- *Supercritical and ultra-supercritical pulverised coal combustion:* The efficiency of a steam cycle is largely a function of steam pressure and temperature. Typical subcritical steam cycles, as in the vast majority of today's power plants, operate at 163 bar pressure and 538°C. With supercritical designs,

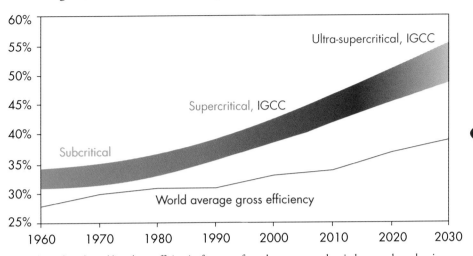

Figure 5.17: **Thermal Efficiency of Coal-Fired Power Generation**

Note: The multi-coloured line shows efficiencies for state-of-art plants on a net electrical output, lower heating value basis.

pressure is typically 245 bar and temperature in excess of 550°C, *i.e.* above the critical point at which water turns to steam without boiling. In ultra-supercritical designs, even higher temperatures are used, sometimes exceeding 600°C. More expensive materials are required, but the impact of this higher capital cost on the overall economics of the plant is, to some extent, balanced by the increased efficiency, which brings fuel and fuel-handling cost savings. Supercritical technology has become the norm for new plants in OECD countries and is increasingly so in China. At end-2006, 6.5% of all coal-fired capacity was supercritical in China. Supercritical plants are planned in India. Commercial ultra-supercritical plants are in operation in Japan, Germany and Denmark. Research into materials taking place today aims to push efficiencies to over 50%.

■ *Circulating fluidised bed combustion:* CFBC plants can be designed for a wide variety of coals and particle sizes. Because coal is burned at low temperatures and in a staged manner, they produce low NO_x compared with conventional pulverised coal (PC) boilers. In addition, operating temperatures are ideally suited for *in-situ* capture of SO_2. The efficiency of CFBC plants is similar to PC units. At present, the largest operating CFBC unit is 320 MW. CFBCs are now available commercially at a scale that allows them to be used in supercritical mode. The first supercritical CFBC unit (460 MW) is currently undergoing construction in Poland, and is scheduled to operate in the first half of 2009. However, relatively low operating temperatures mean CFBCs may not be practicable for ultra-supercritical plants, which operate at steam temperatures much higher than 550°C.

- *Integrated gasification combined-cycle:* IGCC combines coal gasification with a combined-cycle power plant. Coal is gasified under pressure with air or oxygen to produce fuel gas which, after cleaning, is burned in a gas turbine to produce power. Exhaust gas from the gas turbine passes through a heat-recovery steam generator or boiler to raise steam for a steam turbine which generates extra power. Only four successful IGCC plants have so far been built: two in Europe and two in the United States. At high temperatures, efficiency can be as high as 41%, or even higher with the latest gas-turbine models. For IGCC to establish itself in the market, further development to bring down costs and improve operational flexibility is necessary. A number of plants are being built in China and Japan, and several others are being considered elsewhere. IGCC has inherent advantages for emission control, as gas clean-up takes place before combustion of the fuel gas, using relatively small equipment, and solid waste is in the form of a vitrified slag. If CCS becomes an established mitigation measure, then CO_2 capture from an IGCC plant is technically easier than post-combustion capture from a conventional steam plant.

- *Other technologies:* A number of other technologies and hybrid systems are at the research and development stage, notably in the United States and Japan. Integrated gasification-fuel cell combined cycle involves combining a fuel cell and the combined cycle component of IGCC to generate power. Efficiency could reach 60%.

While clean coal technologies have made significant progress in the last decade or so, there are still considerable challenges in exploiting the remaining potential, particularly for low-grade coals. For high-moisture coals, a cheaper and more efficient drying system is needed together with a reliable system for feeding these coals into a pressurised gasifier. For high-ash coals, the main challenge is to overcome fouling problems in gasification and combustion. For all types of coals, gas clean-up at higher temperature is needed to obtain higher efficiency in IGCC units. Considerable research to address these problems is under way. For emerging technologies – especially CCS – the main challenge is to lower costs and demonstrate reliable operation. The addition of CCS equipment increases significantly the capital cost of capacity for all coal technologies, not least because thermal efficiency is lower (Figure 5.18).

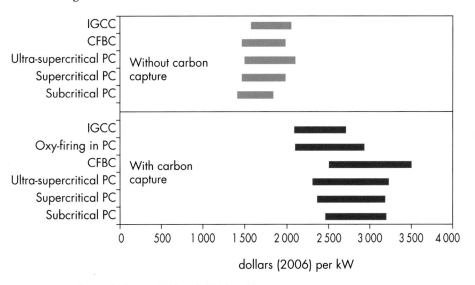

Figure 5.18: **OECD Coal-Fired Power Plant Investment Costs**

Sources: IEA and EPRI databases; MIT (2007); IEA (2006c).

CHAPTER 6

ENERGY POLICY RAMIFICATIONS

HIGHLIGHTS

- The emergence of China and India as major players in global energy markets makes it all the more important that all countries take decisive and urgent action to reconcile the need to meet rising demand for energy services worldwide with ensuring energy security and protection of the environment. The rest of the world needs to engage constructively with China and India to address this common challenge in a mutually beneficial way.

- Many of the policies available to alleviate energy insecurity can also help to mitigate local pollution and climate change, and vice versa. In many cases, those policies bring economic benefits too, by lowering energy costs – a "triple-win" outcome. An integrated approach to policy formulation is therefore essential. The right mix of policies to address both energy-security and climate concerns depends on the balance of costs and benefits, which vary among countries.

- Collective responses to these global challenges are needed. No major energy consumer can be confident of secure supply if supplies to others are at risk. And there can be no effective long-term solution to the threat of climate change unless all major energy consumers, including China and India, contribute. There are large potential gains to IEA countries, on the one hand, and to China and India, on the other, from policy co-operation.

- Collaboration between IEA countries and developing countries is contributing to speeding-up the widespread deployment of new technologies. A portfolio approach to developing new technologies is indispensable. Energy efficiency needs to play a central role. Clean coal technology, notably CO_2 capture and storage, is one of the most promising routes for mitigating emissions in the longer term – especially in the United States, China and India, where coal use is growing the most. China and India stand to benefit from experience of best practice in other countries.

- Given the scale of the energy challenge facing the world, a substantial increase in public and private funding for research, development and demonstration is called for. Public budgets for energy research and development in IEA countries fell sharply in the 1980s and the early 1990s and have barely recovered since. The financial burden of supporting research efforts will continue to fall largely on IEA countries.

Addressing Energy Security and Climate Challenges

Energy developments in China and India over the coming decades will have profound consequences for the world. Rising energy use in these two emerging economies will inevitably add to the concern already felt by *all* import-dependent countries that energy supply is insufficiently secure and that energy-related pressures on the environment, both at the regional and global levels, must be addressed. Yet the economic development that is driving change in the energy sector in these two countries is an increasingly important motor for the world economy, bringing material benefits not only to their citizens but also to most other countries through increased trade (see Chapter 3). The desire of all developing countries, China and India among them, to improve the lives of their people is a legitimate aspiration that the rest of the world should accommodate and support.

In the absence of rigorous new action by governments, the twin threats to energy security and climate will, indeed, grow with rising global demand for energy services – regardless of what happens in China and India. But the inexorable emergence of these two countries as major players in global energy markets makes it all the more vital that all countries, IEA members and non-members, including China and India, take decisive and urgent action to reconcile the need to meet rising demand for energy services worldwide with ensuring energy security and protecting the environment.

Both energy insecurity and climate change stem largely from growing consumption of fossil fuels. As a result, many of the policies that can alleviate one of those threats can also help to address the other – a happy coincidence often referred to as "win-win". Where such policies bring economic benefits too, which is often the case where they are aimed at promoting more efficient energy use, they may be considered "triple-win". Which policies and measures fall into those categories depends on national circumstances. In view of the overlaps, an integrated approach to policy formulation – both within and between countries – is essential.

Energy-related greenhouse-gas emissions, mainly CO_2, can be lowered in one, or a combination, of several ways:

n *Improving energy efficiency:* Energy efficiency refers to the ratio between the output of an energy service – such as light, heat or mobility – and the input of energy. Technology improvements or changes in energy-use practices can bring about an increase in the energy efficiency of all types of equipment that produce or use energy to provide a given energy service. To the extent that this results in less use of fossil fuels, greenhouse-gas emissions will be reduced. The more carbon-intensive the fuel, the bigger the emissions savings. Many energy-efficiency investments have a rapid payback.

- n *Conserving energy:* Conservation refers to going without or using less of an energy service, and therefore saving on the energy that would be needed to provide it. Examples include switching off the light when leaving a room or walking and cycling instead of driving. When fossil fuels are used to provide the service, directly or indirectly, conserving energy lowers emissions.

- n *Switching to less carbon-intensive fossil fuels:* Coal emits almost 75% more carbon per unit of energy contained in the fuel than natural gas and about one-third more than oil. So switching from coal to oil or gas and from oil to gas reduces emissions per unit of energy consumed.

- n *Switching to "zero-carbon" energy sources:* Renewable energy sources, produced in a sustainable manner, and nuclear power produce no CO_2 emissions in operation, even though the construction of the plant and the production and processing of fuel can be a source of emissions.

- n *Capturing and storing CO_2 emissions:* Technology already exists to capture and store the CO_2 emitted when fossil fuels are burned, either before or after combustion. The CO_2 can be stored in geological formations such as depleted oil and gas fields, unmineable coal beds, salt cavities or saline aquifers.

All these approaches to reducing emissions can also contribute to energy security, insofar as they lead to less reliance on imported energy that may be vulnerable to disruption or to more supply diversity. For example, improving the efficiency of cars by cutting fuel consumption reduces emissions and the need to import oil. But, in some cases, there are trade-offs: switching from coal to gas in power generation will cut emissions but may increase reliance on imported gas. Likewise, encouraging the development of indigenous hydrocarbons might cut imports but, by contributing to higher global oil supply (albeit marginally), could lead to lower prices and increased overall consumption, thereby driving up emissions. Coal-to-liquids, which is under development in China, would also reduce imports, but would drive up emissions (see Spotlight in Chapter 11). Nonetheless, many of the policies and measures that governments have adopted or plan to adopt do fall into the category of "win-win".

Policies that aim to prevent or respond to short-term supply emergencies have little direct impact on greenhouse-gas emissions but are essential to energy security. Emergency-response mechanisms form a central pillar of IEA countries' short-term energy-security policies (see Box 4.1 in Chapter 4). Deepening the political dialogue with producing countries through multilateral and bilateral channels can also help to address short-term concerns over supply security. To the extent that such dialogue helps pave the way for investment in export infrastructure in producing countries, long-term security would be enhanced too, though any resulting increase in supply would, of course, have consequences for greenhouse-gas emissions.

The right mix of policies to address both climate and energy-security concerns depends on the balance of costs and benefits. The net costs and benefits – economic, environmental and social – of such policies are extremely difficult to measure, because of the difficulty of expressing them in monetary terms and the uncertainties surrounding the relationship between emissions and climate change and the eventual economic and social impact. They undoubtedly vary markedly across countries, because mitigation costs and the costs of climate change are far from uniform. Strategies for achieving a more sustainable energy system must, therefore, not only be politically feasible but also be tailored to local conditions (there are often significant hurdles to the adoption and implementation of policies that, in principle, could yield considerable net benefits). In most developing countries, including China and India, worries about the availability of energy to support economic growth and about local environmental problems – notably air and water pollution – are likely to remain the primary motivations for actions aimed at curbing fossil-energy use.

The most recent work carried out on the economics of climate change suggests that capping concentrations of CO_2 in the atmosphere, involving a progressive slowing of the growth in emissions, stabilisation and then a downturn, could be achieved at reasonable cost. The Intergovernmental Panel on Climate Change (IPCC), in its latest report on the economics of climate change, estimates that if cost-effective measures are adopted immediately, stabilisation of the CO_2-equivalent concentration of all greenhouse gases at between 445 parts per million and 535 ppm (equating to an average increase in global temperatures of between 2.0° and 2.8°C) would cost at most 5.5% of global GDP by 2050 (IPCC, 2007).[1] Stabilisation of greenhouse-gas concentration at this level would require emissions to peak by 2015 and then fall by 30% to 85% by 2050 compared with the 2000 level (see Chapter 5). Making less severe cuts in emissions could cost considerably less. Stabilisation at between 535 ppm and 590 ppm by 2050, which would cause temperatures to rise by 2.8° to 3.2°C, would cost just over 1% of GDP (median estimate). The cost of stabilising concentrations rises with the concentration level, because the marginal cost of mitigation measures increases once the cheaper options available in the early years have been exploited.

---

1. By comparison, the *Stern Review* estimates that, taking into account the full ranges of both effects and possible outcomes, climate change in a business-as-usual case would reduce welfare eventually by an amount equivalent to a reduction in consumption per head of between 5% and 20% (Stern, 2006). The *Review* estimates the annual costs of stabilisation at 500-550 ppm of CO_2-equivalent at around 1% of GDP by 2050.

The results of our Alternative Policy Scenario lend support to the IPCC estimates. The emissions savings in that scenario – which is likely to be consistent with stabilisation of greenhouse-gas concentration at around 550 ppm (see Chapter 5) – would be achieved at a net financial benefit (negative cost) to society (IEA, 2006). This is because the higher capital spending by consumers to improve energy efficiency – the main contributor to lower emissions – is more than offset by savings in their fuel bills over the lifetime of the equipment. The payback times on these investments are typically very short, especially where energy is priced efficiently: for example, about two years for commercial lighting retrofits or replacing incandescent light bulbs with compact fluorescents. These benefits are in addition to those associated with improved energy security and lower CO_2 emissions, as well as reduced local and regional air pollution.

In China, India and other developing countries, payback periods based on economically efficient prices are generally shorter than in the OECD, because there is more potential for replacing inefficient equipment. In China, for example, paybacks range from less than one year for improved industrial motor systems to just over three years for more efficient cars (see Chapter 11). They can be much longer in India, where energy prices are often heavily subsidised. The removal of electricity subsidies in India would reduce the average payback period on a range of investments in more efficient appliances and equipment by almost a fifth (see Chapter 18). In industry, where electricity is not subsidised, paybacks on more efficient motor systems are already less than one year.

A particular challenge for China and India relates to policy implementation. Policy makers recognise the need for action and have already taken high-level decisions on policies aimed at addressing concerns related to surging energy use, including goals on energy intensity and mandatory efficiency standards. But these policies have not always been translated into firm action. One reason is a shortage in both countries of skilled personnel to devise practical measures and administer their effective implementation. In addition, there are often conflicts between the goals and interests of the central, provincial, state and municipal authorities. In June 2007, in response to a failure to meet energy and pollution targets, the central government released a new plan aimed at improving the implementation at different levels of government of measures to save energy and curb emissions.

The need for all countries to curb the growth in fossil-energy demand, to increase geographic and fuel-supply diversity and to mitigate greenhouse-gas emissions is more urgent than ever. The primary scarcity facing the planet is not of natural resources or money, but of time. The projections of the Reference and High Growth Scenarios leave no doubt about the scale of the challenge. We do not have the luxury of ruling out any of the options for moving the global energy system onto a more sustainable path. The IEA has carried out a considerable amount of work on identifying cost-effective strategies for more sustainable and secure energy development in response to the G8 Gleneagles Plan of Action and on a renewed mandate from IEA Ministers in May 2007.[2] Improving energy efficiency is central to these strategies.

As most of the investment that will be needed will have to come from the private sector, it is essential that governments put in place an appropriate policy and regulatory framework – at both national and international levels. They need to plan far enough ahead to give investors as much certainty as possible about the future policy landscape, while retaining flexibility to adjust policies as required in order to meet policy aims. The most cost-effective approach is likely to include market-based instruments that place an explicit financial value on CO_2 emissions. Regulatory measures, such as standards and mandates, will also be required, together with government support for long-term research, development and commercialisation of new technologies.

The difficulties in agreeing on an equitable sharing among countries of the burden of reducing global greenhouse-gas emissions and in putting into place a harmonised international system for determining a carbon value are evident. But the cost of failure to act could be considerably greater (IPCC, 2007; Stern, 2006). Both energy security and the climate are, at least to some extent, global public goods, whose safeguarding requires a collective response. The success of that response will depend jointly on the participation of China and India and the leadership of OECD countries. This was acknowledged by the G8 leaders, meeting in Heiligendamm in June 2007, who invited China and India, as well as other major emerging economies, to join the G8 countries in taking strong and early action to tackle climate change.[3] They also called on the IEA to continue to support their national efforts to promote energy efficiency worldwide through appropriate advice and to make proposals for effective international co-operation.

Policy Co-operation with China and India

IEA countries worry that their efforts to counter the growing insecurity of energy supplies risk being eroded by the impact of China's and India's rapidly

2. See www.iea.org/G8/index.asp. IEA Ministers, at their meeting in Paris in May 2007, called on the IEA to promote the development of efficiency goals and action plans at all levels of government, making use of sector-specific benchmarking tools to bring energy efficiency to best-practice levels across the globe. They also invited the IEA to evaluate and report on energy efficiency progress in IEA member and key non-member countries, and to continue to work towards identifying truly sustainable scenarios and least-cost policy solutions for combating climate change. The press communiqué, and details of all IEA activities in these areas, can be found at www.iea.org.
3. Heiligendamm G8 Summit Declaration, page 15. Available at:
www.iea.org/G8/docs/declaration_2007.pdf.

increasing need for oil and gas. The IEA's energy supplies will only be secure if those to China and India are too, and vice versa. And there can be no effective long-term solution to the threat of climate change without bringing China and India – along with other major energy consumers – into a global agreement. Effective implementation of the IEA countries' own policies and measures aimed at addressing energy-security and climate-change concerns is essential, but far from sufficient.

Enhanced co-operation between IEA and emerging economies generally could contribute to meeting both these goals. It could also bring broader economic benefits to IEA countries by facilitating exports of advanced energy technology. Similarly, China and India would garner the benefits to economic growth of enhanced collective energy security and smaller changes in climate. China and India could also boost their exports of the innovative energy technologies that are being developed to address their domestic energy challenges.

IEA countries have long recognised the advantages of co-operation with China and India, and this is reflected in a range of co-operative activities between the IEA and China and India, together with other multilateral and bilateral agreements. China and India, in turn, recognise the benefits they can derive from such co-operation and have generally responded in a highly positive manner to IEA overtures. Both countries take very seriously the threat to their energy security and the costs of worsening local pollution caused by rising fossil-fuel use. They also recognise the long-term threat posed by climate change. The imperative to step up this co-operation will increase with the rising importance of China and India in global energy markets and the growing threats posed to energy security and the global climate. As part of its broad programme of outreach activities with non-member countries, the IEA continues to deepen its dialogue and strengthen collaborative activities with both countries (Box 6.1).[4]

Emergency preparedness is an important focus of IEA co-operation with both China and India. Both countries are developing emergency oil stocks (see Chapter 4). The IEA has made available information and has shared its experience about creating and using such stocks and intends to co-ordinate future emergency-response policies. The IEA established co-operative programmes on oil and energy security with China in 2001 and with India in 2004. Within these programmes, emergency-response simulation exercises for oil-supply disruptions were organised with participants from China, India and south-east Asian countries in 2002 and 2004. They have been invited to

4. At present, a country must be a member of the OECD before it may apply for membership of the IEA. More details about IEA's outreach activities with China and India can be found at www.iea.org/Textbase/work/2006/gb/.

Box 6.1: **Co-operative Activities Between the IEA and India and China**

Co-operation between the IEA and China is formalised in a 1996 memorandum of understanding and a 2001 agreement on a Framework for Energy Technology Co-operation. The IEA and India signed a Declaration of Co-operation in 1998. The key objectives of IEA co-operation with China and India are to contribute to the development of the energy sector in both countries. Core areas of activities are energy statistics, indicators and data management, energy security, energy efficiency and environmental issues, market reform and pricing, technology co-operation and studies of the power, coal and gas sectors.

The Agency has, over the past decade, organised with the Chinese and Indian authorities a number of technical-level meetings, seminars and workshops involving experts from member countries. Perhaps most significantly, China and India have begun to send delegations to observe selected meetings of the IEA's Governing Board and committees, which oversee the activities of the Agency. In 2007 the IEA provided training to nearly 150 energy statisticians in China on international practices. The two countries have also collaborated with the IEA on a number of in-depth studies. For example, these events are intended to promote understanding and communication on both sides on a range of energy issues. Indeed, the preparation of this *Outlook* benefited from the results of two workshops on energy prospects and policy challenges held in Beijing and New Delhi in March 2007.

participate in the next emergency-response exercise in 2008. More recently, the IEA has assisted in training Chinese and Indian officials in emergency preparedness statistics.

China's and India's oil security – like that of all consuming countries – is increasingly dependent on a well-functioning international oil market. Market transparency is a vital component. To this end, the IEA is working with the Chinese and Indian authorities to improve their oil data collection and reporting, including through the Joint Oil Data Initiative, which aims to improve the availability, quality and timeliness of monthly oil market information.[5] Neither China nor India yet reports information on the stocks held by private or national oil companies, though China has recently started to provide data on monthly changes in total stock levels. Better reporting of stocks data and improving the reliability and timeliness of oil data generally would add significantly to market transparency and predictability, bringing global benefits to consuming and producing countries alike.

5. Available at www.jodidata.org..

Co-operation is a two-way street; IEA countries have much to learn about energy developments in China and India and how these will affect their own energy markets. In return, the latter can profit from IEA experiences and best practices, notably in building institutional capacity and designing effective policies. Technology collaboration with the IEA, which has been gathering momentum in recent years, provides an additional mutually beneficial mechanism for accelerating the development and deployment of cleaner, more efficient technologies (see below). Both China and India are keen to learn from the industrialised countries about ways of curbing emissions and to take advantage of technological advances, notably in the area of end-use efficiency, renewables, clean coal in power generation and carbon capture and storage. The effectiveness of co-operation hinges on bringing the right stakeholders into the process. The success of many programmes depends on effective implementation at the provincial, state or municipal levels. In recognition of their importance, in 2006, the IEA declared China and India, together with Russia, to be the priorities of the Agency's outreach programme.

6

A number of IEA member countries have developed bilateral and multilateral mechanisms and programmes, in some cases working through the Agency, aimed at assisting and co-operating with China and India on a range of energy issues (Box 6.2). Four IEA countries – Australia, Japan, Korea and the United States – work with China and India on promoting clean, more efficient technologies through the Asia-Pacific Partnership on Clean Development and Climate, launched in 2006. China and India also co-operate on energy-related issues with non-IEA countries through regional organisations, such as the Asia-Pacific Economic Co-operation and the East Asia Summit. In 2006, China itself initiated five-party talks on energy with India, Japan, Korea and the United States. China and India, with the IEA, participate in the Renewable Energy Policy Network for the 21st Century (REN21), a global multi-stakeholder forum for promoting the development and deployment of renewable energy technology. China and the IEA are also members of the Global Bioenergy Partnership, a forum to promote research, development, demonstration and commercial production and use of biomass for energy.

The industrialised countries have obvious long-term economic and political interests – and a moral duty – in helping India deal with energy poverty (see Chapter 20). Many poor households continue to rely heavily on inefficient and polluting traditional fuels and stoves to meet their energy needs for cooking and heating, because they cannot afford modern commercial forms of energy or because it is simply not available. Developed countries, through multilateral organisations such as the IEA and bilateral co-operation, can help in many different ways, including through financial support and technical assistance (IEA, 2004 and 2006). China, which has made great strides in improving its own population's access to modern energy, has valuable experience to share with India and other developing nations.

Box 6.2: **Bilateral Co-operation between IEA Members and China and India: Three Examples**

The **United States** co-operates with China through the US-China Energy Policy Dialogue, the US-China Oil and Gas Industry Forum, the Peaceful Uses of Nuclear Technologies Agreement and the Joint Coordinating Committee on Science and Technology. The United States also launched a new bilateral US-India Energy Dialogue in 2005 aimed at identifying concrete actions that the two countries can take to help India address its energy challenges, through increased trade and investment in cleaner domestic energy production, energy efficiency and diversified imports of energy. The promotion and development of clean coal technologies and carbon sequestration in power generation remain a key focus of current United States initiatives with China and India. In 2005, India and the United States reached an agreement to co-operate on civilian nuclear technology.

The **European Union** has also established co-operative agreements with China and India on energy and climate change. The EU-China Energy and Environment Programme, which runs from 2003 to 2008, promotes sustainable use of energy in China, focusing on energy-policy reform, energy efficiency, renewables and natural gas. The EU-China Partnership on Climate Change, launched in 2005, provides a mechanism for political dialogue on concrete actions in the areas of climate change and energy. Achievements include an agreement with the United Kingdom to set up a near-zero-emissions coal-fired demonstration power plant using carbon capture and storage in China. The EU-India Action Plan, also agreed in 2005, includes an initiative on climate change, focusing on clean coal technology and clean development mechanism (CDM) projects.

Japan is strengthening energy co-operation with China and India. In April 2007, following a meeting between the prime ministers of Japan and China, deals were reached on energy co-operation, notably in the fields of energy efficiency, clean coal use, renewables and nuclear power, as well as on commercial collaboration between Chinese and Japanese companies. The Japan-China Ministerial Energy Policy Dialogue was also launched. The first Japan-China Comprehensive Energy Conservation and Environment Forum was held in May 2006 in Tokyo and a second one was held in September 2007 in Beijing. Energy co-operation with India, under the Japan-India Strategic and Global Partnership launched in 2006, embraces the promotion of energy efficiency, strengthening institutional capacity, developing CDM projects, co-operation in clean coal technologies and helping India develop its oil emergency response capability and emergency stocks. The first Japan-India Energy Dialogue was held in April 2007 in Tokyo and a second one was held in July 2007 in Delhi.

Technology Co-operation and Collaboration

The development and deployment of cleaner, more efficient energy technologies serve the common objectives of energy security, environmental protection and economic growth. Existing technologies can take us some of the way down the path towards more sustainable energy use – a central finding of the Alternative Policy Scenario. But technological breakthroughs that change profoundly the way we produce and consume energy will almost certainly be needed to achieve a truly sustainable energy system in the long term.

A portfolio approach to technology development is indispensable. Carbon capture and storage technology is one of the most promising options for mitigating emissions, with particular promise for the period beyond 2030 – especially in China and India, where coal use is growing fastest, and in the United States and other countries that will remain dependent on coal for decades to come (see Chapter 5). Advanced nuclear reactors and renewable energy technologies could pave the way for a wholesale shift away from fossil fuels in the longer term. International co-operation, including collaboration on emerging energy technologies, can make a big contribution to improving the effectiveness of public and private spending on research and development, and to facilitating the deployment of new technologies around the world. The involvement of China and India is increasingly important to the success of such co-operation.

Governments have a central role to play in pushing technology advances, by directly supporting research, development and demonstration, by encouraging private companies to invest in technological development and by facilitating the international commercialisation of new technologies. In the case of basic science, governments are normally called upon to support the entire cost of research. With technologies that are close to commercialisation, private companies are normally expected to take on much or even all of the cost. Public budgets for energy research and development in IEA countries fell heavily in the 1980s and the early 1990s and have barely recovered since (Figure 6.1); private-sector spending is also thought to have fallen sharply. Given the scale of the energy challenge facing the world, a substantial increase in public and private funding for research, development and demonstration is called for. A greater share of funding may need to be directed to the demonstration of emerging technologies, notably carbon capture and storage (CCS) and other clean coal technologies. The financial burden of supporting research efforts will continue to fall largely on IEA countries.

Although they are increasingly installing state-of-the art energy facilities, China and India continue to use supply- and demand-side technologies that are generally less advanced than those being deployed in IEA countries. This reflects differences in market conditions, including the availability of financial

Figure 6.1: **Public Energy Research and Development Funding in IEA Countries**

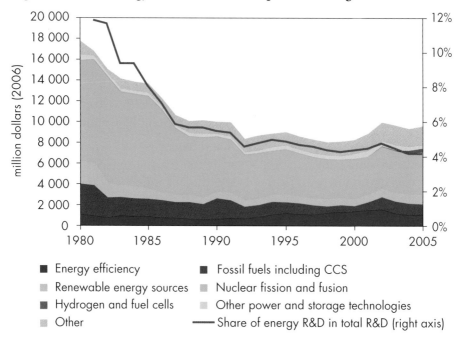

Source: IEA R&D database.

incentives and regulatory requirements, as well as the availability of the most up-to-date technologies and knowledge of how best to use them. In some cases, trade restrictions – notably import tariffs – impede market penetration. Most appliances and equipment that either use energy or are used in the production of energy are subject to tariffs in both China and India, though in both countries tariffs have lowered in recent years. In China, for example, the government requires all new wind power projects to have a 70% minimum domestic content, and levies higher import duties on pre-assembled turbines than on assembled components and individual parts. Tariffs on biofuels in both countries also impede the development of the sector and raise costs. Similarly, import barriers in other countries discourage the export of Chinese and Indian equipment, which is often cheaper and, in some cases, more advanced. Worries about protection of intellectual property rights can discourage firms from entering Chinese and Indian markets. Removing such barriers, within the framework of multilateral and unilateral negotiations over international trade and investment rules, could give a major boost to the rate of deployment of new technology worldwide, bringing mutual energy-security, environmental and economic benefits.

International collaboration on developing new technologies and improving existing ones, involving China and India and other developing countries, will continue to make an important contribution to the rapid and widespread deployment of new technologies. Technology collaboration has been a core activity of the IEA since its establishment in 1974. The IEA brings together policy makers and experts through working parties and expert groups, and provides a legal framework for international collaborative research projects, known as Implementing Agreements (Box 6.3). Initially, collaboration was mainly limited to member countries. In recent years, the participation in collaborative projects of major non-member countries, including China and India, has expanded considerably. China currently participates in four Implementing Agreements: on fusion materials, the IEA Clean Coal Centre, multi-phase flow sciences and hydropower. India participates in the agreements on Greenhouse Gases Research and Development, the Clean Coal Centre and Demand-Side Management. Both countries are discussing participation in more than a dozen other agreements. At the request of the G8 and as part of the G8 Gleneagles Programme, the IEA, with the support of the World Bank and other international and non-governmental organisations, launched in 2005 a major initiative to engage the "big five" developing countries – Brazil, China, India, Mexico and South Africa – and Russia more fully in international energy technology collaboration, including the Agency's Network of Expertise in Energy Technology (NEET).

Box 6.3: **IEA Implementing Agreements and Technology Network**

IEA Implementing Agreements – a legal contract between two or more IEA countries on technology collaboration – form the core of the IEA's Technology Co-operation Programme. They bring together experts in energy technology research, development, demonstration and commercialisation around the world. There are currently 40 active agreements covering advanced, clean exploitation of fossil fuels, the optimisation of new and renewable energy, hydrogen, fusion power and the application of best practice in efficient energy end use in transport, buildings and industry. The scope, strategic plan and work plan of each agreement, which is subject to approval by the IEA Committee on Energy Research and Technology and the Governing Board, must be consistent with the IEA's Shared Goals. Non-member countries and private organisations can participate.

A network of IEA working parties and expert or *ad hoc* groups, involving several hundred people, links the centres of expertise provided by the Implementing Agreements and policy-making bodies in the energy technology field. The network provides a platform for exchanging national experience, through which delegates can learn from each other and thereby help increase the effectiveness of national and international approaches to accelerating development and market penetration of promising technologies.

Several international collaborative activities involving China and India take place outside the framework of IEA Implementing Agreements. For example, both countries participate in the $1-billion FutureGen project being developed by the US Department of Energy. FutureGen aims to create a zero-emissions coal-fired power plant that will produce hydrogen, capture the carbon dioxide emitted and store it underground. The successful deployment of this technology could transform future energy supply, making it possible to use coal resources in a more environmentally sustainable way and lowering reliance on less secure hydrocarbons for power generation, transport and other end uses. China has established a complementary domestic programme called GreenGen, led by the Huaneng Group. India and China also collaborate with the United States and the European Union on other coal-based power generation technologies, including integrated gasification combined-cycle and CCS.

Role of the Clean Development Mechanism

The CDM provides an increasingly important conduit for investment in more sustainable energy technology in China and India. The CDM is a flexible mechanism that can be used to help countries meet their commitments to limit their greenhouse-gas emissions under the Kyoto Protocol. The CDM provides incentives for Annex I countries with greenhouse-gas emission commitments to undertake projects to reduce emissions in developing countries, including China and India, which do not have such commitments. The resulting certified emissions reductions can be used in Annex I countries to assist in complying with their national targets.[6] The marginal emission-abatement cost in developing countries is often much lower than in Annex I countries, so the CDM can help the latter reduce the overall cost of meeting their commitments. The mechanism also provides a means of transferring advanced technology and/or resources to developing countries, with positive knock-on effects for the more widespread deployment of cleaner, more efficient technologies in the longer term.

China and, to a lesser extent, India have emerged as major recipients of CDM investments. China initially lagged behind Latin America in developing CDM policies and institutional arrangements because of concerns about the integrity of the Kyoto Protocol and the CDM. However, CDM activity in China has recently grown extremely quickly and the country has emerged as the dominant recipient of CDM-related investments (see Box 9.6 in Chapter 9). Together, China and India account for around two-thirds of all the projected emissions credits for CDM projects for the period through to 2012 that have

6. A more detailed description of CDM can be found at
http://unfccc.int/kyoto_protocol/mechanisms/items/1673.php

been registered, are being validated or have been submitted for validation as of August 2007 (Table 6.1). China alone accounts for more than half of these credits. Of China's credits, about half come from energy-related projects;[7] this share is around two-thirds in India. Of the cumulative energy-related credits to 2012, about one-quarter are from renewable energy projects in both countries. The main buyers of credits worldwide are industrial companies and power generators in the European Union, covered by the EU Emissions Trading Scheme, and in Japan, where there are voluntary agreements on greenhouse-gas emissions reductions.

China has made the CDM an important component of its strategy to make its economic development more environmentally sustainable. It has devoted considerable effort to developing, at national, local and enterprise levels, expertise in identifying and designing CDM projects. It has developed clear institutional structures and implementation strategies aimed at streamlining CDM procedures. A law on "Measures for Operation and Management of Clean Development Mechanism Projects" has been adopted, setting out priorities for CDM investment – energy efficiency, renewables and methane recovery and utilisation. The law also establishes general provisions, licensing requirements and institutional arrangements for project management and implementation. It stipulates that only majority-owned Chinese companies are eligible to participate in CDM projects, which may limit inward investment.

The long-term prospects for CDM hinge on decisions made by the Parties to the UN Framework Convention on Climate Change over greenhouse-gas emissions reductions after 2012 – the end of the commitment period for Annex I countries under the Kyoto Protocol. In practice, demand for CDM credits will depend on the stringency of emission commitments in Annex I countries. CDM will remain important in establishing price signals for least-cost reductions and in enhancing the institutional capacity to estimate, monitor and review the potential for such reductions in developing countries. The potential for CDM could expand significantly, with the possibility of credits being traded internationally were different emissions-trading schemes to be established. Question marks over validation and certification procedures will need to be addressed to ensure that the costs of projects are minimised, that certified emissions reductions would not otherwise occur and that perverse incentives do not arise to augment emissions so as to benefit from sales of emission credits.

---

7. Almost 40% of China's credits and one-quarter of India's come from projects to reduce emissions of hydrofluorocarbon-23 (HFC-23) – a by-product of making hydrochlorofluorocarbon-22 (HCFC-22), substance largely used for refrigeration, which is both a powerful greenhouse gas and an ozone-depleting agent.

Table 6.1: **Status of CDM Projects in China and India, August 2007**

	China total	Of which registered	India total	Of which registered	China and India total	Of which registered	World total	Of which registered
Number of projects	737	107	727	268	1 464	375	2 392	763
Annual emissions reductions (ktCO$_2$/yr)*	206 867	70 412	54 006	27 232	260 873	97 644	376 727	161 806
Biogas	413	0	1 000	241	1 413	241	7 209	1 214
Biomass energy	3 978	928	10 240	3 469	14 218	4 397	25 437	9 089
Coal-bed/mine methane	19 489	3 523	0	0	19 489	3 523	20 484	3 523
Energy efficiency – industry	408	0	2 180	820	2 588	820	3 243	888
Energy efficiency – autogeneration	29 478	2 533	9 043	3 448	38 521	5 981	40 093	6 087
Energy efficiency – other	177	0	1 267	82	1 444	82	2 877	492
Fossil-fuel switching	16 836	0	7 400	3 545	24 236	3 545	25 743	3 910
Hydropower	32 730	2 696	3 514	1 649	36 244	4 345	45 412	7 767
Solar energy	0	0	130	1	130	1	179	43
Wind power	13 037	4 936	5 020	1 824	18 057	6 760	22 450	9 367
Other	90 321	55 796	14 212	12 153	104 533	67 949	183 600	119 426

*In CO$_2$-equivalent terms.
Source: Based on data from the Joint Implementation Pipeline of the United Nations Environment Programme Risø Centre on Energy, Climate and Sustainable Development (available on line at www.uneprisoe.org).

PART B
**CHINA'S
ENERGY
PROSPECTS**

CHAPTER 7

POLITICAL, ECONOMIC AND DEMOGRAPHIC CONTEXT

HIGHLIGHTS

- Because of its spectacular growth over a sustained period, the Chinese economy was already the second-largest in the world in purchasing power parity (PPP) terms in 2006 and the fourth-largest at market exchange rates. It made up 15% of the global economy, up from only 3% in 1980 in PPP terms. GDP growth has averaged 9.8% per year since 1980, accelerating to around 11% in 2006 and the first half of 2007. No other large country has grown as fast for such a long period.

- China is the world's second-largest FDI recipient and third-largest trader. It is the world's largest producer (and consumer) of several commodities, including iron, steel, cement and ammonia. Economic growth has been mainly driven by industry, which now accounts for almost 50% of GDP. Market liberalisation, reform of state enterprises and accession to the WTO have made major contributions to growth.

- High investment rates and rising exports have generated a huge trade surplus and made the economy vulnerable to a downturn in world demand or higher imported raw material costs. Resource-intensive growth has also led to severe environmental degradation and is increasingly contributing to global growth in CO_2 emissions. The government is pursuing a policy of structural adjustment by curbing investment in overheated sectors, cutting energy intensity and boosting domestic consumption, and redoubling its efforts to cut pollutant emissions.

- Rapid growth has cut poverty dramatically, yet per-capita income is still around one-quarter of the OECD average in PPP terms. Income differences are most striking between rural and urban areas but are also stark between different provinces.

- With 1.31 billion people, China has the largest population in the world – a fifth of the global total. China's population is expected to increase to 1.46 billion by 2030. Chinese households will become older and smaller over that period. The retired population will more than double. Urbanisation, largely the result of rural-urban migration, will continue, aided by an easing of government restrictions on movement. China already contains eight cities with populations more than five million, and has 88 cities with between one and five million, though only 42% of the population lives in urban areas. The urban population will increase by 14 million people per year, rising from 40% to 60% of the total.

The Political Context[1]

Established in 1949, the People's Republic of China is characterised by the pre-eminence of the Communist Party of China (CPC) in all central and local state organs. With about 67 million members, the CPC is the largest political party in the world (OECD, 2005a). The highest level of state authority is the 2 985-seat legislature, the National People's Congress (NPC). Members are elected every five years, via a tiered system, to represent 31 provinces, autonomous regions and province-level cities. They have the power to enact laws, to set policy, to appoint the president and to ratify the selection of the Premier. The NPC also appoints members of the Supreme People's Court. Outside the annual session of the NPC, a 153-member Standing Committee undertakes its work. The Chinese authorities describe the political process in China as one based on consensus-building: support must come from below in order for directions from above to take effect. One platform for building consensus is the Chinese People's Political Consultative Conference. This is an advisory body, made up of members of the Communist Party and other political parties, which provides for political dialogue and consultation.

The president is the head of state and often also serves concurrently as head of the Central Committee of the CPC. The Premier is the head of government and of the highest executive body, the State Council. In addition to vice premiers and state councillors, the State Council also includes heads of ministries, who are nominated by the Premier, vetted by the NPC and appointed by the president. Ministries and other agencies are charged with implementing policy, but also have a role in deciding policy and drafting regulations and laws to be considered by the NPC. These structures are all replicated at provincial, municipal and county levels.

The CPC plays an important role throughout government. Most senior officials are members of the party and of its Central Committee, which is overseen by the Political Bureau and its Standing Committee. The party establishes the aims of, and the philosophical framework for, the work of government. Through the Party Committees, it wields considerable power over appointments to high-level positions, both in government agencies and in organisations outside the government, including public enterprises.

Regional autonomy has been strengthened considerably by reforms enacted since the early 1990s. Local governments have taken advantage of the opportunities these reforms offer to foster business activity within their spheres of influence, which has helped unleash vigorous economic expansion. The

1. Part B of this *Outlook* was prepared in close co-operation with the Energy Research Institute (ERI) in China and benefited from discussions held at the *World Energy Outlook-2007* workshop organised by the ERI in Beijing on 26 March 2007.

central government and the CPC retain powers of strategic direction over key sectors, including energy. Over the past 25 years, the private sector has grown, state enterprises have gradually become more managerially independent and authority over smaller enterprises has been devolved to local governments.

The Economic Context
Economic Structure and Growth[2]

Growth in China's real gross domestic product averaged a phenomenal 9.8% per year since 1980, accelerating to around 11% in 2006 and the first half of 2007. In purchasing power parity (PPP) terms, the Chinese economy was the second-largest in the world in 2006, with GDP of $10 trillion, or 15% of global GDP (see Box 3.1 in Chapter 3). At market exchange rates, it was the fourth-largest economy, accounting for 5.5% of world GDP – just behind Germany in 2006. China's per-capita GDP is still low, at around one-quarter of the OECD average (measured in PPP terms). Economic growth has reduced poverty dramatically by raising personal incomes. Industrialisation has been the main source of growth. In 2006, industry contributed 49% of China's GDP, an increase of four

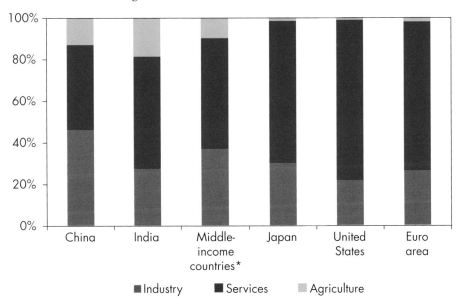

Figure 7.1: **Sectoral Share of GDP, 2004**

*An exact definition can be found at: http://go.worldbank.org/D7SN0B8YU0.
Source: World Bank (2007a).

2. The macroeconomic assumptions underpinning the Reference, Alternative Policy and High Growth Scenarios are outlined in Chapters 9, 11 and 12 respectively.

percentage points over 2001.[3] In 2004, industry's share was markedly higher than that of other middle-income, developing and OECD countries (Figure 7.1). The services sector accounts for 39% and agriculture for 12%.

The Chinese economy has become increasingly integrated into the world economy, particularly since China's accession to the World Trade Organization (WTO) in 2001. China's shares of world trade and foreign direct investment (FDI) have accelerated within the last five years. China is the world's third-largest trader, after the United States and Germany, with the sum of imports and exports reaching $1.8 trillion in 2006. China is the leading destination for FDI, with inflows amounting to $108 billion (see Figure 3.5 in Chapter 3). Its share of global steel production doubled to 31% between 2000 and 2005, while its share of world telecommunications equipment jumped by 14 percentage points to 20% (Table 7.1).

Table 7.1: **China's Importance in the World Economy**
(% of world total)

	1980	1995	2000	2005
GDP (2006 $ PPP)	3.2	9.1	11.3	14.5
GDP (market exchange rates)	2.9	2.5	3.8	5.0
Trade	0.9	2.7	3.6	6.7
Foreign direct investment*	1.0	13.0	7.0	12.0
Ammonia production	17.0	27.0	29.0	30.0
Steel production	8.2	13.0	15.5	31.2
Cement production	9.0	33.6	37.4	46.6
Telecommunications equipment	–	–	6.7	20.4

*Includes Mainland China and Hong Kong.
Sources: IEA Secretariat calculations based on IMF, CEIC, ADB, IISI, UNCTAD and WTO databases.

The increase in China's own domestic consumption and investment, relocation of many manufacturing processes from other parts of the world to China, and burgeoning exports all underpin the dramatic increase in the production of energy-intensive goods. China benefits from having cheaper labour, lower land costs and faster factory construction than most other countries. China's sheer market size also gives rise to economies of scale in the production and distribution of goods. The surge in cement production reflects a rise in the rate of fixed investment in infrastructure and real estate, which has

3. China's National Bureau of Statistics (NBS) revised GDP statistics sharply upwards in January 2006, on the basis of the findings of the 2004 Economic Census. It was the first time that a comprehensive survey of the non-agricultural economy was carried out, raising the estimate of GDP by 17%. The share of the services sector in GDP was also revised upwards.

risen by more than 20% annually over the past few years.[4] Output of services is also rising rapidly, as China seeks to move up the global value chain. In 2004, China surpassed the United States to become the world's biggest exporter of information technology goods (OECD, 2006a).

Exports and investment together accounted for about 80% of GDP in 2006. The export sector grew by 30% per annum over 2002-2006, exceeding the increase in imports. As a result, the external surplus reached $178 billion, or 7% of GDP, in 2006, boosting foreign exchange reserves to above $1.3 trillion by mid-2007 – the biggest in the world (Figure 7.2). The share of exports in GDP increased to 37% in 2006 from 20% in 2001. China has engineered a gradual appreciation of its currency, the yuan, since 2005, when a system pegging the yuan to the US dollar was replaced by one tying the yuan to a basket of currencies. The yuan had appreciated against the dollar by around 8% by mid-2007.

China's public spending, which is officially 19% of GDP, could be above 30% because of huge off-budget spending[5] by local governments (OECD, 2006b). A large part of local public spending goes to investment, especially in infrastructure, with outlays amounting to 9% of GDP in 2002. A relatively small portion of public spending – less than 4% of GDP – went to health and

Figure 7.2: **Foreign Exchange Reserve Holdings at mid-2007**

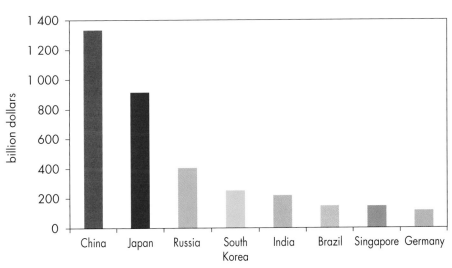

Sources: IMF and central bank websites.

4. Infrastructure and real estate investment made up 60% of urban fixed asset investment, and manufacturing 27%, in 2005.
5. Tax collection is highly centralised in China, while public expenditure is decentralised, so local governments often resort to off-budget spending to avoid deficits (OECD, 2006b).

education. However, the government has boosted public health and education budgets, particularly for rural areas. Private consumption now accounts for 37% of GDP, compared with 50%-70% in most other Asian economies, though it rose by 12% per annum on average over 2002-2006. Urban spending grew most rapidly.

Box 7.1: **China's 11ᵗʰ Five-Year Plan**

The government's 11ᵗʰ Five-Year Plan (2006-2010) aims to build a "harmonious socialist society". Growth of 7.5% per year between 2006 and 2010 is planned to prevent the overheating of the economy caused by over-investment. The government aims to cut energy use per unit of GDP by 20% and pollution by 10% by 2010. Other important goals include setting up a "new socialist countryside", reducing rural-urban inequality by spreading development gains from the coastal to the inland area and strengthening the services sector. The planned growth pattern is designed to benefit the poor and make more efficient use of natural resources. The government recognises that certain targets are proving difficult to achieve, in particular with respect to energy intensity, and is strengthening efforts to achieve these goals. The 11ᵗʰ Five-Year Plan for Energy is described in Chapter 8.

Main Drivers of Growth

Economic growth stems from increased inputs of labour and capital but also, and more importantly in the long term, from productivity gains. Productivity has benefited from the government's decision, in the late 1970s, to give freer rein to market forces, leading to the movement of labour and capital into more productive sectors and faster technological progress. China's growth has been mainly driven by rapid capital accumulation, increased productivity resulting from the steady movement of labour from agriculture into industry and services, and growth of the private sector. Growth in productivity averages 3-4% troughout the reform period (Bosworth and Collins, 2007; OECD, 2005b). In 2005, productivity growth accounts for almost two-thirds of GDP growth (Figure 7.3). Investment has contributed most of the rest.

Reforms of state enterprises were launched in the mid-1990s, with many transformed into corporations with a formal legal business structure and then listed on domestic and overseas stock exchanges. China's accession to the WTO has sharpened up competitiveness through increased competition on domestic markets and enhanced access to foreign markets. These factors have led to a more efficient allocation of labour and capital. Moreover, foreign companies investing in China have combined world-class management skills and technology with local labour to increase exports and improve the overall dynamism of the economy. Meanwhile, in response to foreign competition, domestic companies have been increasing the share of sales revenue devoted to research and development (Hu *et al.*, 2005).

Figure 7.3: **Source of Chinese Growth**

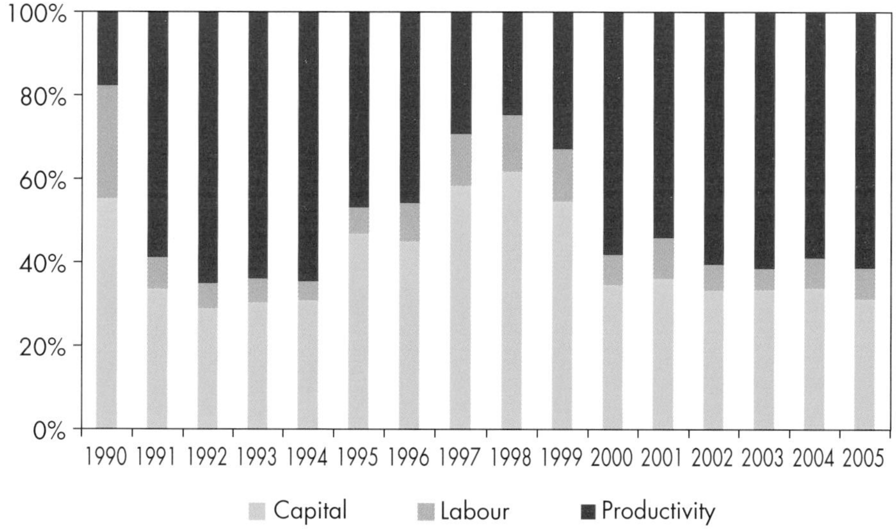

Capital ▪ Labour ▪ Productivity

Source: Aziz (2006).

Further productivity gains are on the way. The structure of China's trade is shifting to higher-value goods and services and new products are emerging. This trend will be reinforced by government efforts to increase the share of research and development spending in GDP to 2% by 2010 and by foreign companies' investment in local research. In addition, the government is placing ever greater emphasis on encouraging the private sector. A constitutional amendment in 2004 reinforced private business rights and a property law, designed to protect private property from arbitrary seizure, was adopted in March 2007. These moves, along with other measures to deregulate the economy, are expected to enhance the growth of domestic private firms. Moreover, there is still much scope for shifting around 170 million surplus labour[6] from agriculture to industry and services. Some 40% of China's labour force is still employed in farming, whereas the figure is less than 5% in most developed countries.

6. Agricultural labour productivity is 80% lower than industrial labour productivity in China (OECD, 2007a).

Capital accumulation was boosted by the introduction of the "open-door" policy in 1978, which has enabled China to attract huge FDI flows to the manufacturing export sector, and by local government investment in infrastructure. The share of fixed investment in GDP rose from 34% in 2001 to 52% in 2006. Investment, particularly in heavy manufacturing industry, is increasingly financed by the corporate savings of domestic private firms as their productivity and profitability rise (Figure 7.4).

Figure 7.4: **Share of Chinese Investment in GDP by Source**

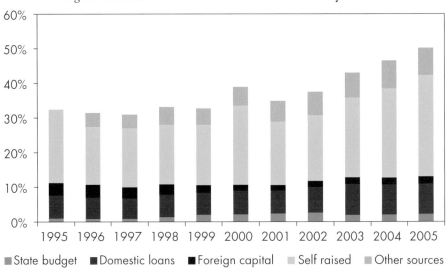

Sources: CEIC database and National Bureau of Statistics, China.

The planned adjustment of the economy towards services and more labour-intensive industry will accelerate urbanisation, which is still relatively low at 41% in 2006 (compared with an OECD average of 75%). This process will involve substantial investment in infrastructure and real estate. Continued reform of the financial sector and corporate governance is expected to improve the efficiency of capital allocation and facilitate investment.

Better education and training will also drive growth. In 1986, China introduced nine years of free education and it has made great strides in improving literacy, notably in rural areas. The government is committed to increasing education spending in rural areas and also in tertiary education. China now produces 3.1 million bachelor graduates a year and is actively encouraging the return of overseas Chinese students with good qualifications to set up their own business.

Economic Challenges and Prospects
Adjustment of Economic Structure

China's economic growth in recent years has relied predominantly on fixed investment in heavy manufacturing industry and exports of industrial goods. As a result, there is a mismatch between the rapid expansion of production capacity and the rate of expansion of domestic demand. This makes the economy vulnerable to a downturn in export demand. Moreover, capital is not always allocated efficiently. Local governments are responsible for 10% to 20% of total investment (World Bank, 2006). This is in part financed by "packaged loans", which are usually borrowed by a company owned by a local government. Until recently, the central government measured the performance of local government in terms of GDP growth, giving local officials an incentive to boost GDP via investment in infrastructure projects and in local industrial champions. The large number of non-performing loans bears testimony to the amount of capital that has been invested by local governments in uneconomic projects.

Continued heavy reliance on investment and exports to generate growth could exacerbate trade frictions, augment pressure to revalue the yuan and increase the risk of deflation. Yet a number of factors are discouraging domestic consumption. In the absence of a well-developed social security system and with rising household spending on education and health care, concerns about future costs have led Chinese citizens to increase precautionary saving, which has depressed consumption. In addition, excessive liquidity, the result of high savings and foreign speculation on a possible revaluation of the yuan, has driven up house prices, further reducing the share of household income available for consumption of other goods.[7] Moreover, surplus labour has kept the rise in wages below that in labour productivity, limiting domestic demand.

Over-dependence on exports of manufactured goods also makes China vulnerable to further increases in the cost of imported raw materials. Rising domestic energy prices, which are gradually approaching international energy prices, are starting to affect corporate investment decisions by squeezing profit margins. In 2005, energy bills accounted for more than 10% of total costs in the industrial sector, while operating margins were around 8-9%. Energy costs also make up a significant share of household spending, so a further rise in energy prices will dampen consumption.

The central government is determined to cool off an overheated investment sector and combat rising inflationary pressures. The People's Bank of China – the central bank – has raised the lending rate and bank reserve requirements

---

7. Consumption data in China do not include housing spending, which is classified as investment. This understates significantly the true level of consumption, as house purchase is an increasingly important part of household expenditure.

several times since the beginning of 2006.[8] The National Development and Reform Commission has used a range of administrative controls on overheated sectors. A decree was jointly issued by several ministries in April 2006 calling on banks to stop signing packaged loans with local governments. In addition, the central government recently provided for local environmental improvement to be a crucial indicator for assessing local government performance.

Environmental degradation is a major and growing problem that is forcing China to adjust its economic structure in favour of less resource-intensive economic activities. Most of China's electricity is produced from coal and most coal-fired plants are far dirtier than those found in OECD countries. Environmental charges are generally too low to reduce pollution significantly. Many pollutants, including nitrogen dioxide and mercury, are largely unregulated. The central government is increasingly taking environmental considerations into account in policy making, but enforcement of most resulting legislation by the provinces and local authorities, who typically focus more on economic goals, is often poor. Air pollution is estimated to cost China in the range of 3% to 7% of GDP each year (OECD, 2007b). China contains 20 of the world's 30 most polluted cities (World Bank, 2007b). Acid rain, water shortages and desertification are all pressing concerns, while global climate change – to which China is increasingly contributing – carries further economic and social costs.

Poverty and Inequality

Economic growth in China has been impressively fast, but it has been uneven – across sectors, across regions (see Chapter 13) and across sections of the population. Breakneck rates of economic growth since the late 1970s have helped to lift hundreds of millions of people out of poverty (Figure 7.5). Employment has increased roughly twice as fast as population, thanks to an increase in the proportion of the population which is of working age and an increase in the participation rate of women. A new urban middle class has sprung up in China. Although their incomes are modest by OECD standards, urban middle-class Chinese nevertheless constitute the world's biggest market for many products, from toothpaste to mobile telephones[9], and the second-biggest market (after Japan) for electronic goods and information technology.

---

8. Many of China's biggest state enterprises and private enterprises have large retained earnings, which are reinvested, dampening the impact of higher interest rates.
9. There are 420 million mobile phones in China today, a number that is increasing by around 4 million per month.

Will Government Efforts to Adjust China's Economic Structure Work?

There is a growing consensus in China and externally on the need for China to "adjust" its economic growth towards more labour-intensive industries and services and away from resource- and capital-intensive manufacturing. This would make growth in the use of energy and other natural resources less intensive, less damaging for the environment, more evenly distributed and driven more by domestic demand (World Bank, 2007c). Such a change could raise the long-term rate of growth in GDP and raise productivity. Private consumption, government consumption and exports of services and high-value manufactured goods could grow faster to offset an investment slow-down. New investment opportunities could come through sustained gains in domestic consumption, as opposed to exports.

The Chinese government is committed to adjusting the structure of the economy through a series of measures, consistent with the 11th Five-Year Plan. In 2006, the government cut income taxes, abolished agricultural taxes and introduced subsidies to grain producers in an effort to boost consumption. Other measures (detailed earlier in this chapter) aim to reduce China's propensity for over-investment. The government's 2007 Work Plan, presented at the annual session of the NPC in March 2007, includes several new initiatives. These include unification of the corporate income tax rates for both domestic and foreign businesses at a low level, a property law and big increases in government spending on education, health and rural development. The government has also introduced measures to reduce corporate and government savings and to redirect funds to households. Reform of the household registration system has begun and a labour contract law was passed in June 2007. That could lead to an increase in industrial wage rates, thereby stimulating consumption.

These policies will undoubtedly affect the pattern of growth in the long term but there are significant barriers to their implementation. Local governments in the less developed parts of the country are not always willing or able to put reforms into practice, because of the sheer scale of the administration involved and strong pressure to replicate the success of industrialisation in the coastal provinces. The central government has been developing various incentives and penalties to consolidate the heavy manufacturing sector, which has become fragmented as a result of local efforts to promote industrial champions.

Yet pockets of extreme poverty remain. Recent Chinese estimates (Shu, 2006) suggest that less than 3.7% of China's total population is currently poor but that figure is based on the official poverty line, which is meant primarily to measure the minimum income needed to meet basic needs, such as food and clothing. Using the World Bank's international poverty standard of $1 per day per person (in PPP terms), it is estimated that at the end of 2006 China had about 105 million poor people – 8% of the population (World Bank, 2007a). Using the $2-a-day standard, the number of poor in 2006 is estimated at a little over 340 million people, or 26% of the population. While poverty is still widespread in urban areas among recent migrants, the long-term challenge for China lies in eradicating rural poverty.

Figure 7.5: **Number of Poor***

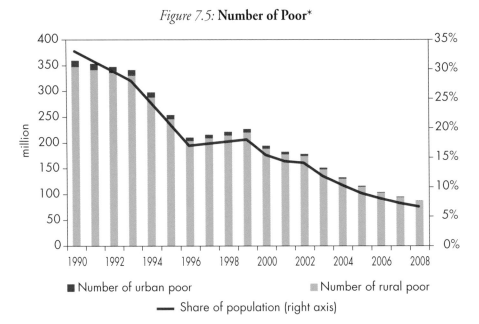

* Measured using the World Bank's "one-dollar-a-day" definition (a poverty line of $1.08/day in 1993 PPP).
Sources: IEA estimates and World Bank (2007c).

Clearly, industrialisation and urbanisation have been key drivers of poverty reduction. Urban poverty is less prevalent than in most other developing countries – even allowing for possible undercounting of recent migrants to urban areas. The registered urban unemployment rate in 2006 was 4.1%, suggesting that cities have been able to absorb workers migrating from rural areas. While the urban share of the population has grown (see next section), the urban share of the poor has decreased in recent years and currently stands at 7%. Alleviating poverty therefore hinges on rural development and the expansion of rural health insurance and social protection programmes. China's central government recognises this and is increasing support for rural areas and agriculture by 15%, to about $51 billion, in 2007.[10]

China's poor are increasingly concentrated among those less able to participate in economic activity – the elderly, the disabled and, above all, those living in remote rural areas. From 1978 to 2004, the national Gini coefficient – a measure of income inequality – increased from 0.31 to 0.47 (World Bank, 2007a). Inequality is now at a level comparable to the South American average. The richest 20% of the population accounts for over half of income in China, while the poorest 20% earns just 4% (World Bank, 2007a). Growing income inequality is compounded by geographical disparity, since cities in general and coastal provinces have progressed much faster than the average. The ratio of rural incomes to urban incomes has changed from an average of 1:2 in the 1980s to about 1:6 at present. This does not take into account the higher cost of living in urban areas but nor does it consider the better access to energy, health care, education and social support enjoyed by many urban citizens. In 2005, GDP per person in coastal areas was $2 787 while that of inland areas was just $1 229 (see Chapter 13).

Demographic Trends and Prospects

With 1.31 billion people in 2005, China has the largest population in the world, accounting for one-fifth of the global population. It is expected to retain that status until just after the end of the projection period (when it is set to be overtaken by India). China spans about the same geographic area as the United States, yet its population is well over four times greater. Population density is much higher in the coastal region than inland (Figure 7.6).

Among Chinese provinces, Henan has the biggest population, with 96 million people in 2005, followed by Shandong with 95 million people and Sichuan with 84 million people. The population of the coastal provinces totals 517 million. Historically, population growth in the interior has outstripped that of the coast, due to lower income levels, a higher degree of dependence on agriculture and a greater number of exceptions to family planning laws. Over the past decade, this has changed, as income levels rise in the interior provinces, rich coastal residents make use of the freedom to pay a fee to have more than one child and migration increases from the interior to the coast as the household registration system known as *hukou* becomes less restrictive (see below).

Our population growth assumptions for China and the rest of the world have been revised for this *Outlook*, based on the latest UN projections (UNPD, 2006). Population is projected to grow at an annual average rate of 0.4% per

10. *The Economist*, Special Survey of China, 29 March 2007.

year over the projection period, reaching 1.46 billion in 2030. The rate of growth is significantly lower than that of other developing regions. Chinese population growth has slowed considerably over the past thirty years because of a sharp fall in fertility rates – itself a result of family planning and the choices of a more urbanised population. Population grew by 1.5% per year in 1980-1990, but by only 0.9% per year in 1990-2005.

Rapid urbanisation is expected to continue, with the urban share of the population reaching 50% by 2015, before levelling off at around 60% in 2030 (the OECD average is 75%). We assume that the shares of the coastal and inland regions in total population will remain constant. As population growth slows, the average age of the population will rise and household size will decline.

Figure 7.6: **Distribution of Population and Major Cities in China**

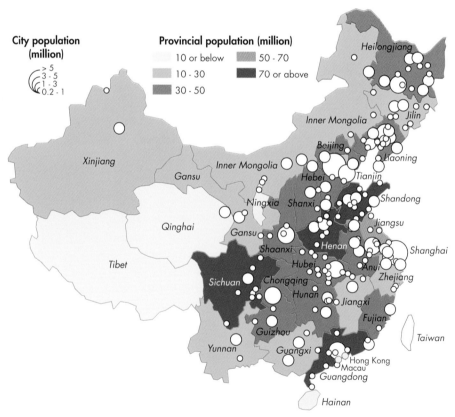

The boundaries and names shown and the designations used on maps included in this publication do not imply official endorsement or acceptance by the IEA.

Sources: CEIC (2007), National Bureau of Statistics.

Urbanisation

China has undergone rapid urbanisation in recent decades. Cities have, on average, doubled in size over the past 20 years. China has more than 660 cities[11] (UNFPA, 2007), 120 of which together contribute 75% of the country's GDP (World Bank, 2007b). Along with Shanghai and Beijing, whose populations are over 10 million, China contains eight cities with more than five million people and 88 cities with 1 to 5 million (Table 7.2). Four Chinese cities rank among the 30 most populous in the world today. Cities in the coastal provinces, notably Shanghai and Beijing, have generally grown much faster since 1980 than inland cities. Urbanisation averages 49% in the coastal provinces and 35% inland. Only a quarter of inland provinces reach the national urbanisation rate of 40%.

Table 7.2: **Populations of the Top 20 Urban Agglomerations* in 2005**
(thousands)

City	Population	City	Population
Shanghai	14 503	Chengdu, Sichuan	4 065
Beijing	10 717	Xi'an, Shaanxi	3 926
Guangzhou, Guangdong	8 425	Harbin, Heilongjiang	3 695
Shenzhen, Guangdong	7 233	Nanjing, Jiangsu	3 621
Wuhan, Hubei	7 093	Guiyang, Guizhou	3 447
Hong Kong (SAR)	7 041	Dalian, Liaoning	3 073
Tianjin	7 040	Changchun, Jilin	3 046
Chongqing	6 363	Zibo, Shandong	2 982
Shenyang, Liaoning	4 720	Kunming, Yunnan	2 837
Dongguan, Guangdong	4 320	Hangzhou, Zhejiang	2 831
Total			**110 978**

*This table uses the United Nations Statistics Division (UNSD) definition: "The *de facto* population contained within the contours of a contiguous territory inhabited at urban density levels without regard to administrative boundaries. It usually incorporates the population in a city or town plus that in the suburban areas lying outside of but being adjacent to the city boundaries."
Source: UNSD online databases (2007).

The fact that China has kept some controls on internal migration has had the advantage of allowing quite rapid urbanisation to take place without the creation of massive slums.[12] Urbanisation generates huge demand for

11. This is according to the UN definition of a city (see note to Table 7.2). According to a broader definition, for example including non-*hukou* construction and services workers, China has nearly 200 cities with more than one million inhabitants, while the population of Beijing is 15 to 17 million.
12. Because of the scale of their populations, China and India still account for 37% of the world's slums (UNFPA, 2007).

infrastructure, housing and services. Between 1949 and 1990, 300 million people were provided with urban housing. Urbanisation also affects energy use: annual energy consumption per person in cities is much higher than in rural areas, mainly because incomes are higher. China is still at a lower level of urbanisation than OECD countries.

China's urban population will grow strongly as a result of natural increase, the transformation of rural areas into urban areas[13] and continuing rural-urban migration. Migration is the most important factor (Sicular *et al.*, 2007), even more so than in other developing countries. The urban population is projected by the United Nations to grow by almost 2% per year over the period to 2030. According to UN projections, the rural population will drop to 574 million, or 40%, by 2030. Within ten years, more than half of the Chinese population will live in cities (UNFPA, 2007).

Population migration is theoretically controlled by the central government in China through the *hukou* system. Established in the 1950s, this created a two-tier system covering urban and rural areas, largely to control migration to cities. *Hukou* migrants – those who change residence with official approval – actually make up a minority of Chinese migrants. Non-*hukou* migrants tend to move from poorer areas, often in the interior, to richer coastal areas. They do not have easy access to schools, health care, and other basic services.

There is growing support for reform, as the current system is widely regarded as unfair. The government is concerned that liberalisation should not result in an uncontrolled influx to the cities, putting excessive strain on infrastructure and services (low-lying cities already face the potential adverse effects of global warming (see Box 11.1). In practice, constraints on movement are gradually being eased, as part of the process of economic reform. This, combined with a persistent income gap between rural and urban areas, will stimulate further rural-urban migration. Migration will increasingly be intra-regional and intra-provincial, as coastal mega-cities become saturated and the urbanisation drive shifts to second- and third-tier cities.

Ageing

China is experiencing a profound demographic change. The one-child policy instituted in the early 1970s, restricting most urban families to one child, has been a major contributor to Chinese economic growth. But the downside is that China might well get old before it gets rich. The current fertility rate of 1.7, the same as that of the Netherlands, is well below the replacement rate (around 2.1). At the same time, life expectancy has reached 71 for men and

---

13. This process, known as peri-urbanisation is one of the drivers of structural change in the economy, as rural workers shift from agriculture to manufacturing and services.

74 for women – levels seen in middle-income countries such as Brazil or Saudi Arabia (UNFPA, 2007).

China's age dependency ratio[14] has been rising since the mid-1980s and this trend is set to accelerate. By 2030, the share of the population above the age of 60 will more than double to 23.8%, equivalent to the level of Japan in 2000 (Figure 7.7). Most of the elderly will live in urban areas. The number of workers for every dependant (elderly person or child) is projected to fall from 2.1 at present to 2.0 by 2015 and to only 1.4 by 2030. The working-age population as a proportion of the total population is projected to peak as early as 2010 (Dunaway and Arora, 2007). Age-related illnesses, such as cancer and cardiovascular disease, will become more prevalent, requiring an increase in health spending. Rapid ageing will also increase the burden on the public pension system.

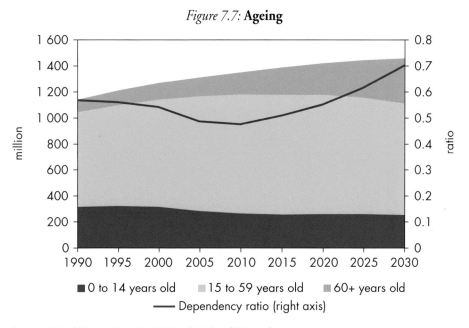

Figure 7.7: **Ageing**

■ 0 to 14 years old   ■ 15 to 59 years old   ■ 60+ years old
—— Dependency ratio (right axis)

Sources: United Nations Statistics Division (2007) and IEA analysis.

14. The ratio of those aged 0 to 14 and 60 plus to the population aged between 15 and 59.

CHAPTER 8

OVERVIEW OF THE ENERGY SECTOR

HIGHLIGHTS

- In less than a generation, China has changed from being a minor and largely self-sufficient energy consumer to become the world's second-largest and fastest-growing energy consumer and a major player on the global energy market. Demand for all forms of energy has been rising very fast in the last few years.

- The increase in China's energy demand between 2002 and 2005 was equivalent to Japan's current annual energy use. Some 105 GW of new power plants, almost all of which are coal-fired, were built in 2006 alone – a rate of increase for which there is no precedent worldwide. Growth in energy demand has been driven by booming industrial output of manufactured goods for domestic and export markets and building materials for domestic construction.

- China's energy imports have risen sharply, raising concerns about energy security. In 2006, its net oil imports reached 3.5 mb/d – the third-largest in the world after the United States and Japan. In response to growing concerns about energy security, the Chinese government is building emergency stocks, as most OECD countries have already done. Over the past two years, China also started to import natural gas in the form of LNG and, in 2007, became a net importer of coal for the first time.

- After having fallen during most of the 1980s and 1990s, energy intensity rose between 2002 and 2004, threatening the sustainability of economic expansion. Despite some successes in countering environmental damage from energy production and use, emissions of air pollutants in China remain very high and energy-related greenhouse-gas emissions are rising rapidly. Strong policy action is still needed to address these issues, aimed in particular at encouraging more efficient energy use throughout the energy system. Necessary measures include economic instruments, as well as higher standards for industry, buildings and transport.

- China already has an energy regulatory and institutional framework in place and is actively seeking ways to enhance it to meet current and future challenges. In its technical and policy approaches to expanding energy supply and improving the efficiency of transformation and use, China is taking note of developments abroad and adapting lessons to its own context. This can be seen clearly in its reforms of the gas and electricity sectors.

China's Energy Sector

In less than a generation, China has moved from being a minor and largely self-sufficient energy consumer to become the world's fastest-growing energy consumer and a major player on the global energy market. Soaring energy use is both a driver and a consequence of the remarkable growth in the country's economy – especially in heavy industry. For many years, China was able to meet its rapidly growing energy needs entirely from domestic resources, so its impact on global markets was minimal. That has changed dramatically in the last decade and national concerns about supply security have grown in parallel. Rising fossil-fuel use has worsened already acute local pollution and driven up greenhouse-gas emissions, casting doubts on the sustainability of China's pattern of development.

Coal is the backbone of China's energy system. It meets just over 60% of the country's primary energy needs, providing most of the fuel used by power stations and much of the final energy used by industry, commercial businesses and households (Figure 8.1). In fact, coal's importance in the overall fuel mix has been growing in recent years, due to booming demand for electricity, which is almost 80% coal-based. Oil demand has been growing quickly, with its share of primary demand reaching 19% in 2005. Because of the continued use by so many rural households of fuelwood and crop wastes for cooking and heating, biomass remains an important source of energy. Still, its share of primary demand is only half what it was two decades ago. Natural gas and the country's many hydropower projects constitute just 2% each. Nuclear power provides less than 1% of primary energy. Other renewables, while growing very rapidly, continue to represent a small share.

Figure 8.1: **Total Primary Energy Demand in China, 2005**

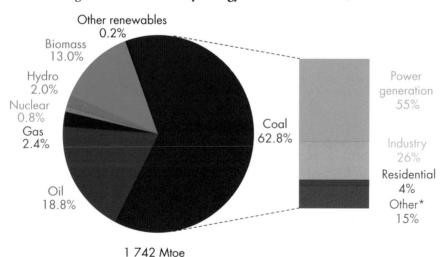

1 742 Mtoe

* Includes other energy sector, transport, services, agriculture, non-energy use and non-specified.
Note: Totals and shares are different from those shown in Chinese statistics (see Box 8.1).

In the 1980s and 1990s, energy demand grew more slowly than gross domestic product (GDP). Between 1980 and 2000, primary energy demand rose by a factor of two, while GDP increased six-fold. As a result, energy intensity – measured as primary demand per unit of GDP – improved by nearly 6% per year during this period (Table 8.1). This was in stark contrast to most other developing countries, which experienced an increase in intensity during the early stages of their development. But between 2002 and 2004, energy demand grew much faster relative to GDP, causing energy intensity to rise. This reversal was driven mainly by surging electricity demand (met largely by increased coal use) and by the manufacture of metals, building materials and chemicals for infrastructure and for consumer goods for domestic and export markets. Between 2000 and 2005, primary energy demand unexpectedly jumped by 55% while GDP increased by 57%. Indeed, in the three years to 2005 alone, energy demand rose by 44% – almost entirely as the result of coal-based electricity generation and associated thermal losses. That increase – 530 million tonnes of oil equivalent – is equal to total primary energy use of Japan in 2005. Since 2005, intensity has once again begun to fall, thanks in part to strong government action to rein in energy-demand growth.

8

Table 8.1: **Key Energy Indicators for China**

	1980	1990	2000	2005	2006**
Total primary energy demand (Mtoe)	604	874	1 121	1 742	n.a.
Oil demand (mb/d)	1.9	2.4	4.7	6.7	7.1
Coal demand (Mtce)	447	763	899	1 563	n.a.
Gas demand (bcm)	14.3	15.3	27.7	50.6	60.7
Biomass and waste demand (Mtoe)	180	201	214	227	n.a.
Electricity output (TWh)	313	650	1 387	2 544	n.a.
Primary energy demand/GDP (index, 2005=100)	342	207	101	100	n.a.
Primary energy demand per capita (toe per person)	0.6	0.8	0.9	1.3	n.a.
Oil imports (mb/d)*	–0.2	–0.4	1.4	3.1	3.5
Electricity demand per capita (kWh)	318	570	1 093	1 940	n.a.
Energy-related CO_2 emissions (Gt)	1.4	2.2	3.1	5.1	n.a.

* Negative numbers denote exports.
** Preliminary estimates based on NBS (2007a).

Power generation currently accounts for just under 40% of total primary energy use in China – a slightly higher share than that of the rest of the world. This reflects both the relatively large share of electricity in final demand and the low thermal efficiency of power stations – despite some improvement in recent years. In 2006, nearly 90% of new power-generation capacity was coal-fired, compared to 70% in 2000 (China Electricity Council, 2006).

Box 8.1: **China's Energy Statistics**

The primary source of energy statistics for China is the National Bureau of Statistics (NBS). The memorandum of understanding between China and the IEA also covers statistics. The NBS provides detailed statistics in both physical units and energy units, with a detailed consumption breakdown, in January of each year. However, although China is trying to match its definitions to the Asia-Pacific Economic Cooperation energy reporting format (very similar to the IEA's), some definitions still differ from the IEA methodology and considerable work is needed before Chinese statistics can be entered into the IEA databases.

Historically, one of the main problems with energy statistics submitted by China was the presence of a huge discrepancy between coal supply and demand, which led to enormous statistical differences. One of the reasons behind this discrepancy is the lack of proper data on stock changes (this also applies to crude oil and petroleum products). Therefore, the IEA Secretariat draws on additional sources to estimate Chinese coal production, based on demand-side statistics. There are also discrepancies in biomass, hydro and other statistics. Because of a lack of regular and proper surveys on the use of biomass, traditional biomass is excluded from primary energy use in the energy balance submitted (though in the *China Energy Statistical Yearbook* there are some biomass data for firewood, stalks and biogas, in addition to the energy balance). As for hydro, IEA statistical methodology calculates the primary energy equivalent of hydroelectricity by assuming 100% conversion efficiency. The NBS of China, on the other hand, assumes 35% conversion efficiency for hydroelectricity.

Over the past year, especially, the IEA Secretariat has worked closely with the Chinese authorities to improve China's energy statistics (see Chapter 6). For this *Outlook*, we have also made full use of external data sources – international organisations, research institutes and others. Great attention has been paid to compiling reliable energy statistics at provincial level (Box 13.1). Much higher quality and more comprehensive data have resulted. In August 2007, the NBS organised, together with the IEA, a major workshop on energy statistics. Over 130 statisticians from all the provinces attended the 10-day event – a clear indication of Chinese commitment to improve the coverage, timeliness and the quality of its energy statistics.

Industry has long been the largest final user of energy and in recent years its share has gone back up to levels seen a decade ago, reaching 42% of total final consumption in 2005. The rapid growth of industrial energy use has depressed the shares of other sectors. The residential sector is the next-largest, at 30%. Vigorous growth of transport demand – from a relatively small base – has raised its share from 5% in 1980 to 11% in 2005. The commercial sector, even though its share has increased, accounts for only 4% of final demand, the same as agriculture.

Until the 1960s, when China began exploiting its first large domestic oilfields, the country relied almost entirely on biomass and coal for the needs of its mainly rural population and its expanding industrial sector. Rising oil and coal output, like other material and financial resources, was assigned to industry, to the detriment of other sectors. By the late 1970s, China was producing almost twenty times as much commercial energy as it had in the years after the founding of the People's Republic three decades earlier. Yet two in five rural households still had no access to electricity and many homes remained unheated in winter – partly a result of policy restricting winter heating to northern districts. The launch of major economic reforms at the end of the 1970s led to a renewed surge in investment in energy-supply infrastructure, including a major expansion of power-generation capacity and the electrification of almost every community and household in China.

China is playing an increasingly important role in world energy markets. In the first six months of 2007, even as it remained a major coal exporter, China was a net importer, mainly of steam coal – to meet only a small share of domestic demand but amounting to a big addition to demand for internationally traded steam coal. China started to import natural gas in liquefied form in 2006 and volumes are set to rise gradually in the coming years as new import capacity is added. A net oil exporter until the early 1990s, China is now the world's third-largest oil importer behind the United States and Japan, though it is the sixth-largest oil producer. Of its 7.1 mb/d of oil consumption in 2006, 3.5 mb/d, or close to half, was imported. Its participation in oil and gas trade is set to grow significantly.

China's per-capita energy use remains less than 30% of the average of OECD countries, but it is higher than in most other developing regions because of the importance of heavy industry (Figure 8.2). Rising household incomes have also pushed up energy use in buildings (in line with housing space per person) and transport. The urban construction boom that has resulted in large part from rural-urban migration has been accompanied by a switch from solid fuel (biomass and coal) heating to coal-based district heating. In some areas, there has been a further move away from district heating to direct use of natural gas and oil for central heating, and electricity for air conditioning.

Figure 8.2: **Per-Capita Primary Energy Demand in China, India and Other Selected Countries, 2005**

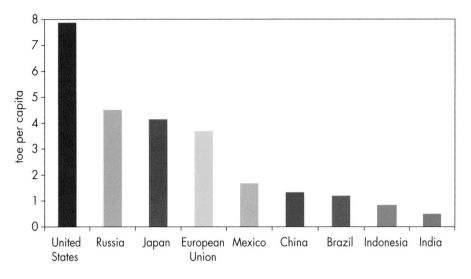

The speed of demand growth has led to some temporary mismatches between supply and demand. Higher than expected industrial growth in 2003 and 2004, combined with a surge in the numbers of air conditioners, led to shortfalls in power supply in large parts of the country, mainly in summer. In the short term, much of this gap was filled by new diesel generators, leading to an unexpected jump in oil use in 2004 – and subsequent sluggish growth as new coal-fired power stations eventually came on line, making up the shortages. The magnitude of the spurt in the construction of power plants – amounting to 105 GW in 2006 alone – was without precedent anywhere. While overall power supply and demand are evening out, some areas continue to suffer periodic imbalances and associated power cuts. Regional differences are pronounced, both in energy production and consumption (Figure 8.3). Most of China's coal is concentrated in a few inland provinces, while the largest centres of demand for coal, and for the electricity that is overwhelmingly generated by coal, are in the coastal provinces. Coal-fired power stations are increasingly built close to mines or waterways and some new rail capacity continues to be built to handle the large additional volumes of coal that need to be moved. Still, trucks and ships have met much of the incremental need for coal transport. In this way, rising coal demand has contributed to growing oil demand.

Figure 8.3: **China's Energy Production and Consumption by Province, 2005**

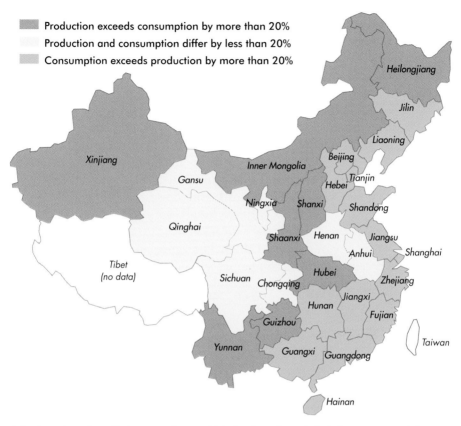

- Production exceeds consumption by more than 20%
- Production and consumption differ by less than 20%
- Consumption exceeds production by more than 20%

Heilongjiang
Jilin
Liaoning
Xinjiang
Inner Mongolia
Beijing
Gansu
Tianjin
Hebei
Ningxia
Shanxi
Shandong
Qinghai
Shaanxi
Henan
Jiangsu
Anhui
Shanghai
Tibet
(no data)
Sichuan
Chongqing
Hubei
Zhejiang
Jiangxi
Hunan
Fujian
Guizhou
Yunnan
Guangxi
Guangdong
Taiwan
Hainan

8

The boundaries and names shown and the designations used on maps included in this publication do not imply official endorsement or acceptance by the IEA.

Sources: NBS (2007b); IEA analysis

The state maintains extensive direct control over the energy sector (though some energy companies are also highly influential with the government). In line with the ownership and economic regulatory reforms since the 1980s, the central government has relinquished ownership of smaller enterprises to local governments and allowed non-state enterprises to enter certain segments of the energy industry. A prime example of this latter trend was the flourishing of small non-state coal mines, but these mines are now the subject of a closure and consolidation programme (though how quickly it is implemented is uncertain given the pressure on local governments to promote employment). Small mines have been crucial in meeting periodic upswings in demand. Shenhua Group, a state-owned company, is the leading coal company with production of 137 Mt in 2006. Large state-owned companies still account for almost half of total coal output, though 90% of coal mines in 2005 were small and belonged to towns and villages.

The oil and gas industry and the power sector are dominated by large shareholding enterprises, most now with overseas listings for their most profitable subsidiaries, but in which the government is the largest stakeholder, via the State-Owned Assets Supervision and Administration Commission. Appointments for top posts are guided by the government, but operationally these large corporations are increasingly independent. The Chinese oil and gas sectors are dominated by the big three: China National Petroleum Corporation (CNPC), China Petroleum and Chemical Corporation (Sinopec) and China National Offshore Oil Corporation (CNOOC). Together they accounted for 88% of China's oil and 94% of gas output in 2005. The three companies have opened their equity to private investment by successfully carrying out Initial Public Offerings. In the electricity sector, State Grid and China Southern Power Grid are in charge of trans-regional grids, while China Huaneng Power Corporation leads the power generation sector with 43 GW of installed capacity, or 8% of the total capacity of China in 2005. The aggregate installed capacity of the top five central power companies amounts to 36% of the nation's generating capacity, while there are also several large local companies, including Shenergy Group in the Shanghai area.

Energy Administration and Policy

Responsibility for making and implementing energy policy in China is shared among a number of different bodies at national and local levels. The State Council is the country's highest governing body. Energy policy-making responsibilities are held by the National Development and Reform Commission (NDRC) – the ministry-level agency responsible for overall management of the economy. The NDRC has policy, regulatory and administrative functions, such as making development plans and issuing project approvals. The Energy Bureau within the NDRC takes the lead in formulating energy-supply policy, while other NDRC departments have responsibilities for energy efficiency, pricing and regulation of industrial sectors. The National Energy Leading Group of the State Council, established in 2005 and headed by Premier Wen Jiabao, has authority to co-ordinate among ministries and other government agencies to achieve energy policy goals (Figure 8.4). It is supported by the Office of the National Energy Leading Group (ONELG), which has responsibility for advising on energy strategy and co-ordinating the drafting of energy legislation. ONELG is currently undertaking a broad set of studies to support the establishment of a long-term energy strategy to 2030.

Figure 8.4: **Organisation of Energy Policy Making and Administration in China**

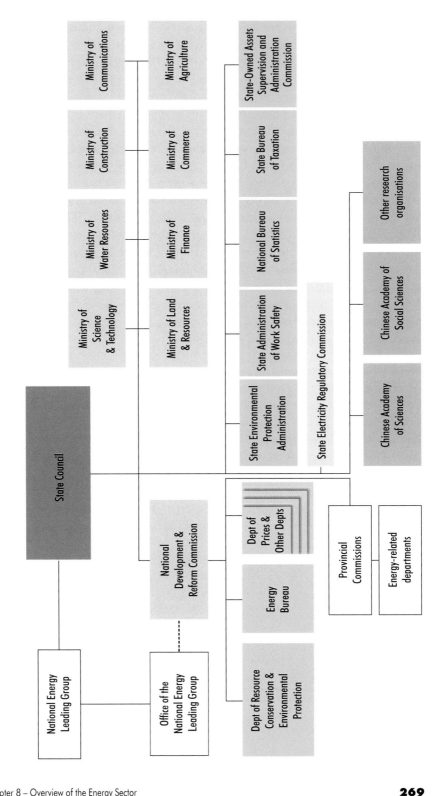

A variety of other government agencies have important roles. The State Administration of Coal Mine Safety governs mine safety. The Ministry of Land and Resources oversees the exploitation of fossil-fuel resources and reserves, except hydropower resources, which are the preserve of the Ministry of Water Resources. The Ministry of Construction oversees building construction, including the granting of building permits and enforcement of building codes. The Ministry of Communications devises plans for transport infrastructure development and proposes energy saving policy and standards for the transport sector. The Ministry of Agriculture is responsible for rural energy, including much renewable energy development. Energy trade is overseen by the Ministry of Commerce. The State Electricity Regulatory Commission plays a growing role in overseeing power-sector reform and competition. The Ministry of Finance and the State Bureau of Taxation are responsible for taxes and fees and are closely involved in any reforms that involve financial incentives. The State Environmental Protection Administration regulates environmental standards at industrial facilities; it has recently begun to exert more influence over the development of energy projects. Finally, nearly everything that is decided in Beijing must be carried out by the provincial and municipal counterparts of all these agencies. Local capacities vary widely.

The national government's Five-Year Plans are the main framework point for policies relating to energy and are an essential touchstone for justifying specific projects and actions (see Box 7.1 in the previous chapter). Energy-sector details of the 11th Plan, which covers the period 2006-2010 (NDRC, 2007a), were released by the NDRC in early 2007 (Box 8.2). In addition to specifying key supply infrastructure projects, the plan sets goals for energy efficiency, environmental protection and research and development. The NDRC and other agencies periodically issue a variety of other policy statements, as well as directives and circulars focusing on specific issues. Local governments often issue directives on energy matters as well, in line with policy set in Beijing.

In response to the rapid evolution of the domestic energy market and changing policy priorities in the current plan, the government is preparing a comprehensive new energy law, which will set out the principles that subsidiary sectoral laws should follow. It may also reorganise the powers and responsibilities of various government agencies regarding energy, possibly involving the re-establishment of a Ministry of Energy, which was abolished in 1993. The drafting and approval process could take some time. In the meantime, regulations governing detailed matters of energy-sector management requiring urgent action, such as the operation of the newly established emergency petroleum reserve, will continue to appear, as required.

Box 8.2: **Energy Goals in China's 11th Five-Year Plan**

Saving energy and expanding domestic supplies are given priority in the 11th Five-Year Plan for Energy, covering the period 2006-2010. The central goal is to reduce energy intensity – energy consumed per unit of GDP produced – by 20% in 2010 compared to 2005. That national target has been broken down into provincial targets imposed as part of performance targets for provincial leaders. Preliminary data for 2006 show a slight fall in energy intensity, reversing the upward trend of the recent past, but not fast enough to meet the 2010 target. Other objectives are diversifying energy resources, protecting the environment, enhancing international co-operation and ensuring a stable supply of affordable and clean energy in support of sustainable economic and social development. The plan also maps out the institutional and legislative changes needed to realise these goals, including pricing reforms.

For all energy sectors, the plan establishes specific goals. These include target shares for each major fuel in the primary energy mix in 2010: 66.1% for coal, 20.5% for oil, 5.3% for natural gas, 0.9% for nuclear power, 6.8% for hydropower and 0.4% for other renewables. Keys to expanding energy supply are identified, including which coal-mining areas and transport infrastructure to prioritise, which river systems to develop and how many small wind turbines to install in villages. The plan also sets efficiency targets for power generation and industrial processes (see below). The number of small coal mines is to be cut from 22 000 to around 10 000, and early indications are that implementation may be more robust than in previous closure campaigns. Priorities for technology development are set, ranging from those technologies already beginning to enter markets, such as better methods of coal washing and new nuclear reactors, to more speculative options, including hydrogen systems and exploitation of gas hydrates. Target shares outlined above are not directly comparable with projected *WEO* fuel shares, because of methodological differences (see Box 8.1).

8

Energy Policy[1] Challenges and Uncertainties

The energy-related issues that China faces are familiar the world over, although the magnitude and the speed of change the country is experiencing are unprecedented. As in any country, China's energy-policy challenges are largely framed by national socio-economic policy goals. Paramount among them is the need to sustain rapid economic development and growth in output, but in a

1. The policies mentioned in this section are not meant to be exhaustive. More detailed sectoral policies can be found in Chapter 11 as well as in the *World Energy Outlook Policy Database* at www.worldenergyoutlook.org.

way that is more equitable, more environmentally sustainable and, therefore, less energy-intensive than has so far been the case.

Chinese leaders have long recognised the importance of reducing energy intensity. Technical efficiency improvements have been sought through many channels, but big changes in energy use relative to economic output may require major changes in the structure of the economy. Despite rapid transformations within sectors, the share of industry, and especially of heavy industrial activities such as iron, steel, cement and, increasingly, aluminium production, has not diminished. In the meantime, migration to cities, the growth of rural centres into new cities and gradually rising incomes in rural areas have led to much greater household energy use. Rising aspirations, a policy of further urbanisation and greater personal mobility are setting the foundation for even higher demand in the future. The major energy policy challenges and uncertainties which arise and the potential impact on energy trends are described below. Our energy demand and supply projections are set out in detail in Chapters 9 to 13.

Security of Energy Supply

China is generously endowed with energy resources, particularly coal and renewables like hydropower, wind, biomass and solar. However, while China possesses significant oil and gas resources and continues to enjoy some success in exploration within its territory, its reserves are insufficient to match the projected growth in demand through to 2030. The prospect of a continuing rise in oil imports has led the government to seek to curb the growth in demand, through measures to promote energy efficiency and conservation (see below), to diversify geographic supply sources and to secure preferential access to foreign resources. Measures to use energy more efficiently will also contribute to meeting environmental goals. The government has encouraged its national oil companies to invest in developing oil resources abroad – a policy referred to as "going out". Today, those companies have investments in several countries, including Sudan, Kazakhstan, Nigeria and Angola, though the effectiveness of this means of enhancing energy security is open to debate – see, particularly, Chapters 4 and 10.

In addition, the government, with the encouragement of the IEA and its member countries, is enhancing its ability to respond effectively to a major supply disruption, notably through the development of an emergency oil reserve. China has begun filling the first phase of the reserve, though how the system will be run, including the roles of both government-owned and commercial stocks, has yet to be decided. Recent statements indicate that the government intends to have reserves equivalent to 30 days of net imports by 2010 – an ambitious goal, particularly if oil prices remain at current, let alone higher, levels.

China is becoming increasingly dependent on oil imports, so the attraction of technologies that can convert coal into various liquid fuels is clear. The technology is available and the economics have become more favourable in recent years, due to surging oil prices. Some such fuels can be used directly as transport fuel, while others ease oil demand by substituting for petrochemical feedstock. For example, coal gasification can replace oil gasification to produce syngas for fertilizer manufacturing, while methanol is a basic building block for plastics, paints and construction materials. Methanol can also be blended with gasoline or used to produce dimethyl ether (DME), which in China is being promoted for blending with LPG. Integrated coal gasification and methanol and DME production seem particularly promising.

In 2006, NDRC announced a plan to invest more than $128 billion in the development of alternative coal-based synthetic fuels and chemical feedstock to ease oil import dependence. Seven production centres have been identified, notably Xinjiang for coal-to-liquids and eastern Inner Mongolia for methanol. About 30 coal-based projects are currently at the planning or construction stages in China.

Coal-to-liquid projects are attracting particular attention. Shenhua Group is building the first direct coal liquefaction plant of 60 000 b/d, with a planned investment of $3 billion. The company is also planning at least two additional projects of roughly the same size. There are about twenty coal-to-oil projects under construction or under consideration, involving total investment of $15 billion and estimated capacity of 16 million tonnes of oil.

The government has recently become more cautious about coal-to-chemical and coal-to-liquid plants because they consume, and often pollute, large quantities of water (see Spotlight in Chapter 11). Also, the CO_2 implications will be onerous unless take-up of CO_2 capture and storage technology takes place. Alternatives to coal-based fuels include bioethanol, biodiesel and hydrogen. However, those options have drawbacks of their own. Although there are still barriers to be overcome, prospects for widespread commercialisation of alternative coal-based fuels over the projection period are good, particularly if conventional fuel prices remain high and stricter transport fuel quality standards are introduced as planned.

Environmental Issues

The Chinese government is acutely aware of the serious environmental ramifications of the country's energy system. These include pollution of the air, water and soil caused by the production, transformation, transportation and burning of fossil fuels, the loss of soil fertility and deforestation caused by overuse of biomass, the disruptions to ecosystems and communities caused by the construction and operation of hydroelectric facilities, the disposal of waste from coal and nuclear power plants, water shortages caused by over-extraction for coal production and growing emissions of greenhouse gases.

Despite improvements in fuel quality, changes in combustion and emissions control technologies and the relocation of some industrial activities, air pollution remains a major problem in China. The government has stepped up efforts to address the problem, with a degree of success in some cities. Most cities have forced factories to move to less populated areas and pushed for fuel-switching from coal to gas where possible, though growth of emissions from vehicles is offsetting emissions reductions in the industrial and household sources that have been the main contributors in the past. The State Environmental Protection Administration (SEPA) has temporarily halted some large projects for failing to comply with impact-assessment requirements, has introduced new emission standards for vehicles and has begun to set standards for total allowable emissions from stationary sources in addition to those for concentrations of pollutants in emission streams. The stringency of environmental regulation and enforcement is likely to increase. The government's overall objective of reducing the energy intensity of the economy, to the extent that it is successful, will also contribute to curbing pollution. Total emissions of major pollutants, including SO_2, are targeted to fall by 10% below the 2005 level by 2010; meeting the energy-intensity reduction target of 20% over the same period (see below) would reduce SO_2 emissions by 8.4 million tonnes.

The country's heavy reliance on coal is the primary cause of energy-related environmental degradation. In coal-producing areas, washeries and especially coking plants are major culprits. Some progress has been made in reducing particulates, SO_2, NO_x and other air, water, and solid waste discharges from industrial facilities and power plants, with the installation of pollution-control equipment and greater application of waste recovery and reuse, offsetting the impact of rising coal use. But coal-related pollution from the latter remains a big challenge. Pollution caused by road traffic is also getting worse as car ownership and freight expand. The sheer number of new vehicles and growing congestion on the ever-expanding road network threaten to overwhelm technical and regulatory progress on energy efficiency and emissions.

Diversification to other energy sources – especially nuclear power and non-hydro renewables – is a high priority, alongside better ways of using coal. But even if targets for nuclear (40 GW by 2020) and non-hydro renewables (over 60 GW by 2020) are achieved, they will represent a small fraction of total installed capacity for many years to come (NDRC, 2007b). Hydropower projects are still the largest alternative to coal and capacity of 300 GW by 2020 is planned, yet while the 22.5 GW Three Gorges Project has encouraged further development, the barriers remain considerable – high costs for dams and long-distance transmission, opposition to resettlement and ecological impacts, and complex interregional and transnational water rights issues.

China's heavy reliance on coal is the main cause of the recent escalation in CO_2 emissions, which has outpaced the growth in energy use. Local and regional impacts of energy use have long been considered more urgent than climate change and water quality issues are even more pressing than air quality. But concern about CO_2 is growing, as awareness among government agencies and the general public has spread about the probable adverse impacts of climate change on, for example, water availability and agricultural production. China set up its National Climate Change Co-ordination Office in 1998. In June 2007, it released its first national action plan to respond to climate change and constituted a National Climate Change, Energy Conservation and Emissions Reduction Leading Group, headed by the Premier (see Box 11.1). Nonetheless, China, along with other developing countries, maintains that it is the responsibility of already industrialised countries, which have higher per-capita emissions and which are overwhelmingly responsible for the current high atmospheric concentrations of greenhouse gases, to take the lead in mitigating emissions.

Energy Efficiency

Long held up as a high priority by policy makers, energy efficiency has attained even greater prominence over the past few years as the main approach to addressing both energy security and environmental concerns. The Energy Conservation Law, providing for the proper use of energy resources and the promotion of energy-saving technology, was enacted in 1998. The NDRC released a *Medium and Long Term Plan for Energy Conservation* in late 2004 (NDRC, 2004), which set out specific energy efficiency improvement targets for the industrial, transport and building sectors. It is closely related to the Ten Key Projects of Energy Conservation. These provide guidance on technological measures to achieve the 20% energy intensity reduction target between 2005 and 2010, which is a feature of the 11th Five-Year Plan for Energy. National programmes are now in place and there is very strong political will behind their implementation. In addition to changes in economic structure (see Chapter 9), the energy plan calls for major industrial equipment to reach the "international level" of the 1990s by 2010 and for the performance of appliances and motor vehicles to attain the equivalent of current "international levels" by then (Table 8.2).

Table 8.2: **Selected Targets for Improvements in Energy Efficiency in the 11th Five-Year Plan for Energy**

Indicator	Unit*	2000	2005	2010
Power generation (coal-fired, gross)	gram coal-equivalent / kWh	392	370	355
Raw steel (total)**	kgce / t	906	760	730
Raw steel ("comparable")**	kgce / t	784	700	685
Average of 10 kinds of non-ferrous metals	tce / t	4.81	4.67	4.60
Aluminium	tce / t	9.92	9.60	9.47
Copper	tce / t	4.71	4.39	4.26
Oil refining	kgoe / tonne factor	14	13	12
Ethylene	kgoe / tonne	848	700	650
Synthetic ammonia, large plants	kgce / t	1 372	1 210	1 140
Caustic soda	kgce / t	1 553	1 503	1 400
Cement	kgce / t	181	159	148
Tiles	kgce / m²	10.04	9.9	9.2
Railway freight transport	tce / Mt-km	10.41	9.65	9.4
Coal-fired industrial boilers (operational)	efficiency (%)	65		70-80
Small and medium power generation units (design)	efficiency (%)	87		90-92
Wind turbines (design)	efficiency (%)	70-80		80-85
Pumps (design)	efficiency (%)	75-80		83-87
Air compressors (design)	efficiency (%)	75		80-84
Room air conditioners	energy efficiency ratio (EER)	2.4		3.2-4.0
Refrigerators	energy efficiency index (EEI) %	80		62-50
Household cookstoves	efficiency (%)	55		60-65
Household gas water heaters	efficiency (%)	80		90-95
Average automobile fuel economy	litres / 100 km	9.5		8.2-6.7

* To convert from tce to toe, multiply the value in tce by 0.7.
** Two types of intensity are calculated for the steel intensity, one adjusting for differences between product structures of different plants (comparable) and one not (total).
Source: NDRC (2007).

One of the key initiatives is the Top 1 000 Enterprises Energy-Efficiency Programme, under which contracts and targets have been drawn up with the thousand largest companies in terms of energy intensity. The overall goal is to achieve savings of 100 Mtce compared to the expected 2010 energy consumption of these businesses. In its early stages, the programme is facing some implementation difficulties, as expertise in energy-saving opportunities in industry is developed. The Top 1 000 Enterprises are distributed around the country as shown in Figure 8.5.

Figure 8.5: **Number of "Top 1 000" Enterprises by Province**

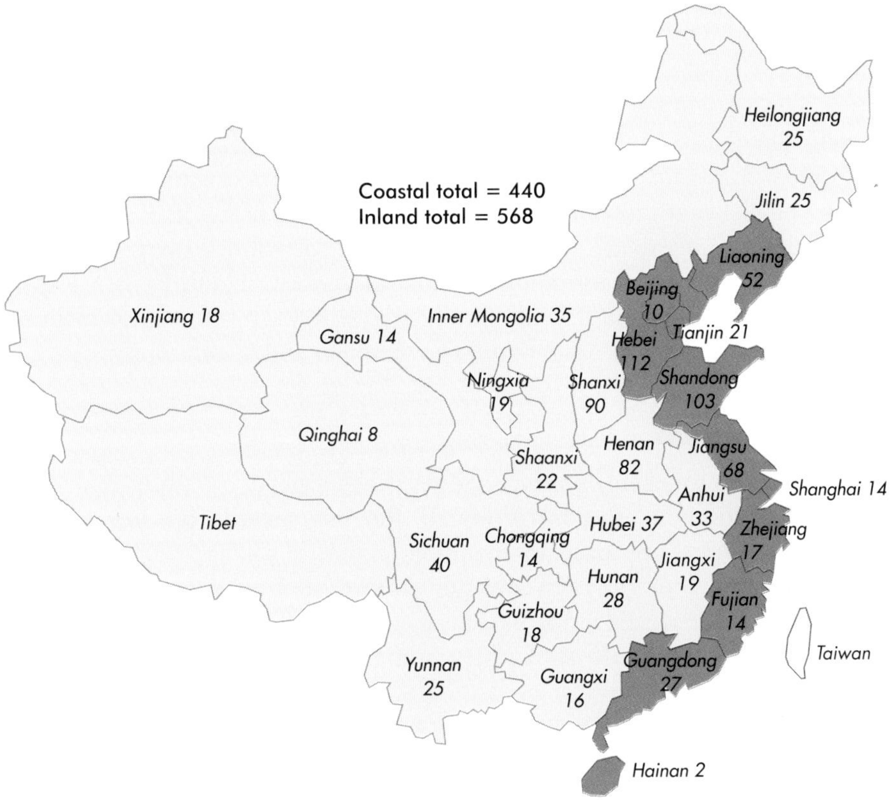

Coastal total = 440
Inland total = 568

Heilongjiang 25
Jilin 25
Liaoning 52
Beijing 10
Hebei 112
Tianjin 21
Shandong 103
Shanxi 90
Shanghai 14
Jiangsu 68
Anhui 33
Zhejiang 17
Henan 82
Shaanxi 22
Hubei 37
Jiangxi 19
Hunan 28
Fujian 14
Taiwan
Guangdong 27
Guangxi 16
Guizhou 18
Yunnan 25
Sichuan 40
Chongqing 14
Tibet
Qinghai 8
Ningxia 19
Gansu 14
Inner Mongolia 35
Xinjiang 18
Hainan 2

The boundaries and names shown and the designations used on maps included in this publication do not imply official endorsement or acceptance by the IEA.

Source: NDRC (2006).

In addition to efficiency in end uses, efficiency in energy production and transformation is also a top priority. Efficiency potential is unsurprisingly greatest in the largest energy-intensive sectors – power generation and industry. Within the industrial sector, steel, building materials (cement, brick and tile, flat glass, ceramics), chemicals and non-ferrous metals are particularly important. With huge investments in new plants over the past decade, important progress has been made in introducing more efficient equipment and processes. Yet China still hosts a broad range of technology, from among the world's best to among the worst, and much can be gained by closing the worst facilities and improving the way the remainder are used.

In August 2006, the State Council issued a statement on strengthening energy conservation work, and it circulated a major document by NDRC and other ministries in June 2007 called *Comprehensive Action Plan for Energy Saving and Emissions Reduction*, designed to ensure action at all levels of government. One area with considerable scope for improvement is the enforcement of existing energy-efficiency regulations for public procurement, energy-efficiency standards and labelling, and – most importantly – for energy codes for buildings.

Improvements in technical efficiency usually proceed very slowly and require either retrofits to existing equipment and buildings or high-quality new installations. China has had the opportunity to take advantage of surging rates of investment in recent years to install a huge amount of new equipment, raising average efficiencies. However, in markets where speed of installation has been paramount, in order to take advantage of prevailing demand, systems are still not necessarily designed with operating efficiency as a priority. Similarly, there is wide recognition that technical and administrative capacity for implementing efficiency needs strengthening, both in government and in enterprises.

Market Reforms

China's markets for energy products are at various stages of evolution, as reforms progress at different speeds. Although the government has adopted an increasingly liberal approach to economic policy, it has approached energy-market reform conservatively, on the grounds that energy is a strategic commodity. Price-setting remains a sensitive issue and subsidies remain large in some cases, albeit in an effort to achieve important policy goals. Although coal prices at the local and national levels are determined largely by market processes, access to and the cost of long-distance transport is still subject to government regulation, particularly by the Ministry of Railways. The oil market has seen gradual moves towards liberalisation, but consumers remain insulated from global markets and high regulatory barriers protect the incumbent national oil companies from international competition. Natural gas development is hindered by inefficient pricing and it takes time to develop liquefied natural gas (LNG) and long-distance pipeline gas import projects.

The freeing of coal prices without concomitant liberalisation of electricity prices (wholesale or retail) has been a point of great contention, particularly during the 2003-2004 power shortages. The price-setting mechanism for power now allows some flexibility, but many power generators are trying to protect themselves by buying upstream into coal supplies and some upstream companies such as Shenhua Coal Group have growing electricity businesses. The separation of generation from transmission and distribution businesses was completed soon after 2002. This has set one of the preconditions for the introduction of wholesale and retail competition, but much work remains in detailing how the grids will be regulated, including pricing of grid services (IEA, 2006). The issue of plant dispatch, for instance, is a thorny one. Currently, local control over dispatch often means in practice that preference goes to locally owned plants, which may be smaller and less efficient, while newer plants that are more efficient and have better pollution control may be left idle. Future decisions about feed-in tariffs for preferred generators, for example those using renewable energy sources, will affect investments and operation and influence future demand for different fuels.

Since 1990, in most years the prices of most energy products have risen faster than those for other industrial commodities. Prices have become increasingly reflective of costs and subsidies have been progressively squeezed out (Figure 8.6), although underpricing of oil is high today as prices lag the sharp rise in international prices. In general, high energy prices relative to the prices of other goods and services help to explain falling energy intensity in the 1990s. But higher prices did not prevent the rise in intensity from 2002 to 2004. The government has shown an increasing willingness to use the financial levers at its disposal to influence energy use, despite thorny issues arising from deliberately raising retail oil prices. The resource tax on coal was recently raised, albeit by a small amount, to encourage more efficient extraction from currently producing fields. A proposed tax on motor fuels, intended in part to constrain rising demand for oil imports, awaits a final decision. Gradual moves have been taken

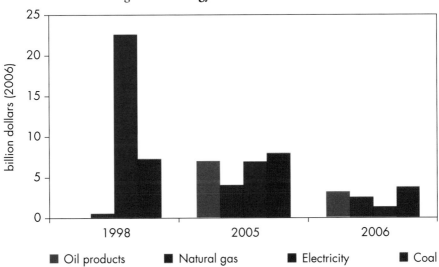

Figure 8.6: **Energy Subsidies in China***

* Energy subsidies were calculated using a price-gap approach, which compares final consumer prices with reference prices that correspond to the full cost of supply or the international market price, adjusted for the costs of transportation and distribution.
Source: IEA analysis.

to bring domestic oil prices closer to those prevailing on global markets. Rebates on exports of energy products and energy-intensive manufactured goods have been removed, while tariffs on imports of similar goods are being selectively dropped in order to help reduce domestic energy demand. Reforms have long been considered that would make electricity prices more responsive to demand. More recently, mechanisms for incorporating external costs into energy prices are under discussion. Future tax and pricing policies will help determine how quickly efficiencies are exploited along the whole chain of energy supply and use.

SPOTLIGHT

How Big are China's Energy Subsidies?

Until very recently, substantial energy subsidies remained in place in China (Figure 8.6). For example, China has long protected end-users, such as farmers who use diesel for tractors and irrigation pumps, from the effects of oil price increases by keeping domestic oil product retail prices significantly lower than those on the international market. The preferred method has been to squeeze refiners' margins through control of refinery gate prices. One effect has been to suppress the incentive to improve end-use efficiency that would flow from exposing consumers to higher market prices.

After thirty years of gradual convergence with world prices, progress has been much faster since 2005. We estimate that total consumption subsidies (net of taxes) in China amounted to around $11 billion in 2006. This is a reduction of 58% compared to 2005 (see *WEO-2006* for information on the methodology used).

In absolute terms, coal is today the most heavily subsidised form of energy, followed by oil products. Nominal subsidies to oil products and coal have fallen sharply, despite rising international prices – the result of even faster increases in wholesale and retail prices. Subsidies to oil products amounted to $3 billion in 2006. Consumption subsidies on road-transport fuels have now been largely eliminated, but some subsidies to household heating and cooking fuels remain. In percentage terms, under-pricing is biggest for natural gas and coking coal. On average, consumers pay a little over half the true economic value of the gas they use. Some subsidisation is also present in the LPG and steam coal markets.

One of the major reforms contained in the 11th Five-Year Plan concerns the system of energy pricing and taxation. It will involve a further upward adjustment of oil product and natural gas prices (along with subsidies for renewable energy sources).

Access to Modern Energy

China has alleviated energy poverty on a scale and at a pace seen nowhere else. In the early 1980s, a large fraction of the country's mainly rural population had barely enough fuel for basic cooking and heating needs and no access to electricity. Thanks to rising incomes and a rigorous policy of making modern energy services available to the entire population, all but around 10 million households now have some access to electricity; and LPG, biogas and natural gas are distributed in a growing number of towns and cities. Improved stoves are also widely available though, unfortunately, not always used as intended: frequently, improved biomass stoves that burn efficiently and vent smoke outside the home are left idle in favour of small coal briquette stoves that have the advantage of convenience but are often unvented, worsening indoor air quality.

Box 8.4: **Access to Electricity in China**

Electrification in China is a remarkable success story. China's electrification rate reached 99% in 2005. Solid fuel use for cooking and heating is still common in rural areas but such fuels are now used in conjunction with electricity and to a declining degree as incomes rise. Nevertheless, because of the country's large total population, the small fraction without connections still numbers about 10 million people. There is also uncertainty about the quality of service: a household that uses a single light bulb is considered just as connected as one that can run a full set of household appliances.

Nevertheless, the spread of electricity in China dominates the global electrification picture. Between 1990 and 2005, the number of people without electricity fell from 2 billion to 1.6 billion worldwide. Excluding China, the number of people without electricity has *grown*. To reach the relevant Millennium Development Goal, the number would need to fall to less than one billion by 2015.

Electrification rates vary greatly between countries. With similar GDP (PPP) per capita to China's, Ukraine has almost 100% electrification while Gabon has just 48%. India seems to be making good progress but, at an estimated rate of 62% in 2005, it has a long way to go. China's success has distinct features that make it difficult for other countries to replicate. First, other developing countries will have to depend more on funding from international partners than China did. Second, schemes providing subsidies and low-interest loans to households, which ensured rapid take-up in China, are difficult to implement in an efficient and equitable way. Third, China's programme benefited from the cheap cost of everything from light bulbs to hydro turbines and generators. This is rarely the case in other developing countries, which do not have China's manufacturing capacity.

Yet there may be some transferable strategies. For example, the Chinese government put in place a supportive institutional framework by establishing the electrification goal as part of its poverty alleviation campaign in the mid-1980s and maintaining an unwavering commitment to it. This facilitated mobilisation at the local level. The plan prioritised the building of basic infrastructure and the creation of local enterprises. Numerous local networks based on hydropower were gradually incorporated into the provincial grids by the Ministry of Electric Power and later by State Power. The most remote communities benefited from programmes of decentralised electrification, often with renewable energy. But the most important lesson for other developing countries and the international community may be that electrified countries reap great benefits, both in terms of economic growth and human welfare. In this respect, China stands as an example.

Although rural electrification programmes have been particularly successful (Box 8.4), the level of supply of electricity to a large proportion of the rural population is modest. The availability of clean-burning gaseous fuels is also very limited in rural areas and they are not affordable to many families. These issues are addressed in the 11th Five-Year Plan, which promotes the development of off-grid renewables, biogas and solar thermal technologies, among others, supported by technical outreach, grants and credit facilities. For cooking and hot water, large integrated biogas programmes have been carried out through the rural energy centres of the Ministry of Agriculture. Nevertheless, many millions of rural households will continue to rely on traditional biomass to meet much of their energy needs, even as China's cities swell with citizens whose patterns of consumption are converging with those of their counterparts in already developed countries.

CHAPTER 9

REFERENCE SCENARIO DEMAND PROJECTIONS

HIGHLIGHTS

- China's primary energy demand is projected to climb from 1 742 Mtoe in 2005 to 3 819 Mtoe in 2030 in the Reference Scenario. China becomes the world's largest energy consumer after 2010, overtaking the United States. China's energy demand grows strongly, at an average rate of 6.6% per year through to 2010, driven, particularly, by heavy industry. In the longer term, demand slows, as the economy matures, the structure of output shifts towards less energy-intensive activities and more energy-efficient technologies are introduced. Growth over the whole projection period averages 3.2% per year.

- Energy used to produce goods for export – equipment, textiles and chemicals – accounts for as much as one-quarter of Chinese energy demand. This share has increased since 2001, as exports have surged. The trend reverses in the medium term, as an increasing amount of fuel goes towards meeting domestic demand for such energy services as personal mobility and heating.

- Successful implementation of policies aimed at shifting the structure of the economy towards lighter industries and improving energy efficiency slows the pace of industrial energy use after 2015. The fuel mix in industrial energy consumption changes radically, shifting from coal to electricity. Oil demand for transport almost quadruples over 2005-2030, making up more than two-thirds of incremental Chinese oil demand. The vehicle fleet grows by some 230 million between 2006 and 2030, to reach almost 270 million. Fuel economy regulations, adopted in 2006, nonetheless temper growth in oil demand.

- China is experiencing an unprecedented construction boom, with 2 billion m² of new building space added every year. Per-capita residential and commercial energy consumption grows by nearly 40% by 2030. Rising income and urbanisation underpin strong growth in housing and use of appliances. On average, 14 million people migrate to cities each year, driving up demand for energy for space heating and cooling. Electricity and natural gas are expected steadily to replace biomass and coal in the residential sector.

- The projected rise in energy demand has major implications for the local and global environment. China's SO_2 emissions are projected to increase from 26 Mt in 2005 to 31 Mt in 2015, before levelling off to 30 Mt by 2030. NO_x emissions rise from 15 Mt in 2005 to 21 Mt in 2030. China's energy-related CO_2 emissions will exceed those of the United States in 2007, making it the world's largest emitter. They reach 11.4 Gt in 2030. However, China's per-capita emissions remain much lower than those of the United States in 2030, not even reaching current average OECD levels.

Key Assumptions

The energy projections under each scenario, for China as well as all other regions, rest on a number of key assumptions, especially those regarding the economy, demographic trends, energy prices and technology (see Introduction). The Reference Scenario provides a vision of how China's energy markets would be likely to evolve over time if no new government policies were to be introduced during the projection period, thus providing a series of reference points for the consideration of new policy options. Our population assumptions, which are the same for all three scenarios described in this *Outlook*, are detailed in Chapter 7. The other main assumptions for the Reference Scenario are detailed below.

Economic growth is the main driver of demand for energy services. Energy projections are, therefore, highly sensitive to underlying assumptions about GDP growth and the structure of the economy. In China, in contrast with most other countries, energy demand has not always grown predictably with GDP. For example, primary coal demand grew steadily between 1971 and 1996, but fell between 1997 and 2001 – despite continuing rapid economic growth. Coal demand growth restarted in 2002, surging in 2003 and 2004 by around 20% per year. Demand for other fuels has soared relative to GDP in recent years.

Despite higher oil prices since 2002, the Chinese economy has continued to grow very strongly. There is considerable potential for further increases in productivity and GDP, especially in the near term (see Chapter 7). The government has a stated goal of increasing GDP four-fold between 2000 and 2020 and aims at growth of 7.5% per year on average to 2010 in the 11th Five-Year Plan. Recent economic performance has surpassed the government target by a wide margin, with GDP surging by more than 10% per annum from 2002 to 2006 and exceeding 11% in the first half of 2007.

China is expected to continue to grow rapidly, provided economic reforms continue and favourable global economic conditions persist. Nevertheless, GDP growth is expected to slow gradually over the projection period as the economy matures, population levels off and the dependency ratio[1] increases. In the Reference Scenario, the Chinese economy is assumed to grow at 7.7% per year between now and 2015. Over the entire projection period, growth is assumed to average 6% per year (see Table 2 in the Introduction). The coastal region's economy is expected to continue to grow more rapidly, averaging 6.1% per year through to 2030. The share of the services sector in GDP is assumed to increase steadily over time, from 40% in 2005 to 47% in 2030, partly thanks to government policies aimed at structural adjustments to economic

---

1. The ratio of those aged 0 to 14 and 60 plus to the population aged between 15 and 59.

growth (Table 9.1). It is possible that China will manage to sustain its current high growth rate pattern for a longer period than we assume in this scenario. The implications for energy and the environment of faster economic growth are analysed in a High Growth Scenario (see Chapter 12).

Table 9.1: **Key Assumptions for China's Energy Projections in the Reference Scenario**

	1990	2005	2015	2030
Services share of GDP (%)	31	40	43	47
Population (millions)	1 141	1 311	1 387	1 457
Urbanisation (%)	26	40	49	60

Sources: IEA Secretariat; UNPD (2006).

Energy prices in China are assumed to move in line with international prices (see Introduction). China's ongoing market reforms, which are making domestic energy prices more sensitive to international price movements, are assumed to persist. Subsidies to energy were reduced significantly in 2006 and 2007, but underpricing remains (see Chapter 8). We assume that the Chinese government will gradually phase out all energy subsidies over the projection period.

The pace of technological innovation and deployment affects the cost of supplying energy and the efficiency of its use. In general, it is assumed that the end-use technologies available in China become steadily more energy-efficient. However, the pace of change varies for each fuel and each sector depending on the potential for efficiency improvements and the state of technology development and commercialisation. The rate at which technologies are actually taken up by end-users also varies, mainly as a function of how quickly the current and future stock of energy-using capital equipment is retired and replaced. How efficiency improvements are expected to reduce energy intensity – the amount of energy used per unit of output – in power generation, industry and transport is summarised in Table 9.2.

In addition to being a large and growing domestic economy, China is also an export-oriented economy, with very high ratios of trade and investment to GDP. It is also an economy which is undergoing rapid transformation, as market reforms are introduced, a new middle class emerges and consumption patterns change. Projecting trends in energy demand, therefore, involves an analysis of the changing structure of the economy, as well as traditional modelling of domestic consumer demand for goods, services and mobility.[2]

2. For this *WEO* we have greatly expanded the modelling framework for China and increased the degree of sectoral, technological and regional disaggregation.

Table 9.2: **Energy Intensity in Selected Power-Generation Technologies and End-Use Sectors in the Reference Scenario**
(Index, 2005 = 100)

			2015	2030
Power generation	Coal-fired	heat rate	95	85
	CCGT	heat rate	96	86
Industry	Iron and steel	toe per tonne	93	86
	Cement	toe per tonne	88	83
Transport	Cars	l/100km	86	68

It should also be remembered that China's energy consumption per person is still low: 1.3 tonnes of oil equivalent in 2005. This is only about three-quarters of the world average, and 28% of that of OECD countries. Because of the huge population, a small change in per-capita consumption means very large volumetric changes.

Included in this year's *World Energy Outlook* is a special analysis of current and future energy trends in the coastal region of China (see Chapter 13). The coastal region has been and will continue to be the main driver of energy use in China, so its role in China's energy future is of particular interest.

Primary Energy Demand

In the Reference Scenario, China's primary energy demand is projected to register an average annual growth rate of 5.1% between 2005 and 2015 and of 3.2% over the period 2005 to 2030 as a whole. That will take China's energy demand from 1 742 Mtoe in 2005 to 2 851 Mtoe in 2015 and 3 819 Mtoe in 2030 (Table 9.3). Without biomass, primary demand grows from 1 515 Mtoe in 2005 to 3 592 Mtoe in 2030. The growth rate of 3.2% over the entire projection period is slower than the 4.3% per year seen between 1980 and 2005. China's energy intensity continues to decline very rapidly, by 2.6% per year between 2005 and 2030. The Chinese government has adopted a number of policies to meet the target of the 11th Five-Year Plan of reducing energy intensity by 20% by 2010 compared with 2005 (see Chapter 8). While energy intensity has begun to fall, preliminary estimates for 2006 show that the intensity improvement in that year fell short of the required trajectory (Box 9.1).

Table 9.3: **China's Primary Energy Demand in the Reference Scenario** (Mtoe)

	1990	2005	2015	2030	2005-2015*	2005-2030*
Coal	534	1 094	1 869	2 399	5.5%	3.2%
Oil	116	327	543	808	5.2%	3.7%
Gas	13	42	109	199	10.0%	6.4%
Nuclear	0	14	32	67	8.8%	6.5%
Hydro	11	34	62	86	6.1%	3.8%
Biomass	200	227	225	227	–0.1%	0.0%
Other renewables	–	3	12	33	14.4%	9.9%
Total	**874**	**1 742**	**2 851**	**3 819**	**5.1%**	**3.2%**
Total excl. biomass	673	1 515	2 626	3 592	5.7%	3.5%

* Average annual rate of growth.

Box 9.1: **Recent Energy Trends in China**

In February 2007 the National Bureau of Statistics of China (NBS) released the *Statistical Communiqué of The People's Republic of China on the 2006 National Economic and Social Development*. Also, in July 2007 the NBS issued revised GDP and energy intensity figures by province. These 2006 data are useful to understand the current momentum of China's energy sector and the data are incorporated in our analysis. Highlights include the following:

- Real GDP increased strongly in 2006, by 11.1% over 2005. The growth rate of the agriculture, manufacturing and mining, and services sectors is 5%, 13% and 10.8%, respectively.
- Output of heavy industries continued to grow rapidly in 2006. Crude steel production increased by 19.7% from 2005 to 423 million tonnes, and cement production by 15.5% to 12 billion tonnes. Production of ethylene increased by 24.5% from 2005 to 9.4 million tonnes.
- Total energy consumption increased by 9.3% over 2005. Coal consumption increased by 9.6%, crude oil consumption by 7.1% and natural gas consumption by 19.9%.
- Overall energy intensity fell by 1.3% – well below the government's target of 4% per year, though reversing the previous rise in energy intensity. All the provinces, except Beijing which reduced intensity by 5.25%, fell short of this target rate of energy intensity improvement (see Chapter 13), highlighting the considerable challenges faced in implementation of supporting measures.

Coal remains the dominant fuel in China's energy mix (Figure 9.1). Coal consumption is expected to grow most rapidly in the near term, boosting coal's share of total primary energy demand by three percentage points to a peak of 66% around 2010, before falling back to 63% by 2030. The power sector remains the main coal user throughout the projection period, accounting for more than two-thirds of the incremental demand. Industry use also grows significantly, in particular in the near term, during which output of heavy industries grows rapidly. After 2010, coal use for coal-to-liquids (CTL) plants is expected to rise rapidly, reaching 72 Mtoe in 2030.

China's oil consumption increases from 6.7 mb/d in 2005 to 11.1 mb/d in 2015 and 16.5 mb/d in 2030 – an average growth of 3.7% per year. More than two-thirds of the increase comes from the transport sector whose share in total oil demand rises sharply, from 35% in 2005 to 55% in 2030. The country's oil-import dependence increases sharply, with imports increasing from 3.1 mb/d in 2005 to 13.1 mb/d in 2030. In 2000, imports were only 1.4 mb/d. China is expected to overtake Japan to become the world's second-biggest oil importer, after the United States, around 2010. It will import as much as all 27 EU member states combined in 2030.

Use of natural gas increases faster than any other fossil fuel over the *Outlook* period, at an annual rate of 6.4%; but the share of natural gas in total demand reaches only 5% in 2030 – up from 2% in 2005. In the residential and services sectors, the share of natural gas in total energy use rises from 3.3% in 2005 to 10% in 2030; in the industrial sector, its share rises from 2.6% to 4.6%; in the power generation fuel mix, the rise is from 1% to 3.6%. The National Development and Reform Commission (NDRC) released in September 2007 a new policy on the use of natural gas, giving priority to urban consumption and prohibiting new gas-fired power plants in coal-abundant areas and the use of natural gas as feedstock for methanol production (NDRC, 2007a).

The share of nuclear power in total energy demand rises from 0.8% in 2005 but, despite rapid growth, does not exceed 2% by 2030. The share of hydro slightly increases, from 2% in 2005 to 2.3% in 2030. The share of biomass drops from 13% in 2005 to 6% by 2030, reflecting a continuing shift towards greater use of modern forms of energy. The share of other renewables[3] increases steeply, but still only accounts for 0.9% of primary energy demand in 2030.

Power generation accounts for 53% of the increase in China's energy demand over the *Outlook* period. Its share of primary demand increases from 39% in 2005 to 46% in 2030. Thanks to improvements in power station efficiency, the rate of growth in power-sector energy demand (3.9%) is lower than that of final electricity demand (5.1%). Coal dominates inputs to power generation,

3. Other renewables data in 2005 were estimated from various sources as they are not available in IEA statistics.

with an 89% share in 2005. By 2030, its share declines to 84%, as use of gas, nuclear power, biomass and other non-hydro renewables rises. Oil's share declines from 3% to less than 1%; the share of hydropower also falls marginally.

Figure 9.1: **China's Primary Energy Demand in the Reference Scenario**

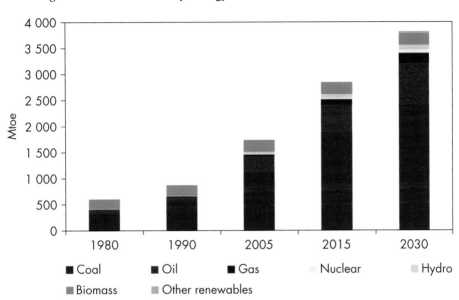

Final Energy Consumption

Total final energy consumption increases at a similar rate to primary energy demand, more than doubling between 2005 and 2030. At 3.0%, the annual rate of growth is lower than that from 1990 to 2005, when it averaged 3.5%. Total final consumption of coal grows by 4.4% per year on average between 2005 and 2015, before easing off after 2015. Most of the increase in coal use to meet final energy demand comes from industry. Final gas consumption increases nearly four-fold by 2030. Final oil demand rises by 4.0% per year, driven mainly by transport. Oil accounts for 96% of total energy for transport in 2030, an almost unchanged share. Electricity use increases three-and-a-half times between 2005 and 2030, with its share in final energy consumption rising from 15% to 26%. The use of biomass and waste declines, mainly with households switching to modern fuels. Other renewables, including wind and solar technologies, grow rapidly, but their combined share in total final energy consumption reaches only 0.7% in 2030.

In absolute terms, industry is the single biggest element in the growth in final energy demand over the projection period and remains the largest energy consumer in 2030 (Figure 9.2). However, its share in final demand excluding biomass declines from 53% in 2005 to 47% in 2030, as economic growth becomes more consumption-driven and personal mobility increases. The share of transport in final demand excluding biomass, on the other hand, rises sharply – from 13% in 2005 to 21% in 2030, while that of residential and services sectors increases from 17% to 20%.

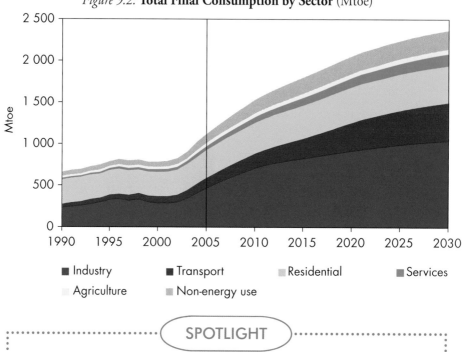

Figure 9.2: **Total Final Consumption by Sector** (Mtoe)

How Much of China's Energy Use Goes to Making Export Goods?

A significant share of the energy used in China, as an export-oriented economy, goes to manufacturing its exported goods. The energy embedded in these goods can be calculated using input-output analysis, which traces the distribution of fuels, raw materials and intermediate goods to and from industries throughout the economy (Leontief, 1936; Miller and Blair, 1985).

We estimate that the energy embedded in China's domestic production of goods for export was 452 Mtoe in 2004, or 28% of the country's total energy consumption. However, China also imported goods with an embedded energy content of 190 Mtoe in 2004, equivalent to 12% of Chinese energy demand. In 2001, the amount of energy embedded in exported goods was only 197 Mtoe, or 18% of total energy use. Some 40% of this energy in 2004 was in industrial and commercial equipment and consumer appliances, 12% in clothing and textiles, 10% in chemical, rubber and plastic products, and 9% in other manufactures. This share of energy use embedded in exports is high in China compared with other countries. For example, the comparable figures in 2001 were 6% in the United States, 7% in the European Union, 10% in Japan and 20% in Korea. The difference reflects variations in the amount and type of exports and the efficiency of energy use. On the basis of carbon intensities and trade data, we estimate that the energy-related CO_2 emissions embedded in China's domestic production for export were 1.6 Gt in 2004, or 34% of China's total emissions. These results are comparable with those of other studies (Shui and Harriss, 2006; Wixted *et al.*, 2006)

These estimates should, nonetheless, be treated with caution. It is very difficult to calculate embedded energy because of outsourcing – trade in intermediate goods, rather than final products – and the aggregation of industrial sectors. Trade is accounted for in output terms; if a component is produced in one country, shipped to another to be attached to another component, and finally assembled in a third country, the component may be counted several times in the statistics. For example, in the case of electrical machinery, a sector with particularly large economies of scale, trade in parts and components accounts for 80% of total exports from the sector. Outsourcing is almost 40% more prevalent in East Asia than in the rest of the world (Yeats, 2001; Gill and Kharas, 2007).

The share of total energy use and carbon emissions embedded in exports is expected to fall over the projection period, as the economy becomes more oriented towards the domestic market. An increasing share of equipment and manufactures is expected to be used domestically, rather than exported.

Industry

The industry sector has driven the surge in China's final energy demand over the past two decades, as output has grown strongly. In 2005, demand in the industry sector, at 478 Mtoe, accounted for 42% of total final energy consumption, which is significantly higher than the level in 1990 (36%) and the OECD average (22%).[4] This reflected national and provincial policies to boost industrial investment and production. Three sectors – iron and steel,

---

4. The projections described in this section are based on IEA statistics, which do not include fuels used for blast-furnace gas and non-energy use in the industry sector. In addition, electricity is converted differently in Chinese statistics, resulting in totals that are considerably higher than reported by the IEA.

chemicals and petrochemicals, and non-metallic minerals – accounted for two-thirds of China's industrial energy use in 2005. The energy intensity of industrial processes, measured by energy consumption per unit of value added, has improved significantly since the 1990s. Over the period 1990 to 2002, industrial value added grew at an average rate of 10.5% per year while energy consumption grew by only 2% per year. However, the downward trend in industrial energy intensity recently reversed: from 2002 to 2005, industrial value added grew by 12% per year, while energy consumption grew by 16% per year. This reflects recently surging output from energy-intensive heavy industry as well as increasing energy intensity in non-metallic minerals and other sectors.

Energy demand in industry is expected to continue to grow strongly to 2015, at 5.7% per year, with rapid growth in the output of heavy industries. Though the rate of growth of energy demand slows over time, industrial energy consumption reaches 1 046 Mtoe, accounting for 44% of final energy consumption in 2030 (Table 9.4). The deceleration in the rate of growth of industrial energy demand results primarily from a shift in the economic structure from heavy industry towards less energy-intensive lighter industry and services, as a result of policies aimed at structural adjustment of the economy (see Chapter 7). The share of the three major heavy industries – iron and steel, non-metallic minerals, and chemicals and petrochemicals – in industry value added is expected to rise from 20% in 2005 to 21% in 2015, then fall to 18% in 2030. Their share in industrial energy demand falls significantly, from 66% in 2005 to 52% in 2030.

Table 9.4: **Industrial Energy Demand in the Reference Scenario** (Mtoe)

	1990	2005	2015	2030	2005-2015*	2005-2030*
Total energy	242	478	833	1 046	5.7%	3.2%
Iron and steel	42	132	260	273	7.0%	2.9%
Non-metallic minerals	56	109	157	142	3.7%	1.1%
Chemicals and petrochemicals	38	74	119	127	4.9%	2.2%
Other	106	163	298	504	6.2%	4.6%
CO_2 emissions (Mt)	800	1 430	2 186	2 373	4.3%	2.0%

* Average annual rate of growth.

Energy intensity is expected to decline as a result of the closure of small, less efficient plants, investment in new, more efficient plants, retrofits to existing facilities, and increased recycling of scrap steel and waste material. The China Medium and Long Term Energy Conservation Plan targets a reduction in energy intensity of 9% in steel production[5] by 2020, 19% in cement, 17% in ammonia and 14% in ethylene (NDRC, 2004). We project energy intensity to decline in line with the target for iron and steel but not for non-metallic minerals or chemicals and petrochemicals, where the targets are much more ambitious. Saving energy in the chemicals sector depends on shifting towards processes based on natural gas rather than coal. The limited scope for new supplies of gas will restrict efficiency gains.

Industrial coal consumption is projected to grow by 4.7% per year to 2015 and 2.1% per year over the full period to 2030. Coal remains the dominant fuel for industry throughout the *Outlook* period, but its share in Chinese industry energy use drops from 59% in 2005 to 45% in 2030 (Figure 9.3). This share is still large compared with other countries; coal accounts for only 8% of industrial energy consumption worldwide and only 3% in the OECD. The use of electricity is expected to grow by 8.5% per year to 2015 and by 5% per year over the period 2005 to 2030. The share of electricity in total industrial energy consumption jumps from 24% in 2005 to 38% in 2030. A gradual shift from resource-intensive products towards higher value added products, as well as the shift to electric arc furnaces from blast furnaces in iron production, drives the projected shift from coal to electricity. Oil and natural gas retain relatively small shares of industrial energy consumption.

9

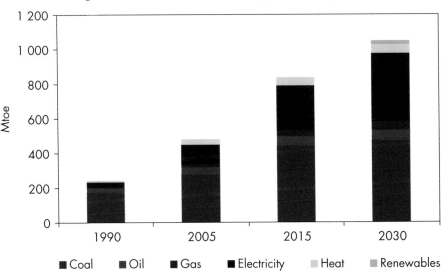

Figure 9.3: **China's Industrial Energy Demand by Fuel**

5. This energy intensity reduction is in comparable energy consumption terms; it includes energy savings from steel scrap recycling.

Iron and Steel

The iron and steel sector is the largest industrial consumer of energy in China, accounting for 28% of total industrial energy use in 2005. It grew by 14.5% per year between 2000 and 2005, while steel production grew by more than 20% per year in the period. China is currently the world's largest producer of steel, accounting for 34% of world steel production in 2006 (World Coal Institute, 2007). The average efficiency of medium and large plants in China is lower than that of plants in other countries using advanced technologies, partly because scrap steel is not available for recycling and because ore is of poor quality (IEA, 2007).[6] Moreover, less efficient small-scale blast furnaces drag the average down still further.[7] Nevertheless, current efficiency levels are far better than those prevailing in the early 1990s.

It is projected that energy demand growth for iron and steel production will continue, but slow to 7% per year from 2005 to 2015 and to 2.9% per year over the *Outlook* period. By 2030, coal continues to dominate the energy mix, at 72% or 196 Mtoe, but electricity gains seven percentage points, from 17% in 2005 to 24% or 64 Mtoe in 2030, as the use of electric arc furnaces increases. The share of iron and steel in total industrial energy consumption increases slightly in the near term but falls back to 26% by 2030.

The rapid increase of growth in energy use for iron and steel in the short term and slow-down in the longer term is due to several factors. Current sharp output growth is expected to continue to around 2010, mainly driven by booming domestic construction, but this is expected to slow down over the *Outlook* period. In addition, major energy-intensity improvements are expected. By 2030, China's energy intensity falls by 14% from today's levels, as more scrap steel becomes available (boosting recycling) and more efficient technologies are deployed (IEA 2007). The Chinese government is speeding up this process by requiring the closure of small inefficient plants. It aims to consolidate smelting companies, so that the top ten companies' production will comprise more than 50% of national output by 2010 and more than 70% by 2020 (Steel Business Briefing, 2005). There are currently 6 686 steel enterprises, 58% of which are in the coastal region.

---

6. Chinese ore contains less metal than ore in other countries (US Geological Survey, 2007). Imports of iron ore have increased dramatically in China recently, reaching a record 326 million tonnes in 2006 (CEIC, 2007).
7. Efficiency data for small-scale plants are not available. We estimate the energy efficiency gap between the average and the best plant in China to be about 20% (IEA, 2007). Coking coal used to produce coke and, as a by-product, blast-furnace gas are included in the transformation sector and not in the industrial sector.

Non-Metallic Minerals

The non-metallic minerals sector[8] is the second-largest industrial energy consumer in China, accounting for 23% of total industrial demand. Energy use by the non-metallic materials sector grew by 4.5% per year between 1990 and 2005. Although, non-metallic minerals production doubled over the last three years, production growth is expected to slow markedly, not doubling again before 2030 (NDRC, 2007b). Energy use by the sector is, accordingly, projected to slow down to 3.7% per year to 2015, then decline further. This decline is due to the stabilisation of per-capita output, in line with the slow-down of new building and infrastructure construction, and energy intensity improvements. The share of the non-metallic minerals sector in total industry demand is projected to drop to 14% by 2030. Coal still dominates the energy mix for this sub-sector, at 77%, with electricity remaining around 12% of the energy mix to 2030.

Cement accounts for a large part of energy demand in the non-metallic materials sector. China's cement production, at 46% of worldwide production, is eight times that of the second-largest producer, India (IEA, 2007). China has made significant gains in the energy efficiency of cement production, mainly through the use of clinker substitutes, granulated blast-furnace slag, fly ash and a variety of other by-products from steel production and coal power plants. Clinker production nonetheless remains comparatively inefficient (IEA, 2007). In China, clinker substitutes are both used in cement production and added directly into concrete.

Chemicals and Petrochemicals

The chemicals and petrochemicals sector produces three main types of products: ammonia, methanol and various petrochemicals, including ethylene, used to produce synthetic polymers, and is the third-largest industrial energy consumer in China. Energy use by the chemicals sector grew by 5.1% per year between 2000 and 2005. It is projected to slow to 4.9% per year from 2005 to 2015 and to 2.2% per year over the full period to 2030. In the short term, demand will be boosted by the increasing use of methanol as a fuel additive for gasoline. The later slow-down in demand reflects limits in agricultural growth, which curb ammonia needs. The chemical sector's share of industrial energy consumption drops from 15% in 2005 to 12% by 2030. Currently, coal represents an unusually high 38% of this sector's energy mix, while electricity comprises 29%, heat 18% and natural gas 8%. Coal-based production requires considerably more energy than gas-based production, which is more prevalent in the rest of the world.

---

8. Includes cement, glass, lime, brick and other ceramic building materials.

China is the world's largest producer of ammonia, which is mainly used to make fertilizer. Production in 2005 reached 44 Mt, or 30% of the world total (ADB, 2006). An estimated 70% of ammonia output is based on coal, 20% on gas and 10% on oil (Figure 9.4). This is in stark contrast with North America, where production is based solely on gas, and Western Europe, where around 90% is based on gas and 10% on oil. Coal-based processes in this sector typically use 70% more energy than gas-based processes (International Fertilizer Industry Association, 2006). China's ammonia output is forecast to increase by only 25% between 2005 and 2030, remaining largely coal-based (on the assumption that natural gas prices remain high). China's methanol production, which totalled 5 Mt in 2005, is set to grow fast with rising demand for road-fuel blendstock (China Petroleum and Chemical Industry Association, 2007). Demand for petrochemicals for synthetic polymers is also expected to grow strongly. For strategic and economic reasons, the bulk of this demand is expected to be met from domestic production, involving the use of refinery by-products, which are likely to become increasingly available as the transport sector grows, or from methanol produced from coal.

Figure 9.4: **Global Ammonia Production by Feedstock, 2005**

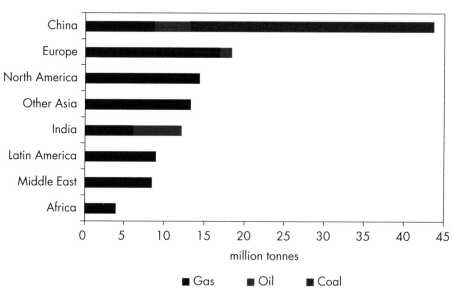

Source: International Fertilizer Industry Association, 2006.

Transport

Energy demand in the transport sector in China is expected to grow by 5.5% per year over the *Outlook* period in the Reference Scenario (Figure 9.5). Growth slows progressively from an annual average 7.6% from 1990 to 2005, to 7.0% in 2005-2015 and 4.4% in 2015-2030 (Table 9.5). Demand will be spurred by rapidly expanding passenger- and freight-vehicle fleets. The transport sector, where the potential for growth is enormous, will increasingly dominate China's oil demand. In 2030, 55% of Chinese oil use is for transport, up from 35% in 2005. Over 30% of the world's growth in transport energy demand over the projection period occurs in China.

Figure 9.5: **China's Transport Energy Demand by Mode in the Reference Scenario**

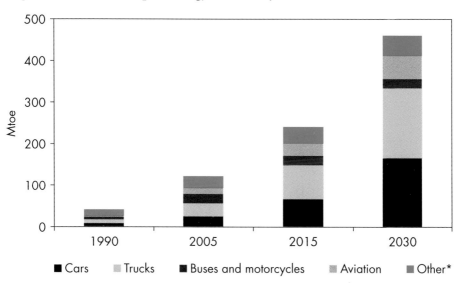

■ Cars ▨ Trucks ■ Buses and motorcycles ▨ Aviation ■ Other*

* Includes rail, pipeline transport and navigation,

Table 9.5: **Transport Energy Demand in the Reference Scenario** (Mtoe)

	1990	2005	2015	2030	2005-2015*	2005-2030*
Total energy	**41**	**121**	**240**	**460**	**7.0%**	**5.5%**
Road	22	78	170	356	8.1%	6.3%
Cars	*7*	*24*	*66*	*164*	*10.7%*	*8.0%*
Trucks	*9*	*31*	*82*	*168*	*10.0%*	*7.0%*
Other	18	43	69	104	4.8%	3.6%
CO_2 emissions (Mt)	*121*	*337*	*664*	*1 255*	*7.0%*	*5.4%*

* Average annual rate of growth.

Road

Road energy use increases more than four-fold, accounting for over 80% of the overall growth in transport energy use. Aviation, navigation and rail use make up the rest. Light duty vehicles (LDVs) – cars and sports utility vehicles (SUVs) – alone account for half the increase in road transport energy use. Gasoline represents 90% of LDV fuel demand. Trucks also see rapid growth in their energy use, almost entirely in the form of diesel.[9] In recent years, an important reason for increasing truck use has been bottlenecks in the rail capacity to transport coal. These bottlenecks are expected to be gradually eliminated. Overall, diesel use in road transport is expected to grow more quickly than gasoline use, due to the gradual phase out of gasoline in trucks and to relatively lower fuel-economy improvements in trucks compared with cars. By 2030, diesel is expected to account for around half of road oil demand, up from 37% in 2005. The share of oil in road transport fuels is expected to increase over time, as alternative fuels – mainly biofuels and natural gas – remain confined to niche markets in those provinces which enact specific policies and build the necessary infrastructure (Box 9.2). Alternative fuels represent only 3% of road transport fuel in 2030.

The vehicle stock in China has increased almost seven-fold since 1990, from 5.5 million vehicles to almost 37 million in 2006.[10] The most spectacular increase has been in cars. China surpassed Germany in 2004 and Japan in 2006 to become the second-largest car market in the world. We project the total number of vehicles on Chinese roads to increase by some 230 million over the *Outlook* period, reaching 270 million in 2030 (Table 9.6). LDVs will make up the lion's share of the new vehicles, going from 22 million to more than 200 million. China overtakes the United States as the largest car market in the world around 2015 (Figure 9.6). Growth in the high- and middle-income population, rapid infrastructure development and the emergence of cheap domestic brands are the main factors behind growth in vehicles sales over the next decade (Box 9.3). In the longer term, growth is expected to slow as demand in the major cities in the coastal area begins to reach saturation. In some cases, delays in expanding the road infrastructure may constrain vehicle ownership. As heavy industry is gradually relocated outside city centres, cars and trucks will account for an increasing proportion of local pollution.

---

9. In 2000 the Chinese government mandated that as of 2010 all trucks should run on diesel, as a measure to increase efficiency and contain demand growth.
10. These figures do not include more than 60 million two-wheelers currently in use on Chinese roads.

Box 9.2: **Prospects for Alternative Transport Fuels in China**

The Chinese government supports the use of alternative fuels, notably biofuels, compressed natural gas and coal-based liquids (see Box 8.3). In 2005, ethanol consumption was 1 billion litres (0.5 Mtoe). Production and consumption is concentrated in the few provinces that have been granted financial support by the Chinese government for production of ethanol for E10, a gasoline blend with 10% ethanol (see Chapter 10). The subsidy is estimated at around 2 000 yuan per tonne. The government has set non-mandatory targets of 10 million tonnes (6 Mtoe) for ethanol and 2 Mt for biodiesel (1.9 Mtoe) for 2020. However, we do not expect ethanol to become a major fuel for transport in the long term, because of supply constraints. Domestic production will be limited by concerns about the security of food supplies, water availability and by increasing competition from imports from Brazil. In the Reference Scenario, biofuel use remains limited, reaching only 2% of road fuel consumption in 2030.

Natural gas-powered vehicles, mainly buses, used 0.1 Mtoe of gas in 2005. China already has a fleet of more than 110 000 gas-powered buses and taxis in use in more than ten cities – exceeding the government's target of 100 000 vehicles by 2010. The small number of filling stations and limited availability of natural gas supply will be the main constraints to the further development of gas as an alternative fuel. The market is projected to reach 1.7 Mtoe in 2030.

Table 9.6: **Evolution of Key Indicators for Transport in China**

	1990	2000	2005	2015	2030
Fuel shares (%)					
Oil*	74	91	95	96	96
Biofuels	–	–	–	1	2
Other	26	9	5	3	2
Vehicles on road (million)					
Light-duty vehicles	2	9	22	81	203
Trucks	4	8	13	34	66

* Includes coal-based liquid fuels.

Figure 9.6: **New Car Sales in China**

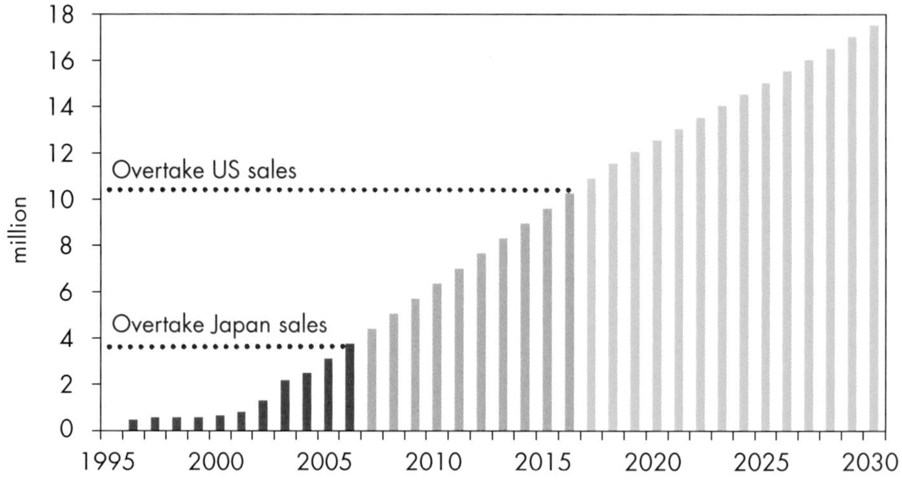

Sources: IEA analysis; Global Insights (2006).

Box 9.3: **China's Car Market**

In 2006, the passenger-car stock in China was 17 million. From 2000 to 2006, sales grew at an average rate of 37% per year, reaching 4.4 million in 2006. The Chinese market used to be skewed towards large cars, its demand originating mainly from institutions and taxis. The sales profile has been gradually changing towards smaller cars, with private car ownership becoming the main driver of growth. In addition, in April 2006, the government reformed the consumption tax with the aim of promoting sales of vehicles with small engines and discouraging ownership of larger sedans.

Domestically assembled cars account for more than 96% of car sales, imports remaining a niche luxury market. Domestic production has expanded rapidly, displacing imports. In only a few years, the Chinese car market has changed from being dominated by a few foreign brands, notably Volkswagen and General Motors, to a market with more than twenty manufacturers. They can be categorised as follows:

- *Sino-foreign joint ventures:* Foreign companies are allowed to have a maximum stake of 50% in a Chinese plant, so all foreign companies operate through joint ventures. Every major foreign car manufacturer now has a plant in China, accounting for more than 60% of car sales in 2006.

- *State-owned companies:* These include companies like First Auto Works, Dongfeng and Shanghai Automotive Industry Corporation, which rely on international joint-venture partners for technology but aim to develop their own-brand vehicles in the near future. They accounted for more than 20% of car sales in 2006.

- *New private Chinese companies:* These recently formed companies, which include Chery, Geely and Great Wall, operate without a formal foreign joint-venture partner. Their main challenge is to build up their vehicle design and development capabilities without violating intellectual property rights. They accounted for 15% of car sales in 2006.

Chinese companies have a price advantage, offering models similar to those of foreign car manufacturers, but at a price 50% to 70% lower. But Chinese companies will find it more difficult to comply with new environmental, efficiency and safety regulations, at least in the short term.

The spectacular growth in the overall vehicle stock in China in recent years masks huge differences between the provinces (Table 9.7). Vehicle ownership in Beijing, one of the richest cities, is 133 vehicles per 1 000 people – five times more than the national average and equal to that of South Korea in 2000. Hunan, the province with the lowest vehicle ownership, has only 9% of the level of vehicle ownership of Beijing. Differences in personal income, infrastructure development and urban planning explain the regional variations. Inland provinces have a higher share of trucks and tractors than coastal provinces, where cars represent two-thirds of vehicles.

Local policies also affect car ownership. Shanghai, the wealthiest region of China, has a vehicle ownership of only 52 vehicles per 1 000 people, reflecting policies that restrict the number of driving licences and promote public transport. In 2005, the average vehicle ownership in coastal provinces, at 35 vehicles per 1 000 people, was twice as high as in inland provinces (see Chapter 13). Growth in the inland provinces lags approximately eight years behind the trend of the coastal area.

Future vehicle growth will continue to be led by the 11 coastal provinces. Together, they account for 70% of the vehicle-stock increase between now and 2030, by which time ownership – at 260 vehicles per 1 000 people – will be close to that of South Korea. Vehicle ownership in the inland provinces steadily increases throughout the *Outlook* period, but they lag behind the coastal provinces.

National fuel-efficiency standards for cars, SUVs and minibuses are being introduced in two phases. In the first phase, the standards were applied on 1 July 2005 to new models and a year later to models already in production. A second implementation phase, involving tougher standards, will take effect in 2008 (Figure 9.7). The standards set maximum fuel consumption levels for 16 weight classes, using the New European Driving Cycle. The

Table 9.7: **Vehicle Ownership by Province in China, 2005**

	Vehicles on road (thousands)	Vehicle ownership (per thousand people)	Share of LDVs (%)	Per-capita income ($)
Inland	**13 527**	**17**	**63**	**1 229**
Anhui	805	13	54	1 050
Chongqing	469	16	56	1 312
Gansu	337	13	57	891
Guangxi	591	12	64	1 046
Guizhou	468	12	56	634
Heilongjiang	859	22	67	1 725
Henan	1 522	16	65	1 350
Hubei	862	15	62	1 365
Hunan	783	12	63	1 231
Inner Mongolia	658	27	58	1 952
Jiangxi	484	11	56	1 125
Jilin	653	24	71	1 594
Ningxia	156	26	51	1 216
Qinghai	122	22	58	1 196
Shaanxi	632	17	68	1 182
Shanxi	1 074	31	63	1 490
Sichuan	1 380	16	70	1 075
Tibet	71	25	44	1 084
Xinjiang	564	27	59	1 549
Yunnan	1 036	23	60	933
Coastal	**18 069**	**35**	**71**	**2 787**
Beijing	2 097	133	90	5 354
Fujian	698	19	64	2 222
Guangdong	3 730	40	66	2 909
Hainan	164	19	64	1 292
Hebei	1 982	28	61	1 762
Jiangsu	1 923	25	75	2 929
Liaoning	1 349	31	66	2 269
Shandong	2 470	26	67	2 394
Shanghai	952	52	80	6 157
Tianjin	677	63	80	4 239
Zhejiang	2 029	41	71	3 281

Source: CEIC (2007).

standards are most stringent for heavier classes (An and Sauer, 2004). In the Reference Scenario we assume that the standards are prolonged and tightened progressively. In 2030 a new car in China will have, on average, the same efficiency as an EU model in 2012.

China is also tightening vehicle emission standards. In most provinces, they are currently at the level of Euro II standards, which came into force in the European Union in 1996. Euro III standards have been adopted in Beijing and Guangzhou. Current plans call for adoption of Euro III for LDVs by July 2007 and Euro IV from July 2010 (Global Insight, 2006). However, significant challenges face Chinese refineries in meeting demand for high-quality fuels. New passenger vehicle crash standards will come into force in 2009.[11] Because of the external standards to which they have been built, foreign brands will have a competitive advantage over local brands in complying with both emission and crash standards.

Figure 9.7: **International Comparison of Fleet Average Fuel Economy Standards as of July 2007**

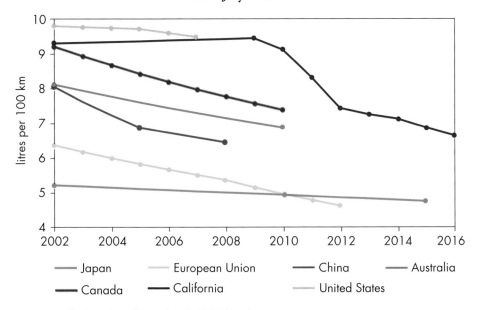

Sources: IEA database; An and Sauer (2004); ICCT (2007).

11. In 2005, there were more than 450 000 traffic accidents with 100 000 fatalities (Global Insight, 2006).

Other Modes

In 2005, aviation, navigation and rail combined accounted for nearly one-third of energy use in transport. Oil accounted for 85% of the fuel consumption in these modes. Together, navigation and rail, which are much more efficient than road transport, accounted for 80% of total traffic volume. Aviation is expected to be the fastest growing of these modes over the projection period, as China more than doubles its share of global aviation oil consumption, from 6% in 2005 to 12% in 2030. Aviation fuel consumption quadruples between now and 2030, growing at an average rate of 5.4% per year.

Aviation fuel demand in China – for domestic and international air routes – grew on average by 12% per year from 1990 to 2005. Faced with this booming growth in air travel, the Chinese transport authorities recognised that it was necessary to restructure the domestic route network to establish a more efficient hub and spoke system. Beijing, Shanghai and Guangzhou were established as the three major mainland hubs. They are expected to expand rapidly to become major Asian hubs, accounting for much of the increase in traffic. Increasing wealth, particularly among the growing middle class, will drive Chinese outbound air travel, while increasing interest in China as a tourist destination will drive inbound travel.

Residential

In 2005, residential energy consumption accounted for 30% of total final energy consumption. Per-capita consumption was around 0.26 tonnes of oil equivalent, compared with an OECD average of 0.62. The gap is due to differences in income, climate, housing type, fuel choice and social habits. Energy demand in the residential sector is expected to rise by 1.1% per year over 2005-2030 in the Reference Scenario. The fuel mix is expected to change markedly (Figure 9.8). The share of biomass, which is mostly used for rural space heating and cooking, is projected to drop from around two-thirds in 2005 to 36% in 2030, as more people switch to commercial fuels and the urban share of the population rises.[12] Excluding biomass, residential sector energy demand grows by 3.8% per year over the projection period. Coal, mainly used for heating (especially in the north), drops from 14% to 10% as more district heating and natural gas become available. Electricity and gas will see the biggest gains in market share. Natural gas use increased at a rate of 13% per year from 2000 to 2005 (Box 9.4) and is projected to continue to rise, albeit at a slower rate, averaging growth of 6.3% per year through to 2030. Electricity demand grows at 5.9% per year, boosting its share of residential energy use from 8% now to 24% in 2030. The residential sector accounts for

---

12. The use of biogas, which makes up a small share of biomass use, is being promoted by the government. The Medium and Long Term Renewable Energy Development Plan (NDRC, 2007c), released in August 2007, targets raising the use of biogas to 44 bcm by 2020.

nearly one-fifth of the increase in total electricity demand for final consumption in 2005-2030. Delivered heat sees its share in the residential fuel mix double by 2030. The use of solar water heaters has been rapidly expanding in the last two-and-a-half decades[13] and this increase in use is expected to continue, along with use of other renewables in the residential sector.

Heating, cooling and appliances will account for most energy end uses in urban households. It is estimated that heating in urban areas of northern China already makes up 40% of energy use in buildings (Tsinghua University, 2007). This share is set to grow, as the government is actively promoting collective central heating, in particular district heating boilers and combined heat and power (CHP) plants.[14] The total urban area covered by collective central heating systems more than doubled from 1.1 to 2.5 billion square metres in 2000-2005 (NBS). Moreover, heating has been extended rapidly southwards from the cold north to the "hot summer, cold winter" region,[15] where electric heaters and air conditioners with a heating function provide the majority of heating to urban households in the winter. Recently, extensive use of air conditioners has exerted huge pressure on urban electricity supply in the summer. For example, air-conditioning load accounts for 40% of peak load during the summer in Shanghai (Kang, 2005).

Figure 9.8: **Residential Energy Consumption by Fuel, 2005 and 2030**

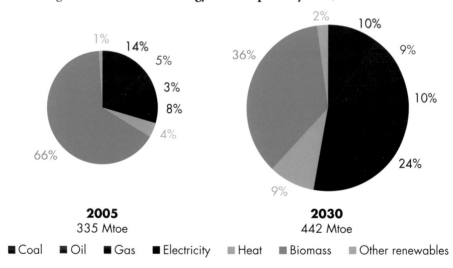

2005
335 Mtoe

2030
442 Mtoe

■ Coal   ■ Oil   ■ Gas   ■ Electricity   ■ Heat   ■ Biomass   ■ Other renewables

13. China is the world leader of solar water heaters with 63% of global installed capacity in 2005 (REN21, 2006). For example, Rizhao, a city of three million inhabitants in Shandong province has over half a million square metres covered by solar water-heating panels.
14. In the 11th Five-Year Plan (2006-2010), the government aims to achieve 40% collective central heating provision in the cities by 2010, up from 30% today.
15. In this region, no heating or cooling infrastructure was available until recently, although there are 60-80 days per year when the temperature falls to less than 5ºC and 15-30 days when temperatures exceed 35º. Cities in this region include Shanghai and Chongqing.

Residential natural gas consumption amounted to around 11 bcm, or 22% of China's total gas consumption, in 2005. The use of natural gas in urban areas has been growing rapidly in recent years with the construction of distribution infrastructure nationwide under policies to shift households from solid fuels to gas and electricity. It is estimated that gas was supplied to more than 140 cities in 2005, a jump from only 60 cities in 2003. The number of cities with gas distribution is expected to rise to 270 in 2010 (Merrill Lynch, 2007). The number of urban residents with access to gas more than doubled, from 32 million in 2001 to 71 million, or 14% of the urban population, in 2005. The coastal area accounts for more than half of this growth. Some inland provinces, including Sichuan and Xinjiang, which are close to gas supply sources, have also seen a rapid increase in access to gas. The strong growth potential for natural gas in the residential sector results from the government's policy of prioritising this use (NDRC, 2007a). In addition, the price of natural gas is lower, on a heat-content basis. The particular residential uses of gas vary between regions. In Beijing, heating accounts for 60% of total residential gas consumption, while cooking and hot water showers are the primary use in Chengdu.

Buildings Stock and Efficiency

China is currently in an unprecedented construction boom, with 2 billion square metres of new commercial and residential buildings added every year. Urban residential living space increased by 50%, from 9.3 billion square metres in 2000 to almost 14 billion m^2 in 2005. A rapid increase in both the urban population and in floor space per capita explains this growth. In 2005, urban Chinese had an average living area of 26 m^2, an increase of 6 m^2 over 2000. A massive programme of privatisation of state-owned apartments and the introduction of mortgage lending by domestic banks, which started in the late 1990s, have contributed to the increase. Access to housing has also been boosted by the rapid development of the compulsory housing provident fund and by a recent government decision to build more houses for low-income households. The Ministry of Construction has set a target of 35 m^2 per capita for urban residents and 40 m^2 for rural residents for 2020 – official benchmarks for achieving a "well-off society". These targets are expected to be met.

We project that 800 million square metres of new urban residential floor space will be built annually to 2030 to accommodate the growth in urban population and the demand for larger dwellings. The proportion of the population living in cities is assumed to grow steadily from 40% to 60% over the projection period, adding around 14 million new urban residents each year. We project residential floor area per capita to rise to 38 m² for urban residents and 41 m² for rural residents as living standards increase (Figure 9.9). Average household size, which dropped from 4.5 people in 1985 to 3.5 in 2005, is projected to continue to diminish, to 3 in 2030. The trend towards smaller households is expected to be fastest in the coastal urban area, where two- to three-person households are emerging rapidly.

Figure 9.9: **Residential Floor Area per Capita and Household Size in China**

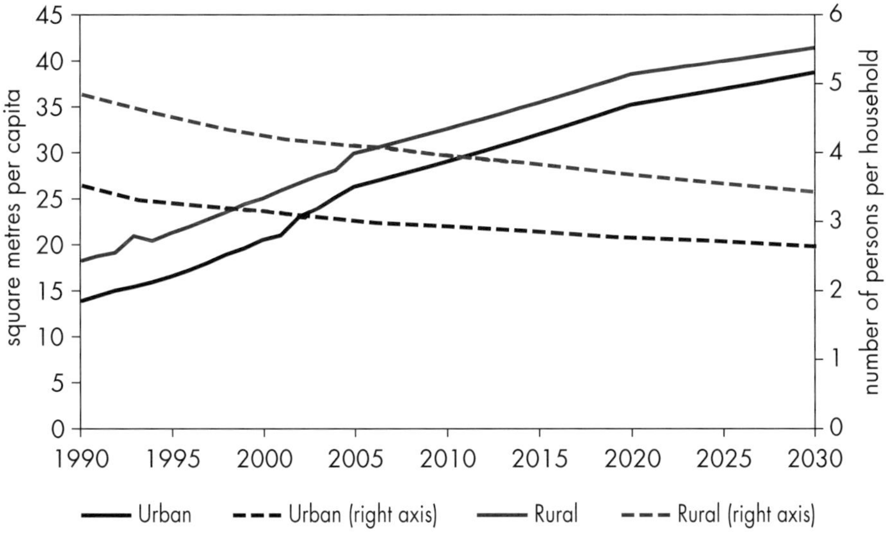

Building codes and standards were established in 1986 and revised in 1995. They divide the country into several zones, according to climate. The standards were initially applied for the northern cold-winter heating zone and were extended to hot-summer cold-winter zones in central China in 2001 and to hot-summer warm-winter zones in southern China in 2003. A bill setting national standards for residential buildings, which aims to harmonise the three building standards, is under consideration. Energy efficiency in buildings is still low and varied in China: compliance rates with building standards in new buildings are around 60% in the northern region, 20% in the central region and 8% in the southern region.

Appliances Stock and Efficiency

Four major appliances[16] – air conditioners, refrigerators, washing machines and televisions – use about 21% of residential energy (Zhou *et al.*, 2007). Rapid income growth and declining appliance prices have caused ownership of the major appliances to soar in recent years, especially in urban China. For example, hardly anyone owned an air conditioner in the early 1990s, while today on average about 80% of urban household do – a doubling of ownership every three years. The ownership growth rates of some appliances have begun to slow as they reach saturation levels. By 2005, most urban households already had one or more colour televisions. Urban ownership of washing machines and refrigerators is also approaching saturation. On the other hand, rural ownership of appliances is still less than half that of urban areas, except for televisions.[17]

Appliance ownership is expected to continue to increase steadily, though growth rates are likely to slow. It is expected to reach saturation in urban areas by around 2020 or earlier, and by about 2030 in rural areas. For example, ownership per 100 households[18] in 2030 is projected to reach 100 for washing machines and 83 for refrigerators (Figure 9.10).

Figure 9.10: **Major Appliance Ownership in China, 1985-2030**

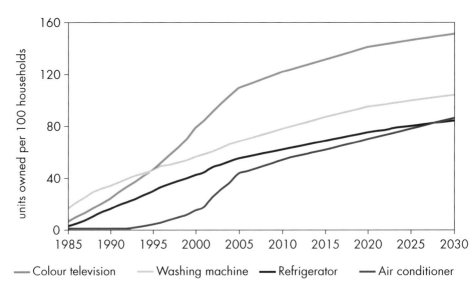

16. As income rises, many households start to own electronic equipment such as home theatre systems which consume more energy than a household refrigerator. This contributes to the rise of appliance electricity consumption.

17. In 2005, 84% of rural Chinese households owned a colour television (NBS), while overall ownership exceeded 100%.

18. This indicator allows for the possibility of more than one unit of appliance owned per household, so it can exceed 100.

Appliance efficiency improvements will offset part of the impact on residential electricity demand of rising appliance ownership. Chinese appliances are relatively inefficient compared to similar products in OECD countries. For example, the most popular Chinese refrigerator has a volume of 220 litres and uses on average 1.2-1.3 kWh per day, compared with around 0.8 kWh per day for a European refrigerator of similar size with class A labelling. However, the government is adopting minimum efficiency standards and has launched various labelling programmes. The gradual phase-out of subsidies to residential electricity will also encourage more rational and efficient use of appliances.

Other Sectors

Energy demand in the *services* sector accounted for 4% of total final consumption in 2005. The value added of the services sector grew at a rate of 12.7% per year, and energy consumption at 8.2% per year, from 1990 to 2000. However, from 2000 to 2005, growth of value added slowed to 9.9% per year, while energy consumption grew by 9.1% per year. We expect the value added of the services sector to grow faster than that of either the industrial sector or overall GDP over the projection period, partly as a result of the government's efforts to adjust China's economic structure.[19] The energy use is projected to grow at 6.9% per year from 2005 to 2015 and slow to 3.3% per year from 2015 to 2030.

Electricity currently accounts for 27% of energy consumption by services in China. This share is projected to increase to 33% by 2015 and 44% by 2030. Coal's share falls from 11% to 3%, while that of gas rises from 8% to 11%. Use of renewables in this sector, excluding biomass and waste, is forecast to grow at 7.1% per year in 2005-2030, reaching a 4% share. Oil currently accounts for 50%, but this share may be inflated by a statistical convention in China that includes some commercial transportation in the services sector.

Energy demand in the *agriculture* sector accounted for 3.6% of total final consumption in 2005, although the sector accounts for 44% of employment (World Bank, 2007). Just over half energy use is oil, with the rest split fairly evenly between coal and electricity. Energy use increases by 1.3% per year over 2005-2030, a much slower rate than in all other sectors. This reflects the declining share of agriculture in China's economy.

19. The 11ᵗʰ Five-Year Plan aims to increase the share of the services sector in GDP by three percentage points to 43% by 2010.

Environmental Implications

Local Air Pollution[20]

The energy trends contained in the Reference Scenario have unsustainable environmental consequences, in several respects. International observers are keenly aware of greenhouse-gas emissions, whereas many rural Chinese are more worried about indoor air pollution (Box 9.5). Perhaps the most immediately striking environmental phenomenon is local ambient air pollution, which carries a high social and economic cost. Twenty of the world's thirty most polluted cities, as ranked by the World Bank, are in China (World Bank, 2007). The cost of health damage alone from air pollution is predicted to rise to the equivalent of 13% of Chinese GDP by 2020 under a business-as-usual scenario (OECD, 2007).

Coal contributes most to air pollution in China. In 2005, coal-fired power plants supplied 78% of China's electricity supply. Rapidly rising coal consumption in the power sector and in industry in China has had a profound environmental impact. Yet coal will continue to play a large part in China's energy future, especially in the Reference Scenario, because most new power generating capacity over the *Outlook* period will be coal-fired (see Chapter 10). Coal combustion accounts for most of China's sulphur dioxide (SO_2) emissions. All coal contains some sulphur, which is released as SO_2 when the coal is burnt. In the absence of sulphur-control technologies or flue-gas desulphurisation (FGD), this noxious gas can cause breathing difficulties and acid rain and precipitation. Air pollution generally, and particularly the adverse impact of coal-fired generation on the environment, is a major force behind China's drive to reduce energy intensity and wish to accelerate the use of nuclear, hydro, other renewables and combined heat and power (CHP) technologies.[21] Measures such as flue-gas desulphurisation at power plants have the same aim of mitigating air pollution.

The Chinese government had already set an ambitious SO_2 reduction target in its 10th Five-Year Plan – a 10% reduction in 2005 compared to 2000 levels. This target was not met (OECD, 2007). China did not meet its target of washing 50% of the coal it burns, and the implementation of FGD has been slow to date. In 2005 only 45 GW out of 389 GW of installed thermal capacity had an FGD unit installed. In the Reference Scenario, China's SO_2 emissions are projected to increase from 26 Mt in 2005 to 31 Mt in 2015 before levelling off to 30 Mt by 2030 (Figure 9.11). The increasing number of

---

20. The projections in this section are based on analysis carried out by the International Institute for Applied Systems Analysis (IIASA) on behalf of the IEA.

21. The successful installation of CHP and combined cooling, heating and power (CCHP) systems can reduce fuel consumption by approximately 25% compared with conventional power plants, bringing about a proportional reduction of pollutant emissions. When fuelled with natural gas rather than coal or oil, SO_2 emissions can be reduced to zero.

FGD units installed, along with the diversification of the power sector, explains the levelling-off of emissions around 2015. Increased availability of natural gas in medium and small cities and replacing coal in coal-fired boilers, will reduce air pollution and price reforms will discourage wasteful consumption, but it is likely to prove difficult to control the SO_2 emissions from dispersed and varied industrial plants. Moreover, the level of fuel diversification seen in the Reference Scenario will not be sufficient to limit SO_2 growth in the short term. China's current ambient air quality trends are inconsistent with its emissions reduction targets, total emissions of major pollutants including SO_2 are targeted to fall by 10%, or 8.4 million tonnes, under the 11th Five-Year Plan for Energy. Much greater emissions reductions will be needed to achieve the targets (OECD, 2007).

Figure 9.11: **SO$_2$ Emissions by Sector in the Reference Scenario**

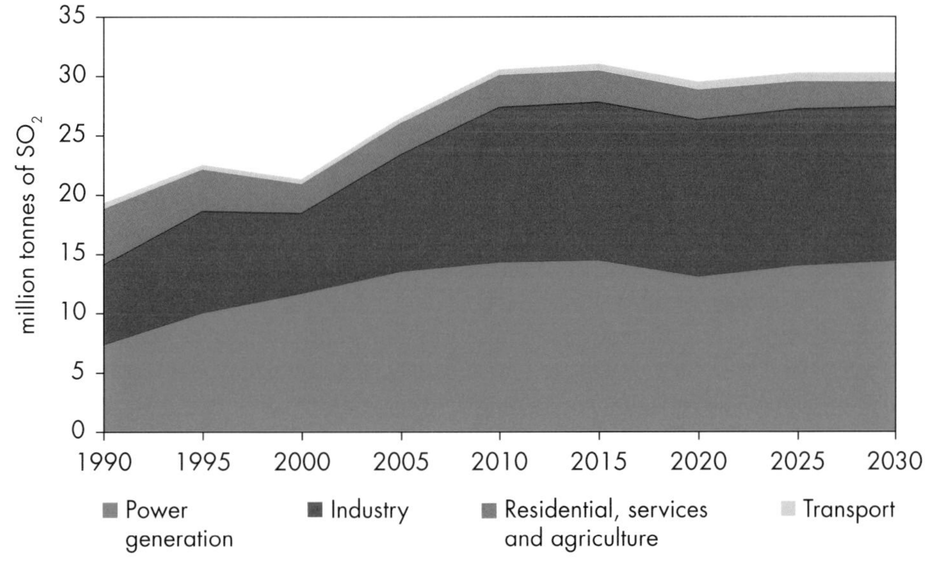

Oxides of nitrogen (NO_x) are also formed during combustion; its release into the atmosphere is linked to environmental impacts including acidification, eutrophication and ozone formation. Transport is the fastest growing sector and rising vehicle ownership does not bode well for local air quality. Slowing or reversing the rise of NO_x emissions from the transport sector presents difficulties because of the large number of individual sources. NO_x emissions rise over the *Outlook* period, from 15 Mt in 2005 to 21 Mt in 2030 (Table 9.8). Emissions of $PM_{2.5}$ continue the declining trend seen since the mid-1990s, reaching 9 Mt in 2030 compared to 14 Mt in 2005.

Box 9.5: **Household Use of Biomass and Coal**

The proportion of the population using coal and biomass (everything from firewood to manure) for cooking is one of the indicators used to assess progress towards the Millennium Development Goals. The World Health Organization (WHO) estimates this share at 80% in China (WHO, 2007). For biomass alone, we estimate the figure to be 37%, or 483 million people. Most of them – an estimated 428 million – live in rural areas. We project the total number of people relying on biomass for cooking to fall to 390 million in 2030.

The use of these fuels in conjunction with inefficient stoves causes indoor air pollution and, in the case of biomass, can have consequences such as local deforestation and soil erosion. The WHO estimates that some 380 000 people in China die prematurely every year because of indoor air pollution from the combustion of solid fuels (WHO, 2007). This is a considerably greater number of deaths than the corresponding figures for deaths from *outdoor* air pollution (around 270 000) or from lack of clean water (95 000). These considerations led to the establishment of the Chinese National Improved Stoves Programme (NISP) in the 1980s to disseminate household stoves with chimneys. The NISP was implemented in a decentralised way in order to diminish bureaucratic hurdles and speed up payments. Provincial and county stove programmes provided for marketing, subsidised training, a subsidy to households and a range of other measures. Centralised production of critical stove components ensured quality control, while local modification of designs ensured that the stoves would meet user needs. China's Ministry of Agriculture estimates that by 1998, 185 million out of 236 million rural households had improved biomass or coal stoves (Sinton *et al.*, 2004). The programme was one of the most successful energy-efficiency programmes in China and perhaps the world's most successful household energy initiative. Financial and institutional support has now tapered off (except for certain programmes targeted on impoverished areas) but the NISP left a positive legacy of private infrastructure for producing and marketing improved stoves.

China's stove industry sells more than 10 million improved stoves per year, is worth about $30 million to the economy and is growing at a rate of 10% per year. From the 1990s onwards, however, there was significant switching away from biomass to coal. As a result, 90% of manufacturers' revenue comes from coal stoves rather than biomass stoves (Spautz *et al.*, 2006). While both biomass and coal can give rise to respiratory illness, coal can also contain large quantities of arsenic, lead, mercury, other poisonous metals and fluorine. Exposure to indoor air pollution from coal fires is associated with a two-fold increased risk of lung cancer among women (WHO, 2006). Further improvements in indoor air quality will require both greater take-up of efficient cooking equipment and better ventilation, as well as faster switching to electricity (via grid connection, but also solar panels and micro-hydropower), piped gas, biogas, modern biomass fuels, such as ethanol gel, and alternative fossil fuels, such as LPG and DME.

Table 9.8: **Emissions of Major Pollutants in the Reference Scenario** (Mt)

	1990	2005	2015	2030
NO_x	7	15	19	21
$PM_{2.5}$	12	14	12	9
SO_2	19	26	31	30

Energy-Related CO_2 Emissions

In the Reference Scenario, China's energy-related carbon-dioxide emissions are expected to exceed those of the United States in 2007, making it the world's largest emitter (see Box 5.2 in Chapter 5). Yet per-capita emissions, at 3.9 tonnes of CO_2 in 2005, are only 35% of those of the OECD. Greenhouse-gas emissions are receiving increasing attention from the Chinese government which, until recently, had been mainly preoccupied with local environmental pollution. In June 2007, the Chinese government published *China's National Climate Change Programme*.

From 1990 to 2005, China's CO_2 emissions grew strongly at an average annual rate of 5.6%, driven by the country's rapid economic expansion. We project emissions to grow by 5.4% annually to 2015 and 3.3% over the period 2005-2030, reaching 11.4 billion tonnes and confirming China's position as the leading emitter. By 2015, China's emissions reach a level 35% higher than that of the United States; in 2030, they are 66% higher. Nonetheless, China's per-capita emissions do not reach even current OECD levels by the end of the projection period (Table 9.9). China accounts for 27% of global emissions in 2030, up from 19% in 2005.

China's carbon intensity is expected to fall by half over the *Outlook* period, as the structure of the economy changes in favour of less energy-intensive activities and

Table 9.9: **China's Energy-Related CO_2 Emission Indicators in the Reference Scenario** (tonnes of CO_2)

	2005	2015	2030
Per capita	3.9	6.2	7.9
Per thousand dollars of GDP*	2.2	1.8	1.2
Per toe of primary energy	2.9	3.0	3.0

* In year-2006 dollars and market exchange rates.

energy efficiency improves. However, as the average carbon content of each unit of primary energy consumption rises slightly, from 2.9 tonnes of CO_2 per toe of energy in 2005 to 3.0 tonnes in 2030, emissions, at 3.3% per year, are projected to grow faster than total primary energy demand, at 3.2%.

The power sector, which is mainly fuelled by coal, contributes most to China's CO_2 emissions. Its share is projected to rise, from 49% in 2005 to 52% in 2015 and 54% in 2030 (Table 9.10). The transport sector's share also increases, from 7% to 11%. Industry's share, by contrast, falls from 28% in 2005 to 21% in 2030.

Table 9.10: **Energy-Related CO_2 Emissions by Sector in the Reference Scenario**
(million tonnes)

	1990	2005	2010	2015	2030	2005-2030*
Power generation	652	2 500	3 589	4 450	6 202	3.7%
Industry	800	1 430	2 014	2 186	2 373	2.0%
Transport	121	337	486	664	1 255	5.4%
Residential and services**	479	468	550	622	715	1.7%
Other***	191	365	585	709	903	3.7%
Total	**2 244**	**5 101**	**7 223**	**8 632**	**11 448**	**3.3%**

* Average annual growth rate. **Includes agriculture sector. *** Includes other transformation and non-energy use.

Box 9.6: **China and the Clean Development Mechanism**

The Clean Development Mechanism (CDM) is a mechanism set up to help countries meet their greenhouse gas commitments under the Kyoto Protocol while also contributing to external development. China is now the dominant player on the supply side of the fast-growing CDM market (*i.e.* the host to projects generating credits). The main buyers are companies in the European Union and Japan (see Chapter 6).

China is expected to account for more than half of all the credits to be generated by CDM projects to 2012. China's big market share means that it sets a *de facto* global price floor for Certified Emission Reductions (CERs, each of which corresponds to one tonne of CO_2). By August 2007, China had 737 CDM projects in the pipeline (including all projects registered, at the validation stage or requesting registration), which were expected to generate almost 1.2 billion CERs by 2012. Of these projects, 107 had already been registered, accounting for 391 million CERS in 2012.

China's preferred categories of CDM projects are renewable energy, energy efficiency and methane recovery projects. However, the largest share of registered credits, 72%, comes from projects to reduce emissions of hydrofluorocarbon HFC-23 (Figure 9.12). HFC-23 is a by-product of HCFC-22, which is a potent greenhouse and ozone-depleting gas used largely for refrigeration. China is well placed to provide these credits, because it is a big producer of HCFC-22 and cutting HFC-23 emissions is very cheap in China, at less than $1 per tonne of CO_2 equivalent, or less than a tenth of the value of the CDM credits generated. Projects that reduce N_2O (another powerful greenhouse gas), hydropower and the collection of methane from coal mines and coal beds make up about 11% each of total expected 2012 credits, while energy efficiency in industry accounts for almost 10%, wind power for 6% and landfill gas for 3% (UNEP, 2007).

New projects are being added to the pipeline all the time. Revenue from the sale of CDM credits could contribute as much as 0.5% per year of Chinese GDP in 2030, mainly thanks to technology transfer (OECD, 2007). China will remain an attractive market for buyers of credits because of the economies of scale available, the broad portfolio of eligible projects, growing energy demand and the inefficiencies which exist in its industrial sector. Indeed, its share of global credits would increase even further if new designs of HCFC-22 plants that produce fewer emissions of HFC-23, new supercritical or ultra-supercritical coal plants or carbon capture and storage projects become eligible for emission credits.[22] However, geographical concentration may become an issue; foreign buyers may wish to diversify their portfolios.

9

Coal was by far the leading contributor to China's CO_2 emissions in 2005 and remains so in the Reference Scenario through to 2030 (Figure 9.13). Coal's share of emissions falls only slightly over the next two-and-a-half decades, from 82% to 78%. The share of natural gas increases from 2% in 2005 to 4% in 2030, while oil's share also increases, from 16% to 18%.

22. HCFC-22 use for refrigeration (but, importantly, not as feedstock) is controlled by the Montreal Protocol and is scheduled for complete phase-out by 2030. Equipping HCFC-22 plants with HFC-23 destruction technology would greatly aid efforts to mitigate climate change, as the annual emissions of such plants are typically of the order of several million tonnes of CO_2.

Figure 9.12: **China CDM CO$_2$ Reduction by Project Type**
(Registered CERs, August 2007)

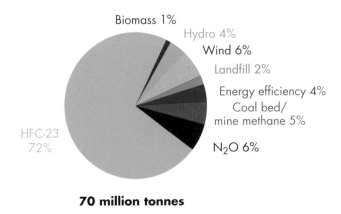

70 million tonnes

Figure 9.13: **Energy-Related CO$_2$ Emissions by Fuel in the Reference Scenario**
(billion tonnes)

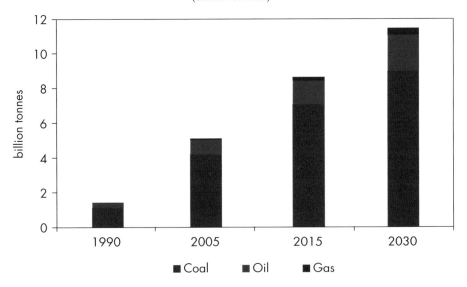

CHAPTER 10

REFERENCE SCENARIO SUPPLY PROJECTIONS

HIGHLIGHTS

- Conventional oil production in China is set to peak at 3.9 mb/d early in the next decade and then start to decline in the Reference Scenario. Non-conventional oil supply from coal-to-liquids plants reaches 750 kb/d in 2030, compensating to a large degree for the decrease in conventional oil production. China's oil import dependence rises to 80% in 2030, from 50% today, as oil imports reach 13.1 mb/d, equal to EU imports in 2030.

- Indigenous gas production more than doubles, thanks to the development of onshore non-associated gas reserves, to reach 111 bcm in 2030, after peaking at 118 bcm in 2020. Despite the increase in indigenous production, gas imports increase substantially to reach 128 bcm in 2030, requiring a huge expansion of pipeline and LNG regasification infrastructure.

- China's coal production is projected to increase strongly to reach 2 604 Mtce in 2015 and 3 334 Mtce in 2030. As more than 90% of Chinese coal resources are located in inland provinces, almost 1 250 Mtce, equal to 36% of Chinese total primary coal demand, needs to be transported to the coastal provinces in 2030. This adds to the pressure on internal coal transport and makes international imports to coastal provinces more competitive. China became a net coal importer in early 2007. Net imports reach 95 Mtce in 2030, representing 3% of its demand and 7% of global trade.

- Over the period to 2030, China needs to add 1 312 GW to its generating capacity, more than the total current installed capacity in the United States. Coal remains the dominant fuel in China's electricity mix. The efficiency of coal-fired generation gradually improves, with the construction of larger, more efficient units and the deployment of supercritical and other clean coal technologies.

- Gas-fired electricity generation grows rapidly, its share in the total generation mix reaching 4% in 2030. The share of nuclear power reaches 3% in 2030. China is already the largest producer of renewable energy in the world and renewable energy is expected to play an even larger role in the future. Capacity additions in renewable energy are mostly in large-scale hydro and wind power. By 2030, hydropower capacity reaches 300 GW and wind power 49 GW.

- Cumulative investment in China's energy-supply infrastructure amounts to $3.7 trillion (in year-2006 dollars) over the period 2006-2030, *i.e.* $150 billion per year. The power sector accounts for three-quarters of the total investment.

Oil Supply

Oil Resources and Reserves

China's proven reserves of oil amounted to 16 billion barrels at the end of 2006, equal to 1.2% of world reserves.[1] Ultimately recoverable resources from discovered fields are estimated at 57.3 billion barrels.[2] We estimate that the volume of oil yet to be produced is close to 29 billion barrels (Table 10.1), to which production from reserves growth and from fields yet to be discovered will be added. The reserves are mainly located in five sedimentary basins: Bohai Gulf (35%), Songliao (22%), Tarim (12%), Junggar (11%) and Ordos (6%). Almost all the reserves are located onshore; only the Bohai Gulf basin is partly offshore (Figure 10.1). The Daqing field in the Songliao basin is by far the biggest in China. It still holds 14% of China's remaining proven and probable reserves, even though it has been producing since 1960. Most other big fields are also mature, having been discovered in the 1960s and 1970s.

Table 10.1: **China's Oil Reserves as of end-2005**

	Onshore	Offshore	On/offshore	Total
Number of fields	724	191	19	934
Proven and probable reserves (Mb)	23 911	4 356	1 168	29 435
Cumulative production to date (Mb)	25 523	1 635	732	27 890

Sources: IHS Energy databases; IEA estimates.

Between 1997 and 2006, 230 oilfields were discovered, adding 7.1 billion barrels to proven and probable reserves (Figure 10.2). While the *number* of new fields is about a third of all previous discoveries, the volume of oil was only 14% of that discovered before 1997. Three-quarters of the oil discovered since 1997 is concentrated in three basins: Bohai Gulf, Junggar and Tarim. The biggest find was the Jidong Nanpu in the offshore Bohai Gulf by PetroChina, reported in 2007, with reserves of 2 800 million barrels – the biggest find in 40 years and the largest offshore discovery ever in China. The most recent study by the US Geological Survey estimated undiscovered resources in 1996 at 16.5 billion barrels (USGS, 2000). Some 9 billion barrels, including discoveries in 1995 and 1996, have been discovered since then, with the possible implication that less than half of that potential remains to be found.

1. *Oil and Gas Journal*, 18 December 2006.
2. Based on data provided to the IEA by IHS Energy.

Figure 10.1: **China's Oil and Gas Resources and Supply Infrastructure**

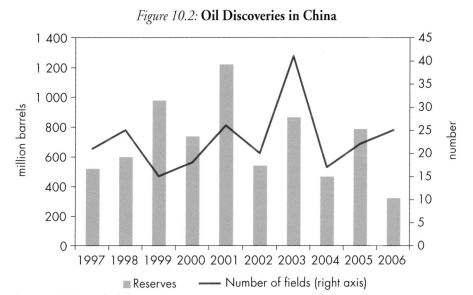

➤	Tanker terminal
★	Existing LNG import terminal
✦	Under const./planned LNG import terminal
☆	Speculative LNG import terminal
●	Main oilfield
●	Main gasfield

KAZAKHSTAN
Karamay
Dushanzi
Almati
Bishkek
KIRG.
Urumqi
Hami
Korla Shanshan
Tarim Basin
Zepu
TAJ.
Islamabad
PAKISTAN
Delhi
NEPAL Kathmandu Thimphu
BHU.
INDIA
Dhaka
BANG. MYANMAR
Hanoi
LAOS Sanya
THAI.
VIETNAM
CAMB.

MONGOLIA
Ulan Bator
Irkutsk
Chita
RUSSIA
Khabarovsk

Daqing
Qianguo
Harbin
Jilin
Liaohe Fushun
Jinzhou Liaoyang
Jinxi Anshan NORTH
Beijing Tianjin Dandong KOREA
Shijiazhuang P'yongyang
Zibo Dalian
Jinan Seoul
Qingdao REP. OF
KOREA JAPAN

Yumen
Ordos Basin
Lenghu
Qaidam Basin
Golmud
Lanzhou
CHINA Xianyang Xi'an
Luoyang
Nanyang
Chengdu Jingmen Wuhan
Yunxi Chongqing Anqing
Baling
Kunming

Rudong
Shangai
Zhejiang LNG

Fujian LNG
Guangzhou Guangdong Taipei
Beihai Hong Kong

PHILIPPINES
Manila

Km
0   900

Vladivostok

——	Existing oil pipeline
-----	Under const./planned/proposed oil pipeline
——	Existing gas pipeline
-----	Under const./planned/proposed gas pipeline
⌘	Refinery

The boundaries and names shown and the designations used on maps included in this publication do not imply official endorsement or acceptance by the IEA.

Figure 10.2: **Oil Discoveries in China**

Reserves — Number of fields (right axis)

Sources: IHS Energy databases; IEA estimates.

Oil Production[3]

Oil production in China was 3.7 mb/d in 2006, of which about 90% was onshore. Output, which was flat last year after gaining about 0.5 mb/d since the beginning of the decade, is concentrated in seven complexes, composed of several neighbouring fields. Most of them are more than 50% depleted. The remaining part of domestic production is therefore fragmented between many small to medium-size fields. The biggest 11 fields, out of a national total of 492 in production, contribute close to half. Production at only one of them, Tahe, has not yet peaked. About half of proven and probable reserves from known fields have been produced. In the Reference Scenario, China's conventional oil production is projected to increase marginally, levelling off at about 3.9 mb/d in 2012 and then declining gradually to 2.7 mb/d in 2030 (Table 10.2), as the largest existing producing fields become almost completely depleted.[4] About one-quarter of China's production by the end of the *Outlook* period is expected to come from fields discovered recently and awaiting development. The Jidong Nanpu field is expected to contribute a significant share, on average 7% of conventional oil production, though the figure is very uncertain, as reserves have not yet been fully assessed. On the basis of partial information, we assume that production will start in 2012 and will reach an average of about 270 kb/d following the build-up phase, with a peak at about 300 kb/d before starting to decline to about 200 kb/d by 2030.[5]

The projected fall in crude oil production is offset to a large degree by increased production from non-conventional sources – notably coal-to-liquids (CTL) plants. The first such plant, being built by Shenhua, is expected to come on stream in 2008. We project that CTL production will reach 250 kb/d in 2020 and 750 kb/d in 2030. The recent increase in oil prices has made CTL production a profitable option. It is expected to be a particularly attractive technology in China because of the availability of cheap local coal (see Chapter 8 on Chinese policy for CTL development).[6]

---

3. Crude oil, natural gas liquids (NGLs) and condensates.
4. These projections are derived from a bottom-up assessment of the top 11 producing fields in 2006 and new oilfield developments in the coming years and from a top-down analysis of longer-term development prospects.
5. In the absence of any further information on geology, this is based on 7 billion barrels in place, estimated ultimately recoverable resources of around 2.8 billion barrels and a 40% recovery factor using primary recovery techniques.
6. The Chinese government has recently expressed doubts about the feasibility of achieving ambitious plans for CTL production, mainly because of uncertainty about production costs, the magnitude of coal imports and the environmental impact, including that on water resources. See the Spotlight in Chapter 11.

Table 10.2: **China's Oil Production in the Reference Scenario** (kb/d)

	2006	2015	2030
Producing fields	915	448	170
Saertu (Daqing)	348	155	40
Xingshugang (Daqing)	131	69	23
Lamadian (Daqing)	64	22	4
Karamay Complex	104	51	15
Tahe Complex	94	73	48
Huanxiling (Liaohe Depression)	15	1	0
Shuguang (Liaohe Depression)	40	16	4
Suizhong 36-1	28	2	0
Ansai	45	32	18
Xijiang 30-2	9	0	0
Chengdao (Shengli)	36	28	18
Other fields and developments			
Discovered fields to be developed	–	989	738
Of which Jidong Nanpu field	*–*	*284*	*200*
Other currently producing fields, new discoveries and reserve additions	2 759	2 403	1 792
Non-conventional oil (CTL)	–	184	750
Total oil production *	**3 674**	**4 024**	**3 450**

* Including condensates and NGLs.
Sources: IHS Energy databases; IEA analysis.

10

Our projections are inevitably subject to a number of uncertainties. The large number of fields currently undergoing appraisal (126) and development (33) means that the average delay between a field being discovered and being brought into production is likely to be longer than was the case in the recent past (typically around seven years). In the Reference Scenario, we assume that only 30 fields can realistically be brought on stream in any given year. In practice, slippage caused by rising upstream costs and a shortage of available rigs may lead to even longer delays, slowing the rise in production in the early part of the projection period. Recovery factors may also turn out to be higher or lower than assumed in our analysis. In particular, the adoption of the latest improved and enhanced recovery techniques could result in higher levels of output from both existing and new fields. All Chinese companies have invited international companies to participate in several of their most complex developments, so they have been

Box 10.1: **Restructuring of China's Oil Sector**

The oil sector in China is dominated by three majority state-owned companies: China National Petroleum Corporation (CNPC), China Petroleum and Chemical Corporation (Sinopec) and China National Offshore Oil Corporation (CNOOC). These companies are subject to policy set by the National Development and Reform Commission (NDRC). All three companies were established in the 1980s. They were given specific sector-based responsibilities. CNPC was put in charge of oil and gas exploration and production activity onshore and in shallow offshore waters; CNOOC was given responsibility for all other offshore regions; Sinopec was given the primary responsibility for refining, petrochemicals and other downstream activities.

A second round of oil-sector restructuring occurred in 1998, with the establishment of CNPC and Sinopec as fully vertically-integrated oil companies and a geographical partition of the domestic market. CNPC took control of all oil activities, both upstream and downstream, in the northern and western provinces while Sinopec did likewise in the southern provinces. CNPC then transferred its major domestic assets to its then wholly-owned subsidiary, PetroChina, and developed an internationally-oriented strategy, especially for upstream activities. Between 2000 and 2002, all three companies opened their equity to private investment by carrying out initial public offerings (IPOs). CNPC raised $3 billion through the sale of a 10% stake in PetroChina, Sinopec sold a 15% stake in its main subsidiary, raising $3.5 billion, while CNOOC sold off 27.5% of its equity. The IPOs attracted considerable interest, including from major international oil companies. BP, with 20% of the listed shares, was the largest purchaser of PetroChina stock. ExxonMobil, BP and Shell together took almost 60% of the Sinopec IPO, while Shell purchased a big stake in CNOOC. The Chinese government still holds majority stakes in all three companies and all international companies are now reported to have sold their shares.

Despite the close control exercised over the sector, Chinese companies make their own decisions about awarding contracts, through bidding or bilateral negotiation, to foreign companies which operate under production-sharing contracts. They have a legal option to take up to a 51% share in any new upstream development.

able to gain access to the latest technology and best practice in development and production. There is obviously enormous uncertainty about the success of future exploration and about production rates from fields yet to be discovered. Our projections are based on USGS resource estimates, which may turn out to be conservative given recent discoveries.

Oil Refining

China's refining capacity is growing fast. Distillation capacity climbed to 6.6 mb/d in 2006 and to about 7.5 mb/d in early 2007. An average of 460 kb/d of additional capacity is expected to be brought on stream annually over the next five years from identified projects, much of it at greenfield refineries (see Figure 10.3). On that basis, total installed capacity will reach 9.9 mb/d in 2012. Some 55% of the new capacity will be built by Sinopec and 27% by PetroChina. The number of refineries should grow from about 100 at present (the exact number of refineries in China is not known with certainty) to approximately 120.

Figure 10.3: **Planned Refining Capacities in China**

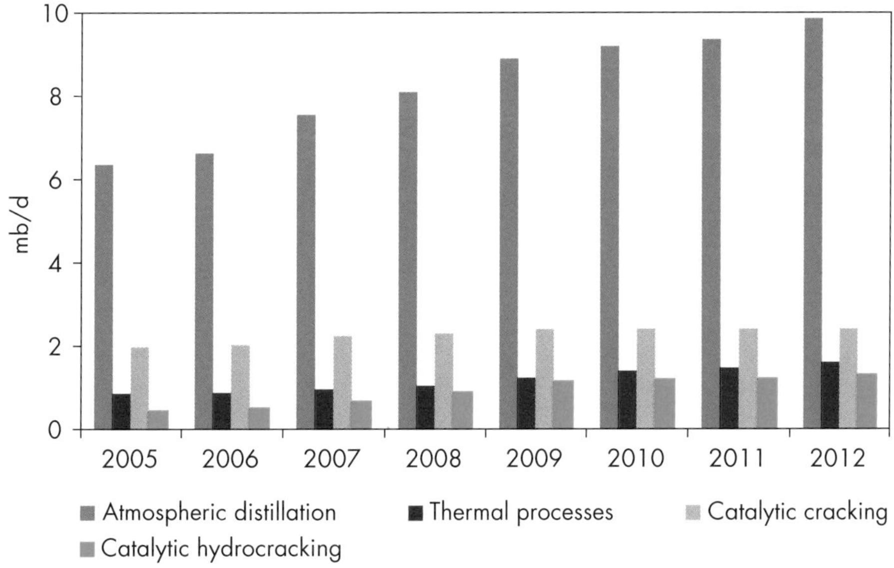

Sources: IEA analysis.

Beyond 2012, up to 2030, refining capacity is assumed to increase in line with oil demand in the Reference Scenario. On this basis, distillation capacity reaches 11.1 mb/d in 2015 and 16.6 mb/d in 2030 (Figure 10.4). The challenge facing Chinese refiners in the years to come is two-fold. First, they need to keep pace with the strongly growing domestic market, which is projected to reach 9.8 mb/d in 2012. Current investment plans suggest they will achieve this, as capacity outstrips slightly the growth in demand. Second, refiners need to adapt their output to the changing structure of oil demand and meet the stricter fuel quality requirements. In the Reference Scenario, the share of gasoline in the total Chinese oil demand is projected to grow from 16% in 2006 to 23% in 2030, while that of middle distillates, including diesel and kerosene, rises from 37% to 42%. Refiners recognise the need to maximise the output of middle distillates: coking capacity is projected to double and hydro-cracking capacity is to be multiplied by a factor of 2.6 between 2006 and 2012, whereas atmospheric distillation capacity will increase by only half. China is a net exporter of gasoline, so relatively little investment is going into catalytic cracking capacity, which is projected to increase by only 20% up to 2012 (see the Oil Trade section below). The projected increase in the share in refineries' crude oil slates of medium to heavy crude supplies from Middle Eastern exporting countries also calls for increased complexity in China's refineries over the projection period.

Figure 10.4: **China's Refining Distillation Capacity in the Reference Scenario**

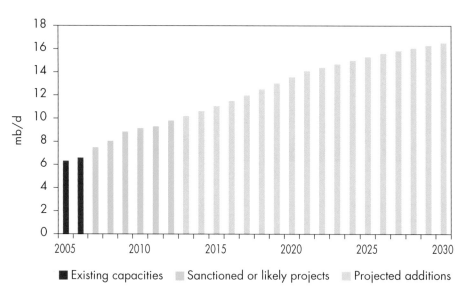

The large expansion of refining capacities which we project in China in the Reference Scenario could be underpinned by the convergence of interests with major oil-producing countries that are concerned about the prospects for oil-product demand. Early in 2007, the Chinese government approved, for the first time, foreign investment in domestic refineries, with ExxonMobil and

Saudi Aramco taking combined stakes in two joint-ventures with Sinopec of 50% and 45% respectively. The entire Fujian project which is due to start in 2009, will fully integrate refining, petrochemicals and marketing of oil. The refinery, which forms part of this project, will have a 160-kb/d refining capacity. In addition, Sinopec plans to build a 200-kb/d refinery at Qingdao, due to be completed in 2008 with Saudi Aramco, and a 200-kb/d refinery at Nansha to be completed in 2011 with Kuwait's KPC. In the case of Qingdao, Saudi Aramco will reportedly take a 25% stake and provide much of the crude supply.

Oil Trade

China imported 3.7 mb/d of oil in 2006, equal to about 50% of its total oil consumption. Of these imports, 2.9 mb/d, or 80%, were in the form of crude oil. The Middle East and Africa supplied almost 80% of China's crude oil imports. The biggest suppliers are Saudi Arabia and Angola, with about a 16% market share each of the import market (Figure 10.5). Russia currently supplies 11% of China's imports, entirely by rail. Preliminary data for 2007 show crude oil imports accelerating to reach more than 3.6 mb/d and the emergence of new suppliers, including Sudan, Kazakhstan and Equatorial Guinea. In 2006, China also imported more than 700 kb/d of refined products, two-thirds of which were fuel oil. China is currently a net exporter of gasoline and naphtha. On a net basis, China imported 3.5 mb/d of oil in 2006.

Figure 10.5: **China's Crude Oil Imports by Origin in 2006**

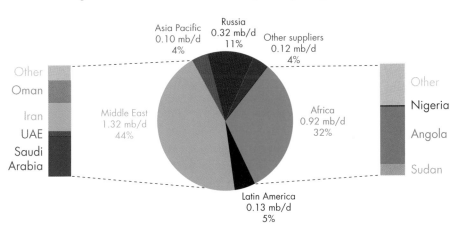

In the Reference Scenario, net oil imports are projected to increase markedly to 5.1 mb/d in 2010, 7.1 mb/d in 2015 and 13.1 mb/d in 2030 as demand rapidly outstrips production (Figure 10.6). In 2030, China will import as much as the European Union. China's dependence on imports rises from about 50% today to 80% in 2030.

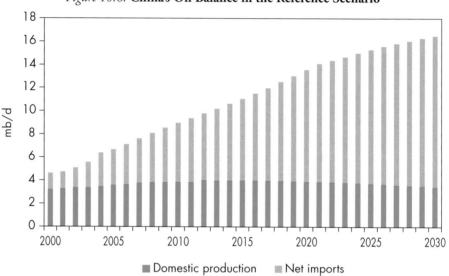

Figure 10.6: **China's Oil Balance in the Reference Scenario**

■ Domestic production   ■ Net imports

China's Oil Security Policies

Oil security has emerged as a central policy issue in China and is increasingly affecting domestic, economic and foreign policy (see Chapter 4). China has adopted a number of policies aimed at mitigating the adverse effects of the increase in oil-import dependence, diversifying the sources and routes of imported oil and preparing for supply disruptions. The heavy reliance on maritime shipping through the Straits of Malacca is a particular worry. China has reduced the proportion of its oil supplies that are shipped from the Middle East by increasing purchases from Africa, Central Asia and Russia. It also now obtains more oil by pipeline from Kazakhstan and by rail and road from Russia, helping to diminish the effects of any disruption to seaborne transportation. There are plans to increase pipeline capacity from Kazakhstan and to build a line from eastern Siberia in Russia and tentative plans for one from Myanmar. China will import Russian oil through the new $13-billion East Siberia-Pacific Ocean pipeline, which has a capacity of 0.6 mb/d and which will come on stream in 2009. China is building storage facilities for emergency oil stocks, with plans to stockpile up to 100 million barrels of oil in the coming years (Box 10.2).

Box 10.2: **China's Emergency Oil Stocks**

After several years of debate, the government decided in 2003 to proceed with building an emergency petroleum reserve. It will be built in three phases. The first phase will provide total storage capacity of about 100 million barrels at four sites: Zhenhai (33 mb) and Aoshan (31 mb) to the south of Shanghai in Zhejiang, Huangdao (19 mb) near Qingdao in Shandong, and Dalian (19 mb) in the north-eastern province of Liaoning. These four storage facilities are all above-ground tank farms. In addition, the Chinese government is contemplating imposing an obligation on industry to create emergency stocks of an as yet unknown size. Each of the second and third phases of the government build-up of stocks is expected to add a further 200 million barrels. A number of sites have been proposed for the second and third phases, but final selections have not been announced. These future phases will make use of both above-ground tank farms and underground caverns.

Construction of the first phase is expected to be completed by the end of 2008, providing storage capacity that, if filled, would be equivalent to 24 days of net imports in the Reference Scenario. The second phase would increase capacity to the equivalent of 61 days of 2010 net imports: so far, the government has announced that it intends the actual stocks held to be equivalent to 30 days of net imports by 2010. The third phase would raise coverage to 75 days of 2015 net imports, when all storage sites are filled. If no new capacity were built thereafter, coverage would fall gradually as Chinese imports continue to grow.

10

The acquisition of equity stakes in oil assets overseas, which began in the late 1980s, forms another plank of China's energy-security policy. This "going-out" policy was at least partly motivated by the ambition of the state companies to increase their reserves, diversify their activities and increase profits, with the ultimate aim of creating internationally competitive world-scale businesses. Chinese companies' equity oil output from overseas assets amounted to about 370 kb/d in 2004, most of it produced by CNPC and CNOOC in Africa and Kazakhstan (Table 10.3). An estimated 40% to 50% of this oil is shipped to China. Total Chinese equity oil production could increase to 1 mb/d by the beginning of the next decade, equivalent to about 10% of the country's total oil needs, though not all of this would actually be physically shipped to China for technical and cost reasons. It is doubtful whether Chinese equity oil investments contribute materially to improving the country's energy security or even whether they are economically competitive as no serious economic or strategic assessment has been carried out so far (see *Spotlight* in Chapter 4).

For example, the current value of Chinese investment in African oil is equivalent to 8% of IOC investment, and 3% of all commercial investment (Downs, 2007).

Table 10.3: **Chinese Oil Companies' Foreign Equity Oil Production, 2004**

Country/company	b/d	Share (%)
Sudan	134 752	36
Kazakhstan	110 452	30
Indonesia	46 941	13
Other	80 225	21
Total	**372 370**	**100**
Of which CNPC (including PetroChina)	*329 810*	*89*
CNOOC	*29 941*	*8*
Sinochem	*8 603*	*2*
Sinopec	*4 016*	*1*

Source: Downs (2006).

Whatever the underlying reasons for the going-out policy and its implications for energy security, Chinese companies certainly benefit from it in competing with other companies to acquire overseas assets. The Chinese state, by far the dominant stakeholder in these companies, sets the financial targets which are significantly less onerous than those of the international oil companies. Moreover, the Chinese companies benefit from privileged terms on loans from Chinese state banks. The Chinese government also helps the national companies to acquire assets overseas through active diplomacy and development aid to resource-rich countries. However, it is not clear that such assistance is a major factor in enhancing the competitiveness of China's NOCs compared to IOCs and other NOCs.

Natural Gas Supply

Gas Resources and Reserves

China's proven reserves of natural gas amounted to 3 720 bcm at the end of 2006, equal to 2% of world gas reserves (Cedigaz, 2007). On the basis of IHS data, we estimate that recoverable, proven and probable reserves from identified fields are approximately 30% higher, at around 5 000 bcm. We estimate that 80% of proven and probable reserves are non-associated gas, of which close to 90% are onshore (Table 10.4). Most production to date has come from onshore associated reserves, which are estimated to be about 35% depleted. Onshore non-associated gas reserves are largely untapped.

Table 10.4: **China's Natural Gas Reserves as of End-2005**

	Onshore	Offshore	On/offshore	Total
Total gas				
Number of fields	745	129	15	889
Proven and probable reserves (bcm)	4 391	549	18	4 958
Cumulative production to date (bcm)	713	41	17	771
Non-associated gas				
Number of fields	306	44	–	350
Proven and probable reserves (bcm)	3 538	445	–	3 983
Cumulative production to date (bcm)	256	32	–	288
Associated gas				
Number of fields	439	85	15	539
Proven and probable reserves (bcm)	853	104	18	975
Cumulative production to date (bcm)	457	9	17	483

Sources: IHS Energy databases; IEA estimates.

Reserves are mainly located in five sedimentary basins: Ordos (27%), Sichuan (23%), Tarim (19%), Bohai Gulf (8%) and Songliao (7%). The remaining 16% are distributed in small reservoirs in about ten basins. The Ordos, Sichuan, Tarim and Songliao basins together hold the bulk of onshore non-associated gas and, therefore, form the core of potential future production. Gas reserves from the Bohai Gulf are mostly associated with oil in mature fields. Bohai Gulf is located closest to the consuming areas of the country and is, unsurprisingly, the most depleted. In the last ten years, a total of 227 natural gas fields have been discovered – equal to about 35% of the number discovered before 1997 (Figure 10.7). Yet the reserves in the recently discovered fields exceed those of all the previously discovered fields. The biggest recent discovery, the Sulige field found in 2000, has proven and probable reserves of 466 bcm.

Figure 10.7: **Natural Gas Discoveries in China since 1997**

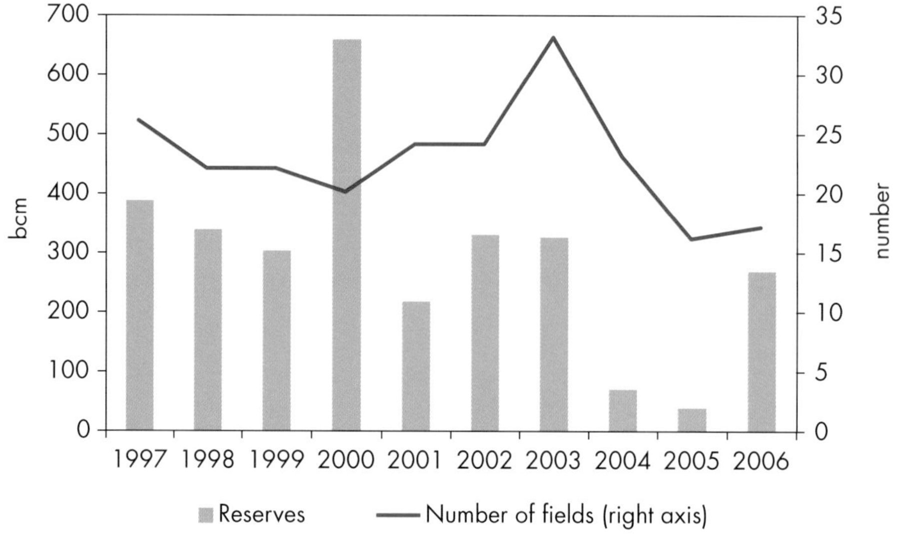

Sources: IHS Energy databases; IEA estimates.

Gas Production

Natural gas production in China totalled 51 bcm in 2005. According to preliminary data, output went up by 17% in 2006, taking it to a record high of 60 bcm. About 60% of production comes from 133 onshore non-associated fields; another 370 fields provide the rest. Offshore fields, which started to come on stream only in the mid-1990s, today contribute about 15%. Some 173 of the 190 currently producing fields were discovered before 1990. Only a small number of gas fields have reached their production peak, suggesting that there is considerable remaining production potential from existing fields over the projection period. A large number of fields, with ample reserves, yet to be developed or which started producing recently (such as Kela 2 and Sulige), will also contribute increasingly to total production.

In the Reference Scenario, China's gas production is projected to reach more than 76 bcm in 2010 and 103 bcm in 2015.[7] Output is projected to reach 118 bcm in 2020 and then start to decline, to reach 111 bcm by 2030 (Table 10.5). The 13 largest existing fields see a collective increase in production of about 29% in the next five years, then reach a plateau of about 30 bcm over the following decade, before going into decline. Discovered fields not yet in production are projected to produce 32 bcm in 2010, 60 bcm in 2015 and 65 bcm in 2030.

7. These projections are based on a bottom-up analysis of the 13 largest fields currently in production (representing more than 40% of total output), and the development of new fields in the coming years, together with a top-down analysis of longer-term development prospects.

Table 10.5: **China's Natural Gas Production in the Reference Scenario** (bcm)

	2005	2015	2030
Existing fields	**20.6**	**30.8**	**12.4**
Jingbian-Hengshan	7.6	23.3	10.0
Yacheng 13-1	3.0	1.5	0.5
Karamay Complex	2.7	0.5	0.0
Dongxin	0.9	0.6	0.3
Sebei-1	1.1	0.7	0.3
Pucheng	0.6	0.2	0.1
Shaping	0.9	1.0	0.3
Xinchang	1.4	1.4	0.4
Zhongba	0.3	0.0	0.0
Tahe	0.5	0.6	0.2
Suinan-Moxi	0.8	0.6	0.1
Pinghu	0.4	0.3	0.1
Jinzhou 20-2	0.4	0.2	0.0
Other fields and developments	**30.1**	**71.9**	**98.4**
Fields awaiting development	–	59.7	65.0
Other currently producing fields and new discoveries	30.1	12.2	33.3
Total	**50.7**	**102.7**	**110.8**

Sources: IHS Energy databases; IEA analysis.

10

The major expansion of China's gas production projected here will call for considerable investment in pipeline and storage infrastructure, particularly since production will be increasingly concentrated in the centre and west of the country, while demand is concentrated in the southern and eastern provinces. The West-East pipeline, built by CNPC and fully completed in 2005, connects the Tarim basin in the remote western Xinjiang Uygur autonomous region to Shanghai. It has an estimated capacity of about 12 bcm per year, which is to be increased to 17 bcm by the beginning of the next decade. Pipeline capacity is expected to be supplemented substantially by 2010, through the addition of a second line with a capacity of 30 bcm per year. The West-East pipeline forms the backbone of an ambitious plan to develop a national gas network, involving 20 000 km of pipelines. Reserves from the Ordos and Junggar basins – and imports from Turkmenistan and Kazakhstan – will also feed into the pipeline later.

Gas Imports

China imported gas – as liquefied natural gas (LNG) – for the first time in 2006.[8] In the Reference Scenario, natural gas imports in China are projected to increase sharply, to 12 bcm in 2010, 28 bcm in 2015 and 128 bcm in 2030 (Figure 10.8). Gas will come in the form of LNG and via pipeline from neighbouring countries. The prospects for imports are, nonetheless, highly uncertain, as they depend critically on the balance of production and demand, which in turn is very sensitive to the relative prices of coal and gas.

Figure 10.8: **China's Gas Balance in the Reference Scenario** (bcm)

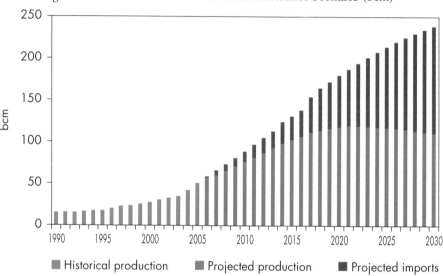

A dozen LNG projects are currently under construction or planned (Table 10.6). An expansion of the existing Shenzhen terminal is due to be completed by 2008. New terminals at Putian and Yangshan are also under construction. Most of the gas will be used in the residential sector. Imports from Russia and Kazakhstan by pipeline are also under discussion. CNPC has signed a memorandum of understanding with Exxon and Rosneft to import 8 bcm per year from the Sakhalin-1 project from 2011, but pipeline routing issues have not been finalised and this project is included in the projections only from 2016. In 2006, Gazprom and CNPC signed a protocol providing for the importation of gas through two new pipelines - one would connect western Siberia to China's Xinjiang region, where the internal West-East gas pipeline starts, and the other would run from eastern Siberia to north-eastern China. The total volumes of gas mentioned in the protocol amount to 68 bcm per year and the proposed first delivery date is 2011. However, we do not include those two pipelines in the Reference Scenario as they are not certain to

8. Imports from Australia into the Shenzhen terminal amounted to 0.9 bcm in 2006.

be built, because of concerns about the environmental impact, the price of the gas (which is expected to be comparable to European levels) and the time scale for development of Russian gas fields and the related infrastructure. A 3 000-km pipeline from Kazakhstan to the Xinjiang province is also planned and is supposed to come on stream in 2009. Initial capacity would be 10 bcm per year, rising to 30 bcm in 2012. Gas is due to come mainly from Turkmenistan, following the deal signed recently by CNPC to receive 30 bcm per year over 30 years, as well as from the gas fields of other Central Asian countries.

10

Table 10.6: **LNG Regasification Terminals in China**

Terminal	Location (province)	Main operator	Status	Start-up	Initial capacity (bcm)
Shenzhen	Guangdong	CNOOC/BP	Operating	2006	5.0
Shenzhen (extension)	Guangdong	CNOOC/BP	Under construction	2008	2.7
Putian	Fujian	CNOOC	Under construction	2009	3.5
Putian (extension)	Fujian	CNOOC	Under construction	2010	3.3
Yangshan	Shanghai	Shanghai LNG (CNOOC, Shenergy)	Under construction	2009	4.1
Subtotal existing and under construction					*18.6*
Qingdao	Shangdong	Sinopec	Feasibility study completed	After 2010	4.1
Ningbo	Zhejiang	CNOOC, Zhejiang Energy Group	Feasibility study	After 2010	4.1
Rudong	Jiangsu	PetroChina	Feasibility study completed	After 2011	4.8
Tangshan	Hebei	PetroChina	Feasibility study completed	After 2010	4.1
Dalian	Liaoning	PetroChina	Feasibility study completed	2012	4.1
Tianjin	Tianjin	Sinopec	Pre-feasibility study	2012	2.7
Beihai	Guangxi	PetroChina	Unknown	After 2010	4.1
Subtotal under consideration					*28.0*
Total					**46.6**

Sources: Company reports and IEA estimates.

In the Reference Scenario, the Turkmenistan-Kazakhstan-China gas pipeline is assumed to come on stream later than announced, by the middle of next decade. In addition, we expect the Shenzhen, Putian and Yangshan LNG plants to be fully operational by 2012. Even taking into account those import projects, we estimate that substantial additional import capacity will be required from the end of next decade, reaching 80 bcm in 2030 (Figure 10.9). This suggests that all the LNG projects at the planning stage reported in Table 10.6, with combined capacity of 28 bcm, will be needed to balance the Chinese gas market from about 2020 onwards. Other LNG or pipeline projects would be necessary after 2025. China has for long been in negotiation with Russia over access to gas, via pipeline or in the form of LNG.

Figure 10.9: **China's Natural Gas Imports in the Reference Scenario**

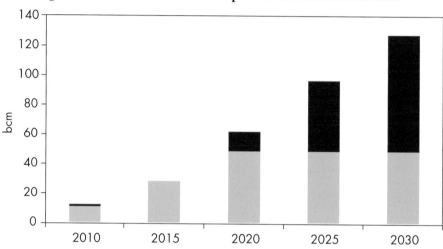

■ Existing, under construction and confirmed capacities
■ Additional capacities required

Coal Supply

Coal Resources and Reserves

China's remaining coal resources are second only to Russia's, totalling 1 003 billion tonnes (General Geological Bureau, 1999). These resources have been defined by exploration and mapping, but only 115 billion tonnes can be regarded as proven reserves, yielding a reserve-to-production ratio of around 50 years at current production levels. More recent assessments conclude that proven reserves could be as high as 192 billion tonnes (Barlow Jonker, 2007). A prospecting programme is currently under way to prove up more resources, using revenues from the competitive tendering of mining rights. China's coal resources lie predominantly outside the major demand centres, which are located in the industrialised east and south-east (Figure 10.10). Some 80% of coal resources lie in the provinces of Shanxi, Inner Mongolia, Shaanxi, Xinjiang, Ningxia, Hebei, Gansu and Qinghai that together make up China's north-west. Shanxi has the most hard coking coal, with 26 billion tonnes, or 38% of China's total resources. Just 6% of coal resources lie in the coastal provinces. To reach consumers, some coal must therefore be transported over very long distances, resulting in congestion on China's national railway system and much higher costs.

Figure 10.10: **China's Coal Resources**

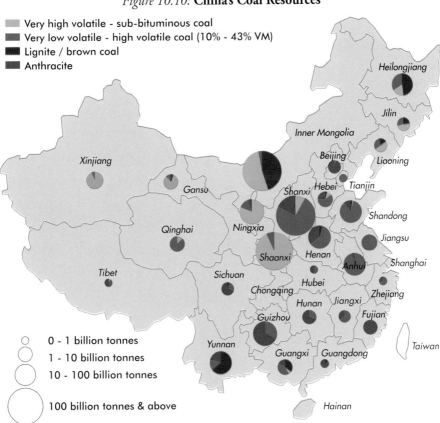

The boundaries and names shown and the designations used on maps included in this publication do not imply official endorsement or acceptance by the IEA.

Source: Beijing HL Consulting (2006).

The most significant coal fields in China date from the Jurassic period and account for approximately 60% of known deposits. They produce good quality steam coal, typically high volatile, with low ash (5% to 10%) and low sulphur (less than 1%). Older coals, from the Carboniferous and Permian periods, are also important, ranging from anthracites to high volatile bituminous coal and much good quality, medium volatile, coking coal (Barlow Jonker, 2005). Ash and sulphur levels can vary enormously, with raw ash of between 20% and 40% not being unusual and sulphur ranging from less than 1% to 5% or more. Almost all coal in China lies deep underground, with little potential for surface mining. Although the average sulphur content of coal mined today is low to medium, it increases with depth in north China, suggesting that it will rise over time. Ash may also increase, as coal from China's deeper seams is less amenable to washing (Minchener, 2004). The average heating value of Chinese coal is below that of internationally traded coals, but comparable with that of coal produced in many other countries: it averages 5 400 kcal/kg in China compared, for example, with 5 600 kcal/kg in the United States (IEA, 2006).

Coal Production

Chinese coal production has surged since the start of the decade in response to strong demand, reaching 1.8 billion tonnes of coal equivalent (2.4 billion tonnes) in 2006. In the Reference Scenario, coal production is projected to increase further to 2 248 Mtce in 2010, 2 604 Mtce in 2015 and 3 334 Mtce in 2030 (Figure 10.11 and Table 10.7). Output of steam coal, which currently accounts for almost 90% of production in volume terms, increases faster than that of coking coal. Shanxi province is expected to continue to dominate coal production, with output from Inner Mongolia, Shaanxi, Ningxia and Guizhou also growing significantly. The coastal provinces produce only 321 Mtce in 2030, an increase of 28% over 2005 levels but only 10% of China's total production, compared to 15% in 2005.

Figure 10.11: **China's Coal Supply**

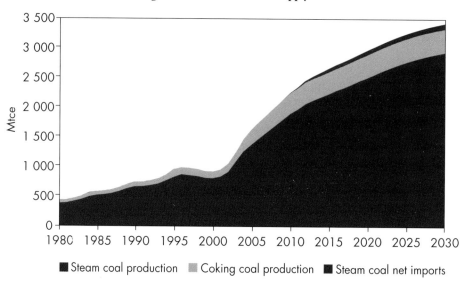

Table 10.7: **China's Coal Production by Type and Region** (Mtce)

	2005	2015	2030
Coastal			
Steam coal	238	259	305
Coking coal	12	11	16
Inland			
Steam coal	1 142	1 978	2 633
Coking coal	244	356	380
Total			
Steam coal	1 380	2 238	2 939
Coking coal	256	366	395

The projected expansion of coal output in the Reference Scenario hinges on the continued restructuring and modernisation of the coal-mining industry and massive investment in the transport infrastructure to move coal to market. Average productivity, at around 500 tonnes per man-year, is extremely low in China's coal mines, mainly because of the very large number of people working in small mines. The manual mining methods used in small mines, which number at least 20 000, often yield just a hundred tonnes per man-year. In contrast, China also has many highly productive coal mines employing large capacity equipment. For example, Shenhua Group operates underground mines on the border of Shaanxi and Inner Mongolia that are considered to be the most efficient in the world. Five mines, Daliuta, Bulianta, Yujialiang, Kangjiatan and Shangwan, have an annual output of over 10 million tonnes each with an overall productivity of over 30 000 tonnes per man-year. In 2006, Bulianta became the first underground mine to have produced over 20 Mt in a single year. At Yujialiang, new longwall mining equipment will shortly be installed to create a 413-metre coal face, the world's widest, to cut panels of coal up to 6 000 metres long (Chadwick, 2007).

Coal demand has been underestimated in successive five-year plans, with the result that the sustained high GDP growth could not be fuelled by flexing the output from the existing and new Key State-Owned Coal Mines (KSOCM), large though this has become. Instead, it has been the huge growth in output from Township and Village Coal Mines (TVCM) that has balanced supply with demand, but at a cost. These mines are dangerous places where to work (see Box 10.3). Beyond that, the very low extraction rate of around 15% means that China is forsaking a large part of its coal resource. The government had planned to close 10 000 TVCMs, but this policy was not fully implemented as

Box 10.3: Coal Mining Safety in China

Modern coal mining, whether at underground or surface operations, poses significant risks to workers who install and operate powerful machinery, often in physically harsh environments, facing the possibility of fire, flood, explosion and collapse. Annual fatalities are one measure of how successfully risks are managed and allow some comparison to be made between the safety of coal mining and that of other industrial sectors. In OECD countries, with long histories of regulation and legislation, mining fatalities are now at a level that makes coal mining a relatively safe occupation. However, this has not always been the case and is certainly not the case in China today, where around 6 000 coal miners are killed each year, predominantly in the thousands of small, inefficient mines. In terms of fatalities per unit of output, China's record in 2004 was very poor, at 3.08 fatalities per million tonnes compared with 0.03 in the United States and 0.24 in India. In 2006, Chinese fatalities fell to 2.04 deaths per Mt.

meeting demand took priority and local resistance proved difficult to overcome. Nevertheless, the intent remains to reduce the total number of small mines to 10 000 by 2010 (NDRC, 2007a).

The Chinese government is acutely aware that its coal industry needs to be modernised and expanded in order to meet future demand. It is giving priority to the exploitation of reserves in Shanxi, Shaanxi and the western part of Inner Mongolia ("Mengxi" area). In the longer term, China could exploit the so-called "western backup coal reserve region", comprising Xinjiang, Gansu, Ningxia and Qinghai. The government has declared that the coal sector is one of seven sectors in which enterprises are to remain under state control, as a minimum through majority shareholdings. Industry consolidation, further mechanisation and the removal of heavy social burdens are NDRC's priorities in the effort to raise productivity and improve resource efficiency, with a target to raise the average recovery rate from 46% achieved in 2005 to 50% by 2010 (NDRC, 2007a).

To this end, NDRC aims to establish six to eight large, highly productive coal-mining companies, each with an annual output of 100 Mt or more, and eight to ten with an output of around 50 Mt each, all located in Shanxi, Shaanxi, Inner Mongolia and Ningxia (NDRC, 2007a). By 2010, 75% of coal output is expected to come from large or medium-sized mines. To reinforce this policy, NDRC announced in October 2006 that only mines producing more than 300 000 tonnes per annum would be considered for approval, a threshold already applied in the major coal-producing areas. An estimated 229 projects, being developed by the top 66 mining companies, could be producing a total of 829 Mt of coal each year by the end of 2011 (Barlow Jonker, 2007). The Shenhua Group, a vertically integrated company, is seen as a role model, employing modern corporate management systems and achieving high levels of productivity. Vertical integration in other companies would promote the policy of "west-east power transmission" from mine-mouth power plants located in the Mengxi area.

Coal Transport

Our Reference Scenario projections imply a continuing need to expand China's inland coal transport infrastructure. Shipments from inland to coastal provinces will need to increase from 507 Mtce in 2005 to 1 060 Mtce in 2030 for steam coal and from 117 to 182 Mtce for coking coal. At present, coal is transported to consumers in China by rail, road, inland waterways and coastal vessels. Shanxi is a key coal supplier to other provinces, exporting around 300 million tonnes, mainly by rail, with the Datong to Qinhuangdao Port (Daqin) line being of particular significance, carrying coal for both domestic and export customers. Despite the remarkable growth in coal production since 2000, China suffered electricity blackouts during 2003-2004 partly due to coal transport bottlenecks.

More than one-half of China's total coal supply is moved by rail. This one billion tonnes of coal accounts for around 44% of national rail freight (Barlow Jonker, 2005). Much coal is transported on older trains, with an average payload of 3 000 tonnes, over routes shared with passenger and other freight trains. There are only two modern rail links dedicated to coal: the 600-km Daqin line and the 588-km line from Shuozhou to Huanghua. Both are highly efficient and carry trains of up to 25 000 tonnes. New investment is being made to expand rail capacity, including the newly expanded Houma-Yueshan link in Shanxi and a potential third dedicated link of 740 km from Baotou in Inner Mongolia to the port at Tangshan. Foreign investment is being encouraged, although passenger routes are likely to be more attractive to investors. The government's Middle- and Long-term Railway Network Construction Plan envisages a range of major investments in new long-distance lines to create separate lines for passenger and freight traffic. If it is implemented, the length of the railway network would reach 100 000 km by 2020. However, rail constraints are likely to persist as coal is increasingly transported over longer distances from the western provinces.

National rail freight costs for coal are high by international standards, particularly on the important west-east routes, where rates are 0.12 yuan/tonne-km ($0.016/tonne-km). In addition to these national rates, coal must often be shipped on local rail links from mines to the national network at roughly double the national per tonne-kilometre rate. *Ad hoc* taxes and rail access fees for unplanned coal movements further complicate the rail transport cost structure in China.

Coastal shipping comprises mainly handysize vessels, chartered at rates that correspond to those in the international freight market. Larger vessels would improve efficiency, but loading facilities for capesize[9] vessels are generally only used for China's coal exports and not for domestic supplies. Around 10% of China's domestic coal supply moves through ports, including Qinhuangdao, for onward shipment to ports such as Shanghai and Guangzhou in the south. In 2005, China's ports handled 370 million tonnes of coal (ABARE, 2006). Expanding port capacity to handle more coal is a priority of the government's Coordinated Seaport Plan, which aims to establish specialised ports for particular commodities. While port capacity is unlikely to become a constraint, there remains uncertainty about the capacity of the rail system to move coal to and from the ports, specialised or not (see above and Minchener, 2007).

Stretches of the Yellow River, Yangtze River and Grand Canal provide an economic supply route for smaller quantities of coal from certain mines. Large trucks of 60-80 tonnes' capacity are used to move coal from mines in Inner Mongolia and Shanxi, but the inefficient and costly transport of coal in small, often overloaded 20-tonne trucks is common throughout China. Fees range from 0.5 to 0.8 yuan per tonne-kilometre, but do not appear to curb demand for journeys of up to 300 km by truck.

Coal Pricing

Prior to 1993, coal prices were administered by the Ministry of Coal and the State Planning Commission. Since then, there has been a gradual move towards free market pricing. Initially, a system of "in-plan" and "outside-plan" prices was used, whereby the National Development and Reform Commission set guideline price bands, within which annual settlements were negotiated for in-plan supplies, these accounting for almost half of total supplies to the power generation sector (Barlow Jonker, 2005). In the case of coking coal, prices are now set by domestic and international market forces. However, with electricity price controls still in place, steam or power coal prices are still agreed in-plan, often below cost, with little price visibility even when "contract" tonnages have been announced. Nevertheless, in 2006 the National Development and Reform Commission announced that coal sales to electricity generating companies would be determined freely, without state involvement.

---

9. Capesize vessels (typically over 150 000 deadweight tonnes) need to sail between oceans via Cape Horn or the Cape of Good Hope as they are too big for the Panama or Suez canals. Handysize vessels (typically 30 000 dwt) are often used for shorter distances.

An outside-plan spot market has existed for a number of years in China, accounting for around half of all coal sales. An important set of marker prices are those quoted at Qinhuangdao, since coal is shipped from there to ports in southern China and also to export destinations. Only four Chinese companies hold coal export licences and government controls, through quotas, taxes and tariffs, mean that traded volumes and prices are not responsive to developments in the international market.

In the future, power coal prices are expected to converge with outside-plan market prices and thus to reflect the full cost of production and supply. China is also expected to remove barriers to coal trade, as encouraged by the World Trade Organization (WTO). The costs of production and transport will then determine supply patterns. With rail transport to Qinhuangdao, port loading and taxes, the FOB cost of coal from Shenhua's mines is currently below $30/tonne, making it very competitive in both domestic and international markets (Figure 10.12). As China continues to consolidate its state-owned mines to create larger, more efficient mining companies like Shenhua,

Figure 10.12: **Coal Prices in China Compared with International Markets**

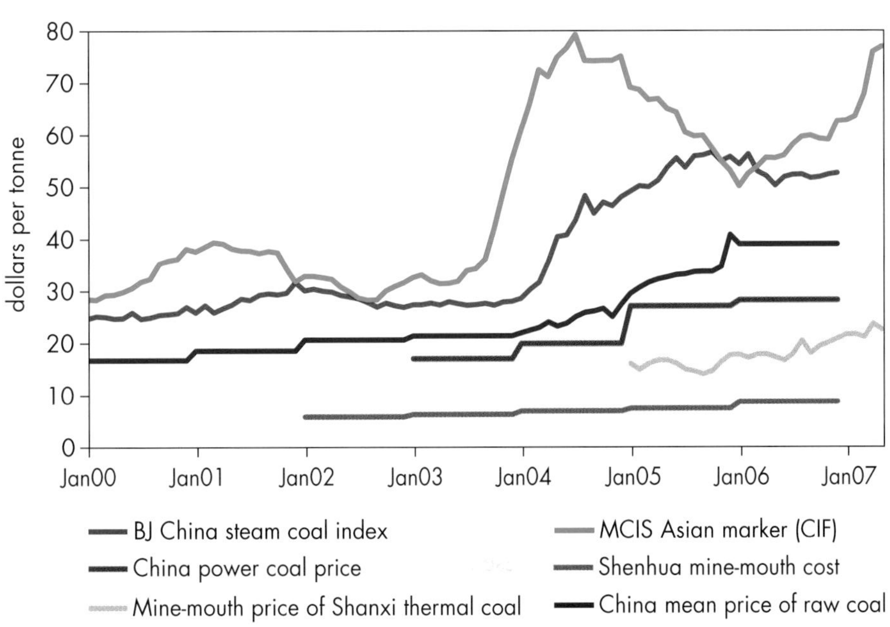

Note: The BJ (Barlow Jonker) China Steam Coal Index is a weighted average of steam coal export prices (FOB) calculated from Chinese customs returns that includes both spot and contract trades of all coal qualities. The MCIS Asian marker price reflects delivered (CIF) prices to ports in Japan, Korea and Chinese Taipei.
Sources: McCloskey Coal Information Services (MCIS), Barlow Jonker (BJ), FACTS, Beijing HL Consulting, Shenhua Group and China Coal Resource (www.sxcoal.com).

indigenous coal supply costs are expected to remain competitive in the short to medium term. In the longer term, transport of coal from remote areas in the western provinces, such as Ningxia and Gansu, would add significantly to the cost of coal supplies and become the most significant cost component, despite some scope for productivity improvements, as current over-manning in the rail sector is reduced. Similarly, costs will increase as deeper reserves are exploited. Some 60% of China's coal resources lie more than 1 000 metres underground (Pan, 2005). Experience elsewhere in the world suggests that economic recovery from such depths is unlikely, on the basis of recent coal prices. Given these factors, imported coal from Australia, Vietnam and Indonesia will become more attractive to coastal users in the south and east.

Coal Trade

China exported 71 Mt of coal in 2005, one-quarter below 2003 when exports peaked at 94 Mt. Imports totalled 25 Mt, mainly to meet demand in southern coastal provinces remote from the major coal-producing regions. In the first half of 2007, China's net coal imports stood at 4 Mt (McCloskey, 2007), a sharp reversal from the 25 Mt net exports of 2006. As recently as 2003, net exports were 83 Mt. This sudden swing to become a net coal importer has had a large impact on international coal trade. The swing began with a spate of serious accidents that led to forced production stoppages in August 2003 at all Shanxi mines (except KSOCMs) and a nationwide rise in coal prices (Huang, 2004). Remedial safety work was not completed until early 2004 by which time international prices had doubled. In the absence of strategic coal stocks and until there is more scope for a free market response within China, price volatility, affecting all regions of the world, is likely to remain a feature of coal trade.

China becomes a net importer of steam coal. By 2015 net imports reach 65 Mtce, which further strengthens the price relationship between domestic and internationally traded coal. In 2030, net imports reach 95 Mtce. By 2015, China will also stop being a net exporter of coking coal. With its high-quality coking coal resources, China has no need to import coking coal despite the gradual growth in demand from 2015 to 2030.

China will continue to export coal, mainly to Chinese Taipei, Korea and Japan, which are closer to the big Chinese export terminals than some of the coastal provinces in the south. Steam-coal exports expand from 66 Mt in 2005 to 178 Mt in 2030 and coking coal exports increase from 5 to 16 Mt during the projection period (Figure 10.13). Overall net imports in 2030 reach 129 Mt.

Coal imports to China mainly meet demand in coastal provinces (see Chapter 13). Key suppliers in 2030 will be Indonesia, Australia, South Africa, Mongolia, Vietnam and Russia.

Figure 10.13: **China's Hard Coal Trade** (million tonnes)

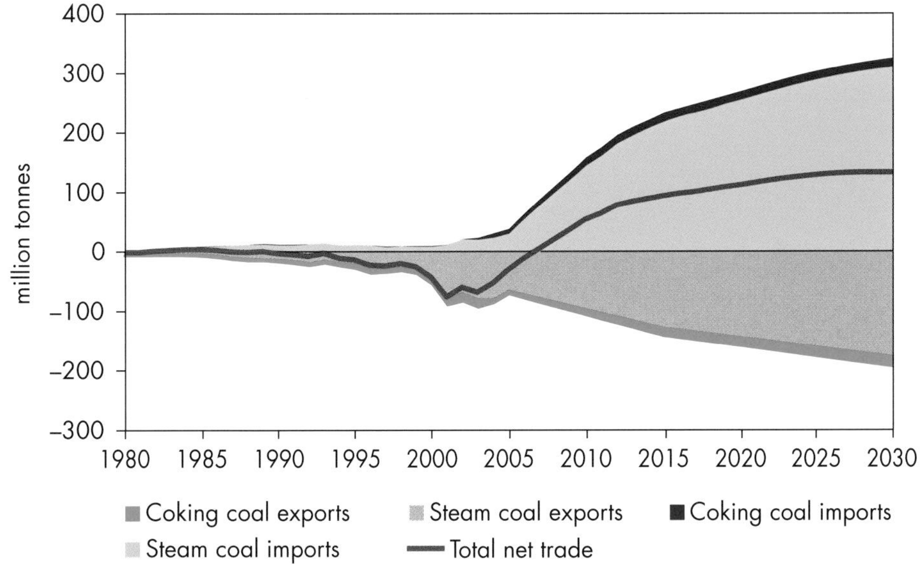

Legend:
■ Coking coal exports   ▨ Steam coal exports   ■ Coking coal imports
▨ Steam coal imports   ── Total net trade

Electricity Supply

Overview of the Power Sector

China has the second-largest electricity market in the world, behind the United States. The country's per-capita electricity consumption is approximately one-fifth of the OECD average. Predominantly coal-fired, total electricity generation reached 2 544 TWh in 2005 and installed capacity 517 GW. Generation is unevenly distributed across the provinces. The ten largest electricity-producing provinces account for 62 % of total generation.

A gradual process of reforms began in the mid-1980s, with the opening-up of generation to investment by parties outside central government. In 1997 most of the assets of the Ministry of Power (nearly all of the grid, as well as 40% of generating capacity) were transferred to the newly formed State Power Corporation. In 2002, the State Power Corporation was split into two transmission companies and five power generation groups. State Grid Corporation of China (SGCC) and China Southern Power Grid (CSG) cover respectively about 80% and 20% of the national market. The five generation entities were initially given around 20 GW of capacity each, with the aim of ensuring that each had less than a 20% market share in any one region. Private investments, often in joint-ventures with local or government-owned corporations, are playing a growing part in generation.

The establishment of the State Electricity Regulatory Commission (SERC) in 2002 marked an important step towards independent electricity regulation. However, the dominant influence of the NDRC on energy policies, planning and project approval is limiting the independence of the SERC. In its 11th Five-Year Plan, the central government announced its intention to improve the market regulation system so as to create a fair and competitive market environment, consolidate the separation of generation from power grids, improve energy efficiency and provide a reliable and high-quality power service at reasonable costs.

Over the last two decades, China's unprecedented economic growth has led to rapid growth in electricity demand. Supply has not always kept pace with demand growth. Power shortages occurred frequently during 2002-2005, notably in the high economic growth regions, such as the provinces of Shanghai, Jiangsu, Fujian, Anhui and Guangdong. A return to a more balanced supply and demand situation was achieved in most regions in 2006, as capacity expanded by more than 100 GW. A few regions still face potential shortages, due to transmission constraints or severe weather conditions. Others, such as the central and north-west regions, have some limited surplus of capacity.

Outlook

Total generation is projected to increase by 4.9 % per year, more than tripling by 2030. In the period 2005-2015, it is projected to grow by 7.8% per year, much faster than the 3.1% average annual rate over the period 2015-2030. At 8 472 TWh, China's generation in 2030 will be comparable to the current level of production in OECD North America and Europe combined. Electricity generation is projected to increase at a slightly slower pace than demand. This is because the combined rate of transmission and distribution losses and own use is projected to decline gradually from the current level of 20% to 16% by 2030.

Coal-fired generation accounted for 78% of total electricity supply in 2005. This share is one of the highest in the world, although lower than in countries such as Australia, South Africa and Poland. Coal will remain the predominant fuel in generation over the projection period (Figure 10.14). Coal-fired generation is expected to increase at an average rate of 4.9% per year.

The expansion of coal-fired generation in China will continue to be based on pulverised coal, with supercritical steam cycle technology expected to play a much greater role in the future, because of its efficiency and emissions advantages. China has made considerable progress in the implementation of state-of-the-art coal-fired generation technologies, by building world-class, larger and more efficient power plants (Table 10.8).

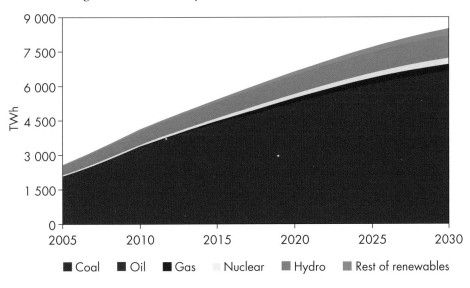

Figure 10.14: **Electricity Generation in China, 2005-2030**

■ Coal   ■ Oil   ■ Gas   Nuclear   ■ Hydro   ■ Rest of renewables

10

Table 10.8: **Coal-Based Power Generation Technology in China**

Technologies	Technological availability	Cost ($ per kW)	Efficiency	Market share in China
Subcritical		500-600	30%-36%	Main base of China's current generating fleet.
Supercritical	Now	600-900	41%*	About half of current new orders.
Ultra-supercritical	Now but needs further R&D to increase efficiency	600-900	43%*	Two 1 000 MW in operation.
IGCC	Now but faces high costs and needs more R&D	1 100-1 400	45%-55%	Twelve units waiting for approval by NDRC.

* Indicates current efficiency. Improvements are expected in the future.
Source: IEA analysis based on data obtained from industry experts.

China added 18 GW of supercritical plant in 2006, bringing total supercritical capacity to about 30 GW. There are about 100 GW of supercritical plant on order, implying that the share of supercritical technology in new capacity will increase significantly over the next few years. The average efficiency of coal-fired generation is expected to improve from 32% in 2005 to 39% in 2030.

Oil plays a limited role in power generation, accounting for 2.4% of the total in 2005. Its share is projected to fall to less than 1% by 2030. Natural gas accounted for just 1% of total generation in 2005. Although gas is not competitive with coal for power generation under current market conditions, China is pursuing policies to diversify the electricity mix and to reduce local pollution, which could boost the share of gas in certain regions. In the Reference Scenario, gas-fired generation is expected to reach 313 TWh by 2030, nearly 4% of total electricity generation.

In recent years, gas-fired generators have suffered from supply constraints and high gas prices. An import infrastructure for LNG is being established as demand increases beyond domestic supply. China received its first LNG shipments at the new Shenzhen terminal in 2006. A second terminal is under construction in the Fujian province, which is expected to be operational in 2009, while construction of a third, in the Shanghai area, started in early 2007, for completion in summer 2009. Gas-fired generation will be concentrated mainly in coastal areas. Its rate of expansion will remain linked to the pace of development of the gas infrastructure and the price of imported LNG – all of which are uncertain.

Nuclear generation amounted to 53 TWh, or 2.1% of total generation in 2005. It is projected to rise five-fold, with its share increasing to 3% of the total by 2030. Installed capacity was 6.6 GW in 2005. Two new reactors were connected to the grid in 2006 and 2007, bringing the total number of reactors in operation to 11 and installed capacity to 8.6 GW.[10] Four reactors with a total capacity of 3.2 GW are under construction. They are expected to be completed by 2010-2011.

The government's target is to have 40 GW in place by 2020, implying that China must add to the plants now operating 31 GW of new plants, as well as 18 GW of nuclear capacity under construction in that year. Although efforts to build more nuclear power plants have been intensified in recent years, the target set by the government seems ambitious given the current level of development, the long construction times and the current global bottlenecks in nuclear component manufacturing, which impose extended delays on delivery. In the Reference Scenario, installed nuclear capacity reaches 21 GW in 2020 and 31 GW in 2030. In this scenario, all new nuclear power plants are assumed to be built in coastal areas.

10. Data are taken from the International Atomic Energy Agency's PRIS database, available at www.iaea.org/programmes/a2/.

China is pursuing a dual objective in nuclear technology: *a)* to adopt a standardised technology for long-term nuclear development and *b)* to develop a home-based technology, so that China becomes self-sufficient in reactor design and construction, as well as other aspects of the fuel cycle. To achieve this, extensive reliance has been placed on technological transfers from leading nuclear technology developers/owners and the accumulation of experience through construction and operation of different reactor designs. China has so far adopted French, Russian and indigenous pressurised water reactors, as well as Canadian pressurised heavy water reactors. The reactor units currently under construction belong to the more advanced Generation II technology. China is currently planning to adopt Generation III technology for the next round of nuclear construction. In December 2006, the Westinghouse AP1000 reactor design was selected for four units to be installed in the Sanmen project in Zheijiang province and the Haiyang project in Shangdong province. Construction is to start in 2009 and the first unit is expected to be operational towards the end of 2013, with subsequent units planned to start up at six-month intervals thereafter.

10

Box 10.4: **Carbon Capture and Storage in China**

As there is no commercial large scale power plant today equipped with carbon capture and storage (CCS) technology, this option is not considered in the Reference or Alternative Policy Scenario. However, the 450 Stabilisation Case (Chapter 5) demonstrates that quick deployment and future development of CCS is needed for a truly sustainable energy future. To achieve the objectives of the 450 Stabilisation Case, China would also need to deploy CCS widely.

China sees CCS as a future technological option for greenhouse-gas emissions abatement and is willing to join international efforts for its development. International co-operation programmes have been initiated with APEC, Canada, the European Union, the United Kingdom, the United States and others (Torrens, 2007). CCS appears in China's 11th Five-Year Plan under the National High Technologies Programme and in the National Medium- and Long-Term Science and Technology Plan Towards 2020. Early opportunities for CCS implementation in China have been documented in an IEA Greenhouse Gas R&D Programme report (IEA GHG, 2002). Twelve such projects would together reduce annual CO_2 emissions by 15 Mt. A summary of the prospects for CO_2 storage is presented by Li *et al.* (2005). Storage estimates vary widely, from 150 Gt to 2 000 Gt. Current experimental projects include:

- A micro-pilot ECBM (Enhanced Coal-Bed Methane Recovery) project in Shanxi province.[11] Initial results indicate a four-fold increase in the performance of the CO_2-ECBM recovery process and that CO_2 storage in high-rank anthracite coal seams is possible in the Qinshui basin (Jianping, 2005).
- A demonstration project at the Yantai IGCC Plant (with the option of future CCS and hydrogen production) (Shisen, 2006). The 300-400 MW demonstration power plant will burn high-sulphur (2-3%) bituminous coal and is planned for 2010. It will closely follow the China Huaneng (CHNG) Greengen first stage plan for a 250 MW IGCC plant. The second phase of the Greengen will have a 400 MW IGCC and CO_2 separation / H_2 power and is planned for operation in 2015.

Thanks to the extensive knowledge base that exists in oil and gas in China, including enhanced oil recovery applications (China ranks first in the world in terms of the proportion of oilfields using EOR), CO_2-EOR could gain early implementation (Qian *et al.*, 2006). CO_2 injection was in use in Daqing between 1990 and 1995 and has been used in Subei.

China is the largest producer of hydroelectricity in the world, producing 397 TWh in 2005. Hydropower is expected to rise to 1 005 TWh in 2030, but its share of total power output will fall from 16% to 12%. China is actively engaged in the development of other sources of renewables to generate electricity, mainly wind power, biomass and solar photovoltaic. Generation from these sources is expected to reach 263 TWh in 2030, about 3% of total electricity. Renewable energy is discussed in more detail later in the chapter.

Combined heat and power (CHP) accounted for over 11% of total installed generating capacity in 2005. The heat from CHP has been mainly used in China in the industrial sector and for central heating in northern cities. Coal remains the predominant fuel, with a small amount of oil use and natural gas now beginning to be used in this application. Efforts are being made to encourage gas-fired CHP schemes. A dozen pilot projects of gas-fuelled trigeneration are being undertaken in Shanghai and Beijing. The potential for CHP is significant, mostly concentrated in Beijing, Tianjing, regions in the Yangtze River Deltas, including Shanghai, Jiangsu and Zhejiang provinces, where direct coal combustion is now forbidden in many cities. The Pearl River Delta regions also have good potential. Power generation from CHP plants is projected to reach 611 TWh in 2030.

11. China's coal-bed methane resources total more than 30 trillion m³ of gas in place (Lako, 2002). Although more analysis is needed to arrive at a representative figure for China, the typical ratio for CO_2 in ECBM is two molecules of CO_2 for one molecule of CH_4 (methane).

Capacity Requirements

In the past twenty years, China has achieved an impressive development of its electricity infrastructure. Installed power generation capacity increased from 66 GW in 1985 to 517 GW in 2005 and 622 GW in 2006.[12] Over 90% of the capacity increase in 2006 was coal-fired. As noted earlier, over 100 GW of new capacity was added in 2006. This was the largest year-on-year increase ever recorded in China or, indeed, in any nation in the world. There has also been significant investment in transmission and distribution as the generation base develops and more load is connected to the system.

This rapid pace of increase in capacity in both generation and the network is expected to continue. Over the projection period, generation investments will lead to capacity additions of 1 312 GW, more than the current installed capacity in the United States (Figure 10.15). Installed capacity will reach 1 775 GW by 2030, nearly as high as the current installed capacity of the United States and the European Union combined.

Figure 10.15: **China's Generating Capacity Additions in the Reference Scenario, 2006-2030**

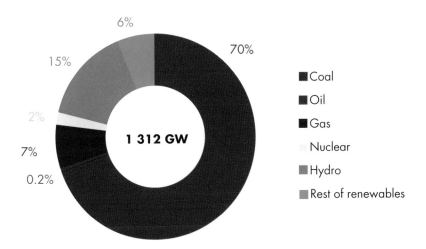

Growth in investment in the electricity infrastructure has underpinned a growing share of electricity in total energy end use. Network expansion has provided greater access to electricity, especially for consumers in rural areas. According to official statistics, by 2005, the electrification rate in China had reached 99%, compared to 73% on average for developing Asia as a whole.

12. Historical capacity data are from the China Electricity Council's website (www.cec.org.cn).

Most of the new generating capacity will be coal-fired, even though the adverse impact of coal-fired generation on the environment is driving China to accelerate the use of nuclear and renewables (mainly hydropower, wind and biomass). The projected increase in coal-fired capacity is equal to about 1.5 times the current installed coal-fired capacity in all OECD countries. The new coal-fired plants are expected to be concentrated in Shanxi, Shaanxi, Inner Mongolia, Guizhou, Yunnan, Henan, Ningxia and Anhui, areas with convenient and economical access to the coal resources. Hydro capacity is projected to reach 300 GW by 2030. Wind power capacity will reach 18 GW by 2015 and 49 GW by 2030, mostly onshore.

With China's generation assets largely under the control of the state, generation investments have been made primarily by state-owned or provincially-owned entities, backed by government funding. Since the structural reforms initiated in the mid-1980s, private investments have played an increasing role.

China is investing in the transmission networks and interconnections, as well as regional power grids. Lack of adequate transmission in some areas has prevented low-cost generation in one province or region from reaching a neighbouring area. In recent years, China has developed interconnections linking the six major regional grids in order to increase capacity to transfer power from the country's resource-rich west to the energy-hungry east, optimising the distribution and use of its existing power resources. Pilot projects are being undertaken for 1 000 MW high-voltage transmission lines. Transmission investments accounted for about 40% of total investment in the power sector in 2006. State Grid reports that, by the end of 2006, the transmission network of 220 kV and above extended over 282 000 km and that 40 000 km of transmission line of 220 kV and above is to be constructed. Some key power transmission projects, such as the West-East Power Transmission Project, are progressing smoothly. China's 2006-2010 plan for grid expansion focuses mainly on the construction of ultra-high-voltage (UHV) lines of 750 kV or higher. Last year, State Grid started construction of China's first UHV transmission line. The 1000 kV alternating current pilot project will link the south-eastern part of Shanxi province with Jingmen city in Hubei province. State Grid also plans to develop an UHV grid connecting the northern, central and eastern regions by 2020.

Electricity Pricing

Historically, until at least 2002, coal was sold to generators at prices below market values. Prices were determined annually at an Annual Coal Procurement Conference run by central government authorities. In 2004, the NDRC adopted a new scheme linking coal prices for electricity generation to on-grid wholesale power prices. The scheme allows generators to pass through to consumers approximately 70% of any increase in coal prices. An increase of 5% or more triggers an automatic adjustment to wholesale electricity prices. With coal prices nearly tripling in the last five years, this reform has saved generators from a financial crisis. Planned retail pricing reforms include a mechanism to adjust end-use prices to reflect fuel cost increases. In the long run, the pricing system is expected to be further reformed to make electricity prices fully cost-reflective and to give timely and adequate signals to consumers and investors.

As wholesale power markets develop, the relationship between coal and power prices may be weakened, as electricity prices will eventually be mainly determined by market forces through the bidding system into the power pool. Coal prices are now to some extent largely determined by markets, but the system of "allocation", under which producers and major users agree contracts on an annual basis, has features which undermine this.

China adopted in the 1960s a so-called Catalogue System for consumer prices that allows for cross-subsidisation between various categories of customer. It allows for preferential treatment for heavy industry, chemical plants, agriculture and irrigation – in terms of both the allocation and price. Time-of-day variations apply to all tariffs, except for those for residential customers and irrigation. However, the differentials are low, being designed to support industry rather than provide incentives for efficient use of energy. Electricity rates vary considerably across the country (Figure 10.16). Each province and major municipality may amend the Catalogue price to suit its own circumstances and policy goals, and may add additional fees. Rates are generally lower in central and western China than in the south and east. Each province has the same rates for each category of customer, regardless of location within the province. For each category, current rates consist mainly of a bundled per-kWh energy charge, plus a capacity charge for large industrial customers.

Thus, power markets in China are not yet structured to provide well-developed market-based price signals. China has tested competitive power pricing in Shanghai and five other provinces, but that pilot programme covered less than 10% of the electricity generated in those areas. Until now, power sector investors have had the security of sales contracts based on a cost-plus pricing regime. The price reform policy seeks to allow the wholesale market to determine tariffs on the generation side, while the government will regulate transmission and distribution prices as well as the relative prices to end users. Sufficiently high electricity prices are needed to attract the necessary investments in power infrastructure.

Figure 10.16: **End-Use Prices by Region and Province, 2006**

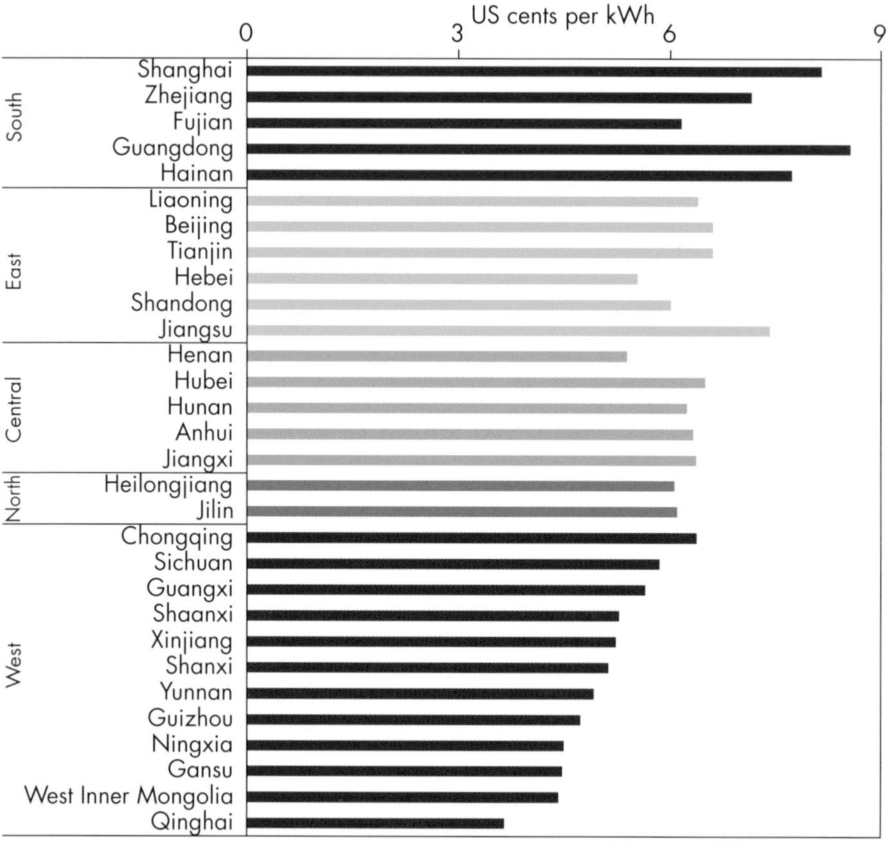

Source: NDRC (2007b).

Power Generation Economics

The costs of alternative generation options have been assessed, on the basis of key parameters related to fuel prices, capital costs, capacity factors and discount rates. The analysis reveals that, in the current Chinese context and without a price on CO_2 emissions, coal is likely to be the most competitive electricity supply source, followed by nuclear and advanced coal. Gas turns out to be the most expensive option, with costs ranging from 4.7 to 7.7 US cents per kWh (Figure 10.17). Coal can provide electricity at costs as low as US cents 2.8 per kWh. The construction cost of nuclear power plants in China is assumed to be in the range of $1 500 to $1 800 per kW. The construction cost of supercritical coal-fired power plants is expected to be in the $600-$900 per kW range.

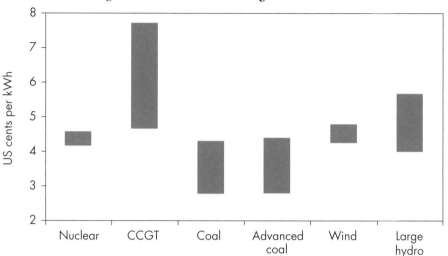

Figure 10.17: **Plant Generating Costs in China**

Source: IEA analysis.

The construction of nuclear reactors is highly capital intensive. Current construction costs of nuclear reactors in China range from $1 769 per kW for Qinshan III-1 to $2 069 per kW for Daya Bay. Construction time is a key element affecting the capital cost of nuclear power.

The capital costs of wind power projects in China are lower than those in North America and Europe because of the lower cost for equipment, land and installation. In China, companies undertaking wind power projects are not required to purchase the five-year warranty on equipment that is standard practice in the European Union and the United States. Thus, investors assume all the risk after two years of operation, which banks will accept in China but not elsewhere. Developers may also opt for equipment with lower-quality steel in order to lower investment requirements. Turbines are also relatively new and untested in China, thus potentially yielding slightly more unpredictable output.

On financing, Chinese domestic wind power developers can borrow up to 80% of the project costs, while there is a 66% limit on borrowing by foreign investors. This tends to lower the return on equity of foreign investor-owned projects. Furthermore, to qualify for clean development mechanism (CDM) credits, projects are required to be at least 51% Chinese-owned, which forces international investors to hand over control of the project to a Chinese partner. Most wind power developers expect that higher electricity prices will be offered to wind projects, as the government moves towards a pricing regime more designed to support renewables and more cost-reflective. Currently, the rate of

10

return to foreign investors on wind projects is around 2% to 4%, which explains their lack of enthusiasm. Domestic investors earn much higher returns, of at least 8%, depending on the particulars of the project.

Renewables

Renewable energy accounted for about 15% of China's total primary energy consumption in 2005. The main renewable energy source now is biomass, mainly used for cooking and heating in rural households. In the electricity sector, hydropower is the main renewable energy source, accounting for 16% of total generation in 2005. Solar thermal heating is also well developed. The Chinese government plans to significantly expand the use of renewable energy in the future, for electricity and heat production and for making transport fuels.

China's *biomass* consumption, at 227 Mtoe in 2005, is the largest in the world. It is almost entirely traditional biomass. Only 3.3 Mtoe were used in power generation in 2005. China's main biomass resources comprise agricultural wastes, scraps from the forestry and forest product industries, and municipal waste. Agricultural wastes are widely distributed across the country. Among them, crop stalks suitable for energy production represent a potential of about 105 Mtoe yearly. In the mid to long term, the forestry sector could provide a yearly potential of 210 Mtoe. Wastes from the processing of agricultural products and manure from livestock farms could, theoretically, also contribute another 80 billion cubic metres of biogas per year. Municipal waste could provide some 16 billion cubic metres of landfill gas. The government target calls for 5.5 GW of biomass-fired generating capacity by 2010 and 30 GW by 2020.

Non-food-grains biofuels are seen as an important means of helping to meet fuel demand in the transport sector. China has so far established two ethanol fuel production bases, with a total yearly production capacity of over 1 Mt. Production of biofuel in China has reached about half a million tonnes yearly.

In the Reference Scenario, total biomass consumption remains broadly unchanged through to 2030. However, the utilisation pattern changes considerably. Traditional biomass consumption falls to 159 Mtoe by 2030. By contrast, demand for electricity and heat from biomass, including industrial on-site generation, is projected to increase, from 8 TWh in 2005 to 110 TWh in 2030, requiring 3.3 Mtoe and 38 Mtoe of biomass fuel respectively. Demand for biofuels reaches 8 Mtoe.

China's economic *hydropower* potential – some 1 750 TWh – is the highest in the world (WEC, 2007). The resources are located mainly on the Yangtze, Lancang, Hongshui and Wujiang rivers. Further hydro development will be undertaken because of its economic advantages and advantages in reducing gaseous emissions. Hydropower is projected to increase from 397 TWh in 2005 to 1 005 TWh in 2030, but its share in total generation will fall from 16% to 12%. The government's 300 GW target is met by 2030 in the Reference Scenario.

The huge Three Gorges Dam on the Yangtze River in Hubei province, when fully completed in 2009, will have a total installed capacity of 18.2 GW, by far the largest hydro generating facility in the world. There are 14.7 GW currently in operation. Construction has recently started on two other very large hydropower plants: the Xiluodu project, located along the Jinsha River in south-western China, which, when completed in 2015, will have a total capacity of 12.6 GW; and the Xiangjiaba project, in Sichuan province, which is projected to be completed also in 2015, with a capacity of 6 GW. Small-scale hydropower plants are widely used. About one-third of China's counties rely on small-scale hydropower as their main power generation source, with a total installed capacity of 50 GW.

Wind power capacity was 1.3 GW in 2005 and doubled in 2006. By the end of that year, there were 91 wind farms in operation in 16 provinces, equipped with 3 311 wind turbines. Besides large wind farms, over 200 000 stand-alone small-scale wind turbines (with an installed capacity of 40 MW) provide electricity to households in remote areas. With its large land mass and long coastline, China has relatively abundant wind resources. Estimates by the China Meteorology Research Institute, based on measurements done at ten metres above ground, indicate a potential of 253 GW for onshore wind power. The institute estimates offshore wind resources to represent an exploitable potential of about 750 GW.

The government's target for large-scale wind turbines is 5 GW in 2010 and 30 GW in 2020. In the Reference Scenario, wind power capacity is projected to reach 49 GW in 2030 and wind power to account for 1.6% of China's electricity supply. Wind power development will need to be accompanied by investment in grid expansion and transmission upgrades.

The domestic manufacturing industry – including joint-ventures – accounted for 45% of the wind-turbine market in 2006. At present, over twenty manufacturers are established in China. By comparison with what is available internationally, there are still gaps in Chinese design and manufacturing capacity for large wind turbines.

Competition between foreign and local manufacturers and suppliers of wind turbines and related components may put downward pressure on wind turbine prices or may force some manufacturers to cut back on quality in order to remain in business. Domestic turbine manufacturers are being supported by a government requirement that more than half of the equipment in the first phase of a wind project must be made in China, with, in the later phases, the domestic manufacturing share increased to 70%. This discrimination against foreign manufacturers and developers is compounded by the prospects of low returns on investment, given low electricity prices and a lack of supportive pricing mechanisms such as feed-in tariffs, and by a lack of flexibility in implementing wind power projects (*e.g.* site selection and wind farm size are determined by the government). Furthermore, there is a lack of information and maps about wind resources.

By the end of 2005, China's installed capacity of *photovoltaic* systems was about 70 MW, of which approximately 50% was used to supply electricity in remote rural areas without grid connection. Since 2000, China's domestic PV industry has grown rapidly, achieving annual PV module production capacity of approximately 300 MW at the end of 2006.

The future potential is very large, as most areas benefit from high solar radiation. The national targets are 300 MW installed by 2010 and 1.8 GW by 2020. PV technology can be expected to make significant advances beyond 2020 along with cost reductions. In the Reference Scenario, China's PV capacity is expected to reach 9 GW in 2030.

China is the world leader in *solar thermal* systems for heating and hot water supply. About 75 million m² of solar collectors are installed in China at present, about half the world total (IEA SHC, 2007). This technology is already cost-effective. The success of the past is likely to continue. The national target for 2010 is 150 million m² and 300 million m² for 2020 (NDRC, 2007c). In the Reference Scenario, the target is expected to be met around 2025.

Policy Framework

In order to encourage the development of renewable energy, China introduced the Renewable Energy Law, which came into effect on 1 January 2006. The law provides for the compulsory connection to the grid of power plants producing electricity from renewables. It stipulates that all energy offered, which is generated from renewable sources, must be purchased and that utilities must provide grid-connection services and related technical support. The law provides a subsidy of yuan 0.25 ($0.032)/kWh for biomass-fired projects, but does not have preferential pricing policies for wind power projects. Instead, the standard price for wind power is determined through competitive tendering. There is no minimum wind power price: each wind project receives an individual on-grid price which varies significantly; from yuan 0.382 ($0.051)/kWh to yuan 0.79 ($0.105)/kWh. There are two types of tendering procedures, one through the central government and another one through the provincial/local government. The law also stipulates that an element to covering construction costs of related power grid facilities may be included in the electricity price.

China designated biofuels as a priority in both its Medium- and Long-Term Development Programme for Renewable Energy[13] and the 11th Five-Year Plan. In fact, China has been subsidising biofuels research and development since the 1980s and carried out trials using ethanol, biodiesel and fermented methane gas as long ago as 1986. Bioethanol (E10) was chosen as an appropriate gasoline replacement and standards were introduced in 2001. Pilot schemes began the following year in five cities. By 2005, bioethanol (E10) was available at petrol stations throughout Heilongjiang, Jilin, Liaoning, Henan and Anhui provinces. Two years later, E10 is also available in Hebei, Jiangsu, Shandong and Hubei provinces and it is expected to spread to more provinces as time goes by. Four bioethanol producers have so far been approved by the government. Sinopec and PetroChina also participate in production and distribute bioethanol through their retail networks. Biodiesel is the other key biofuel, but the industry is relatively less developed. Many projects are being planned, but the relevant standards (necessary in order for biodiesel to be sold in petrol stations) are still awaiting approval.

As the Chinese government works to roll out bioethanol use across the country, at increasing distances from maize- and wheat-producing provinces where stale stocks can be used for production, it is promoting technologies which use cassava, sweet potato, sugar cane, various wood materials and other inputs in an effort to ensure supply. The NDRC suspended the use of edible grains for fuel production at the end of 2006 because of concerns over livestock feed prices and food security. The Ministry of Agriculture then published an Agricultural Biofuel Industry Plan (2007-2015) that aims to develop by 2010 a number of new crop bases capable of meeting growing ethanol and biodiesel demand without competing with the food sector.

The Chinese government set production targets of addtional 2 million tonnes of non-food-grain bioethanol and 200 000 tonnes of biodiesel by 2010, which would be expanded up to 10 million tonnes of bioethanol and 2 million tonnes of biodiesel by 2020 (NDRC, 2007c). Biofuels in the Reference Scenario reach 1.3 million tonnes in 2010 and 5.7 million tonnes in 2020, well below these targets. Issues such as import availability, depletion of food reserves and water resources, difficulties in scaling up energy crop production, the availability of waste oil from restaurants (in the case of biodiesel), and the growing financial burden of subsidisation are likely to persist (IEEJ, 2006). For both biofuels, adequate provision in the distribution networks of Sinopec and PetroChina

10

13. On June 7th, 2007, the State Council reviewed and passed the Medium and Long Term Development Programme for Renewable Energy. The government made it clear that the development of biofuels should not endanger arable land, consume food in great quantity or damage the ecosystem.

will be essential. The Agricultural Plan recognises that new crop production technologies and new crop hybrids will be necessary in order to meet the Five-Year Plan targets, as well as stronger support policies for non-grain fuel ethanol.

Investment

The cumulative investment needed to underpin the projected growth in energy supply in China is $3.7 trillion (in year-2006 dollars) over the period 2006-2030 in the Reference Scenario (Figure 10.18). This corresponds to annual investment of $150 billion. Required investments are lower in the Alternative Policy Scenario (see Chapter 11).

Figure 10.18: **China's Energy Investments in the Reference Scenario, 2006-2030**

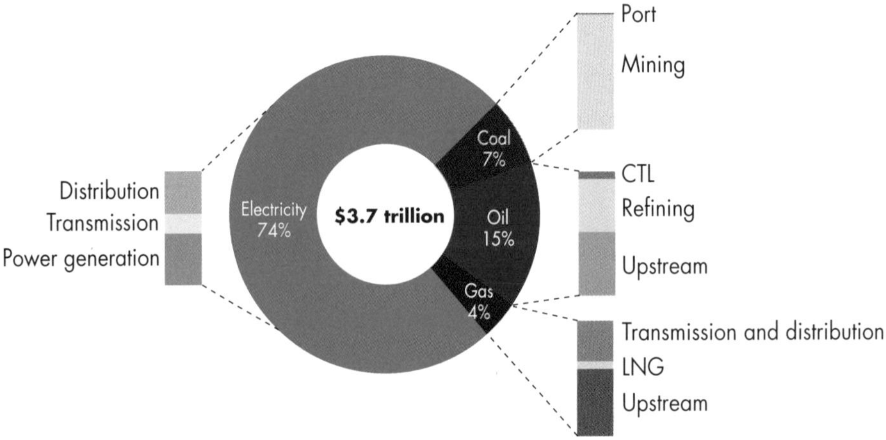

Note: Oil includes investment in biofuels.
Source: IEA analysis.

Cumulative oil and gas investments amount to $715 billion over the projection period. Upstream oil investment accounts for $260 billion, or more than $10 billion per year. Expansion of the oil refining sector adds another $247 billion. Investments in CTL amount to $41 billion. Cumulative gas investments are projected at $168 billion, or $7 billion per year. Exploration and development of new fields will account for 58% of that. LNG investment makes up $11 billion, or 6% of total gas investment.

The Chinese coal industry is currently very profitable. Over the last few years, capital for massive new investment has been raised with relative ease. However, the Reference Scenario requires yet more substantial investment. Cumulative coal investment to 2030 amounts to $251 billion, 42% of the world total. Just 1% of that is spent on ports, the rest on mining itself. Foreign investment could help China achieve its objectives for the sector, but experience to date suggests some significant hurdles remain before the sector can be considered attractive to foreign investors (ADB, 2004). Uncertain validity of good title under evolving Chinese law, unfair and inconsistent enforcement of health, safety and environmental laws and regulations, lack of transparency in the allocation and valuation of coal reserves, weak or missing transport infrastructure and restrictive export controls are the main concerns cited by potential investors. In addition, investment risk is heightened where provincial governments seek to link coal mine investments with much larger downstream investments in power generation, CTL, chemicals and coking plants. To date, only two coal mines have been developed by foreign investors: Antaibo and Daning, both in Shanxi and with a combined capacity of 19 Mt per year (UNDP/World Bank, 2004).

In the Reference Scenario, China's total investment in the electricity sector will account for a quarter of the world's total. Cumulative investments in generation, transmission and distribution over the period to 2030 will amount to $2.8 trillion. Generating capacity needs investment of $1 255 billion over the *Outlook* period, while transmission and distribution require $1 510 billion. Financing these huge investment requirements in the power sector is going to demand funding from both public and private sources. Private finance is expected to play an increasing role in generation investment, but transmission and distribution remain the responsibility of the central government. Transmission investments have been significant in recent years, accounting for about 40% of total investment in the power sector in 2006.

10

CHAPTER 11

ALTERNATIVE POLICY SCENARIO PROJECTIONS

HIGHLIGHTS

- The results of the Alternative Policy Scenario demonstrate that China can move onto a more sustainable economic and environmental path through stricter enforcement of existing policies and introduction of the new ones now being discussed. There is a net economic benefit for Chinese energy consumers and for China as a country – even before the energy-security and environmental implications are taken into account.

- In 2030, the energy savings are comparable to Africa's current consumption. Energy demand nonetheless increases by around 90% between 2005 and 2030. In addition to energy-efficiency improvements along the entire energy chain, realisation of the government's objectives for structural change in the economy is pivotal. It accounts for more than 40% of energy savings. Demand for coal and oil is reduced substantially. In contrast, demand for other fuels – natural gas, nuclear and renewables – increases.

- Coal demand is reduced by 23% in 2030. Close to 40% of the savings comes from reduced electricity use – to which industry contributes two-thirds – which reduces the need to burn coal to generate power. Improved power-generation efficiency accounts for another 30%. More efficient industrial applications and increasing reliance on lighter industries accounts for most of the remainder. Increased reliance on coal-to-liquids only marginally offsets the savings. In this scenario, China remains self-sufficient in coal.

- More efficient use of energy has positive environmental benefits. In 2030, SO_2 emissions are 20% lower, compared with the Reference Scenario. NO_x emissions are stabilised after 2010. An associated benefit is the dramatic reduction in CO_2 emissions, by an impressive 2.6 gigatonnes. In fact, in the Alternative Policy Scenario, CO_2 emissions stabilise soon after 2020.

- Oil demand grows on average by 2.8% per year, reaching 13.4 mb/d in 2030 – 3.2 mb/d less than in the Reference Scenario. Two-thirds of the oil savings originates from the transport sector, notably from the increased fuel efficiency of new vehicles and faster introduction of alternative fuels and vehicles. In 2030, oil imports are 9.7 mb/d, or 3.4 mb/d, lower than in the Reference Scenario. The Chinese oil import bill over the *Outlook* period is $760 billion lower.

- The majority of the measures have a very short payback period. In addition, one dollar invested in more efficient electrical applications saves $3.50 on the supply side. China's efforts to improve the efficiency of vehicles and electrical appliances will impact not only on domestic energy efficiency but also, because China is a net exporter of these products, on global energy efficiency.

Background and Assumptions

China's energy development, like that of most of the rest of the world, is on an unsustainable path. The Reference Scenario projections demonstrate very clearly that, without new government policies and measures or technological breakthroughs, the country's energy needs will continue to grow very fast. This trajectory of rising demand would drive up its dependence on imports of oil, natural gas and coal, add to upward pressure on international energy prices and worsen already dire problems of local pollution. Continuing heavy reliance on fossil fuels would also push up emissions of greenhouse gases and the adverse effects of climate change in China and elsewhere. In short, unchecked growth in energy use poses a serious threat to China's future prosperity and the well-being of the Chinese people. It also carries serious implications for the rest of the world.

Chinese policy makers take these challenges very seriously and have formulated a range of policies to respond to them. Chief among these are interventions aimed at diversifying the country's energy sources, improving energy efficiency and restructuring the economy away from highly energy-intensive activities. Some have already been implemented and are taken into consideration in the Reference Scenario. Other more ambitious actions are still under discussion. The Alternative Policy Scenario takes these into account, presenting a picture of the extent to which they can address China's energy-security and environmental challenges. The benefits of stronger policy action are potentially very large, provided there is effective implementation and strict enforcement on the ground.

Most of the initiatives that China has already adopted are set out in the 11th Five-Year Plan. More than in any other country, energy policy and economic policy in China are inextricably linked. One of the main planks of Chinese economic policy is to effect a fundamental change in the structure of the economy towards less energy-intensive industries and services, with the aim of reducing energy needs per unit of value added produced.[1] Accordingly, the 11th Five-Year Plan contains a target to reduce the country's energy intensity by 20% between 2005 and 2010. It also sets out targets for the share of each energy source in the overall primary energy mix.

Most of the specific policies set out in the plan are included in the Reference Scenario. However, it is not yet clear how some of those are to be implemented or enforced. In these cases, we have assumed that they are not fully implemented in the Reference Scenario. The cloud of uncertainty over implementation is dense in some areas. For example, the plan contains a strong commitment to favour natural gas over other fossil fuels, yet few concrete measures or incentives have so far been announced. In many cases, the

---

1. We call this aim "structural adjustment" in the rest of this chapter.

uncertainty stems from how policies prepared by central government are to be translated into firm action at the local government level. An obvious example is the central government's long-standing intention to reform pricing and introduce taxes on sales of fuel to final consumers. What form this reform will take, the level of taxation and the responsibilities for applying the tax and collecting the revenues are still under discussion.

In the Alternative Policy Scenario, we assume the government at all levels takes stronger action to ensure that policies and measures are implemented fully, are enforced effectively and are supplemented by new measures where necessary. For example, it is assumed that structural change within the economy is more vigorous than in the Reference Scenario and that switching to natural gas is actively promoted. Detailed assumptions for each policy are described below in the sections on each sector.[2]

The results of the Alternative Policy Scenario demonstrate that China can move onto a more sustainable economic and environmental path through stricter enforcement of existing policies and the introduction of new policies already under discussion. In net terms, this result comes at a negative financial cost – *i.e.* net benefit – to energy consumers and to China as a country – even before the energy-security and environmental implications are taken into account.

Key Results

Energy Demand

In the Alternative Policy Scenario, the stricter enforcement of existing policies and the implementation of new policies to promote energy diversification and savings significantly curb the growth in energy demand. Primary demand in 2030 is reduced by about 15%, relative to the Reference Scenario (Table 11.1). This saving is comparable to Africa's entire energy consumption in 2005. The average rate of growth of China's demand is reduced to 2.5% per year, against 3.2% in the Reference Scenario. Demand nonetheless increases by around 90% between 2005 and 2030.

Most of the energy savings in the Alternative Policy Scenario in the short term come from stricter implementation of the central government's policy of closing small and inefficient industrial facilities and power plants and their replacement by plants using modern technologies. In the longer term, structural economic change increasingly drives the faster improvement in energy intensity, alongside more widespread use of efficient energy production

2. The Alternative Policy Scenario also takes into account new policies and measures in all other regions. These policies are assumed not to affect international oil and gas prices, but do lead to lower international coal prices than in the Reference Scenario, with consequences for China's coal and electricity prices and demand. See the Introduction for more details about the methodology and global assumptions underlying the Alternative Policy Scenario.

and consumption technologies. By 2030, structural change accounts for 43% of energy savings, and energy-efficiency improvements and fuel switching for the rest (Figure 11.1). Energy-intensity improvements average 3.3% per year in 2005-2030 (against 2.6% in the Reference Scenario). Energy intensity is reduced by 20% in 2013 relative to 2005. Most of the overall savings in the Alternative Policy Scenario occur in 2015-2030, when more capital stock is added or replaced.

Table 11.1: **China's Primary Energy Demand in the Alternative Policy Scenario** (Mtoe)

	2005	2015	2030	2005-2030*	Difference from the Reference Scenario in 2030	
					Mtoe	%
Coal	1 094	1 743	1 842	2.1%	−556	−23.2
Oil	327	518	653	2.8%	−155	−19.2
Gas	42	126	225	6.9%	25	12.6
Nuclear	14	44	120	9.0%	53	79.4
Hydro	34	75	109	4.8%	23	26.4
Biomass and waste	227	223	255	0.5%	28	12.4
Other renewables	3	14	52	11.9%	19	57.4
Total	**1 742**	**2 743**	**3 256**	**2.5%**	**−563**	**−14.7**

* Average annual rate of growth.

Primary demand for coal and oil is reduced substantially compared with the Reference Scenario. In contrast, demand for all other fuels – natural gas, nuclear and renewables – increases. Coal accounts for 78% of energy savings in the Alternative Policy Scenario in 2030. Coal consumption in 2030 is an eye-catching 23% lower than in the Reference Scenario. Policies directed towards the industrial sector have the most effect – both through structural change and improved energy efficiency. More efficient industrial applications and increasing reliance on lighter industries directly contribute 22% of all the savings in coal use (Figure 11.2). Close to 40% comes from reduced electricity demand – to which industry contributes two-thirds – which cuts the need to burn coal to generate power. More efficient coal-fired power plants and fuel-switching account for another 30%. Coal inputs to coal-to-liquids (CTL) plants increase, marginally offsetting the reductions in other sectors. Despite the overall fall in coal use relative to the Reference Scenario, coal demand still increases by about 70% between 2005 and 2030.

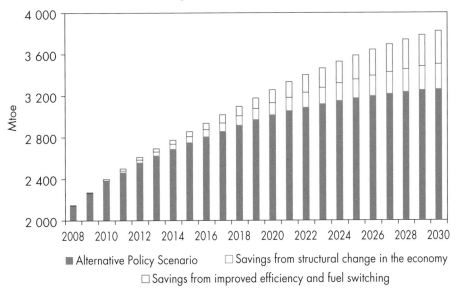

Figure 11.1: **China's Primary Energy Demand in the Alternative Policy Scenario and Savings Relative to the Reference Scenario**

■ Alternative Policy Scenario □ Savings from structural change in the economy
□ Savings from improved efficiency and fuel switching

Figure 11.2: **Savings in China's Primary Coal Demand in the Alternative Policy Scenario Relative to the Reference Scenario, 2030**

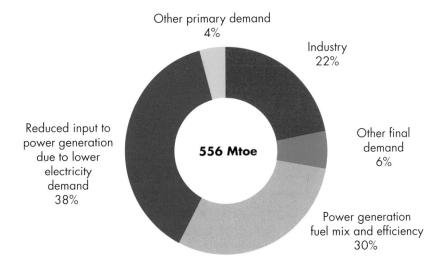

Other primary demand
4%

Industry
22%

Reduced input to
power generation
due to lower
electricity
demand
38%

556 Mtoe

Other final
demand
6%

Power generation
fuel mix and efficiency
30%

Oil savings are also significant, at 19% in 2030, in the Alternative Policy Scenario. Oil demand grows on average by 2.8% per year, reaching 13.4 mb/d in 2030 – 3.2 mb/d less than in the Reference Scenario. By 2015, oil demand is around 60% *higher* than in 2005 – a saving of 0.5 mb/d compared with the

Reference Scenario. More than two-thirds of the oil savings over the whole projections period arise in the transport sector, notably through the increased fuel efficiency of new vehicles and faster introduction of alternative fuels and vehicles (see the section below on Transport). Most of the rest comes from savings in oil use in industry and building.

Contrary to coal and oil, demand for natural gas is *higher* than in the Reference Scenario, because of policies that encourage its use in the residential sector and in power generation in provinces without abundant coal reserves. By 2015, gas demand is 20 bcm, or 15%, higher than in the Reference Scenario and 30 bcm, or 13%, higher by 2030. The rate of growth over the entire projection period is 0.5 percentage points higher than in the Reference Scenario.

A number of Alternative Policy Scenario policies promote the use of renewables and nuclear power. In all, primary demand for energy from non-fossil fuel primary sources is 30% higher under the Alternative Policy Scenario. Nuclear power accounts for 43% of this increase, reaching levels 36% higher than in the Reference Scenario in 2015 and 79% higher in 2030. Hydropower grows more quickly in the Alternative Policy Scenario: by 4.8% per year to 2030, compared to 3.8% in the Reference Scenario. Biomass accounts for 8% of primary energy demand in the Alternative Policy Scenario, compared to 6% in the Reference Scenario. There are two opposing forces at work here: the use of biomass increases in combined heat and power (CHP) plants and for the production of biofuels for transport – in both cases reaching levels in 2030 more than twice as high compared with the Reference Scenario – while households' use of non commercial biomass for cooking and heating drops. Other renewables – wind, geothermal and solar energy – are 57% higher than in the Reference Scenario.

Final energy demand in total is 4.3% lower in 2015 and 17.2% lower in 2030 in the Alternative Policy Scenario than in the Reference Scenario. Electricity demand is 2% lower in 2015, with a gap between the two scenarios projected to grow to 14.5% by 2030, as energy efficiency measures take effect and more capital equipment is replaced. Heat demand is 5% lower in 2015 and 18% lower in 2030, mainly thanks to stricter building codes.

Implications for Energy Markets and Supply Security

In the Alternative Policy Scenario, conventional oil production from Chinese fields is assumed to remain at the same levels as in the Reference Scenario, decreasing to around 2.7 mb/d in 2030.[3] There is, however, an increase in production of oil from non-conventional sources – mainly CTL – and in the output of biofuels, spurred by government policies aimed at limiting oil import

---

3. This is because international oil and gas prices are assumed to be the same in the Alternative Policy Scenario as in the Reference Scenario.

dependence. CTL production reaches 1 mb/d in 2030. This is still below the target of 1 mb/d by 2020 set by Chinese companies, but is 250 kb/d above the Reference Scenario level. The production of biofuels increases to 19 Mtoe by 2030, twice as high as in the Reference Scenario.

SPOTLIGHT

How Much Would More Coal-to-Liquids Increase China's CO_2 Emissions?

In the Alternative Policy Scenario, CTL production reaches 1 mb/d in 2030, emitting between 140 and 250 Mt of CO_2, depending on the production process used. This level of emissions is one-third higher (40 to 60 Mt of CO_2) than in the Reference Scenario, because of the higher level of CTL production. The 140 Mt estimate corresponds to direct coal liquefaction, a process with 60% energy-conversion efficiency, while the 250 Mt figure corresponds to indirect liquefaction, which reaches only around 40% efficiency.

CTL emissions of CO_2 per unit of fuel produced are five to seven times higher than in a conventional refinery. Even so, in 2030 their share in total Chinese emissions in the Alternative Policy Scenario is only 2%, compared with less than 1% in the Reference Scenario.

More worrying are the implications for water. Water needs will be between 350 and 550 million cubic metres per year. Most of the currently planned CTL projects are located near to coal resources, notably in Inner Mongolia and Shanxi, provinces which already face serious water shortages. Priority is at present given to supplying households, irrigation for agriculture and existing power facilities. It may prove to be very difficult for new CTL projects to obtain sufficient water supplies.

China's oil imports continue to increase over the period, albeit at a slower rate. They will still rise significantly – by 6.6 mb/d from 2005 to 2030, reaching 9.7 mb/d in 2030, but are well below the level of 13.1 mb/d in the Reference Scenario. The difference is equivalent to the current combined production of Indonesia and the United Arab Emirates. China's degree of dependence on oil imports differs markedly between the two scenarios. In the Alternative Policy Scenario, the share of imports in total demand rises from 46% in 2005 to 72% in 2030 – seven percentage points lower than in the Reference Scenario. Imports from the Middle East would probably fall the most. The slower growth in oil imports would significantly reduce the level of emergency oil stocks China would need to hold.

The most striking difference concerns coal. While in the Reference Scenario we project China as a growing net importer of coal over the *Outlook* period, in the Alternative Policy Scenario China remains largely self-sufficient in coal. Net imports of coal peak around 24 Mtce in 2015 and decline to 4 Mtce in 2030. Coking coal exports increase, as domestic demand is significantly lower than in the Reference Scenario. By 2030, Chinese coal demand is 23% lower than in the Reference Scenario. However, it still reaches 2 632 Mtce, 68% higher than today.

Unlike oil and coal, natural gas import needs *increase* in the Alternative Policy Scenario, compared with the Reference Scenario (Figure 11.3). As for oil, production from Chinese gas fields is assumed to be the same as in the Reference Scenario, increasing from 51 bcm in 2005 to 111 bcm in 2030, but because gas demand goes up faster, imports rise even faster, reaching 158 bcm in 2030 – 30 bcm, or 24%, more than in the Reference Scenario. By 2030, these additional gas imports would require the construction of 6 additional terminals, were imports to be all in the form of LNG. While financing these terminals would not be difficult, securing affordable LNG supplies might be.

Figure 11.3: **China's Net Energy Imports in the Reference and Alternative Policy Scenarios, 2030**

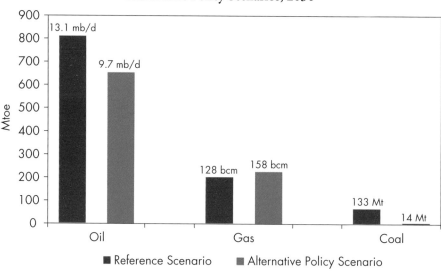

Environmental Implications
Local Air Pollution[4]

In the Alternative Policy Scenario, emissions of SO_2, NO_x and particulate matter ($PM_{2.5}$) are lower than in the Reference Scenario (Figure 11.4). SO_2

4. The projections in this section are based on analysis carried out by the International Institute for Applied Systems Analysis (IIASA) on behalf of the IEA.

emissions began to rise again in 2002, as coal consumption spiked. Coal use accounts for 70 to 80% of SO_2 emissions. In 2005, SO_2 intensity in China was almost three times higher than the average level in the OECD (OECD, 2007). In the Alternative Policy Scenario, thanks to reduced coal use, mainly in industry and power generation, SO_2 emissions peak around 2010 and decline afterwards. In 2030 they are around 20% lower than in the Reference Scenario. NO_x emissions stabilise around their 2010 level and are 20% lower in 2030 compared with the Reference Scenario. As in most other countries, vehicle emissions are the main source of urban NO_x pollution. Road transport policies, including a shift to mass transportation and encouraging the use of alternative fuels, allow NO_x to stabilise, after a sharp rise to 2010. Particulate matter, even more than in the Reference Scenario, continues to fall, following a trend that started in the mid-1990s.

Figure 11.4: **China's Local Pollution Trends in the Reference and Alternative Policy Scenarios**

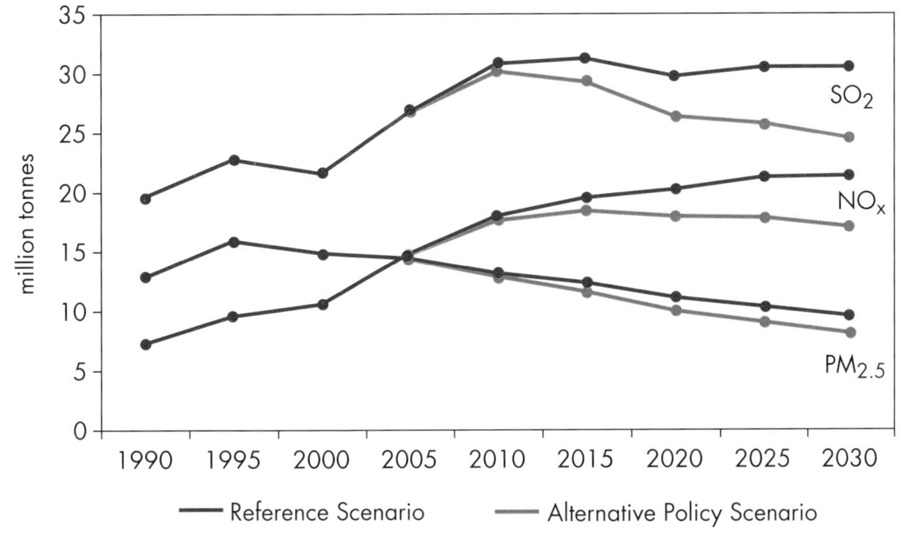

Energy-Related CO_2 Emissions

The policies and measures in China analysed in the Alternative Policy Scenario, while largely intended to alleviate growing energy imports and worsening local pollution, have the additional benefit of curbing the growth in the country's energy-related carbon-dioxide emissions. Climate change poses a particularly important threat to China (Box 11.1). Lower overall energy consumption, combined with a larger share of less carbon-intensive fuels in the primary energy mix, yields savings of 22.5% in emissions by 2030, compared with the Reference Scenario. The total avoided emissions are an impressive 2.6 gigatonnes (Gt).

The slow-down in the growth of CO_2 emissions is already apparent by 2015, when savings reach 0.5 Gt, or 6%. But China's emissions remain on an upward path around 2020, stabilising thereafter around 8.9 Gt. Emissions in China still account for 52% of the global increase over the *Outlook* period, despite falling, in absolute terms, more than those of any other country (Chapter 5).

The largest contribution to avoided CO_2 emissions comes from improved energy efficiency and structural economic change, which together account for close to 70% of total savings (Figure 11.5). Increased fuel economy in vehicles, stricter building codes, and structural change in the economy account for 41% of savings. More efficient motor systems, and more efficient appliances account for another 28%. Increased use of renewables in power generation and increased use of alternative fuels in transport account for a further 17%, switching from coal to gas and improved coal-fired generation efficiency for 8%, and increased use of nuclear for the remaining 6%.

Figure 11.5: **China's CO_2 Emissions in the Alternative Policy Scenario Compared with the Reference Scenario**

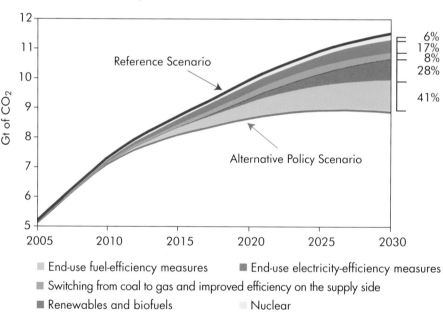

■ End-use fuel-efficiency measures   ■ End-use electricity-efficiency measures
■ Switching from coal to gas and improved efficiency on the supply side
■ Renewables and biofuels   ■ Nuclear

The biggest contribution to emission savings comes from the power sector, where emissions are 1.5 Gt lower than in the Reference Scenario. This sector alone contributes 57% of the saving in emissions in China, thanks to policies aimed at reducing underlying electricity demand, promoting carbon-free

Box 11.1: **Impact of Climate Change on China**

The energy projections in this *Outlook* make climate change an important challenge for the country. China's first National Climate Change Programme (NDRC, 2007), published in June 2007, recognises this and notes that climate change "will bring about significant impacts on China's natural ecosystems and social economic system in the future." This finding echoes those of an IPCC report on impacts of climate change, released a couple of months previously (IPCC, 2007). The issues of most concern are rising sea levels, an increase in the frequency of extreme weather events and glacial retreat in the north-west.

Even without global warming, China's climate presents major challenges. Most of China already experiences seasonal extremes of temperature, precipitation is unevenly distributed and natural disasters have had severe impact. More than one-quarter of China's area is already affected by desertification. Over 18 000 km of coastline and more than 5 000 islands are at risk in the event of a rise in sea level. Any exacerbation of these situations therefore poses a grave threat.

Because of the complexity of the climate system, it is difficult to foresee the regional and local impact of climate change. But there is a consensus among scientists that the repercussions of changes in average temperature will be severe and wide-ranging:

- *Agriculture:* Without effective adaptation measures, agricultural yields are likely to decline and costs to rise. Droughts will become more frequent and longer, further aggravating desertification and reducing productivity; and the frequency of the outbreak of animal disease could increase.
- *Forests and ecosystems:* Effects are already being observed, from shrinking glaciers in the north-west to a thinning of the Tibetan permafrost. Further warming would affect the geographical distribution of forest cover, increase the frequency of insect infestations and disease outbreaks, accelerate the drying-up of lakes and the shrinking of glaciers, and threaten biodiversity.
- *Water:* Further warming would worsen the already declining runoff in China's main rivers and increase the frequency of extreme weather events, such as droughts in the north and floods in the west and south.
- *The coast:* Sea levels have long been rising in China but the trend is accelerating, making adaptation ever more difficult. Hurricanes and storms are likely to become more frequent, aggravating coastal erosion. Groundwater and surface water are likely to become more saline and the homes of millions of people could be flooded.

The greatest danger to human health may be that of more frequent and intense heat waves, which are debilitating in themselves, because of heat stress, but also spread diseases such as malaria and dengue fever.

11

power generation and improving the efficiency of coal-fired generation. Emissions from final use of energy in industry are 0.6 Gt, or 25%, lower in 2030. Transport sector emissions are 0.3 Gt, or 23% lower. Residential, services and agriculture sectors account for the remainder, 0.1 Gt.

Results by Sector

Power Generation

Policy Assumptions and Effects

China's power sector now accounts for almost 40% of total energy consumption and for almost half of total CO_2 emissions. Both these shares are expected to rise in the future, if the government were not to make additional efforts to diversify the electricity supply mix and to reduce CO_2 emissions and local pollution. The Alternative Policy Scenario demonstrates that, if policies to improve the efficiency of the way electricity is used are put in place, electricity generation can be lower by 12% in 2030, compared with the Reference Scenario. Total generation savings in 2030 amount to almost 1 040 TWh and installed capacity is 148 GW lower.

Table 11.2: **Key Policy Assumptions in China's Power Sector in the Alternative Policy Scenario**

Measure	Description	Status	Assumption
Renewable Energy Law	National targets, priority connection, tariffs, renewable energy fund	First introduction Jan. 2006	Greater effort to reach targets
Target for nuclear power	Target for 2020 to have 40 GW in place and 18 GW under construction	Initial stage in place	Greater effort to reach targets
Faster development and deployment of clean coal technologies	More R&D, production of larger, more efficient units	Initial stage in place	Increased efficiency of new power plants
Increased efficiency of existing plants	Measures to increase efficiency of existing plants	Initial stage in place	Increased efficiency of existing power plants
Early retirement of inefficient coal plants	Plans to shut down units less than 50 MW and 100 MW	Initial stage in place	Increase in efficiency of existing stock

Summary of Results

The projected electricity generation mix in 2030 is markedly different from that in the Reference Scenario. While coal continues to be the dominant fuel, its contribution is substantially lower in the Alternative Policy Scenario, contributing 64% of total supply in 2030, as against 78% in the Reference Scenario. Coal-fired generation is cut by around 1 850 TWh, which is close to the total level of coal-based electricity produced in China in 2005. Installed coal-fired capacity is lower by about 350 GW.

Figure 11.6: **Changes in China's Electricity Generation in the Alternative Policy Scenario and Savings Relative to the Reference Scenario, 2030**

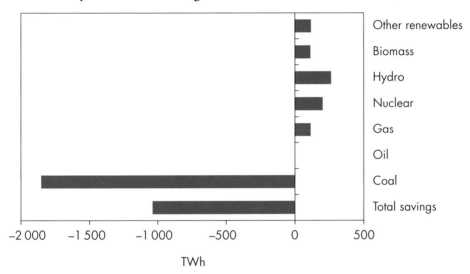

The efficiency of new coal-fired power plants is higher in the Alternative Policy Scenario. After 2015, new power stations are assumed to be as efficient as those built in the OECD. The average gross efficiency increases from 32% in 2005 to 39% in 2030, coming much closer to the OECD average of 42% by 2030. Cleaner technologies include supercritical, ultra-supercritical and integrated gasification combined-cycle plants.

Gas-fired power generation is higher in the Alternative Policy Scenario, reflecting efforts to diversify fuel supplies. Although the generating cost of gas is higher than that of coal, the Chinese government is making efforts to encourage greater use of gas in those provinces where coal resources are not abundant. Natural gas fuels 6% of total generation in 2030, compared with 4% in the Reference Scenario.

Nuclear power rises to 55 GW by 2030, compared with 31 GW in the Reference Scenario, making China one of the largest nuclear power generators in the world with a share of 13% in world nuclear power capacity. The share of nuclear power reaches 6% of total generation in 2030, twice as much as in the Reference Scenario.

Overall, the share of renewable energy in power generation rises steadily, to reach 24% of total electricity generation in 2030, in contrast with the Reference Scenario, where the share of renewables falls from 16% to 15%. The share of hydropower in electricity generation rises to 17% in 2030, up from 12% in the Reference Scenario. There are significant increases, too, in other renewables, notably in wind power and biomass. The increase in renewable energy – other than hydro – is largely driven by the obligations and incentives contained in the Renewable Energy Law. The targets and incentives are summarised in Table 11.3. China is expected to have 311 GW of hydropower in place by 2020, meeting the government target, and 380 GW in 2030. The target for wind power is expected to be exceeded, with wind power reaching 42 GW in 2020 and 79 GW in 2030. Similarly, the target for photovoltaics (PV) is also expected to be surpassed. China is in the process of developing world-class manufacturing industries for wind turbines and solar PV modules, and this is likely to have a strong impact on the domestic electricity market. Installed capacity for biomass is projected to reach 14 GW in 2020, instead of the 30 GW targeted by policy makers. In 2030, however, biomass capacity could reach 39 GW.

Table 11.3: **China's Renewable Electricity Capacity Targets, GW**

Renewable source	Level in 2005	2010 target	2020 target	Potential	Pricing policies
Hydro	117	190	300	400	No premium pricing
Biomass	2.4	5.5	30	n.a.	Feed-in-tariffs premium (0.25 yuan/kWh)
Wind	1.3	5	30	300 onshore and 700 offshore	Competitive tendering
Solar PV	0.07	0.3	1.8	n.a.	Feed-in tariffs based on reasonable production costs and profit

Industry

Policy Assumptions and Effects

The Alternative Policy Scenario assumes more stringent implementation of current Chinese policies to reduce energy consumption in the industrial sector. The main policies are outlined in Table 11.4. These policies have the effect of

Table 11.4: **Key Policy Assumptions in China's Industrial Sector in the Alternative Policy Scenario**

Measure	Description	Status	Assumption
Reduce industrial production compared to the services sector	The 11th Five-Year Plan sets the value added, by 2010, in the services sector as a share of GDP 3 percentage points higher than in 2005. By 2020, value added in services accounts for more than 50% of GDP.	First introduction January 2006	Industry value added loses 7 percentage points of GDP share by 2030. Services comprise 8 percentage points more within GDP in the APS than in the RS by 2030.
Increase industrial energy efficiency by closing inefficient, small-scale plants in energy-intensive industries	For cement production, 250 million tonnes of small capacity will be eliminated; for iron ore production, 100 million tonnes of outdated capacity; all by 2010.	First introduction December 2005	Activity levels for iron and steel and non-metallic minerals grow more slowly compared to the RS.
Taxes	Increases in export taxes for energy-intensive products and decreases in, or elimination of, export credits for steel products.	First introduction mid-2007	
Top 1 000 Enterprises Energy Conservation Programme	To manage energy use at the top energy-consuming 1 008 industrial firms and utilities through energy auditing, reporting, formulating of goals, incentives and investments.	Initial stage in place	
China Medium and Long Term Energy Conservation Plan	Energy-efficiency reduction targets for energy-intensive industries, including iron and steel, cement, ammonia and ethylene, by 2010 and 2020. 2010 targets are consistent with the 11th Five-Year Plan	Initial stage in place	2020 targets are attained

11

reducing industrial consumption, either through structural changes in the economy or via improved energy efficiency. Examples include an increase of 5-10% in taxes on exports of steel and non-metallic minerals and the removal of, or reduction in, export credits for energy-intensive products, like steel products. The Top 1 000 Enterprises Programme, which covers industrial companies that collectively accounted for 33% of Chinese energy consumption in 2004, will, if fully implemented, save 70 Mtoe over five years, starting from 2006 and contribute between a quarter and half of the 20% reduction in energy intensity targeted in the current Five-Year Plan. The policy on industrial structural change is expected to increase industrial energy efficiency through the closure of inefficient, small-scale plants.

Energy efficiency improvements in the Alternative Policy Scenario assume attainment of the 2020 targets for energy efficiency in iron and primary steel, non-metallic minerals and chemicals and petrochemicals which are set out in the China Medium and Long Term Energy Conservation Plan. The energy intensity of iron and steel production (including steel from scrap) is assumed to improve further beyond these policy targets than in the Reference Scenario, thanks to wider availability of scrap steel. An iron to steel ratio of 0.7 is assumed by 2020, rather than by 2030 as in the Reference Scenario. In 2005, it was 0.9. The ratio, nonetheless, remains above that in the United States.

Summary of Results

Industrial energy demand falls by 18% in 2030 in the Alternative Policy Scenario relative to the Reference Scenario. Electricity, heat and fossil-fuel use is lower compared with the Reference Scenario, but biomass and other renewables use is higher. Reduced consumption of coal accounts for 60% of savings, while electricity accounts for 28%, oil for 5% and gas for 4%. Structural change of the overall economy and shifts within the industrial sector towards less energy-intensive production contribute more than 80% of the energy savings. The former is reflected in lower activity levels in the Alternative Policy Scenario. Improved efficiency accounts for the remainder.

Savings in energy use in iron and steel represent the largest share of the savings, resulting from increased use of scrap steel recycling and energy intensity improvements in the Alternative Policy Scenario (Figure 11.7). Blast furnace size and iron ore quality can make considerable differences to energy intensity. The smallest blast furnaces, at less than 100 m³, are 25% less efficient than those larger than 3 000 m³. The largest share of production, 48%, is from blast furnaces sized 300-999 m³, which are 20% less efficient than the largest ones, which themselves make up only 7.1% of production. The "other industries" sub-sector is also significant, especially in electricity savings, because of a shift to manufacturing of lighter, higher value-added products (Table 11.5). Savings in chemicals and petrochemicals are limited, as we do not assume a shift

Table 11.5: **China's Industrial Energy Consumption and Related CO$_2$ Emissions in the Alternative Policy Scenario**

	2005	2015	2030	2005-2030*	Difference from the Reference Scenario in 2030 Mtoe	Difference from the Reference Scenario in 2030 %
Total energy (Mtoe)	478	807	859	2.4%	−187	−18
Iron and steel	132	249	190	1.5%	−83	−30
Non-metallic minerals	109	148	113	0.1%	−30	−21
Chemicals	74	114	103	1.4%	−24	−19
Other industries	163	296	453	4.2%	−50	−10
CO$_2$ emissions (Mt)	1 430	2 048	1 789	0.9%	−584	−25

* Average annual rate of growth.

Figure 11.7: **Industrial Energy Savings in China by Fuel and Industrial Sub-Sector in 2030 in the Alternative Policy Scenario Relative to the Reference Scenario**

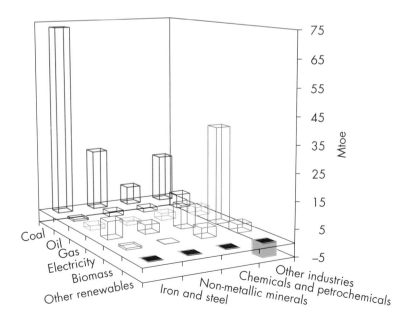

towards gas-based ammonia production. If this process were included, an energy-intensity improvement of more than 20% could be achieved (IEA, 2007). However, we do not expect this to happen, as the supply of natural gas for industry is limited.

Transport

Policy Assumptions and Effects

Over the past few years, the Chinese government has introduced an increasing amount of regulation in the transport sector, with the twin objectives of containing oil import growth stemming from incremental mobility needs and tempering the pollution and congestion that are major consequences of the increase in vehicle ownership (see Chapter 9). A summary of the key policies enacted and proposed is presented in Table 11.6. China introduced mandatory fuel-economy standards for passenger cars and sport utility vehicles (SUVs) in 2006. These will be tightened in 2008. It has also introduced a tax on car ownership that is differentiated according to weight and engine size, to discourage sales of larger and more powerful vehicles. All vehicles also have to comply with scrappage rules. Local governments are supporting, to different degrees, the development of mass transportation; for example, bus and metro rail networks are being expanded substantially in Beijing and Shanghai. Local governments, mainly concerned with curbing pollution, are supporting alternative fuels. A national fuel tax is also under discussion.

In the Alternative Policy Scenario, we assume the following:

■ Fuel efficiency standards are prolonged and tightened. As a result, in 2030, new light-duty vehicles (LDVs) are on average 40% more efficient than 2005 models (compared with 30% in the Reference Scenario), while new trucks are 36% more efficient (25% in the Reference Scenario). In 2030, a new car in China is around 10% more efficient than a new EU model in 2012. New trucks are 15% more efficient than current Japanese new models. These efficiency gains result from improvements in the efficiency of internal combustion engines and the introduction of advanced vehicle technologies, including a higher penetration rate of mild and full hybrid technologies.

■ Cars and trucks are scrapped two years earlier on average than in the Reference Scenario.

■ Public transport develops more quickly than in the Reference Scenario, reducing car usage by 5%.

■ The use of alternative fuels – including coal-based fuels, compressed natural gas, ethanol, biodiesel – is encouraged more than in the Reference Scenario.

Table 11.6: **Key Policy Assumptions in China's Transport Sector in the Alternative Policy Scenario**

Measure	Description	Status	Assumption
Fuel economy standards for LDVs	Phase 1 – maximum fuel consumption standards set for 16 weight classes using New European Driving Cycle, more stringent for heavier classes.	In force for all models as of July 2006	Cars in 2030 40% more efficient than current models. Regulation extended to trucks
	Phase 2 – improvement of 10% over phase 1.	Applies to new models in January 2008, and to all models in January 2009	
Vehicle taxation	Sales taxes for cars, SUVs, medium buses ranging from 3% to 20% of sale price proportional to engine size.	First introduction in April 2006	
Scrappage	Not-for-revenue passenger vehicles to be scrapped after 10/15 years. For-revenues passenger vehicles to be scrapped after 8/10 years (or 500 000 km). Trucks to be scrapped when they reach 10 years (or 400 000 km).	Initial phase in place	Cars and trucks are scrapped 2 years earlier than in the Reference Scenario
Incentive to public transport development and use	Bus Rapid Transit development in major cities including Beijing, Kunming, Shanghai, Tianjin. Subsidy for certain passenger group, *e.g.* students, and low pricing. Favourable terms to prioritise public transport land use.	Initial phase in place	Decrease in private cars and trucks use of 5% in 2030 compared to Reference Scenario
Fuel taxes	Fuel taxes on gasoline and diesel.	Under discussion for implementation before 2010	
Alternative fuels	Support of CNG subsidies on the natural gas refuelling station set-up and land allocation.	Initial phase in place in several municipalities	Biofuel production reaches around 20 Mtoe in 2030
	Biofuels – subsidies for ethanol and biodiesel production. Addition of 2 million tonnes of non-food-grain ethanol in 2010, and 10 million tonnes in 2020. Biodiesel consumption to reach 2 million tonnes in 2020.	In place for ethanol, under discussion for biodiesel	

11

Summary of Results

In the Alternative Policy Scenario, oil savings in the transport sector amount to 2.1 mb/d in 2030, accounting for around two-thirds of the total reduction in China's oil demand, compared with the Reference Scenario (Table 11.7). Oil products still account for the bulk of transport demand in 2030, demonstrating the extent of the challenge of developing commercially-viable alternatives to oil to satisfy mobility needs. Because of the dominant place of road transport in transport energy consumption, new government policies are mainly directed to this sub-sector.

Table 11.7: **China's Transport Energy Consumption and Related CO$_2$ Emissions in the Alternative Policy Scenario**

	2005	2015	2030	2005-2030*	Difference from the Reference Scenario in 2030 Mtoe	%
Road	78	166	283	5.3	−73	−20.5
Cars	*24*	*61*	*121*	*6.8*	*−43*	*−26.2*
Trucks	*31*	*82*	*138*	*6.1*	*−30*	*−17.9*
Other	43	67	84	2.7	−20	−18.9
Total energy (Mtoe)	**121**	**232**	**367**	**4.5**	**−93**	**−20.2**
CO$_2$ emissions (Mt)	**337**	**634**	**961**	**4.3**	**−294**	**−23.4**

* Average annual rate of growth.

As in most other countries, policies that lead to more fuel-efficient vehicles and earlier scrappage produce the largest savings in oil demand. By 2030, those policies combined save 1 mb/d, or around 60%, of the road transport oil savings in the Alternative Policy Scenario (Figure 11.8). Increased use of biofuels and CNG accounts for 14%, modal shifts and reduced fuel consumption in other modes for close to 10%, and other policies – mainly fuel taxes – for the remainder.

Residential

Policy Assumptions and Effects

Strong growth in appliance ownership and residential dwelling space are the main drivers of residential energy use. The relatively poor average efficiency of appliances and of thermal insulation of buildings in China contributes to

Figure 11.8: **Savings in China's Transport Oil Demand in the Alternative Policy Scenario Relative to the Reference Scenario**

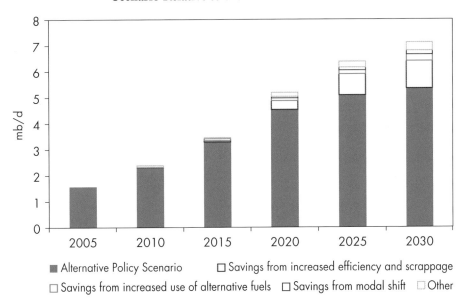

rapidly rising demand. In response to these problems, the Chinese government has introduced a number of mandatory appliances and buildings efficiency standards. Table 11.8 summarises policies that have already been enacted and others that are still under discussion.

China has had minimum energy performance standards for about 20 product groups, including refrigerators, air conditioners, washing machines and fluorescent lamps, since 1989. The recent sharp rise in appliance ownership and usage has prompted the government to adopt a new approach to setting standards. It involves the development of two tiers of standards: one for initial implementation and a more stringent second tier, or *reach*, standard for implementation three to five years later. The lag between the adoption and implementation of the *reach* standards gives manufacturers time to redesign their products[5] and to retool their production facilities, making it easier for them to comply.[6] In the case of room air conditioners, the tier-2 standard will come into effect in 2009, bringing efficiency of mainstream split air conditioners up

5. Much of the production of electrical appliances in China has been geared towards the overseas market. This generates a "spillover" effect for domestic sales, as Chinese exports to OECD countries need to meet the more stringent standards in force there.
6. This practice has been very effective in some OECD countries (Lin and Fridley, 2007). China's standards have, hitherto, typically been implemented within six months of promulgation, giving manufacturers little time to comply. This has resulted in only small incremental improvements in energy efficiency.

Table 11.8: **Policy Assumptions in China's Residential Sector in the Alternative Policy Scenario**

Type of measure	Description	Status	Assumption
Minimum efficiency performance standards – reach standards	Tier 1 mandatory "reach standards" set for refrigerators, air conditioners and colour TVs.	In force for refrigerators as of 2003 and air conditioners in 2005, and TVs as of 2006.	All standards met, strengthened and extended to other appliances
	Tougher Tier 2 "reach standard" for same appliances.	Tier 2 applies to new refrigerators in 2007, air conditioners, and TVs in 2009.	
Energy efficiency labelling	Mandatory labelling for refrigerators, air conditioners and washing machines. Manufacturers must self-declare and send test report of energy efficiency to local energy conservation management and quality inspection departments.	Enacted in March 2005 for refrigerators and air conditioners, and March 2007 for washing machines.	Labelling standards implemented and extended to other products
Building codes and standards	National target to reduce energy consumption in new buildings by 50% by 2010 compared to efficiency of 1980s buildings. Standards for three major climate zones already in place.	Initial stage in place	Standards met and prolonged
Solar thermal	Promotion of building-integrated solar thermal systems in urban areas. Promotion of household solar water heater, solar building and solar cookers in rural areas.	Initial stage in place	Faster deployment of solar thermal

to levels in the United States. China also introduced an energy-efficiency labelling programme, "China Energy Label" for household refrigerators and air conditioners in 2005. Washing machines and central air conditioners have since been added; other appliances, such as flat-screen televisions, are to be included in the future (Fridley *et al.*, 2007). In the Alternative Policy Scenario, it is assumed that efficiency standards and labelling requirements are met and strengthened for the appliances currently covered by *reach* standards. Similar *reach* standards are applied to other appliances with some lags. In 2030, the average new refrigerator is assumed to be 32% more efficient than the 2005 model.[7] Air conditioners will be 35% more efficient than now.

Different building codes and standards are already in place for three regional zones: the heating zone, the hot-summer cold-winter zone, and the hot-summer warm-winter zone, delineated according to winter and summer temperatures. There are also various local standards. The 2007 National Standard for Residential Buildings, which aims to harmonise the current buildings standards, is under consideration. The Alternative Policy Scenario assumes that more stringent building codes are implemented, such that building standards reach today's OECD levels in 2030.

The better enforcement of existing standards, more stringent standards and labelling and tougher building codes which are assumed in the Alternative Policy Scenario ensure faster market penetration of efficient products, so achieving the additional efficiency improvements. For this to happen, it is also assumed that China establishes a set of *implementation and monitoring* systems for appliance energy efficiency standards (that have not yet been adopted) and new building codes. Mandatory and voluntary certification needs to be strengthened and more stringent penalties introduced to ensure the phase-out of inefficient products (Liu, 2006; Jin and Li, 2005). Since some more efficient appliances are more costly in the short term, fiscal incentives such as a reduction of value-added tax for more efficient appliances will need to be offered to consumers and manufacturers. Faster deployment of advanced energy insulation for building and of conservation measures such as combined space heating and cooling systems, is also assumed. Solar water heating in residential buildings has been successfully introduced in rural China and is assumed to be encouraged vigorously in the Alternative Policy Scenario.

Summary of Results

China's residential energy use is 18% lower in 2030 in the Alternative Policy Scenario than in the Reference Scenario. Electricity savings make up 30% of the total. Even so, residential electricity use more than triples, as living standards

---

7. In the Alternative Policy Scenario, the efficiency improvements vary according to the size of the refrigerator. For example, by 2030, refrigerators with volume of 220 litres will be as efficient as European labels A refrigerators.

improve and appliance ownership increases. Large energy savings can be attained through measures to improve appliances efficiency, in view of the rapid growth of appliance ownership and their low efficiency at present compared to OECD models. More stringent efficiency standards for refrigerators and air conditioners alone cut electricity use by 83 TWh in 2020, compared with the Reference Scenario. This is almost equivalent to annual electricity generation by the Three Gorges Dam. By 2030, the saving is equivalent to two such dams (Figure 11.9). Improvements in lighting, water heating, and other appliances bring about savings of around 110 TWh in 2030.

Figure 11.9: **China's Air Conditioner and Refrigerator Electricity Savings in the Alternative Policy Scenario**

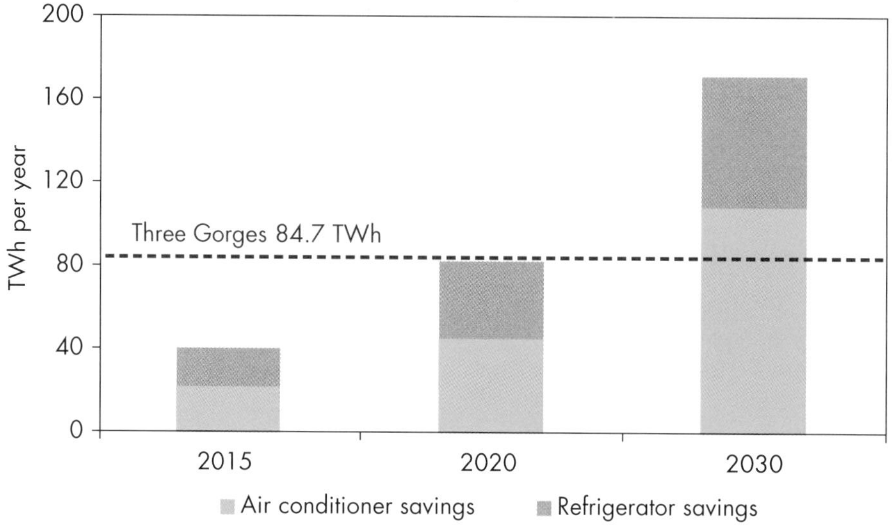

In the Alternative Policy Scenario, coal and oil consumption in China falls by 28% and 16% respectively, compared with the Reference Scenario, as a result of more stringent building codes. Conversely, natural gas consumption is higher, because of policies to contain local pollution that encourage the introduction or expansion of natural gas distribution networks in more cities, compared to the Reference Scenario. The supply of other renewables – mainly solar thermal – is 44% higher than in the Reference Scenario in 2030, accounting for 4% of residential energy demand.

Cost-Effectiveness of Policies

The savings in energy consumption in the Alternative Policy Scenario require a fundamental shift in patterns of investment and spending. Overall, end users invest[8] more, while energy producers invest less. The policies assumed to be

8. The term investment used in this section covers all spending on energy-related equipment, including supply-side infrastructure and energy-using or related equipment and appliances.

implemented in the scenario mean that investment by consumers in more efficient energy-using equipment, over the period 2006-2030, is $308 billion more in total (in undiscounted terms) than in the Reference Scenario. The payback period of the additional demand-side investments is typically very short, ranging from around less than one year for improved industrial motor systems, to less than four years for more efficient cars (Fig. 11.10). Because demand is lower, the need to invest in energy-supply infrastructure is reduced by $385 billion.[9] On average, every additional $1 invested in more efficient energy-using equipment avoids more than $3.5 in investment on the supply side.

Figure 11.10: **Payback Period of Selected Measures in China in the Alternative Policy Scenario**

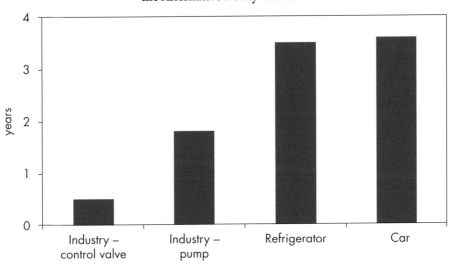

Additional cumulative investment in the industrial sector to achieve the projected energy savings is estimated at only $77 billion over the period 2006 to 2030. Investment in more efficient electrical equipment – mainly motors – accounts for $21 billion. Most measures yield a net financial benefit, as savings in fuel bills over the lifetime of the equipment are larger than the additional investment. The average payback period for the measures considered here is less than three years. The most profitable options are industrial motor systems (see Box 11.2), blast furnaces, continuous casting in the iron and steel sector, and more use of waste heat in the cement sector. The investment cost per unit of energy saved is lower in China than in OECD countries, because of the larger gap in technology between the best available and that in use (see Chapter 9). For example, the difference in the efficiency of the most efficient iron and steel plant and the least efficient is as much as eight times (Tsinghua, 2006).

9. See *WEO-2006* for a detailed description of the methodology.

Box 11.2: **Cost-Effectiveness of Improving Industrial Motor System Efficiency in China**

Industrial motor systems in China consume more than 600 billion kWh of electricity each year, representing more than 50% of industry's total electricity use. Optimisation of motor systems could result in energy-efficiency improvements of 20% or more. Individual industrial components such as motors, pumps, compressors and steam boilers have been improved by manufacturers, but their impact in reducing energy consumption relies on well-designed and optimised systems that use these individual parts efficiently. An optimised system can increase productivity and reliability, as well as save energy. China's Motor Systems Energy Conservation Programme was set up with the intention of establishing a national mechanism to promote motor system efficiency in Chinese industries.[10] Pilot programmes were established by the local energy conservation centres in Jiangsu and Shanghai provinces:

- In Jiangsu, the Sinopec Yangtze Petrochemical Company found that electricity was being wasted by using valves to regulate fluid flow and pressure. Installation of variable speed drives on 34 motors resulted in energy consumption falling from 8.0 kWh per tonne of refined crude oil to 5.8 kWh. This saved 14.1 GWh of electricity and 11 300 tonnes of CO_2 emissions annually. Additional cost savings came from reduced maintenance and prolonged equipment life. Lower noise levels resulted in improved working conditions. The investment cost was paid back within six months.

- In Shanghai, the New Asiatic Pharmaceuticals Company found that the four pumps in the water-cooling system, which use 17 GWh per year, were oversized and unable to respond to seasonal variations in load. Improper pipe configuration and inadequate heat exchanger performance were also found. Appropriate pumps, redesigned pipe configurations and control systems were installed at a cost of 1.2 million yuan ($150 000). This resulted in energy savings of 1.1 GWh, or 49% of system energy usage. Annual cost savings were 660 000 yuan ($82 500). The investment accordingly had payback period of less than two years (Williams, 2005).

The additional cumulative investment in more efficient building shells and more efficient appliances in the Alternative Policy Scenario amounts to $90 billion over the period 2006 to 2030 in the *residential and services* sector – two-thirds of it in electrical equipment, appliances, and solar water heaters. Most measures quickly

10. It was financed by the Chinese government and industry, the United Nations Foundation, the United States Department of Energy and the Energy Foundation. The United Nations Industrial Development Organization, the Lawrence Berkeley National Laboratory and the American Council for an Energy Efficient Economy implemented the programme over 2001-2005.

pay for themselves, as savings in fuel bills exceed the additional investment well within the lifetime of the house, equipment or appliance. More efficient refrigerators are among the most financially attractive options. For example, a refrigerator with a volume of 220 litres consumes about 489 kWh per year, assuming the 2003 MEPS is met. The second-tier reach standard, which is assumed to be implemented in 2007, requires a drop of 10% energy consumption per device to 440 kWh per year. With current electricity prices, the payback period is around three-and-a-half years. More efficient appliances not only save money for manufacturers but also increase the availability of more efficient appliances in other countries. Chinese exports of appliances increased dramatically after 2002. Manufacturers prefer to have a single production line for any single model. It is likely, therefore, that once the standard is established in China, it will also be applied to models for export. As Chinese standards become increasingly stringent, additional energy and financial savings will accrue to energy users in those countries which are China's trading partners (LBNL, 2007).

Additional investment in the *transport sector* amounts to $142 billion, most of which goes to buying light-duty vehicles. Technological advances in road vehicle fuel economy come at a cost of between $150 and $1 800 in 2030 for LDVs compared with the Reference Scenario. Improving vehicle efficiency is cheaper in China than in OECD countries, because the existing fleet there is less efficient and heavier. We estimate the incremental cost to the consumer to improve efficiency by 10% for a medium-weight car is around 1 500 yuan ($185). At current gasoline prices, the payback period would be only about three-and-a-half years – far below the lifetime of the cars. For more powerful cars, for which the cost differential and fuel savings are assumed to be higher, the payback period is only a little longer.

China's cumulative energy-import bill is $684 billion lower over the *Outlook* period in the Alternative Policy Scenario. The savings in oil imports ($760 billion) and coal imports ($47 billion) more than outweigh the increase in the total cost of natural gas imports ($122 billion) (Figure 11.11).

Consumers' additional investment is the consequence of purchasing more efficient, but more expensive cars, industrial motors, appliances and other types of equipment. It reduces Chinese energy demand by 4% in 2015 and 15% in 2030. As a result, significantly less investment is needed in oil, gas, coal and electricity production and distribution. The cumulative reduction in supply-side investment is $385 billion, a fall of 10% compared with the Reference Scenario. Reduced electricity supply investment accounts for most of the overall fall (Figure 11.12). The investment needed in transmission and distribution networks is $345 billion lower. Cumulative investment in power generation is marginally higher, some $30 billion. In the Alternative Policy Scenario the average investment per kWh is higher compared with the Reference Scenario, because of the increased share in the fuel mix of capital-intensive renewables and nuclear power.

Investment in fossil-fuel supply is $70 billion lower in the Alternative Policy Scenario, compared with the Reference Scenario. The increase in natural gas investment – mainly due to additional LNG and transmission lines – is not as great as the decrease in investment for the exploration, development and transportation of oil and coal.

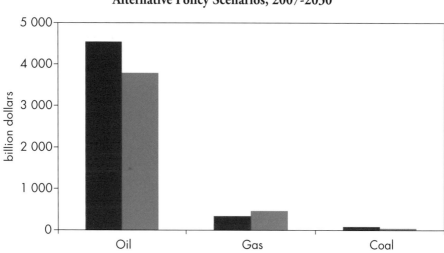

Figure 11.11: **Cumulative Import Bill in the Reference and Alternative Policy Scenarios, 2007-2030**

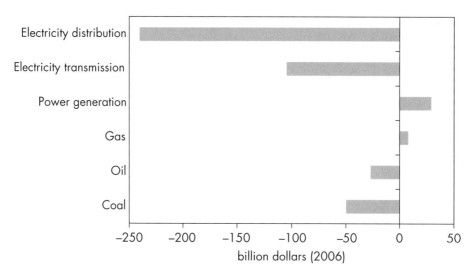

Figure 11.12: **Change in Energy Investment in the Alternative Policy Scenario Compared with the Reference Scenario, 2006-2030**

CHAPTER 12

HIGH GROWTH SCENARIO PROJECTIONS

HIGHLIGHTS

- There is a strong possibility of an even higher rate of gross domestic product growth in China than we assume in the Reference Scenario. The High Growth Scenario analyses the energy-related consequences for China and the rest of the world of China's GDP growing at an annual average rate of 7.5% between 2005 and 2030 – 1.5 percentage points higher than in the Reference Scenario (though GDP grows more slowly than of late).

- China's total primary energy demand in 2030 reaches a level about 23% higher in the High Growth Scenario than in the Reference Scenario. Coal makes up 59% of the difference. Oil demand reaches 21.4 mb/d – 30% more than in the Reference Scenario – about two-thirds of this increase coming from the transport sector, where the vehicle stock reaches 410 million in 2030 (compared with 270 million in the Reference Scenario). Gas demand also grows faster, mainly driven by the power sector. The difference in total primary energy demand is already 10% by 2015 – this is bigger in absolute terms than Canada's energy demand in 2005.

- China relies more on imported fuels in the High Growth Scenario, heightening worries about energy security. Though coal production in China rises by 19% in 2030 with higher coal prices relative to the Reference Scenario, the development of coal mining and the inland transport system does not keep pace with the rapid demand growth and the country's dependence on coal imports rises. Oil imports are 31% higher in 2030, at 17.2 mb/d. China becomes the world's biggest oil importer before 2030.

- The cost of China's energy imports rises sharply in the High Growth Scenario. In total, China's cumulative fuel import bill, at $9.3 trillion (in year-2006 dollars), costs $3.4 trillion more than in the Reference Scenario. Investment requirements in supply infrastructure are $5.1 trillion in year-2006 dollars, $1.4 trillion (36%) more than in the Reference Scenario.

- Energy-related CO_2 emissions increase by 2.6 Gt, or 23%, in 2030 relative to the Reference Scenario. This increase is almost equivalent to the current level of emissions of Russia and Japan combined. In 2030, China's emissions approach those of the OECD in total. Local pollution would also worsen markedly if the government did not respond vigorously.

- On the other hand, higher economic growth would bring substantial social and economic gains to China and benefit the economies of many other nations too. If this growth were associated with stronger policy efforts in China to adjust the structure of the economy and with global efforts to improve energy efficiency, diversify energy sources and mitigate the negative environmental and other consequences of higher energy use (as described in the Alternative Policy Scenario), the net benefits would be yet more substantial.

Background and Assumptions

The rate of China's economic growth is a major source of uncertainty about the country's energy-demand prospects. The projections in the *Outlook* are highly sensitive to the underlying assumptions about GDP growth – the main driver of demand for energy services. Were China's economy to grow significantly faster than assumed in the Reference Scenario, its energy demand could turn out to be much higher by the end of the projection period. Recent experience highlights just how uncertain the outlook is for China's economic and energy-demand growth. The economy grew by 11.1% in 2006, while growth in energy demand over the period 2002-2005 had already averaged 12.9% per year, compared with 3.2% in 1980-2002. As a result, energy consumption in 2006 reached the level that many analysts, only a few years ago, predicted China would reach by 2020.

Under-predictions of energy demand have been largely caused by assumptions about GDP growth that proved to be too low, partly because they failed to take account of the positive impact on the growth of the Chinese economy of trade liberalisation and market-oriented structural reform. Strong export demand and investment – in particular in heavy industry – were largely responsible for the acceleration in the GDP growth rate from an annual average of 8% in 1997-2002 to over 10% per annum from 2002 to 2006. Surging industrial production in energy-intensive sectors is the main reason for the recent reversal in the long-term trend of declining primary energy intensity. The government's 11th Five-Year Plan aims to moderate economic growth to 7.5% per year between 2006 and 2010, but there are few signs as yet that this goal will be attained. Indeed, in the first half of 2007, GDP growth exceeded 11%.

The High Growth Scenario allows us to test the sensitivity of energy demand and supply to an assumed higher rate of GDP growth in China (and India – see Chapter 19) and to analyse the implications for energy trade, investment needs and the environment in China itself and the rest of the world. For China, we assume that the main impetus to growth in this scenario is sustained high investment and continued rapid productivity gains, as the government pushes ahead with reforms to increase the role of the private sector and to open up the economy to foreign investment. China still has a huge labour surplus in the agricultural sector, with relatively low productivity. The movement of labour from the agricultural sector to the industrial and services sectors, and the concomitant urbanisation, could further raise productivity. This would lift millions of people out of poverty, narrow the urban-rural income gap and create a middle class comparable in both size and income to that of the European Union and the United States. Rapidly expanding tertiary education would continue to upgrade China's human capital and contribute to more research and development.

In the High Growth Scenario, we assume that China's GDP grows at an annual average rate of 7.5% in 2005-2030 – 1.5 percentage points higher than in the Reference Scenario. In effect, the slow-down in the rate of growth of the economy is assumed to occur more gradually than in the Reference Scenario. The difference in the growth rate between the two scenarios widens from 1.3 percentage points in 2005-2015 to 1.6 percentage points in 2015-2030. By 2015, China's GDP is 10% higher than in the Reference Scenario. By 2030, it is 42% higher. For the sake of simplicity, the overall economic structure is assumed to be the same as in the Reference Scenario. However, energy prices are higher in the High Growth Scenario, because of higher energy demand from China and India and supply-side constraints. More detail about the methodology used to generate the High Growth Scenario projections can be found in the Introduction and Chapter 3.

12

Energy Demand

In the High Growth Scenario, stronger economic growth raises industrial output, building construction, vehicle and electrical appliance ownership and demand for space and water heating and cooling. All these factors drive up energy demand. Total primary energy demand is projected to grow from 1 742 Mtoe in 2005 to 4 691 Mtoe in 2030, 872 Mtoe or 23% higher than the Reference Scenario in 2030 (Table 12.1). The difference is comparable to energy demand today in Japan and Germany combined. Total primary energy demand grows on average by 4% per annum, 0.8 percentage points higher than in the Reference Scenario.

As in the Reference Scenario, coal remains the dominant energy source in China's primary energy mix in the High Growth Scenario (Figure 12.1). Its share reaches 62% in 2030 – almost the same as in the Reference Scenario. Two-thirds of the additional coal is required for power generation, as electricity demand grows fast and coal remains the cheapest option for power generation. Demand for oil grows faster than for any other fuel in the High Growth Scenario, as demand in the transport sector surges in response to higher incomes. Oil demand grows on average by 4.8% per year, reaching 21.4 mb/d in 2030 – 30% more than in the Reference Scenario. Almost two-thirds of incremental oil demand comes from the transport sector. By 2015, oil demand is already 1.7 mb/d, or 15%, higher than in the Reference Scenario. Natural gas demand grows by 7.8% per year over the Outlook period. It reaches almost 150 bcm in 2015 and 330 bcm in 2030, by which time it is 38% higher than in the Reference Scenario.

Table 12.1: **China's Energy Demand in the High Growth Scenario** (Mtoe)

	2005	2015	2030	2005-2030*	Difference from the Reference Scenario in 2030	
					Mtoe	%
Coal	1 094	2 037	2 910	4.0%	512	21
Oil	327	626	1 048	4.8%	240	30
Gas	42	125	276	7.8%	77	38
Nuclear	14	34	82	7.4%	16	24
Hydro	34	65	100	4.4%	13	15
Biomass and waste	227	235	231	0.1%	4	2
Other renewables	3	13	43	11.1%	10	31
Total	**1 742**	**3 135**	**4 691**	**4.0%**	**872**	**23**

* Average annual rate of growth.

Figure 12.1: **Incremental Primary Energy Demand by Fuel in China in the Reference and High Growth Scenarios, 2005-2030**

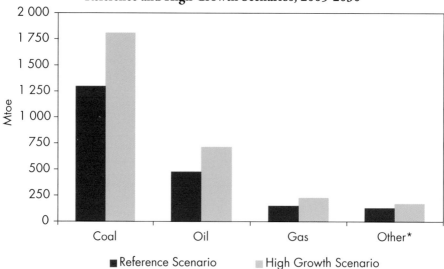

*Includes nuclear, hydro, biomass and waste and other renewables.

Demand for non-fossil energy sources generally grows more rapidly in the High Growth Scenario, but not as fast as demand for fossil fuels. Nuclear energy demand is 24% higher in 2030 than in the Reference Scenario. The Chinese government's target to have 40 GW in place by 2020 is not expected to be met in this scenario, as policies are assumed to be the same as in the Reference Scenario (see Chapter 10). Hydropower is only 15% higher, and faster growth would require a strong push by the government. Biomass use, mainly in the residential sector, drops with higher household incomes, though this is partially offset by stronger demand for power generation and biofuels. Other renewables grow significantly, but their share in total primary demand remains small.

Faster economic growth fosters quicker replacement of old and inefficient capital stock, driving down energy intensity at a brisker pace than in the Reference Scenario. Energy intensity falls on average by 3.2% per year, or 0.6 percentage points more than in the Reference Scenario.

Total final energy consumption in China is projected to grow on average by 3.8% per year in 2005-2030 in the High Growth Scenario, 0.8 percentage points more than in the Reference Scenario. Value added in *industry* grows on average by 7.5% per year, compared with 6% in the Reference Scenario. As in the Reference Scenario, industry is the main driver of energy demand, accounting for 45% of total final consumption in 2030. Industry contributes about half of the difference in final energy demand between the two scenarios (Figure 12.2), most of the difference in final coal demand and nearly three-quarters of the difference in final electricity demand.

12

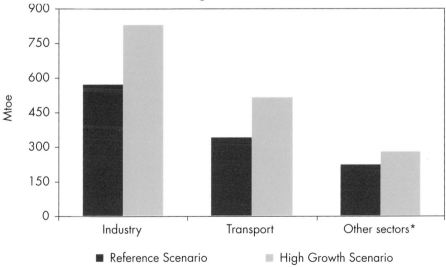

Figure 12.2: **Incremental Final Energy Demand by Sector in China in the Reference and High Growth Scenarios, 2005-2030**

*Includes residential, services and agriculture sectors.

Of the main final sectors, the *transport* sector sees the biggest increase in demand in percentage terms in the High Growth Scenario compared with the Reference Scenario. By 2030, transport energy use is 38% higher. The largest contribution to incremental oil demand comes from this sector, mainly as a result of rapidly rising vehicle ownership. Vehicle ownership, which is closely linked to per-capita income, jumps from 27 vehicles per 1 000 people in 2005 to 285 in 2030, which is four-fifths of Korea's level in 2005. The vehicle stock grows to 410 million in 2030, compared with 270 million in the Reference Scenario (Figure 12.3). The number of passenger cars on the road in China reaches almost 300 million in 2030, more than ten times the current level and almost 50% more than in the Reference Scenario. Oil demand for road transport reaches 9.9 mb/d in 2030 – up from 7.1 mb/d in the Reference Scenario. Demand for biofuels also expands significantly.

Figure 12.3: **Vehicle Stock in the Reference and High Growth Scenarios, Compared with Selected Countries**

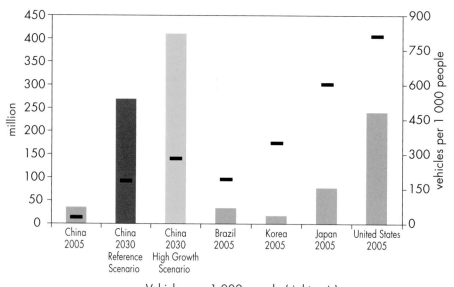

Sources: International Road Federation (2006); NBS; IEA analysis.

In the *residential* sector, building stock and electrical appliance ownership do not grow as fast as economic growth, as they already approach saturation levels in 2030 in the Reference Scenario. However, the switch from coal and biomass use associated with higher incomes boosts demand for oil and gas for space and water heating. Demand in the residential, services and agriculture sectors combined is 9% higher in 2030 than in the Reference Scenario.

Implications for Energy Markets and Supply Security

Higher coal demand in China and India in the High Growth Scenario pushes up global coal demand and, therefore, prices (see Chapter 3). As a result, coal supply in China grows more quickly than in the Reference Scenario. Total coal production is projected to increase to 2 798 million tonnes of coal equivalent in 2015 and 3 959 Mtce in 2030, 19%, or 625 Mtce, more than in the Reference Scenario. This is a volume equivalent to the combined production of Australia, India and Colombia in 2005. Steam coal accounts for 95% of incremental coal supply in 2030. Almost all of the increase in coal production comes from the inland region, which produces 90% of total production in 2030, in particular the Shanxi, Inner Mongolia, Henan and Shaanxi provinces.

Though coal production grows much faster in the High Growth Scenario, it nonetheless fails to keep pace with growing demand. As a result, China's net coal imports rise to reach 199 Mtce in 2030, 106 Mtce (115%) more than in the Reference Scenario (Figure 12.4). The extent of the country's dependence on coal imports is 5% in 2030 – up from 3% in the Reference Scenario. Additional imports of steam coal come mainly from Australia and Indonesia, while Australia and the United States contribute significantly to meeting incremental Chinese coking coal import demand in 2030.

12

Figure 12.4: **China's Net Coal Imports in the Reference and High Growth Scenarios**

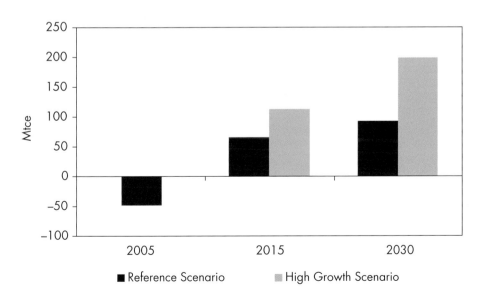

■ Reference Scenario ▧ High Growth Scenario

Higher international prices also boost China's production of crude oil in the High Growth Scenario, to 4.5 mb/d in 2015 and 4.3 mb/d in 2030, which is 830 kb/d, or one-quarter, higher than in the Reference Scenario. Conventional oil production in 2030 increases by 270 kb/d, through the more active use of enhanced oil recovery techniques (Box 12.1). Coal-to-liquids (CTL) production is also boosted by higher oil prices, reaching 1.3 mb/d in 2030 – 560 kb/d more than in the Reference Scenario.

Box 12.1: Prospects for Enhanced Oil Recovery in China

The production profile of China's existing oilfields, as well as of those awaiting development, is a major source of uncertainty. Future rates of production decline (see Chapter 1) will be affected by both geology and economic factors. Currently, only 174 out of a total of 492 fields in production benefit from improved and enhanced recovery techniques. On average, those fields have an expected average recovery rate of 27% of the oil originally in place[1], though rates vary widely, from 10% to 60%.

In the High Growth Scenario, we assume enhanced recovery techniques boost recovery rates on average by 10 percentage points, because the techniques are more widely used with higher oil prices. Table 12.2 shows the results of simulating the implementation of those techniques at the main producing fields where such techniques are not yet used and at 200 discovered fields which are awaiting development and have reserves of more than 20 million barrels (Mb) each. The resulting increase in total production, compared with the Reference Scenario, is more than 360 kb/d in 2015 and 270 kb/d in 2030. Most of the increase would come from fields yet to be developed. It is easier to achieve a high recovery rate using enhanced oil recovery techniques when they are planned and implemented early in the life of the oilfield.

Although our field-by-field analysis does not cover all the 318 fields which are known to be producing through primary recovery, only 61 of them hold initial reserves higher than 50 Mb and their average rate of depletion is close to 40%.

Oil imports in 2015 increase from 7.1 mb/d in the Reference Scenario to 8.3 mb/d in the High Growth Scenario (Figure 12.5). In 2030, China needs to import 17.2 mb/d, which is 4.1 mb/d, or 31%, more than in the Reference Scenario. This increment is larger than Iran's entire production in 2006. China becomes the world's biggest oil importer before 2030. With increased oil imports, China becomes more vulnerable to supply disruptions and needs to spend more on emergency oil stocks to maintain the same forward coverage of oil imports.

1. The total oil content of an oil reservoir.

Table 12.2: **Incremental Oil Production from the Deployment of Enhanced Oil Recovery in the High Growth Scenario*** (kb/d)

	2010	2015	2030
Top 11 existing fields**	140	148	122
Identified non-producing fields	151	214	150
With reserves higher than 50 Mb (31 fields)	*102*	*130*	*89*
With reserves higher than 30 Mb (66 fields)	*36*	*50*	*35*
With reserves higher than 20 Mb (103 fields)	*13*	*34*	*26*
Total	**291**	**362**	**272**

* Relative to the Reference Scenario.
** Xingshugang, Lamadian, Tahe Complex, Huanxiling, Suizhong 36-1, Ansai, Chengdao only. Enhanced recovery is already deployed at the other fields.

Figure 12.5: **China's Oil Demand, Production and Net Imports in the Reference and High Growth Scenarios**

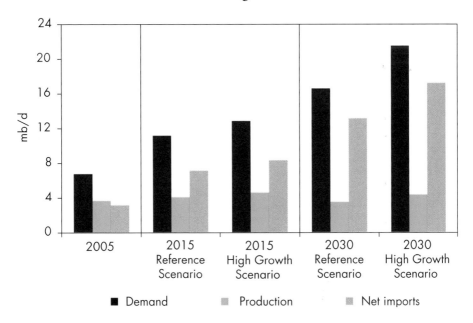

In the High Growth Scenario, with higher gas prices, domestic gas supply exceeds that in the Reference Scenario, but to a marginal extent because of the limited technical scope for capacity expansion. Domestic gas supply reaches 114 bcm in 2030, slightly more than in the Reference Scenario, but only 35% of demand. Accordingly, gas imports jump sharply, from 28 bcm in 2015 in the Reference Scenario to 47 bcm in the High Growth Scenario, and from

128 bcm to 216 bcm in 2030. In order to meet these higher imports, substantial import capacity would be needed, in addition to the Turkmenistan-Kazakhstan-China gas pipeline that is assumed in the Reference Scenario to come on stream by the middle of next decade and the three LNG plants assumed to be fully operational before 2012.

China's electricity generation is projected to reach 10 804 TWh in 2030 in the High Growth Scenario – 28% more than in the Reference Scenario. Installed capacity is 494 GW, or 27%, more, of which 359 GW is coal-fired. In absolute terms, coal-fired power generation expands most in the High Growth Scenario, accounting for 77% of total incremental power generation between 2005 and 2030. Gas-fired power generation expands most in percentage terms, by 13% per year on average in 2005-2030, rising from 1% in 2005 to 5% of the total power generation fuel mix in 2030, compared with 4% in the Reference Scenario. With higher incomes, the shift from coal to gas in the power generation mix is accelerated but it is constrained by higher gas prices.

The cost of China's energy imports rises sharply in the High Growth Scenario. China's cumulative oil import bill in 2006-2030 increases sharply as a result of the much higher imports and higher prices. At $7.1 trillion in year-2006 dollars, it is $2.4 trillion (53%) more than in the Reference Scenario (Figure 12.6). The cumulative gas import bill increases to $2 trillion – $800 billion, or 73%, more. Imported coal costs $220 billion. In total, China's cumulative fuel imports, at $9.3 trillion (in year-2006 dollars), cost $3.4 trillion, or 58%, more than in the Reference Scenario.

Figure 12.6: **China's Cumulative Oil and Gas Import Bill in the Reference and High Growth Scenarios, 2006-2030**

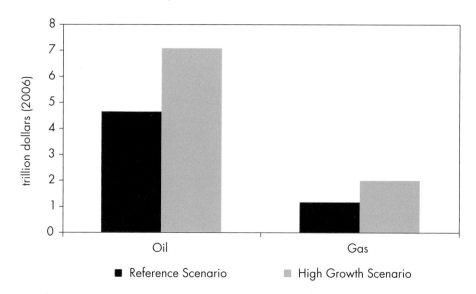

Implications for Investment

To meet the faster projected growth in energy demand in the High Growth Scenario, a total of $5.1 trillion in year-2006 dollars, or almost $200 billion a year, will need to be invested in China's energy-supply infrastructure over 2006-2030. This is $1.4 trillion (36%) more than in the Reference Scenario. The extra investment is needed to build more production and import capacity. As in the Reference Scenario, the electricity sector is most in need of investment, accounting for three-quarters of total energy investment (Figure 12.7). Some 15% of total investment goes to the oil sector, 6% to coal and 4% to gas.

In the High Growth Scenario, cumulative investment in the electricity sector in 2006-2030, in year-2006 dollars, amounts to $3.8 trillion. China's installed generating capacity reaches 2 268 GW by 2030, compared with 1 775 GW in the Reference Scenario. Cumulative investment over 2006-2030 in building generating capacity to meet increasing demand and to offset retirements reaches $1.7 trillion. Investment needs for transmission and distribution exceed those for power plants, totalling $2.1 trillion in 2006-2030 – an increase of 37% over the Reference Scenario.

Figure 12.7: **Cumulative Energy Supply Investment in China in the Reference and High Growth Scenarios, 2006-2030**

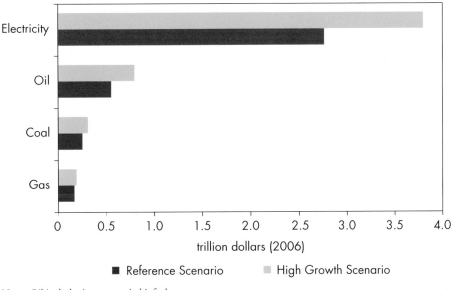

Note : Oil includes investment in biofuels.

As discussed in Chapter 10, power generation investment in China has historically been made primarily by state-owned or provincially-owned entities, backed by public funds. In the High Growth Scenario, China's per-capita electricity generation reaches 7.4 MWh, close to the current level of Korea. The large investment requirements associated with this level of generation would put considerable pressure on government budgets and this, in turn, will increase the pressure to accelerate the pace of market reform, moving towards more efficient market structures, better able to attract private-sector investment.

The projected faster increase in oil prices in the High Growth Scenario leads to the creation of *additional* oil production capacity of 0.7 mb/d relative to the Reference Scenario and the average cost per barrel is higher because of the need for more active use of enhanced oil recovery techniques. Accordingly, cumulative investment needed in the upstream oil industry reaches $384 billion between 2006 and 2030, an amount about 50% higher than in the Reference Scenario. Combined with downstream investment, which rises by 36% to reach $391 billion, China's total oil investment in the High Growth Scenario reaches $775 billion, $228 billion more than in the Reference Scenario.

Total investment in China's *coal* industry in 2006-2030 is $309 billion – $58 billion, or 23%, more than in the Reference Scenario, in line with the rapid increase in domestic coal production and imports. As additional production comes from mines which are deeper and more difficult to exploit, the development costs also rise. Mining accounts for almost all of this investment.

Investment in the gas sector expands by $21 billion, virtually all in the downstream sector. To secure the rapid expansion of gas imports from overseas, investment in LNG facilities increases from $11 billion in the Reference Scenario to $22 billion in the High Growth Scenario. Investment in gas transmission and distribution networks to supply power plants and final consumers increases to $68 billion; $10 billion, or 16%, more than in the Reference Scenario.

Environmental Implications

Local Air Pollution[2]

Greater fossil-energy use in the High Growth Scenario pushes up emissions of various toxic and noxious gases, which worsens air pollution significantly – on the assumption that, like in the Reference Scenario, the government introduces no relevant new policy measures. China is already the largest source of SO_2 emissions in the world because of rising coal use (World Bank, 2007). The even faster increase in these emissions associated with the High Growth Scenario would intensify China's problems with acid rain, particularly in the southeast. The economic cost of the overall increase in pollution could be large. The cost of premature mortality and morbidity related to air pollution was already 1.2% to 3.2% of GDP in 2003 (World Bank, 2007).

---

2. The projections in this section are based on analysis carried out by the International Institute for Applied Systems Analysis (IIASA) on behalf of the IEA.

In the High Growth Scenario, NO_x emissions in 2030 are projected to reach 26 Mt, 22% higher than in the Reference Scenario, and the emissions of SO_2 reach 35 Mt, 17% higher. The Chinese government recognises the fundamental importance of making economic growth and development less resource-intensive and environmentally sustainable, and has already implemented a number of measures aimed at adjusting the structure of the economy (see Chapter 7). It has introduced the "circular economy" concept, which seeks to integrate environmental and economic decision-making. Sustained high growth would reinforce the need for more to be done. Applying the polluter-pays principle in policy-making would help to maximise the cost-effectiveness of anti-pollution policies by giving investors and consumers a financial incentive to limit their pollutant emissions.

Energy-Related CO_2 Emissions

Energy-related CO_2 emissions in the High Growth Scenario are projected to rise to 14.1 gigatonnes by 2030 – 2.6 Gt, or 23%, more than in the Reference Scenario (Figure 12.8). The difference between the two scenarios is almost equivalent to the current CO_2 emissions in Russia and Japan combined. Most of the additional emissions in China come from burning coal, mainly in power stations. Emissions from this source are 1.8 Gt higher in 2030 than in the Reference Scenario. Measured on a per-capita basis, China's emissions reach nearly 90% of the current OECD level.

12

Figure 12.8: **China's CO_2 Emissions in the Reference and High Growth Scenarios**

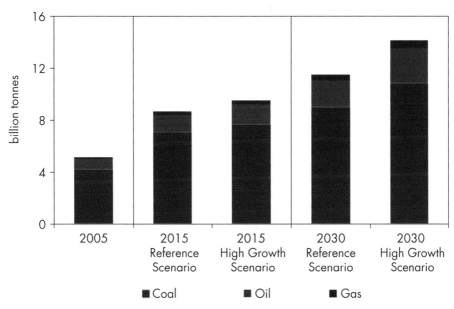

Though higher economic growth would bring greater wealth to China and release millions of people from poverty, the analysis presented here indicates that China's energy security and environment could significantly deteriorate. However, this analysis assumes unchanged government policies. Stronger government action to save energy and mitigate environmental impact, such as that taken into account in the Alternative Policy Scenario (see Chapter 11), would almost certainly be associated with such a level of growth. China's higher economic growth would bring substantial social and economic benefits if it was, indeed, accompanied by stronger policy efforts to modify the economic structure, improve energy efficiency, diversify energy sources and mitigate the negative environmental and other consequences of higher energy use.

CHAPTER 13

FOCUS ON THE COASTAL REGION

HIGHLIGHTS

■ The coastal region of China is the most economically advanced part of the country, and includes clusters of mega-cities in the Pearl River Delta, Yangtze River Delta and Bohai Rim. Eleven provinces located in the coastal region produce 60% of China's GDP and over 90% of exports, with only 39% of its population. In 2005, the GDP of the coastal region amounted to $1.4 trillion (at market exchange rates), bigger than that of Canada.

■ The coastal region has been the main driver of China's increasing energy use because of a high concentration of export industries, investment and urbanisation. The region accounted for 70% of energy demand growth in China between 1996 and 2005. Over the *Outlook* period, export-driven demand is expected to decline relative to expansion of domestic demand in the coastal region, leading a more general change in China's energy-consumption pattern.

■ In the Reference Scenario, coastal energy demand is projected to climb from 954 Mtoe in 2005, which exceeds the current demand of Japan, Korea, Australia, and New Zealand combined, to 2 050 Mtoe in 2030, at an average annual growth rate of 3.1%. The shift towards a more services-oriented economy causes the share of the coastal region in Chinese industrial energy demand to drop from 74% in 2005 to 57% in 2030. Meanwhile, the booming middle-income class in the region increases its consumption of goods and services, boosting demand for mobility, thermal comfort and electrically powered equipment and appliances. Electricity demand and natural gas demand will grow much faster than in China as a whole.

■ The coastal provinces become increasingly dependent upon imported fuels, either from inland Chinese provinces or from the international markets. By 2030, 68% of coastal coal demand is met by supply from the inland provinces and 15% from abroad, with the remaining 17% being produced in the coastal region itself. Gas imports are expected to increase dramatically – reaching 100 bcm in 2030. The West-East pipeline will carry to the coastal region both gas produced in inland regions and imported gas; and LNG imports will surge. Rapid infrastructure development is needed to fuel the coastal economy.

- The magnitude of the region's imports of gas and coal from the international markets is very uncertain. Fuel will need to be brought in to produce electricity. Power plants could be supplied predominantly by domestic coal, or the coast could become more reliant on the global market and import substantial quantities of coal or LNG. Important variables including domestic coal production and transportation costs, environmental regulation and the price of LNG contracts, will greatly affect the share of internationally imported coal and LNG.

Background and Assumptions

The coastal and inland regions differ widely in terms of economic conditions and trends, demographics, energy-resource endowments and energy supply and demand balances. The coastal region – comprising the eleven provinces and municipalities of Liaoning, Beijing, Tianjin, Hebei, Shandong, Jiangsu, Shanghai, Zhejiang, Fujian, Guangdong and Hainan (Figure 13.1) – is the most economically advanced part of China, enjoying greater wealth and better living conditions than the inland region. It is also the region that underpins economic growth in China; it is more urbanised and its favourable location close to the sea has underpinned the development of export industries. However, it is poor in energy resources and relies largely on supplies from outside the region – either from other Chinese provinces or from abroad. The coastal region includes the large cities of the Pearl River Delta, Yangtze River Delta and Bohai Rim, which face urban energy-supply security challenges as they continue to expand.

This chapter provides an analysis of the importance of the coastal region in China's energy market, today and in the future, based on the results of an extensive effort to collect energy, economic and demographic data and to build a detailed model of the region. This exercice is intended to demonstrate how some regional factors may play into the evolution of China's energy system; it is not intended as an exhaustive analysis of complex interactions among regions. We present here the projections derived from the Reference Scenario. The tables in the appendix to this chapter give more information.

In 2005, the GDP of the coastal region amounted to $1.4 trillion (at market exchange rate), bigger than that of Canada. The region generated 60% of China's GDP, yet has only 39% of its population. It contributed 70% of China's economic growth between 1996 and 2005. Coastal per-capita GDP, at $2 787 at current prices and market exchange rate, is more than double that of inland China (Table 13.1). Urbanisation rates are also much higher in the coastal provinces, at 49% compared to 35% in the inland region.[1] Urbanisation

---

1. The enormous regional economic disparities in China have an urban-rural dimension, as well as the coastal-inland (or east-west) divide (see Chapter 7).

Figure 13.1: **Provinces and Regions of China**

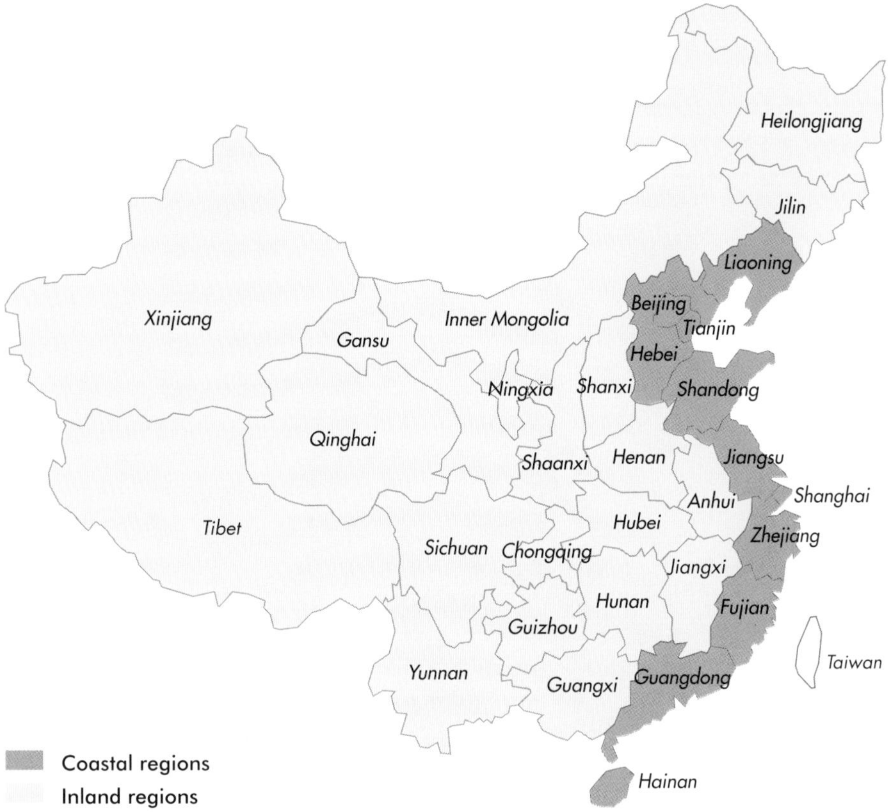

Coastal regions

Inland regions

The boundaries and names shown and the designations used on maps included in this publication do not imply official endorsement or acceptance by the IEA.

Box 13.1: **Provincial Energy Statistics and Modelling of the Coastal Region**

Comprehensive energy balances by fuel and by sector were compiled for each of the 11 municipalities and provinces comprising the coastal region of China. The data cover the period 1996 to 2005 and are based on the *China Energy Statistical Yearbook* and the provincial statistical yearbooks, published by the National Bureau of Statistics. Following the IEA methodology of compiling non-member countries' energy balances, the 11 provincial energy balances were integrated into one single regional balance. This exercise was undertaken for the first time this year. Several challenges emerged in the compilation of the balance, especially in ensuring its coherence with the entire country energy balance. In consultation with the Energy Research Institute of China, the final energy balance for the coastal region of China underwent detailed data verification in order to prepare the most reliable basis possible for the projections.

Table 13.1: **Economic Indicators by Province in China, 2005**

	GDP per capita, current prices, US$	Share of urban population in total*, %	Share of national exports, %	Share of services in GDP, %
Coastal provinces	**2 787**	**49**	**92**	**40**
Liaoning	2 269	54	3	40
Beijing	5 354	80	2	69
Tianjin	4 239	65	3	41
Hebei	1 762	33	2	33
Shandong	2 394	42	6	32
Jiangsu	2 929	54	16	35
Shanghai	6 157	90	11	50
Zhejiang	3 281	34	11	40
Fujian	2 222	39	5	38
Guangdong	2 909	57	32	43
Hainan	1 292	47	0.1	42
Inland (rest of China):	**1 229**	**35**	**8**	**38**
Total China	**1 713**	**40**	**100**	**40**

* IEA estimates based on UN urbanisation rate and do not necessarily match Chinese statistics.
Sources: CEIC (2007); UNPD (2006); IEA analysis.

rates affect energy demand in many ways, for example residential urban consumers own more appliances and consumer electronics than those in rural communities.

The main reason for these disparities is that investment for construction, infrastructure and industry has centred in the coastal regions. Export industries, one of the main drivers of the Chinese economy so far, has gone almost entirely to the coastal provinces. Jiangsu, Shanghai, Zhejiang and Guangdong alone account for 70% of China's exports. The share of industry in GDP in 2005 averaged 51% in the coastal region as a whole, compared with 45% in the inland region. Industry in the coastal region grew by 11.1% per year in 2000-2005, compared with 10.1% per year nationally. The share of services in GDP– which has also grown strongly – is only slightly higher in the coastal region, averaging 40% compared with 38% for inland areas. The services sector on the coast grew by 10.1% per year from 2000 to 2005, compared to 9.9% per year nationally. Agriculture accounts for a much lower share of GDP on the coast (8%) than inland (17%).

The Reference Scenario energy projections for coastal China (set out in the next section), as for China as a whole and all other *WEO* regions, rest on key assumptions regarding the economy, population, international energy prices and technology. One basic assumption of this scenario is that no new government policies are adopted by the central, provincial or municipal governments beyond those already adopted (but not necessarily implemented) by mid-2007. On this basis, GDP in the coastal region is expected to continue to grow more rapidly than in the inland region[2], averaging 8.0% per year in the ten years to 2015, and then at a rate only slightly higher than the inland provinces in 2015-2030. The coastal region grows at 6.1% over the entire projection period. The share of services in GDP is assumed to increase, from 40% in 2005 to 48% in 2030, as the economy matures and economic structural adjustment policies take effect (Table 13.2). The proportion of the Chinese population in the coastal region is assumed to remain constant at 39% through to 2030, because of measures to deter too much internal migration such as the programmes already in existence to develop the central and western inland regions. This coastal share of the population has risen only slightly since 1990, when it was 38%.

Table 13.2: **Key Macroeconomic and Population Assumptions for China's Coastal Region**

	1996	2005	2015	2030
Services share (%)	37	40	43	48
Population (millions)	463	517	547	575
Urbanisation (%)	37	49	60	72

Sources: UNPD (2006); CEIC (2007); IEA estimates and analysis.

13

Energy Outlook
Primary Energy Demand

Primary energy demand in the coastal region in 2005 amounted to 954 Mtoe, 55% of total Chinese demand and exceeding demand in OECD Pacific. Demand grew at a rate of 11% per year between 2000 and 2005, faster than the national average rate. The growth in coastal energy demand over that period was 80% larger than the total demand in Korea in 2005. The region was responsible for about 60% of the Chinese coal, oil and electricity demand in 2005 (Figure 13.2). Coal has always been the dominant fuel, holding a 66% share in primary demand in 2005 and growing at 12.2% per year from 2000 to 2005. Oil accounted for 22% in 2005. Use of natural gas and nuclear power more than tripled between 2000 and 2005. Hydropower saw growth of only 5.1% per year, less than half the rate nationally.

---

2. Coastal GDP grew by 9.9% per year from 2000 to 2005 – 1.2 percentage points higher than in the inland region.

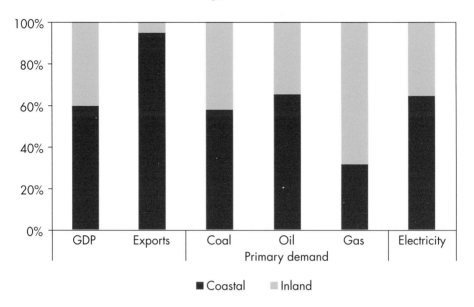

Figure 13.2: **Share of the Coastal Region in China's Economy and Energy Demand, 2005**

Growth in primary energy demand in the coastal region is projected to slow markedly over the *Outlook* period, averaging 3.1% per year (Table 13.3). It grows by 4.5% per year from 2005 to 2015 and by 2.2% per year from 2015 to 2030. Coal remains the dominant fuel in the coastal energy mix, as it does nationwide, demand for coal rising by 2.9% per year in 2005-2030. Coal's share of total primary energy demand jumps from 66% in 2005 to 70% in 2010, but then drops back to 63% by 2030. The slowing of coal growth reflects the movement of heavy industry to neighbouring provinces to the west of the coastal region and an increase in the use of natural gas in the residential and services sectors.[3]

The coastal region's oil consumption increases by 3.4%, from 4.4 mb/d in 2005 to 10 mb/d in 2030. More than 80% of the increased demand comes from the transport sector. By 2030, the coastal region still represents 61% of national oil consumption, slightly down from 65% in 2005, as oil demand from the industrial sector grows more slowly than in the inland region to 2015 and actually declines from 2015 to 2030.

3. On 30 August 2007, The National Development and Planning Commission released a policy of natural gas utilisation, prioritising the residential and services use of natural gas (NDRC, 2007).

Table 13.3: **Primary Energy Demand in China's Coastal Region in the Reference Scenario** (Mtoe)

	2000	2005	2015	2030	2005-2015*	2005-2030*
Coal	356	633	997	1 285	*4.7%*	*2.9%*
Oil	123	214	328	490	*4.4%*	*3.4%*
Gas	4	13	55	131	*15.1%*	*9.6%*
Nuclear	4	14	32	67	*8.8%*	*6.5%*
Hydro	4	6	6	8	*1.4%*	*1.3%*
Biomass	74	74	58	62	*–2.4%*	*–0.7%*
Other renewables	0	0.1	3	7	*35.9%*	*17.8%*
Total	**566**	**954**	**1 479**	**2 050**	***4.5%***	***3.1%***

* Average annual rate of growth.

Primary demand for natural gas in the coastal region increases at an annual rate of 9.6% to 2030. Growth stems mainly from power generation and the residential and services sectors. The share of gas in total primary energy demand increases more than four-fold, from 1.4% in 2005 to 6.4% in 2030. Physical constraints on the availability of gas are expected to discourage more rapid growth in the use of gas in final sectors.

Nuclear power generating capacity continues to be located solely on the coast over the *Outlook* period. The share of nuclear power in the primary energy mix of the coastal region more than doubles, from 1.5% in 2005 to 3.2% by 2030. Support for nuclear power is heightened by the relative scarcity of non-fossil energy resources in the coastal provinces. The 11th Five-Year Plan calls for raising installed capacity to 40 GW.

The share of biomass in the coastal region's total primary energy demand drops from 7.8% to 3% by 2030, as the trend continues towards greater use by households of modern forms of energy and alternative fuels. The share remains much lower than in the rest of China, because the region is more urbanised. The share of other renewables increases steeply and reaches 0.4% of primary energy demand in 2030. Hydropower remains a marginal source of energy, because of a lack of further potential on the coast.

The 11th Five-Year Plan sets energy intensity reduction targets for each province (Figure 13.3). The target percentage reduction in energy intensity by 2010 compared to 2005 levels is 19.6% for the coastal region and 20% for China as a whole. The challenges to meeting the targets vary widely between the provinces, even within the coastal region. For example, Beijing may

Figure 13.3: **Provincial Energy Intensity Targets, 2005-2010**

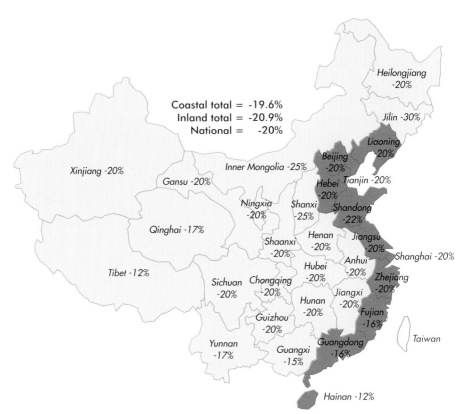

Coastal total = -19.6%
Inland total = -20.9%
National =   -20%

Heilongjiang -20%

Jilin -30%

Liaoning -20%

Beijing -20%

Xinjiang -20%

Inner Mongolia -25%

Gansu -20%

Hebei -20%

Tianjin -20%

Ningxia -20%

Shanxi -25%

Shandong -22%

Qinghai -17%

Shaanxi -20%

Henan -20%

Jiangsu -20%

Shanghai -20%

Tibet -12%

Hubei -20%

Anhui -20%

Zhejiang -20%

Sichuan -20%

Chongqing -20%

Jiangxi -20%

Hunan -20%

Fujian -16%

Guizhou -20%

Yunnan -17%

Guangxi -15%

Guangdong -16%

Taiwan

Hainan -12%

The boundaries and names shown and the designations used on maps included in this publication do not imply official endorsement or acceptance by the IEA.

Source: State Council (2006).

exceed its target, because of the relocation of a major steel company to Hebei province (coastal) and a major coking and chemical plant to Shanxi province (inland). But these changes make it more difficult for Hebei and Shanxi to meet their targets. In the Reference Scenario, the average energy intensity of the coastal region continues to decline rapidly, by 2.8% per year over 2005-2030. In the Reference Scenario, energy intensity declines by 15% between 2005 and 2010. Preliminary energy intensity data for 2006 show that while intensity has declined overall, most areas have further to go before meeting their targets. In only two provinces – Beijing and Fujian – did the actual decline in intensity in that year meet or exceed the targeted rate of decline (Figure 13.4). The coastal region is the key driver for economic growth in China and will lead the rest of the country from an investment-driven economy to a consumption-driven one. Investments in energy efficiency on the coast are crucial because they will have spillover effects for the rest of China. The rising incomes on the coast will result in an increase in appliance and vehicle ownership and hence energy demand; however, the policies and technologies to raise energy-efficiency standards will initially be most effective in the coastal region and will then spill over into the inland region as incomes rise there. There is a danger, however, that restructuring in the coastal region may push heavy industry inland and so slow any declines in energy intensity there.

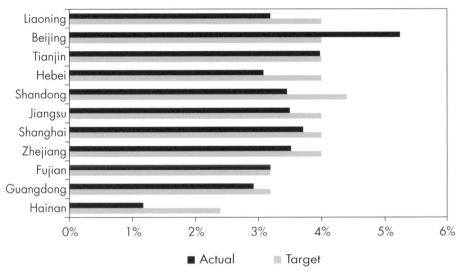

Figure 13.4: **Energy Intensity Reduction Target* and Outcome by Province in China's Coastal Region, 2005-2006**

* Average annual reduction in intensity needed to meet the 2010 target.
Sources: State Council (2006), NBS (2007), IEA analysis.

Energy Demand by Sector

Total final energy consumption increases at a slightly slower rate than primary energy supply, doubling by 2030. The rate of growth is nonetheless markedly lower than that in 2000-2005. Total final consumption of coal grows by 3% per year on average between 2005 and 2015, and then declines by 1.1% per year between 2015 and 2030. Final gas consumption increases more than five-fold, with residential demand accounting for 64% of the increase and industry 25%. Final oil demand rises by 3.8% per year, with oil accounting for 98% of total energy for transport in 2030 (slightly more than the share nationally). Use of biomass and combustible waste drops by 30% between 2005 and 2030. Figure 13.5 shows the contribution of the coastal region to China's incremental final energy demand. For all fuels, except coal, biomass and other renewables, the coastal region contributes the greater part of the increase in energy demand. In the period 2015-2030, the coastal region's coal demand actually declines; however, this effect is offset by coal demand growth in the inland region.

Figure 13.5: **Contribution of the Coastal Region to China's Incremental Final Energy Demand by Fuel in the Reference Scenario, 2005-2030**

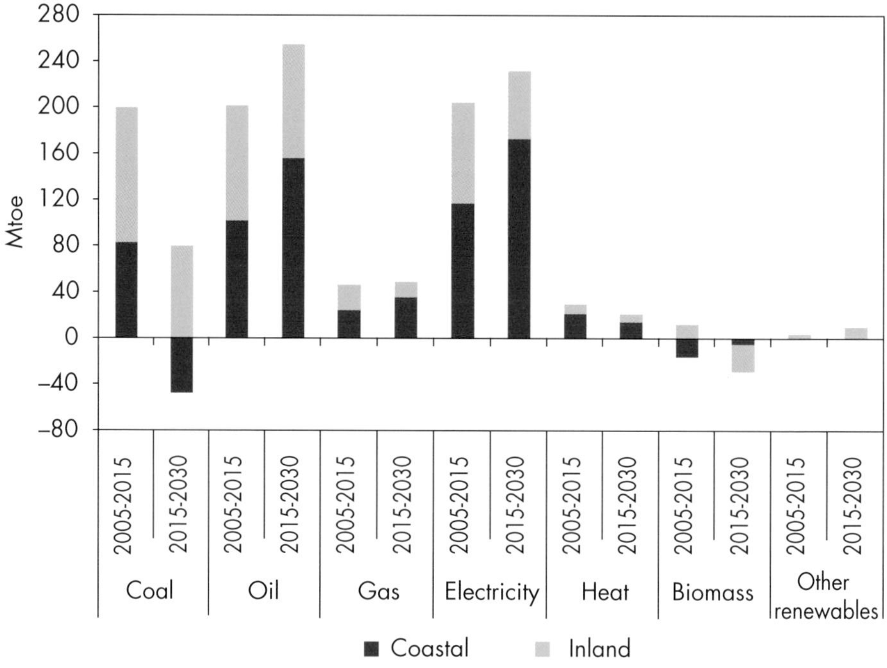

Note: Coastal consumption data for other renewables are not available.

Power Generation

Electricity generation is projected to reach 4 769 TWh in 2030 (Table 13.4). It grows by 8.1% per year between 2005 and 2015, and then slows to 3.9% per year from 2015 to 2030. The share of coal in the generation fuel mix remains high, at 84% in 2030 – more or less the same share as now. Coal-fired generation continues to grow strongly through 2015 and slows thereafter. Hydropower remains the most important renewable source, but its share of generation declines. Other renewables – notably bioenergy, onshore wind and, to a lesser extent, solar photovoltaics – grow much more rapidly. More than 80% of generation from solar photovoltaics is located in the coastal region throughout the *Outlook* period.

Table 13.4: **Electricity Generation Fuel Mix in China's Coastal Region** (TWh)

	2000	2005	2015	2030	2005-2015*	2005-2030*
Coal	575	1 054	2 332	4 016	8.3%	5.5%
Oil	44	52	52	43	0.0%	–0.7%
Gas	1	4	67	249	31.6%	17.6%
Nuclear	17	53	123	256	8.8%	6.5%
Hydro	51	65	75	90	1.4%	1.3%
Biomass and waste	1	3	6	33	5.6%	9.7%
Other renewables	–	1	29	82	35.8%	17.8%
Total	**688**	**1 233**	**2 684**	**4 769**	**8.1%**	**5.6%**

* Average annual rate of growth.

To meet demand growth over the *Outlook* period, the coastal region needs to increase electricity generating capacity from 172 GW in 2005 to 773 GW by 2030 (Figure 13.6). Steam coal is projected to account for 76% of incremental capacity. Onshore wind emerges as a new source of generation towards the end of the projection period, and represents 3% of generating capacity in 2030. The majority of onshore wind is located on the coast, but the inland region catches up by 2030.

13

Figure 13.6: **Power Generating Capacity in the Coastal Region** (GW)

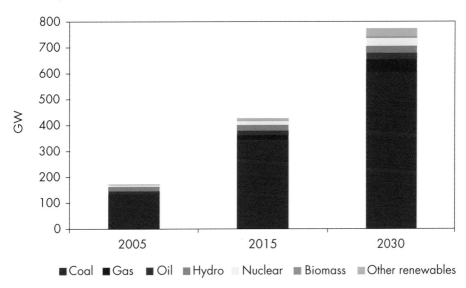

Industry

Industry is today the main final consumer of energy in the coastal region, accounting for 55% of total final consumption in 2005. This share is projected to decline, reaching 46% in 2030. Some 44% of China's 1 000 largest enterprises – and 58% of steel producers (CEIC, 2007) – are currently located in the coastal region. The bulk of new heavy industrial capacity (such as iron and steel, chemicals, and non-metallic minerals) is expected to be located in the inland provinces, resulting in a sharp slow-down in the growth of industrial energy consumption in the coastal region. The share of the industrial sector in coastal GDP is assumed to decline from 51% in 2005 to 49% by 2030. Also, as is the case nationally, there is a shift in the balance of output from heavy industry towards lighter manufacturing. These two effects slow annual coal demand growth to 3.3% per year to 2015 and it declines by 1.2% per year from 2015 to 2030. However, the shift in investment towards lighter industry causes electricity demand to continue to grow rapidly, averaging 6.9% per year from 2005 to 2015, before dropping to 2.8% per year from 2015 to 2030.

Transport

Close to 72% of incremental oil demand for transport in China will come from the coastal provinces. This growth will be driven by road transport. Vehicle ownership has been relatively higher in the coastal region than in the inland region. In 2005, vehicle ownership in the coastal area was 35 vehicles per 1 000 people, double the inland average of 17. Cars accounted for 71% of on-road vehicles, a higher share than inland. Car ownership varies greatly among coastal provinces, reflecting relative wealth, transport infrastructure and transportation policy (Figure 13.7). Passenger-car ownership in the coastal region is projected to climb from 30 cars per 1 000 inhabitants in 2006 to 260 in 2030. Inland ownership rates will reach around 70 in 2030, a level achieved in 2015 in coastal China. By 2030, three-quarters of all the cars on Chinese roads and 64% of new car sales are expected to be in the coastal region. Policies aimed at further improving and encouraging public transport in the coastal provinces will be needed to limit the growth in oil demand and imports.

Driven by rising vehicle ownership, transport energy demand in the coastal region quadruples over the *Outlook* period, growing on average by 5.9% per year. The region accounts for 69% of national oil demand by the transport sector in 2030, up from 62% in 2005. By contrast, electricity use for rail transport, which is currently concentrated on the coast, will grow faster inland. Oil demand for aviation, currently split evenly between the coastal and inland regions, grows faster on the coast through to 2030.

Figure 13.7: **Car Ownership and GDP per Capita in Selected Provinces in China, 1994-2005**

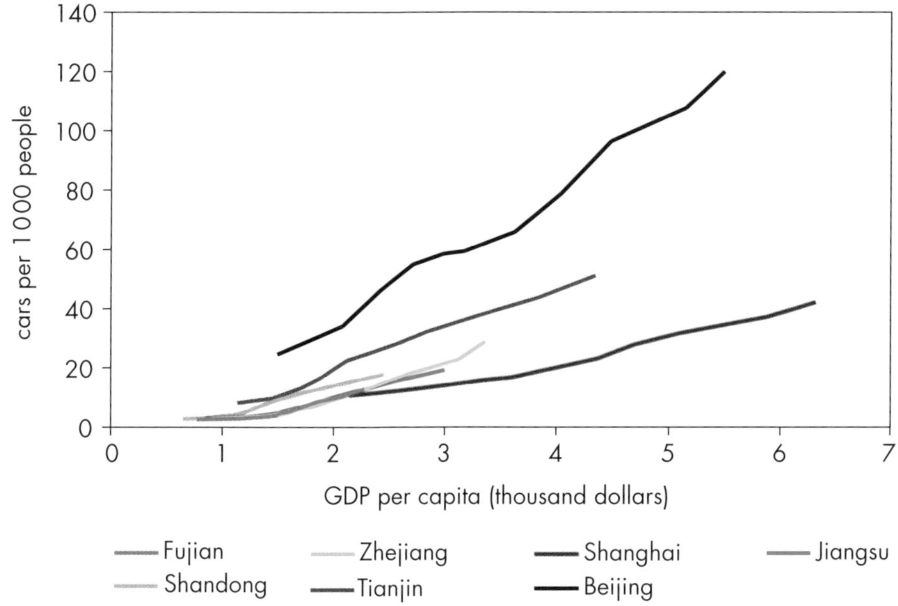

Sources: CEIC (2007); IEA estimates.

Residential Sector

The greater wealth and higher urbanisation rates in the coastal region result in higher appliance penetration rates than in the rest of the country. In 2004, 70% of coastal households owned a washing machine, compared with only 62% of inland households (CEIC, 2007). The same is true for other household appliances like televisions, refrigerators, and air conditioners. In addition to these basic appliances, greater wealth leads to longer hours of use and more purchases of other electronic appliances. By 2030, the average coastal income is 2.4 times that of inland income. Floor space per capita, in both rural and urban areas, remains slightly higher in the coastal region throughout the *Outlook* period. Total residential energy demand in the coastal region remains slightly lower than inland because of lower population, but grows faster – by 2% per year in 2005-2030.

Biomass and waste currently account for 57% of coastal residential demand, but this share drops rapidly to 17% by 2030, with rapid urbanisation and a shift to modern fuels for heating and cooking. The share of electricity soars from 11% in 2005 to 38% in 2030. Electricity demand in 2030 is almost six times that in 2005, because of increases in appliance ownership; it grows by 7.3% per year. Demand for natural gas grows by 12.2% per year, with increased per-capita floor space. Only 17% of the coastal population currently has access to natural gas. Access varies widely between provinces, with 56% of the population in Beijing having access and only 0.5% in Jiangsu. Access is set to grow with the expansion of grids to distribute imported liquefied natural gas (LNG) and the recent government decision to give priority to the residential use of gas (Figure 13.8).

Figure 13.8: **Share of the Population in China's Coastal Provinces with Access to Natural Gas and Planned LNG** (bcm)

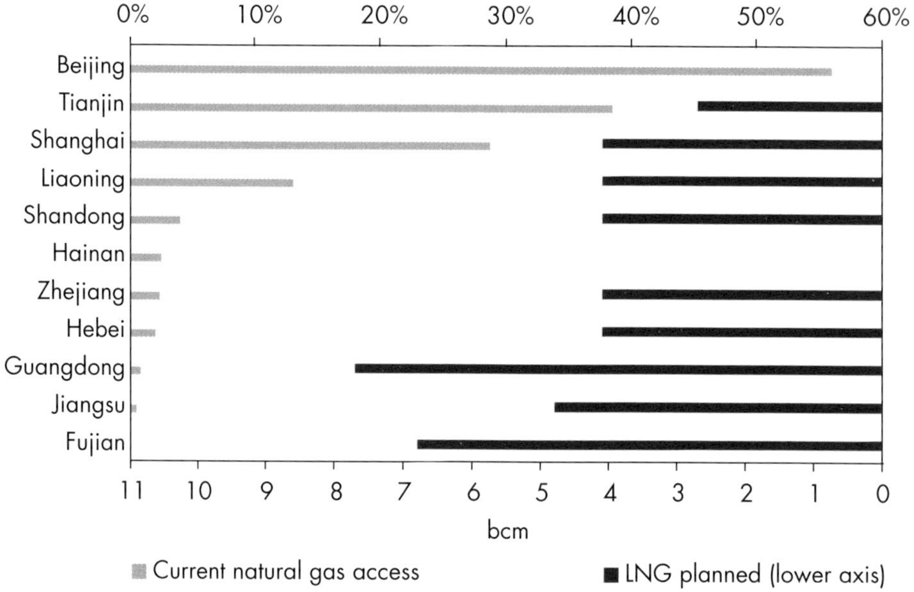

Sources: China Statistics Yearbook (2006); IEA estimates.

Other sectors

The increasing share of services in GDP results in the use of energy in this sector increasing from 13 Mtoe in 2005 to 68 Mtoe in 2030, a growth rate of 8.3% per year to 2015 and 6% per year thereafter to 2030. Today, oil accounts for 35% of total energy demand in the services sector and electricity for 34%. However, by 2030, electricity growth outpaces that of coal and oil, becoming by far the dominant energy source. Agricultural energy use increases from 17 Mtoe to 19 Mtoe in 2030, or 0.5% per year. Oil retains the dominant share in agriculture, gaining 2 percentage points to reach 64% in 2030.

Energy Supply

The coastal region of China has limited energy resources and cannot meet its own energy demand. Coal production is concentrated in the inland provinces of Shanxi, Inner Mongolia, Henan and Shaanxi, where the resources are largest (see Chapter 10). The coastal region has only 6% of the country's coal resources, 19% of oil and 11% of natural gas (located mainly offshore in the eastern China and south-western China basins). Hydro resources are also scarce. Where fuel to meet coastal energy demand will be produced is very uncertain and difficult to estimate.

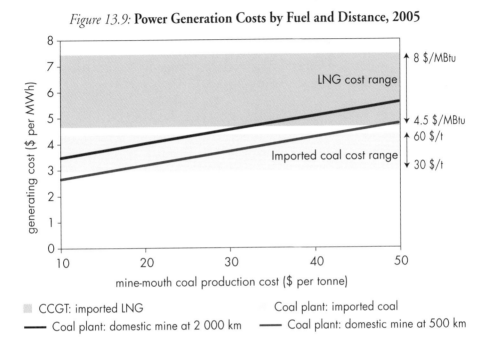

Figure 13.9: **Power Generation Costs by Fuel and Distance, 2005**

Securing future fuel supply for power generation is a major issue for the coastal region, because its own fuel resources are so limited. Power plants could be supplied by domestic coal (requiring additional investments in rail transportation infrastructure) or the coastal region could become more integrated into the global market and import coal or LNG. Figure 13.9 shows coal plant generating costs in 2005, depending on mine-mouth coal costs and the travel distance between mine and plant. It also shows the range of costs for generation with imported LNG or imported coal, in order to illustrate the mine-mouth coal costs required for domestic coal to be competitive with imported coal and LNG. Coal power generation plants using indigenous coal will be competitive for mine-mouth costs up to $42/t for coal mines within 500 km but this cost needs to be below $28/t for mines as far as 2 000 km away (based on the higher-end estimate for imported coal).

The coastal provinces, with a demand of 1 836 Mtce, produce only 321 Mtce of coal in 2030 in the Reference Scenario. This is an increase of 28% over 2005 production and may not be easy to achieve. In 2005, 624 Mtce of coal, representing 69% of coastal demand, were transported from inland provinces to the coastal region. An increasing share of the coal consumed in the coastal region is expected to come from abroad. By 2030, 68% of coastal coal demand is met by supply from inland provinces and 15% from abroad, with the remaining 17% from the coastal region itself. The main external coal suppliers in 2015 are Indonesia, Australia, South Africa and Vietnam. By 2030, Australia and Indonesia are called upon for still higher exports, while African regions outside South Africa also contribute to Chinese import requirements.

Coastal oil imports from abroad almost triple, reaching 9 mb/d in 2030. Oil production in the coast is expected to plateau, creating a need to import more oil. Imports meet 94% of demand on the coast by 2030. In 2005, offshore production of natural gas in the coastal provinces reached 6.9 bcm, and it is expected to reach 12 bcm in 2030. The coastal region imported 9.1 bcm, or 57%, of coastal gas demand, from the inland provinces. The share of imports from inland falls to 15% in 2030 as inland gas demand grows. To supplement gas produced inland and supplied through the West-East pipeline, the coastal provinces are planning to boost LNG imports. Guangdong began LNG imports in 2006 (see Chapter 10). Fujian has also just completed a terminal that is expected to begin operation soon. In 2030, the coastal region needs to import around 100 bcm of gas, in the form of LNG or through pipeline gas from neighbouring countries.

Energy-Related CO_2 Emissions

In 2005, the coastal region's energy-related carbon-dioxide emissions represented more than 10% of worldwide emissions, surpassing OECD Pacific in 2004 and the transition economies in 2005. The coastal region represents 55% of China's emissions. From 2000 to 2005, coastal China's CO_2 emissions grew at an average annual rate of 13%, faster than the average for China as a whole of 11%. We project emissions in the coastal provinces to grow by 5.2% annually to 2015 and 3.3% per year over the entire *Outlook* period. The coastal region accounts for 55% of China's total emissions in 2030, as it did in 2005. The coast's per-capita emissions approach OECD levels by the end of the projection period (Figure 13.10).

Figure 13.10: **CO$_2$ Emissions per Capita, 2005-2030** (tonnes per capita)

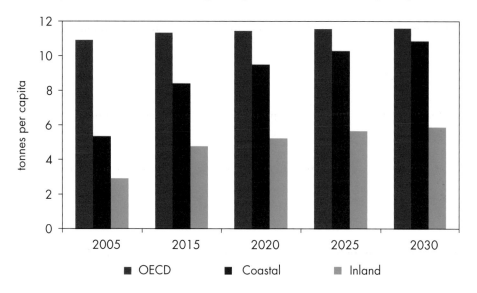

APPENDIX TO CHAPTER 13: CHINA COASTAL REFERENCE SCENARIO PROJECTIONS

For definitions, please refer to Annex A and Annex B at the end of the book.

	Energy demand (Mtoe)				Shares (%)			Growth (% p.a.)	
	2000	2005	2015	2030	2005	2015	2030	2005-2015	2005-2030
Total primary energy demand	**566**	**954**	**1 479**	**2 050**	**100**	**100**	**100**	**4.5**	**3.1**
Coal	356	633	997	1 285	66	67	63	4.7	2.9
Oil	123	214	328	490	22	22	24	4.4	3.4
Gas	4	13	55	131	1	4	6	15.1	9.6
Nuclear	4	14	32	67	1	2	3	8.8	6.5
Hydro	4	6	6	8	1	0	0	1.4	1.3
Biomass and waste	74	74	58	62	8	4	3	−2.4	−0.7
Other renewables	0	0	3	7	0	0	0	35.9	17.8
Power generation	**175**	**344**	**643**	**1 043**	**100**	**100**	**100**	**6.4**	**4.5**
Coal	154	310	568	881	90	88	85	6.3	4.3
Oil	12	12	15	12	4	2	1	1.7	−0.1
Gas	0	1	17	56	0	3	5	30.0	16.5
Nuclear	4	14	32	67	4	5	6	8.8	6.5
Hydro	4	6	6	8	2	1	1	1.4	1.3
Biomass and waste	1	1	2	11	0	0	1	4.8	9.0
Other renewables	0	0	3	7	0	0	1	35.9	17.8
Other energy sector	**72**	**120**	**162**	**205**	**100**	**100**	**100**	**3.0**	**2.2**
Total final consumption	**402**	**643**	**972**	**1 297**	**100**	**100**	**100**	**4.2**	**2.8**
Coal	151	240	323	275	37	33	21	3.0	0.5
Oil	93	170	272	427	26	28	33	4.8	3.8
Gas	5	13	37	73	2	4	6	10.9	7.1
Electricity	59	113	229	402	18	24	31	7.4	5.2
Heat	21	34	55	69	5	6	5	4.9	2.9
Biomass and waste	73	73	56	51	11	6	4	−2.5	−1.4
Other renewables	0	0	0	0	0	0	0	n.a.	n.a.

	Electricity (TWh)				Shares (%)			Growth (% p.a.)	
	2000	2005	2015	2030	2005	2015	2030	2005-2015	2005-2030
Total generation	**688**	**1 233**	**2 684**	**4 769**	**100**	**100**	**100**	**8.1**	**5.6**
Coal	575	1 054	2 332	4 016	85	87	84	8.3	5.5
Oil	44	52	52	43	4	2	1	0.0	-0.7
Gas	1	4	67	249	0	2	5	31.6	17.6
Nuclear	17	53	123	256	4	5	5	8.8	6.5
Hydro	51	65	75	90	5	3	2	1.4	1.3
Biomass and waste	1	3	6	33	0	0	1	5.6	9.7
Wind	0	1	29	69	0	1	1	36.7	17.3
Geothermal	0	0	0	1	0	0	0	36.1	18.2
Solar	0	0	0	12	0	0	0	-0.4	22.1
Tide and wave	0	0	0	0	0	0	0	n.a.	n.a.

	CO_2 emissions (Mt)				Shares (%)			Growth (% p.a.)	
	2000	2005	2015	2030	2005	2015	2030	2005-2015	2005-2030
Total CO_2 emissions	**1 531**	**2 780**	**4 612**	**6 256**	**100**	**100**	**100**	**5.2**	**3.3**
Coal	1 231	2 264	3 607	4 631	81	78	74	4.8	2.9
Oil	289	486	888	1 340	17	19	21	6.2	4.1
Gas	11	31	117	285	1	3	5	14.2	9.3
Power generation	**646**	**1 266**	**2 331**	**3 652**	**100**	**100**	**100**	**6.3**	**4.3**
Coal	607	1 223	2 245	3 482	97	96	95	6.3	4.3
Oil	38	39	46	38	3	2	1	1.6	-0.1
Gas	1	3	40	132	0	2	4	30.0	16.5
Total final consumption	**874**	**1 499**	**2 159**	**2 457**	**100**	**100**	**100**	**3.7**	**2.0**
Coal	624	1 040	1 362	1 149	69	63	47	2.7	0.4
Oil	240	430	720	1 155	29	33	47	5.3	4.0
Gas	10	28	78	154	2	4	6	10.7	7.0

13

POLITICAL, ECONOMIC AND DEMOGRAPHIC CONTEXT

HIGHLIGHTS

- India had the fourth-largest economy in the world, after the United States, China and Japan in PPP terms in 2006. At market exchange rates, India's GDP in 2006 was the thirteenth-largest. India's economic growth has trended upwards over the last three decades, averaging 7% per year since 2000. It was 9.7% in 2006, up from 9% in 2005, mainly thanks to a surge in private investment and in manufacturing. Among the world's twenty largest economies, only China grew faster than India in the two years to 2006.

- Service activities account for a large share of India's economy, compared with most other developing countries. In 2005, they contributed 54% of GDP; industry contributed 27% and agriculture 19%. Despite the relatively small contribution of agriculture to GDP, nearly 60% of the workforce is still employed in farming.

- Services and manufacturing are expected to remain the main drivers of India's economic development. Productivity in India is very low, so the potential for further growth through productivity gains is substantial. The future pace of productivity and GDP growth hinges on structural and business reforms, fiscal discipline and efforts to remove barriers to trade and investment. Infrastructure improvements will be essential to higher productivity in all sectors.

- Poverty remains a huge challenge for India, despite the recent rise in average incomes. Average per-capita GDP (in PPP terms) in 2006 was $3 736 – about an eighth of the OECD average. Per-capita income varies markedly across the country: in Bihar, the poorest state, it is about a tenth of that in Goa, the richest state. Economic growth will reduce poverty but acceptable income distribution will require strong policies to assist the rural sector and the poorest people in urban areas. Greater access to cleaner cooking fuels and electricity must form part of these policies.

- India is home to around 1.1 billion people, about 17% of the world's population. Today, it is the world's second most populous country, after China. It is expected to have the largest population in the world soon after 2030. More than 70% of the population live in rural areas – a higher proportion than in most other Asian countries. The rate of population growth appears to have slowed in many large cities, but the urban population is still expected to nearly double by 2030. The number living in slums today is some 160 million people.

The Political Context

India is a federal republic made up of 28 states and seven union territories.[1] Independent since 1947, it is the largest democracy in the world. The head of state is the president, elected by an electoral college comprising members of both houses of the national parliament and the parliaments of the states. The prime minister heads the government, which has executive power. Ministers are appointed by the president on the recommendation of the prime minister. Legislative power resides with the national assemblies, which consist of an upper house, the Rajya Sabha, and a lower house, the Lok Sabha. Of the 250 members of the Rajya Sabha, 12 are appointed by the president and the rest are chosen by the elected members of the state and territorial assemblies. Two of the 545 members of the Lok Sabha are appointed by the president and the rest are elected by popular vote. General elections to the Lok Sabha are held every five years. Confidence in the stability and integrity of the electoral system is high.

The states have their own elected governments while, with the exception of Delhi and Puducherry, union territories are governed by an administrator appointed by the national government. Some of the state legislatures have two houses, like the national parliament. Each state government is headed by a chief minister, who is responsible to the state legislature in the same way the prime minister is responsible to parliament. Each state also has a governor, appointed by the president, who may assume certain broad powers when directed by the central government. The national government can impose direct presidential rule over the states and has done so in certain circumstances, such as when no party or coalition is able to form a government. Relations between the national and state governments are not always smooth.

India's constitution defines the administrative and legislative relationship between the central and state governments (and union territories). Article 246 lays out the concurrent subjects over which the central and state governments share responsibility. The list of concurrent subjects includes electricity, economic and social planning, education, forests, trade unions and industrial and labour disputes. Policy and laws are set either at the centre or by the states, depending on the subject concerned. In cases where central law is in conflict with state law on a concurrent subject, the central law prevails (provided it is properly directed at national issues).

Hindi and English are the two official languages of communication for the central government. The state governments use their own language, together with English for communication with the central government.

1. See Figure 14.1 for the location of India's states and union territories. The Union Territory of Delhi has an elected chief minister and assembly, although full statehood had not yet been accorded at the time of writing.

In total, India has 23 official languages. India's judiciary is independent of the executive and the legislature. Investors can, accordingly, have some confidence in national and state laws which, in principle, offer considerable protection. However, in practice, the legal system is characterised by very serious delays and there are cases of corruption and undue interference (World Bank, 2006a). There is a huge backlog of court cases. In 2005 the Indian government enacted the "Right to Information Act" in an effort to make information about the working of the country's administration and government more accessible and transparent to its citizens. India has a strong and proactive civil society and a free and vibrant press.

Following the 2004 general election, the government was formed by the United Progressive Alliance (UPA), an 11-party coalition led by the Indian National Congress (INC), with Manmohan Singh as prime minister. Coalition governments are a relatively new phenomenon in India. Since independence, the INC has ruled the country for most of the time. Since 1989, coalition governments have prevailed, reflecting political developments at the state level and the emergence of strong regional parties. At the state level, there has been a trend away from national "all-India" parties, including the INC and Bharatiya Janata Party (BJP), towards smaller, more narrowly-based and caste-oriented regional parties. This trend has made it more difficult to form national governments and to agree on political priorities.

The Economic Context[2]

Economic Growth and Structure

With gross domestic product (GDP) of $4 159 billion in purchasing power parity (PPP) terms in 2006[3], India had the fourth-largest economy in the world, after the United States, China and Japan. India accounted for 6.3% of global GDP and nearly a quarter of GDP in developing Asia. At market exchange rates, India's GDP in 2006, at $887 billion, was the thirteenth-largest in the world, after China, Brazil, Russia and nine OECD countries. On this basis, India represented about 2% of global GDP.

India's economy grew by about 3% per year on average in the 1970s. Growth picked up in the 1980s and 1990s, averaging 5.8% per year. It has accelerated since 2000, averaging close to 7%. India saw growth of 9.7% in 2006,

2. This section has benefited from the discussion among participants of the *WEO-2007* workshop organised in New Delhi on 23 March 2007. The analysis and projections in Part C also benefited from valuable input from The Energy and Resources Institute (TERI) of India.
3. GDP data for 2006 are from the International Monetary Fund.

up from 9% in 2005 and 8.3% in 2004. Among the world's twenty largest economies, only China grew faster than India in the two years to 2006. The government's growth target during the 11th Five-Year Plan (Box 14.1), which runs from 2007 to 2012, is 9% a year. Given the difficulties with meeting targeted growth rates in the past and the considerable infrastructure bottlenecks in India, the *Outlook* assumes that the economy grows at about 7.1% per year over this period in the Reference Scenario.[4] This is about on par with average growth since 2000. In the High Growth Scenario in Chapter 19, India's GDP is assumed to grow by 8.2% per year in 2007-2012.

Box 14.1: **India's Five-Year Plans**

The Indian government implements economic policy through five-year plans, developed, executed and monitored by the Planning Commission. The first five-year plan was introduced in 1951. Performance in meeting plan targets has improved of late, largely as a result of economic reform. For example, the average annual growth rate in 2002-2007 was 7.2%, not far below the target of 8% in the 10th Five-Year Plan and the highest growth rate achieved in any plan period. Traditionally, the rate of growth of GDP has been the central objective. The current plan also sets targets for other dimensions of economic performance, including reversing the deceleration in agricultural growth and providing education and health services to all citizens (Government of India, 2006). The role of the states in meeting targets has been expanded. Many of the focus areas in the 11th Five-Year Plan, such as health, education, drinking water, urban infrastructure and agriculture, are the responsibility of the states, with substantial assistance from the central government.

Average per-capita GDP in PPP terms in 2006 was $3 736 – about an eighth of the OECD average. Per-capita income varies markedly among states (Figure 14.1). In Bihar, the poorest state, it is about a tenth of that in Goa, the richest state. Per-capita GDP has tended to grow faster in the states that were already the most prosperous. Bihar, Uttar Pradesh, Jharkhand, Orissa and

---

4. The macroeconomic assumptions underpinning the Reference, Alternative Policy and High Growth Scenarios and their implications for energy demand are outlined in the Introduction, Chapters 16, 18 and 19.

Figure 14.1: GDP per Capita by State

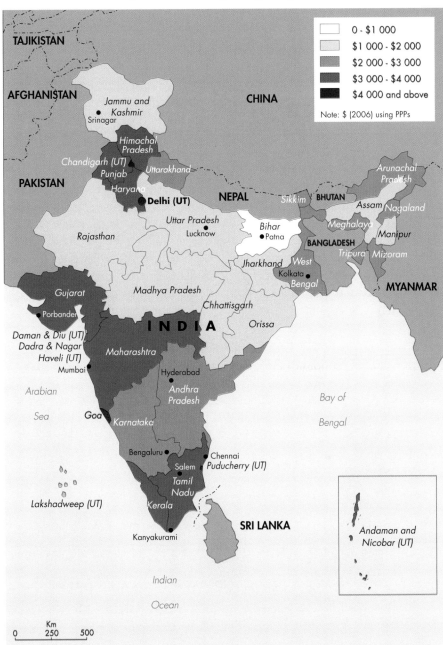

Note: GDP figures are from the Ministry of Statistics, http://mospi.nic.in. Figures for the union territories (UTs) of Dadra & Nagar Haveli, Daman & Diu and Lakshadweep are not available.

Assam are on average poorer than many countries in Sub-Saharan Africa. The union territories of Chandigarh, Delhi and Puducherry have per-capita GDP of more than $6 000 per year, in line with China and Brazil.[5]

Service activities account for a large share of India's economy compared with most other developing countries, including China. In 2005, services accounted for 54% of GDP, industry for 27% and agriculture for 19%. Despite the relatively small contribution of agriculture to GDP, nearly 60% of the workforce is employed in farming (Figure 14.2). Labour is nearly four times more productive in industry and six times more productive in services than in agriculture, where there are estimated to be 160 million surplus workers (McKinsey Global Institute, 2007). Rapid growth of agro-processing industries close to production centres could absorb a significant share of surplus farm labour. In recent years, output gains due to labour migration from agriculture to services and industry have contributed about one percentage point to overall growth (Poddar and Yi, 2007). The gains derive, roughly equally, from agricultural workers moving to industry and to services. With a continuation of this trend, overall productivity and output is set to continue to rise in the coming decades.

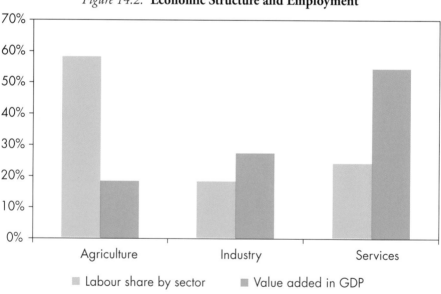

Figure 14.2: **Economic Structure and Employment**

Sources: World Bank database and OECD (2007).

5. See Chapter 20 for a comparison of energy access in Indian states and other countries.

In India, labour productivity, measured as value added per person, is estimated to be about 13% of productivity in the United States (Figure 14.3). There is substantial potential for further growth through labour productivity gains. These will be greater if labour moves to the formal sector of the economy. Productivity gains will be lower if agricultural workers take jobs in the informal sector in Indian cities.[6] India will need to invest more in education and training in order to capture the full potential from productivity gains.

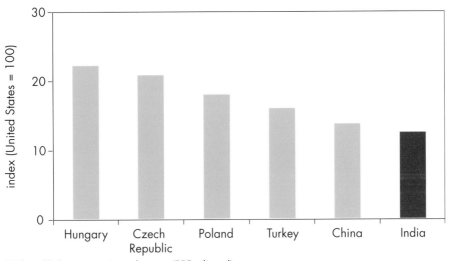

Figure 14.3: **Labour Productivity* in Selected Countries Relative to the United States** (productivity in United States = 100)

* Value added per person in employment (PPP-adjusted).
Source: Van Ark *et al.* (2006).

Drivers of Growth

Capital productivity and total factor productivity increased considerably after 1993 (Figure 14.4), when economic reforms launched at the beginning of the 1990s started to take effect. In the period 1993-2004, increased productivity accounted for well over a third of the 6.5% annual increase in output (Bosworth and Collins, 2007).

The service sector has seen the biggest gains in productivity since the early 1990s. More recently, industry has experienced faster gains: a surge

6. Economic activity in the informal sector of the economy is undeclared and is not included in GDP. If former agricultural workers take up jobs in the informal, non-agricultural sector, the effects on productivity are very modest (OECD, 2007).

in manufacturing productivity was the main reason for higher output growth since 2003, contrary to the conventional wisdom that India's economic growth is largely services-led (Poddar and Yi, 2007). In fact, growth in the two sectors is linked. A quarter of service activity is directly linked to industrial activity, in sectors such as trade, transport, electricity and construction. The underlying factors behind the increase in productivity were an acceleration in international trade, an increased availability of financial services and investments in information and communication technology. The rise in India's annual average growth rate from 6% in the 1990s to nearly 8% since 2003 can be attributed to a rise in the contribution to GDP of manufacturing, trade and banking services.

Figure 14.4: **Sources of Output Growth in India**

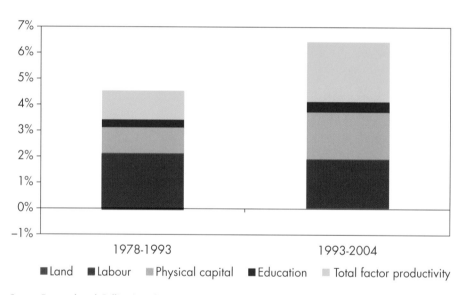

Source: Bosworth and Collins (2007).

Services and manufacturing are expected to remain the main drivers of India's economic development. The pace of GDP growth hinges particularly on structural and business reforms, fiscal discipline and efforts to remove the remaining barriers to trade and foreign investment (see below). The continuing success of the information and technology industry should boost growth, by expanding the pool of technology-skilled labour and by encouraging other domestic firms to increase technology spending. However, the lack of sufficient skilled labour to meet rising demand in this sector has recently begun to constrain expansion in many firms.

Will Economic Growth Solve India's Problem of Rural Poverty?

Poverty remains a huge challenge for India. Rapid economic development has brought an improvement in average living standards, as witnessed by higher life expectancy, lower child mortality and expanded access to clean water. But large numbers of people remain desperately poor. One important dimension of the income gap is the rural-urban divide. During the 1990s, in most states, improvements in urban incomes outpaced those of rural incomes, widening the gulf between rural and urban India; 230 million farmers in India have been largely bypassed by the rapid, urban-led economic growth. Low productivity, public under-investment, inefficient pricing policies, inadequate training and poor maintenance of irrigation systems and road infrastructure continue to characterise the Indian farm sector. Moreover, there are poor regions within otherwise prosperous states. Maharashtra is home to booming and prosperous Mumbai, but at the same time nearly 50% of the population in its rural areas is close to or below the poverty line (World Bank, 2006b).

Economic growth will reduce poverty but needs to be combined with strong policies targeted on the rural sector, including improved access to cleaner, more efficient cooking fuels and technologies. In the 2007/08 budget, the government made plans to fund some of the goals of the Common Minimum Programme, which aims to establish a social welfare scheme. A National Rural Employment Guarantee Scheme has been set up and the Bharat Nirman, the Indian Development Agency, has enjoyed a budget increase of more than half to 0.6% of GDP. The government is also preparing a financial aid plan for indebted farmers in the southern states of Andhra Pradesh, Karnataka and Kerala. Fast and focused implementation of these programmes and plans will be crucial to raising the living standards of India's rural citizens.

The poor quality of education remains a major stumbling block to poverty alleviation and economic growth. Literacy rates have improved, especially among the young, but are still very low in rural areas, at 64% for men and 45% for women. In over one-quarter of rural households, not a single household member can read or write. At 61% in 2006, average literacy in India fell short of the rate in China, where it was 91%.

14

Economic Challenges
Continuation of Policy Reforms

India's impressive economic performance in recent years, after decades of lacklustre growth and underdevelopment, owes a great deal to the economic reforms launched at the beginning of the 1990s. After independence, India pursued import-substitution policies and restricted international trade. The role of the central government was strengthened through regulation and by a wave of nationalisations in the late 1960s and 1970s, causing the public sector's share of GDP to increase steadily.

The disappointingly low rates of growth that resulted from these policies led to pressure on the government to change course. In response to a balance-of-payments crisis, a major programme of economic reforms was adopted in 1991. Industrial and import licensing were progressively abandoned and many public monopolies ended, including those in industry, aviation and telecommunications. Foreign direct investment is now allowed in many sectors. The reforms removed many obstacles to growth and began the process of reintegrating India into the global economy. All national governments since reforms commenced have pursued a similar economic agenda, regardless of political orientation. The commitment of the state governments to reform has varied: it has been strong in Andhra Pradesh and Karnataka, for example, but more tentative in Bihar and Uttar Pradesh.

Notwithstanding the accomplishments of the past fifteen years, much remains to be done. The fiscal deficit and debt to GDP ratios are still high. Economic reforms have largely bypassed the agriculture sector, bringing few changes to the lives of the rural poor. The level of education needs to rise to prepare the growing labour force for employment in the industry and services sectors. Financial sector reforms have to be more aggressive. A major challenge for the Indian government is to ensure that all members of society enjoy the benefits of economic expansion through development policies that create a virtuous circle of growth in investment and income, and increase support for social welfare (see Spotlight on previous page).

International Trade

Rapid growth in international trade has underpinned investment and output growth in industry and services. A gradual reduction in trade barriers following the 1991 reforms gave a substantial stimulus to trade. The economy-wide average tariff fell from 87% in 1990 to 22% in 2005, but it is still high compared to China (12%) and Indonesia (7%).[7] The value of exports (in year-2006 dollars) rose from $25 billion in 1990 to

7. World Bank Group, www.worldbank.org.

$103 billion in 2005. The share in GDP of trade in goods, however, was only 28% in 2005, compared with 64% in China. India currently contributes some 1% to global trade.[8]

According to the World Trade Organization, the share of fuel imports (including mainly crude oil but also mineral fuels and feedstocks, lubricants and related materials) in India's total imports increased to 37% in 2005, up from 29% in 1990, largely the result of an increase in oil prices. Fuel exports have also risen, from less than 3% in 1990 to 12% of total exports in 2005 (Figure 14.5). Exports will rise further if oil-refinery plans go ahead (see Chapter 17). Expansion of refineries in India accounted for almost a quarter of the total increase in world refining from 1996 to 2006. India's trade balance went into deficit in 2003 and the deficit is likely to increase in the near term. However, India's large foreign exchange reserves, which now exceed $200 billion, and its low levels of external debt provide a cushion against any external crisis.

Figure 14.5: **Share of Fuel in Value of India's Imports and Exports**

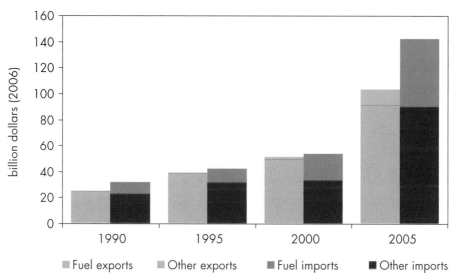

Note: Fuels include mainly crude oil but also mineral fuels and feedstocks, lubricants and related materials (SITC section 3).
Source: WTO database, available at www.wto.org.

8. See Table 3.1 in Chapter 3 for a perspective on India's share in global trade compared with other countries.

South Asia is the least integrated region in the world, with trade among the countries in the region at only 2% of GDP. Trade could increase substantially if more conducive framework agreements were put in place on transport, custom formalities and other trade-related issues (World Bank, 2006c). India has been using the South Asian Association for Regional Cooperation (SAARC)[9], in which it plays a leading role, as a forum for rapprochement with Pakistan. Trade between India and Bangladesh is heavily weighted in India's favour and Bangladesh sees a more balanced trading regime as one condition for enhanced co-operation in other areas. Energy, particularly natural gas, and transport are key areas for co-operation between SAARC countries.

Fiscal Discipline

The persistence of high fiscal deficits is a potential threat to continued rapid growth in India. India's combined central and state government fiscal deficit was around 6.3% of GDP in 2006, while the level of public debt stood at around 75% of GDP in March 2007 (OECD, 2007). The Planning Commission argues that fiscal targets should be subordinated to social spending needs, but the current fiscal deficit and public debt constrain the scope for financing much-needed public investment in infrastructure through borrowing. The large deficit also crowds out credit to the private sector. The government has accordingly proposed measures to cut the deficit, including reducing subsidies, widening the tax net and cutting government employment; but these face stiff opposition in parliament. States, too, mostly face high levels of debt which limit their ability to increase investment through borrowing.

Some progress has nonetheless been made. The central government introduced a value-added tax in 2005 to increase state savings and is committed to abiding by the framework established in the Fiscal Reforms and Budget Management Act, enacted in 2003. The goal is to reduce the fiscal deficit to no more than 3% of GDP by March 2009. Curbs on spending will be crucial to meeting this goal. Some states have begun to increase user charges to limit subsidies and to reform pension systems for their employees.[10] Others, however, are lagging behind. Funding for infrastructure development could be made available by creating favourable conditions for a sustained increase in private investment.

Progress in reducing the fiscal deficit in order to garner the needed funds for infrastructure investment in all sectors will be an important determinant

9. Created in 1985 by Bangladesh, Bhutan, India, Maldives, Nepal, Pakistan and Sri Lanka. Afghanistan joined in April 2007.
10. See IMF (2007) for a review of the fiscal performance of India's states.

of growth in energy demand. How the deficit is cut will also be important. Reducing energy subsidies is important, even though it will make consumers more vulnerable to high international oil prices. Cutting employment in the state-owned coal sector has proven to be very difficult. Privatising the electricity sector has had mixed results in India (see Chapter 17).

Investment and Business Climate

The framework conditions for doing business in India have improved as a result of economic reform. India has simplified business registration, cutting the time required to set up a business from 71 to 35 days. Tax payments have been simplified and the corporate income tax rate reduced. In addition, a Supreme Court decision to simplify the rules governing loan collateral has helped to ease access to credit. Import and export processing times have fallen following the introduction of new risk-management procedures in customs and investor protection has improved as a result of changes in stock-exchange rules. But there is still considerable room for improvements. India is ranked 134th in the World Bank's global rankings of the ease of conducting business (World Bank, 2007).

Compared to China, potential investors in India face greater regulatory hurdles and other constraints on investment. India has accordingly attracted far less foreign direct investment (FDI). In 2005, FDI amounted to $72.4 billion in China, and only $6.6 billion in India.[11] But FDI has increased rapidly in India since 2004, including a three-fold increase in the year to March 2007 (OECD, 2007). The introduction in India of special economic zones (SEZs), which have been successful in attracting foreign investment in China, should increase foreign capital flows even further (Box 14.2). Foreign investment flows currently differ widely between Indian states, since production and labour market reforms have progressed at varying speeds. The richest states attract most of the foreign direct investment and those states with the most FDI also have the highest rates of investment generally. According to the World Bank, six states – Andhra Pradesh, Gujarat, Karnataka, Maharashtra, Punjab and Tamil Nadu – attracted over 66% of total FDI in 2003, down only slightly on their 72% share in the 1990s (World Bank, 2006a). Seven of the least developed states – Bihar, Jharkhand, Madhya Pradesh, Chhattisgarh, Orissa, Rajasthan and Uttar Pradesh – attracted only 13% of FDI in the 1990s.

Capital formation is an important source of economic growth in emerging economies. In India in 2000, the level of capital formation as a percentage of GDP was below that of China and roughly equivalent to that of Japan,

14

11. http://stats.unctad.org.

Box 14.2: **Special Economic Zones in India**

Special economic zones (SEZs) are business locations for which the government offers tax incentives and in which India's strict labour laws do not apply. Export businesses located in SEZs are exempt from tariffs. The Indian government approved over 100 SEZs in 2006 and 2007. Reliance Industries was given clearance to build a SEZ in Haryana and Tata has an SEZ approved in Orissa. The information technology firm, Wipro, has obtained permission to build an SEZ in Andhra Pradesh. The creation of these zones has, however, prompted resistance from the rural communities directly affected. The Indian government is examining ways to compensate people displaced by the zones, but no policies have yet been adopted.

There are doubts about the efficacy of SEZs as a way of fast-tracking economic development and the Reserve Bank of India has recently raised the cost of lending to them. Supporters think that they could contribute to developing the country's comparatively small manufacturing sector. The broader debate is focused on whether the gains in investment (estimated by the Ministry of Commerce to be a potential $22 billion) and employment (estimated by the same Ministry to be some 500 000 jobs over the next few years), will outweigh the loss of revenue from tax concessions (EIU, 2006). The Finance Ministry and Reserve Bank of India have also voiced concerns that the scheme will work only if the SEZs are very large and are built in coastal regions, as in China.

a country with a much more mature economy (Figure 14.6). By 2005, it had risen to 33% in India, but was still some ten percentage points below China. India's high fiscal deficit affects capital formation by reducing private and public-sector investment. In many cases, simply accumulating capital is not sufficient to increase output — it must also be efficiently allocated. This can best be accomplished through market-directed allocation of new investment.

Despite recent reforms, the private sector continues to be burdened with regulatory restrictions, price distortions, deficiencies in legal practice, problems in getting access to finance and severe infrastructure bottlenecks. The ratio of private credit to GDP is under 40% in India, compared with 100% in China and Malaysia. While the largest Indian companies have much better access to finance than in the past, small and medium-sized enterprises and micro-enterprises still find it difficult to borrow. This is partly because lending rates on small loans are capped, which makes banks reluctant to lend to small clients. India's financial sector is still relatively small compared with the size of its economy, but policies to open up the sector, including the entry of foreign banks from 2009, are expected to contribute to improve capital markets.

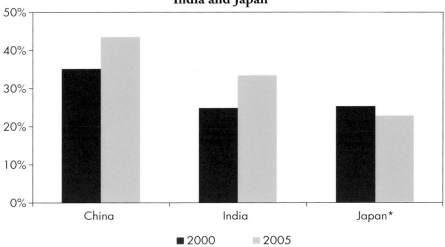

Figure 14.6: **Share of Gross Capital Formation in GDP in China, India and Japan**

■ 2000 ■ 2005

* Second column refers to share in 2004.
Source: World Development Indicators database, available at http://devdata.worldbank.org.

Infrastructure improvements are needed to underpin productivity gains in all sectors. Limitations exist today in roads, ports, power plants and transmission and distribution systems. The government is addressing the constraint in the transport sector. The Golden Quadrilateral Highway, which the National Highways Authority of India estimated to be about 95% complete in July 2007, will connect Delhi, Kolkata, Chennai and Mumbai. The north-south and east-west corridors will be further improved by roads connecting Srinagar in the north to Kanyakumari in the south and by a spur from Salem in the south-east to Porbander in the west. The new facilities will reduce travel times, open up villages and contribute to rural-urban migration. The system will also promote local economic development by attracting business activity along the route.

Private-Sector and Foreign Investment

India's economic growth rests, to a large extent, on a vibrant private sector, particularly in telecommunications, aviation and software. The role of the private sector is expected to continue to expand. The dynamism of India's private sector is visible in overseas direct investment (ODI), which is developing rapidly. From just $509 million in 2000, ODI grew to nearly $1.4 billion in 2005.[12] Liberalisation of investment requirements by the Reserve Bank of India is the main driver. ODI is concentrated in information technology, software and pharmaceuticals, sectors in which the private sector established itself first within India and in which it has a large comparative advantage. However, ODI has

12. See Figure 3.5 in Chapter 3 for a comparison of ODI and FDI in India and other countries.

recently extended into petrochemicals, metals, industrial goods, renewable energy, automotive components, health services and other industries.

Demographic Trends

India is home to around 1.1 billion people, about 17% of the world's population. It is the world's second most populous country after China. Uttar Pradesh is the most heavily-populated state, with some 166 million inhabitants, while Sikkim is the least populated, with about 540 thousand. India's population grew by 1.8% per year from 1990 to 2000 and then decelerated to 1.5% per year from 2000 to 2005. The *annual* increase in population in India is about equivalent to the population of Australia. Some 80% of Indians are Hindu.

Table 14.1: **Demographic Indicators**

	1980	1990	2000	2005
Population (million)	687	850	1 016	1 095
Rural share (%)	77	74	72	71
Economically active population (%)	38	39	39	40
Sex ratio (females per 1 000 males, ages 0-6)	978	995	927	n.a.
Life expectancy at birth (years)	54	59	63	64
Under-5 mortality rate (per 100 000 live births)	173	123	94	74

Sources: Population statistics from UNPD (2007). Other statistics are from http://unstats.un.org/.

Although the fertility rate has fallen sharply in recent decades, mortality rates have fallen even faster. In particular, the under-five mortality rate has more than halved since the 1980s, to 74 per 100 000 live births in 2005 (Table 14.1). While this is lower than the regional average, it is still far above the figure for Latin America (31) and East Asia (33). About 35% of the Indian population is below age 15. The share is 38% in rural areas, reflecting higher fertility rates. As a result of the large proportion of children in the population, India's economically active population was 40% of the total in 2005, compared with 60% for China. Only the Middle East and North Africa have lower rates of economically active population than India.

India's demographic profile is expected to add some 270 million people to the workforce in the period to 2025 (McKinsey Global Institute, 2007). There will be no shortage of future workers, particularly for low to medium-skilled jobs. India is well positioned to reap the benefits of favourable demographics over the long term due to the continued movement of labour from agriculture to industry and services.

This will depend on having sufficient economic flexibility to allow for the creation of new jobs and a higher share of workers in formal markets.

According to the United Nations Population Division, about 780 million people, or 71% of the population, live in rural areas. The country's urbanisation rate of 29% is very low compared with 81% for Korea, 67% for Malaysia, and 42% for China (Figure 14.7). Like other developing countries, India is experiencing strong migration from the rural areas to the cities (the urban population is growing by about 2.3% per year, according to the United Nations Population Division). One result is that slum areas are growing in some cities. The 2001 National Census of India reveals that nearly 160 million people live in slums.[13] Some 300 to 400 families move into Mumbai every day and the city needs at least 1.1 million houses for poorer residents.[14] Indian policy makers hope to slow urban growth through implementation of the National Rural Employment Guarantee Bill, enacted in 2004, which provides a legal guarantee for 100 days of employment every year for rural households with an adult member willing to do unskilled manual work. Even though urban population growth is expected to continue, the rate of urban growth is expected to accelerate over the next two decades in only 12 of India's 56 largest cities (UNPD, 2007). Nonetheless, between 2005 and 2030, India's urban population is expected to increase from 317 million to 590 million (UNPD, 2007). But India will still be less urban in 2030 than Korea, Malaysia and China in 2005.

Figure 14.7: **Percentage of Population Urbanised in India Compared to 2005 Urbanisation Level in Selected Countries**

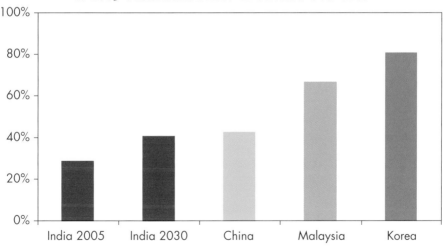

Source: UNPD (2007).

13. See Chapter 20 for a discussion of energy access in slums.
14. "The World Goes to Town", 3 May 2007, *The Economist*.

CHAPTER 15

OVERVIEW OF THE ENERGY SECTOR

HIGHLIGHTS

- Primary energy demand in India was 537 Mtoe in 2005, roughly equivalent to demand in Japan. While energy demand in India is growing fast, at a crisp 3.2% per year in 2000-2005, energy demand per capita – 0.5 toe in 2005 – remains extremely low at about one-tenth of the OECD average.

- Energy-import dependence is growing. In 2005, India imported 70% of its crude-oil requirements and consumed about 3% of world oil supply. LNG imports commenced in 2004 and now make up 17% of total gas supply. India also imports about 12% of its coal supply. Inefficiencies in logistics, the low quality of domestic resources and slow reforms in the coal sector have contributed to recent growth in coal imports.

- India's energy sector is dominated by state-owned companies. Coal India produces 84% of domestic coal. The Oil and Natural Gas Corporation and Oil India are the dominant players in the upstream oil sector, while the Indian Oil Corporation is the largest player downstream. Until 2006, pipeline gas transport was the sole responsibility of the publicly-owned GAIL (India). Most electricity generating capacity is state-owned. Private generation is undertaken mainly by industrial autoproducers and a few independent power producers. Recent reforms have brought more private participation in India's energy sector, particularly oil and gas.

- In 2006, after comprehensive public consultation, the Indian government approved an Integrated Energy Policy, aimed at achieving co-ordinated action among energy ministries, particularly at the federal level. The ability of India's existing institutions to implement planned and proposed policies, which has been lacking in the past, remains unproven.

- Energy prices in India are heavily subsidised. LPG and kerosene subsidies impose an enormous burden on domestic oil companies. Electricity subsidies and theft cause the State Electricity Boards to incur big financial losses. Subsidised natural gas prices are a disincentive to investors, which is a concern given the investment needed for domestic gas exploration and production.

- Building much-needed infrastructure for power generation, oil and gas and energy transport will require the mobilisation of public and private funds within a transparent and predictable investment framework. With India's growing appetite for personal transport, stronger government policies are needed to enhance efficiency and to promote the supply and use of alternative modes of transport and fuels.

India's Energy Sector

India, the slumbering giant, is waking up, with a growing thirst for energy. The average annual volumetric increase in India's primary energy demand over the past five years has been 15 Mtoe. In 2004, it was nearly 30 Mtoe – almost equal to the entire energy market in Greece. Total primary energy demand in India was 537 Mtoe in 2005, roughly equivalent to demand in Japan. Power generation was 699 TWh, not much higher than that of Germany. The rate of growth of energy demand in India in 1990-2000 was a brisk 3.7% per year, slipping to 3.2% from 2000 to 2005. Yet energy demand per capita, at 0.5 toe in 2005, is extremely low (Table 15.1): it is 0.8 toe per capita in Indonesia, 1.3 in China and 4.7 in the OECD. One in six people in the world live in India, but they account for only 5% of world energy demand. Appliance and car ownership levels are much lower in India than in China and electricity demand per capita, 639 kWh in 2005, is a third of Brazil's.[1]

Table 15.1: **Key Energy Indicators for India**

	1980	1990	2000	2005
Total primary energy demand (Mtoe)	209	320	459	537
Oil demand (mb/d)	0.7	1.2	2.3	2.6
Coal demand (Mtce)	75	152	235	297
Gas demand (bcm)	1.4	11.9	25.4	34.8
Biomass and waste (Mtoe)	116	133	149	158
Electricity output (TWh)	119	289	562	699
TPES/GDP (index, 2005=100)	163	142	120	100
Total primary energy demand per capita (toe)	0.30	0.38	0.45	0.49
CO_2 emissions per capita (tonne)	0.43	0.69	0.95	1.05
Oil imports (mb/d)	0.5	0.6	1.6	1.8
Electricity demand per capita (kWh)	174	341	553	639

With GDP per capita rising by about 5.4% per year in 2000-2005 and expected to grow by 6.4% in 2005-2010, the potential for energy demand growth is enormous. But there are challenges. India has vast coal resources, but most of them are of low quality. Indigenous oil and gas reserves are in short supply. Energy imports are growing. Renewable energy holds promise, but, with the exception of traditional biomass and hydropower, its use is very limited today.

India's economy relies heavily on coal, which accounted for 39% of total primary energy demand in 2005. India is the world's third-largest coal user,

---

1. See Chapter 16 for more information.

after China and the United States. Biomass and waste, mostly fuelwood for cooking and heating in rural areas, met 29% of demand, down from 56% in 1980. Oil demand accounted for about a quarter, up from 16%. The share of natural gas has increased recently and is now about 5% of primary demand, the result of a deliberate government policy of diversifying the fuel mix. Coal accounts for nearly 70% of electricity output. Hydropower represents another 14%, though its share in the power generation mix has declined for the past 35 years. Other renewable energy sources, mostly wind and solar, together with nuclear power, account for almost 5% of electricity supply. The contribution of wind has been steadily increasing since the 1990s.

The increasing use of fossil fuels in India has driven up carbon-dioxide emissions. At 1.1 billion tonnes in 2005, they are fast approaching the level of Japan (1.2 billion tonnes). But they are still currently only about a fifth of those in the United States and China. Per-capita emissions in India, at one tonne in 2005, were among the lowest in the world's largest economies and compared with 11 tonnes per capita in the OECD.[2]

India is a large importer of energy, mostly oil. In 2005, it imported 70% of its crude-oil requirements and consumed about 3% of world oil supply. LNG imports commenced in February 2004 and accounted for 17% of total gas demand in 2005. India also imports coal, about 12% of total demand (Figure 15.1).

India's overall energy intensity, measured as primary energy consumption per unit of GDP, has declined significantly since 1980, mainly thanks to the growing share of the services sector in GDP, which is less energy-intensive. Improved efficiency of energy use and a changing fuel mix in the industry sector also contributed. The share of coal in total energy use in industry, at around 41% in 1990, dropped to around 30% in 2005.

India is the world's third-largest coal producer, after China and the United States, with output of 262 Mtce in 2005.[3] It also has the third-largest proven coal reserves, totalling 98 billion tonnes. Coal demand was 297 Mtce in 2005, resulting in imports of about 36 Mtce.[4] Most imported coal is of higher quality than the grades produced indigenously, particularly coking coal for use in the iron and steel industry. Oil consumption has more than doubled since 1990. With domestic production flat for over a decade, India's dependence on crude oil imports has increased from 44% of total primary oil consumption in 1990 to 70% in 2005. Currently, a small fraction of these crude oil imports are exported as refined products. India produced 28.8 bcm of natural gas in 2005, while it consumed about 34.8 bcm, mostly in the power and fertilizer sectors.

15

2. See Figure 5.10 in Chapter 5 for a comparison of per-capita emissions in India and other countries/regions.
3. A tonne of coal equivalent (tce) is defined as 7 million kilocalories (1 tce = 0.7 toe).
4. Imports are not equivalent to the difference between demand and supply because of stocks.

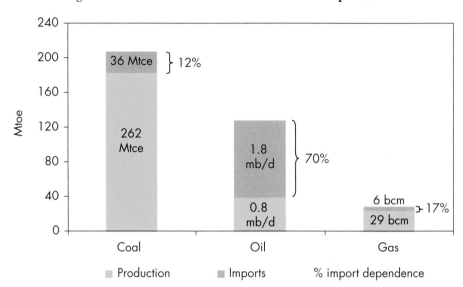

Figure 15.1: **India's Fossil Fuel Production and Imports, 2005**

Imports of LNG, which started in 2004 from Qatar, are set to grow strongly in the coming years as new terminals are built. India's proven reserves of natural gas amounted to 1 100 bcm at the end of 2006 (Cedigaz, 2007). Prospects for expanding domestic production are good after a recent major discovery in the Krishna-Godavari basin and recent announcements regarding gas pricing (see Spotlight below).

India's energy sector is dominated by state-owned companies. Coal India produces 84% of domestic coal and employs some 450 000 workers, making it the second-largest national employer, after the railways industry. Singareni Collieries, a joint undertaking between the central government and the Andhra Pradesh state, accounts for another 9% of production. The remaining 7% is shared between domestic private-sector companies whose production is used exclusively for their own purposes. The coal industry is the only key energy sub-sector that has not seen any fundamental restructuring of its legal and organisational structure in over 30 years. The foreign stake in coal mining remains small, despite the opening-up of the industry to private investment in captive[5] mining projects in 1993. A coal-sector reform bill has been pending since 2000.

By contrast, the private sector is playing an increasingly important role in the oil and gas sectors, following reforms launched in the early 1990s (see below), even though public companies still dominate. The state-owned Oil and Natural Gas Corporation (ONGC) and Oil India Limited (OIL) are the

---

5. Captive mining projects are only able to supply coal to an associated project, *e.g.* steel plant, cement works or power station. The Indian government does not allow these projects to sell coal in the market.

IEA energy statistics for India are based on annual reports and statistical reviews from several ministries, including the Ministries of Coal, Petroleum and Natural Gas, as well as the Central Authority of the Ministry of Power, and the Central Statistical Office of the Ministry of Planning and Programme Implementation. The lack of a centralised information system, including the lack of an official energy balance for the country, leads to difficulties and discrepancies when trying to construct an energy balance from the various sources of information.

Moreover, since India does not submit an energy balance, the IEA has to estimate most of the data for the transformation sector. IEA coal statistics also differ from government estimates as a result of different calorific values. Another complication is that data are not reported on a calendar year basis but for a fiscal year which runs from 1 April to 31 March (for example, 2006 = 1 April 2006 to 31 March 2007).

As a result of analysis for this year's *Outlook*, the IEA's energy data for India have been improved considerably. For example, data on biomass consumption have been revised (on the basis of information from the National Sample Survey Organisation of India and the International Institute for Applied Systems Analysis) and better allocated within the residential, services and agriculture sectors. Biomass consumption in the residential sector has been broken down between urban and rural areas. Electricity demand has also been better allocated across industry sub-sectors. The Energy and Resources Institute of India made a valuable contribution to this work. The projections here are also supported by information from industry and company reports, commercial databases and international organisations, for example data relating to vehicle stocks (including CNG vehicles), industrial output, appliance ownership levels, LNG plants, oil and gas fields and pipeline infrastructure.

In February 2007, the Central Statistical Office hosted a major international meeting on co-operation and harmonisation of energy statistics. The meeting gathered participants from over 30 countries and organisations, including the IEA, as well as participants from all the Indian ministries involved in energy statistics. This was an important initiative towards more internal co-operation and a step towards better harmonisation with international standards and practices.

15

dominant players in the upstream oil sector, while Indian Oil Corporation (IOC), also state-owned, is the largest player in the downstream sector. Pipeline gas transportation and the marketing of gas under the administered price mechanism (APM) is the responsibility of the publicly-owned GAIL (India),

Why Does the Government Need to Reform Gas Prices?

Gas produced by India's public-sector companies is sold at controlled prices under the Administered Pricing Mechanism (APM). Many of these prices are below open market prices. APM gas currently accounts for about 60% of the gas sold in India but its share is expected to decline. Gas that is not covered by APM includes imported LNG, gas produced from fields exploited by companies with production-sharing contracts and gas to be produced from finds made under the New Exploration Licensing Policy (NELP) (Box 15.2).

The current APM price for power and fertilizer feedstocks is $1.90/MBtu. The gas price for CNG vehicles is about $2.40/MBtu. All other APM gas consumers pay more. The prices for non-APM gas are generally based on the price of regasified LNG being imported into India from Qatar under a long-term supply contract. During 2006, spot cargos of LNG were brought into India at prices up to $12/MBtu, which is high by international standards. In 2006, the average price of LNG in the Pacific was around $7/MBtu. Taxes and duties on LNG are levied in India at both the federal and state levels.

In September 2007, the Indian government approved a gas pricing formula for finds under the first six NELP rounds. The formula has three components: a fixed price element, set at $2.50/MBtu, a second element linked to the price of crude oil and a third element which is subject to a bidding process. The new pricing formula yielded a well-head price of $4.20/MBtu, fixed for the first five years, for gas to be supplied by the private company, Reliance Industries, from the Krishna-Godavari basin. Supply is expected to start in mid-2008. The delivered price of gas will include transmission charges, marketing margins, taxes and levies.

The current dualistic price regime appears untenable over the longer term. The APM segment of the market is based on strict quantitative allocations at subsidised prices. Low gas prices to producers act as a disincentive to investment to upgrade mature fields and to explore in deep and ultra-deep waters, where almost all recent gas discoveries have been made. The private share of the gas market is expected to increase in the future. The new gas pricing formula is expected to improve the investment environment for exploration and production, but the government needs to be diligent in allowing gas prices to follow market developments in the future. This could also provide the necessary impetus to move ahead with power-sector reform, particularly for distribution and tariff-setting.

which was created in 1984 to reduce gas flaring, promote gas use and develop midstream and downstream infrastructure. Since December 2006 private investors have been free to develop their own infrastructure. Reliance Industries is the most active private player in the gas sector, both upstream and midstream. Foreign private companies are also active in the upstream oil sector. ONGC Videsh is a wholly-owned subsidiary of ONGC which operates exclusively in foreign markets. Other public and private Indian companies are active in overseas upstream and downstream markets.

The majority of the power sector in India is also publicly owned. In 2005, the State Electricity Boards (SEBs) owned 34 GW of generating capacity (CEA, 2006).[6] Central government-owned public companies, including the NTPC[7], the National Hydroelectric Power Corporation and the Nuclear Power Corporation, controlled about 27% of total capacity. The remaining 27% was owned mainly by industrial autoproducers and independent power producers (IPPs).

The central government has exclusive responsibility for high-voltage bulk inter-state transmission, through Powergrid, a public company. India has interconnections with Nepal and Bhutan, and small power exchanges take place. Transmission within states is in the hands of state transmission utilities. Most distribution is carried out by SEBs. In March 2006, there were ten SEBs, in Himachal Pradesh, Punjab, Madhya Pradesh, Chhattisgarh, Kerala, Tamil Nadu, Bihar, Jharkhand, West Bengal and Meghalaya (CEA, 2007). Several states have unbundled their activities, although the new companies remain under the holding structure of the SEBs. The government aims to develop the national grid, linking the current regional grids, to permit more efficient use of power-generation capacity. The north-eastern region has abundant hydropower resources and the eastern and central states have huge coal reserves, while the main centres of demand are on the coast in the west and south and in the Delhi region. Increased transmission capacity would ease the tight electricity supply situation and avoid an extra burden on the railways' coal transport system, which is over-stretched.

Power shortages and fluctuations in voltage and frequency are a common feature of power supply in India. They arise from insufficient investment in new capacity and the poor performance of existing equipment. The gap between demand and maximum supply nationwide reached 14% in 2006 during peak periods, because of unreliable supply and the limitations of the national transmission network (Ministry of Power, 2007). Underpricing and failure to collect payment from many customers are fundamental causes of these problems (see Chapter 17). Consumers at all levels are affected.

15

6. Total installed capacity in 2005, including utilities and non-utilities, was 146 GW.
7. The National Thermal Power Corporation is officially called NTPC.

Unreliable power damages business and has a high cost for power-intensive and continuous-process industries in particular. Unscheduled power cuts also cause considerable inconvenience to households. Poor reliability has led some consumers to instal generators as a backup. In Delhi, for example, demand for generators, inverters and batteries is increasing by an estimated 20%-25% per year.[8]

As in many developing countries, the distribution sector in India is the weakest part of the power-supply chain. Losses of electricity due to theft and technical factors remain stubbornly high, averaging around 32% to 35% of total generation. They are even much higher in some states. Allowing also for poor bill collection, around 40% to 60% of total potential revenue is lost, depending on the state. The central and state authorities have made some progress recently in improving payment discipline, reforming the regulatory framework and strengthening efforts to improve the financial performance of the SEBs and other publicly-owned power companies to reduce their financial losses.

Energy Administration and Policy

Responsibilities for policy making and implementation in the energy sector are split between five different ministries and several government commissions and agencies (Figure 15.2). The Planning Commission is responsible for assessing energy, capital and human resources in the country, formulating plans for their effective utilisation and appraising progress in meeting targets. Its main function is to formulate India's five-year plans (see Box 14.1 in Chapter 14). Due to administrative bottlenecks and overambitious expectations regarding GDP growth and energy capacity additions, targets are rarely met in practice.

The development and administration of energy policy lies with the various federal ministries and departments:

- The Ministry of Power is concerned with long-term power-sector planning, policy formulation, assigning investment priorities, monitoring the implementation of power projects, training and manpower development, and the enactment and implementation of legislation with regard to thermal and hydropower generation, transmission and distribution. It liaises within the central government, with the SEBs and with the private sector. The Bureau for Energy Efficiency (BEE) is a statutory body under the Ministry of Power, set up under the Energy Conservation Law 2001, to co-ordinate energy efficiency and conservation policies and programmes. The Central

---

8. According to information from the Associated Chambers of Commerce and Industry of India, available at www.assocham.org.

Electricity Regulatory Commission (CERC) is responsible for regulating all activities related to power at the central and interstate level. Its responsibilities include managing electricity trading, regulating interstate transmission and tariffs, generating the tariffs of central utilities and regulating transmission lines. State Electricity Regulatory Commissions (SERCs) deal with licensing, tariffs and competitive issues within each state.

■ The Ministry of Coal has overall responsibility for determining policies and strategies with respect to the exploration and development of coal reserves. It also supervises Coal India and its subsidiaries, as well as Neyveli Lignite Corporation.

Figure 15.2: **Energy Policy Administration in India's Energy Sector**

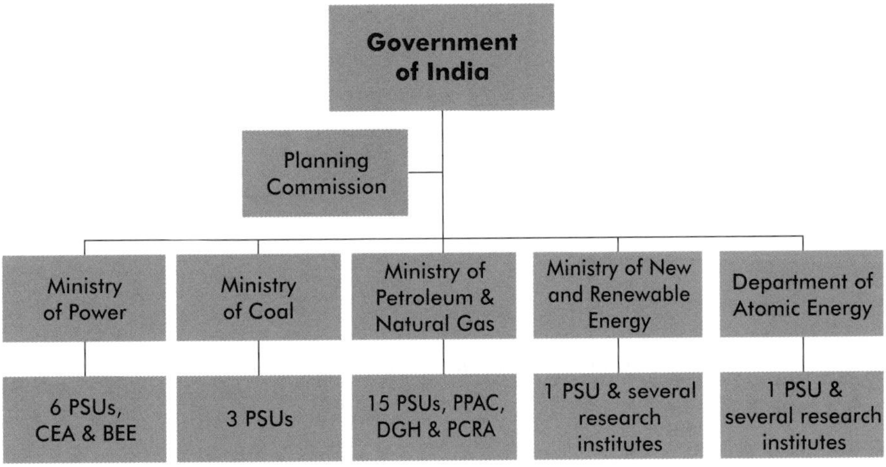

Note: PSU: public sector undertaking; CEA: Central Electricity Authority; BEE: Bureau of Energy Efficiency; PPAC: Petroleum Policy and Analysis Cell; DGH: Directorate General for Hydrocarbons; PCRA: Petroleum Conservation and Research Association.

15

■ The Ministry of Petroleum and Natural Gas oversees the exploration and production of oil and natural gas, their refining, distribution and marketing, and the import, export and conservation of petroleum products and liquefied natural gas. It also has responsibility for development and implementation of pricing policy and for supervising the marketing of biofuels. The Petroleum and Natural Gas Regulatory Board was established in 2006 to set regulations for LNG terminals, refining, transmission and retailing.

■ The Ministry of New and Renewable Energy is responsible for carrying out a national programme to increase wind, small hydro and biomass-based power-generation capacity. It aims to expand the use of renewable energy in urban, industrial and commercial applications and, in remote rural areas, its application, particularly in cooking, lighting and motive power. It is also in charge of policy making in the field of biofuels.

■ The Department of Atomic Energy is responsible for administration of India's nuclear programme.

Other ministries that have influence over energy policy include the Ministry of Agriculture, which handles research and development for the production of biofuels feedstocks, the Ministry of Rural Development, which has responsibility for promoting jatropha plantations for the production of biodiesel, the Ministry of Science and Technology, which supports research into biofuel crops, especially in the area of biotechnology, and the Ministry of Environment and Forestry, which approves and administers clean development mechanism projects in India.

State governments in India have considerable responsibilities in the energy sector, especially in the area of power. The Indian parliament cannot legislate over certain aspects of this sector in the states. In general, as in most federal systems, the states are responsible for implementing national laws, but can also issue state laws and regulations of application in their own territory. As a result, the evolution of power-sector reforms and the level of penetration of renewable energy sources, particularly biofuels, differ widely among states.

Energy-sector reforms started in the early 1990s. The first phase of oil-sector reform involved allowing private and foreign firms to participate in onshore exploration and production through production-sharing contracts.[9] In 1996, a second phase of reforms began, allowing gradual private participation first in refining (1996-1998), then in upstream production (1998-2000) and finally in marketing (2000-2002). In 1997, the government announced a New Exploration Licensing Policy (NELP) to provide a more attractive framework for private domestic and foreign investment in oil exploration (Box 15.2). The government officially abolished the administered pricing mechanism (APM) in 2002 for all petroleum products except kerosene and liquefied petroleum gas (LPG). The abolition was respected for a while but, as international oil prices subsequently rose, the government re-imposed price controls on gasoline and diesel. These controls, plus those on kerosene and LPG, have imposed heavy deficits on downstream oil companies. Some gestures have been made to alleviate these, but the burden essentially remains (see Price and Subsidy Reform below).

Joint-ventures in building oil-product pipelines have been allowed since 2002. Private investors have been free to develop their own gas-pipeline infrastructure since 2006. Foreign direct investment in the gas sector is allowed in exploration

---

9. These are contracts with the government that lock in fiscal terms for the life of the project and ring-fence it from future changes in the general upstream tax regime.

and production, and in liquefied natural gas terminals. In 1998, GAIL, ONGC, IOC and Bharat Petroleum Corporation Limited (BPCL) formed a major joint venture, Petronet LNG, to build and operate LNG import terminals. Qatar's Rasgas is supplying gas to the first plant, Petronet LNG Dahej, commissioned in 2004. A second plant was commissioned by Shell in 2005 at Hazira.[10] Both plants are located in Gujarat.

Box 15.2: **India's New Exploration Licensing Policy**

In response to growing concerns over long-term oil supply and the few discoveries resulting from previous exploration rounds, the Indian government adopted in 1997 a New Exploration Licensing Policy (NELP), based on production-sharing contracts (PSCs). Acreage is awarded under a competitive-bidding process organised by the Ministry of Petroleum and Natural Gas.

The NELP has been in operation since January 1999, when the first round was launched. In the first six rounds, 162 PSCs were signed. In response to an initial lacklustre response from major oil and gas companies, the government removed the ceiling on foreign direct investment in virtually all upstream activities, allowing up to 100% equity by foreign investors. Foreign companies were also given the freedom to sell oil and gas at market prices on the Indian market. In addition, conditions for all deep-water projects have been made very attractive by charging a royalty as low as 5% for the first seven years of commercial production against 10% for other offshore projects and 12.5% for those onshore. The 6th round, launched in February 2006, proved to be the most successful, resulting in 165 bids for the 52 blocks offered and attracting 20 new companies among the 35 foreign companies which submitted bids. The 7th round is expected to be launched in late 2007, with 80 to 90 blocks on offer. The government is considering introducing an Open Acreage Licensing Policy (OALP). This would allow investors to bid continuously for exploration opportunities, with the freedom to choose the areas that interest them.

About 30 oil and gas discoveries have been made since the NELP was adopted, the most significant being the gas discovery in late 2002, by the Reliance-Niko consortium, of almost 10 tcf (283 bcm) of recoverable gas reserves in the Krishna-Godavari basin in Andhra Pradesh State on the east coast. Since then, several more discoveries have been made in the same basin and expectations are high that a second discovery of a size similar to Reliance's find will be made. The largest oil discovery was made by Cairn Energy in Mangala field in Rajasthan in 2003. The find is reckoned to have about 350 million barrels of oil; production is scheduled to start in 2009.

15

10. See Chapter 17 for more information about LNG plants that are planned or under construction.

These reforms have achieved their objective (Table 15.2). For example, private-sector or public-private joint ventures now control 14% of oil exploration and production and more than one-fifth of natural gas production. In 2005, more than one-quarter of India's installed refining capacity was privately owned. Private companies marketed 14% of petroleum products; prior to reform, their market share was nil.

Table 15.2: **Private Participation in India's Energy Sector, 2005**

	Ownership	
	Public (%)	Private (%)
Electricity		
Generation*	76	24
Transmission	100	0
Distribution and end-user supply	87	13
Trade	93	7
Oil and gas		
Crude oil exploration and production	86	14
Natural gas production	77	23
Oil refining	74	26
Marketing	86	14
Coal		
Exploration, production and marketing	93	7**

* Includes industrial autoproducers.
** Includes captive mines.
Note: The first public-private partnership in transmission became operational in 2007 (see Box 15.3).
Source: The Energy and Resources Institute of India.

Foreign investment in the coal sector has been allowed since 1993, but only with approval from the Foreign Investment Promotion Board in cases where a foreign investor takes a controlling equity interest. Foreign companies are allowed to invest in captive coal mining on a case-by-case basis and up to a maximum of 50% of equity. The Indian government has considered the introduction of competitive bidding for coal blocks, but legal constraints in the 1973 Coal Mines Nationalisation Act rule this out. A major failing of the attempt to attract private investment is that Coal India still plays a role in identifying blocks for private participation and naturally retains the best prospects for itself. Without the possibility of selling into a free market (because of the captive restriction) when prices are attractive, investors in the coal sector are likely to remain cautious. The political and social dimensions of reform in the coal sector are more complex than in many other sectors because of the concentration of coal mining and related activities in a small number of states.

The government continues to pursue reform of electricity markets, in order to address chronic problems of under-investment and poor quality of service. With the enactment of the Electricity Act of 2003, India initiated a much-needed overhaul of its power sector. The act consolidates the laws relating to generation, transmission, distribution, trading and use of electricity. It promotes competition and protects the interests of consumers. It also lays out plans to rationalise electricity tariffs. The act, however, does not specify any concrete time frame for elimination of subsidies, which remain very large. Provisions in the act will end investors' obligation to sell to a single buyer.

The act brings some clarity to the roles of different organisations and provides for better management of the regulatory commissions. It also allows for open access to transmission and distribution systems to encourage the development of competitive power markets, and permits private investment in generation and transmission. The act requires the central government to consult with the Central Electricity Authority and state governments in formulating a national electricity and tariff policy. Accordingly, a new National Electricity Policy was announced in 2005 and a National Tariff Policy in 2006. These policies aim to provide everyone with access to reliable electricity supply and to make the power sector commercially viable through cost-reflective tariffs. Progress in implementation of the Electricity Act has varied from state to state. Some states have made progress in separating transmission from generation and developing open access regulations, but more significant progress, especially in grid expansion, will require that all states implement reforms.

In 2006, after comprehensive public consultation, the Indian government approved an Integrated Energy Policy, which lays out recommendations for the main energy challenges facing the country (Table 15.3). Some of the recommendations lack sufficient precision but, since the Integrated Energy Policy was approved, many working groups and committees have been set up to plan the necessary action and evaluate progress. The Expert Committee that drafted the Integrated Energy Policy has acknowledged that greater precision will be needed regarding specific policy measures; but the definition of many of the objectives themselves first needs to be made more precise if they are to be successfully communicated and implemented by appropriately well-directed policies.

Table 15.3: **India's Integrated Energy Policy: Priority Recommendations of the Expert Committee**

Recommendation	Targets/goals
Ensure adequate supply of coal of consistent quality	Make more coal blocks eligible for development by private companies or joint ventures; build infrastructure to facilitate steam coal imports; rationalise coal pricing; amend the Coal Nationalisation Act to facilitate private participation.
Address the concerns of resource-rich states	Allow these states to share in profits; revise royalty rates; create National Policy on Domestic Natural Resources.
Ensure availability of gas for power generation	No new gas-fired capacity to be built until firm gas supply agreements are in place.
Reduce the cost of power	Reduce losses through use of automated meters and separate metering of agricultural pumps; proper setting of cross-subsidy surcharges and wheeling and backup charges; create an efficient interstate and intrastate transmission system; refurbish power stations; generation and transmission projects built on tariff-based bidding.
Rationalise fuel prices	Price energy at trade-parity prices; remove administered pricing scheme.
Enhance energy efficiency and demand-side management	Improve power generation efficiency from 36% to 38-40%; information dissemination; minimum fuel standards.
Augment resources for increased energy security	Carry out surveys of energy resources; enhanced recovery of domestic resources; private-sector involvement.
Use more energy abroad	Invest in captive fertilizer and gas liquefaction facilities.
Enhance role of nuclear and hydropower	Tap thorium reserves; create more hydro storage facilities.

Table 15.3: **India's Integrated Energy Policy: Priority Recommendations of the Expert Committee** *(Continued)*

Recommendation	Targets/goals
Enhance role of renewables	Link incentives to outcomes like energy generation, not installed capacity; enact policies to promote alternatives like plantations, gasifiers, solar thermal and photovoltaics, biodiesel and ethanol; expand equity base of the Indian Rural Energy Development Agency.
Ensure energy security	Maintain strategic oil reserves in line with IEA standard of 90 days; engage in bilateral agreements to reduce supply risk.
Boost energy-related R&D	Set up a National Energy Fund to finance energy R&D.
Improve household access to energy	Provide electricity to all households by 2009/10 (Rajiv Gandhi Grameen Vidhyutikaran Yojana); have more targeted subsidies using debit card systems; improve efficiency of cook stoves and kerosene lanterns; use more distributed generation; increase access to financing for micro-enterprises; involve rural communities in decision-making.
Enable environment for competitive efficiency	Devolve regulatory responsibilities from ministries to state level; regulators should mimic competitive markets.
Address climate change concerns	Enhance energy efficiency in all sectors; increase mass transit; use more renewables and nuclear; invest in clean coal technologies; more research and development.

Source: Planning Commission (2006).

15

Energy Policy Challenges

While acknowledging the steps taken towards a coherent energy policy (see above), this section recapitulates the main challenges which face India in the energy sector, in order to highlight the key areas for action. To meet India's large energy infrastructure investment needs will require the mobilisation of public and private funds. To attract private investment, a transparent and predictable investment framework must be established. Reducing the number of people who do not have access to electricity and the even greater number that use inefficient, polluting fuels for cooking and heating is a huge and pressing challenge. The country's growing appetite for personal transport, which carries the double threats of environmental degradation and energy insecurity, calls for policies aimed at improving efficiency and promoting both alternative fuels and alternative methods of transport. Supply-side and demand-side approaches must go hand in hand. Successful pilot projects need to be scaled up to meet the challenges ahead. Many of the necessary policies have already been proposed or are actively under consideration (see Table 15.3 and Chapter 18). Successful implementation will depend on effective co-ordination of implementing action between the ministries and different departments at the national level and between the central government, the states and union territories, and the municipalities. The Prime Minister has addressed this need through the creation of the Energy Coordination Committee. The Committee is responsible for adopting a systematic and co-ordinated approach to policy formulation and decision-making across the whole energy field. Some areas of energy policy also require integration of India's action into a global framework.

Price and Subsidy Reform

Energy pricing policies to date have resulted in an economically inefficient fuel mix and distorted allocation of energy and financial resources.[11] High subsidies crowd out funds for capital investment. They frequently fail to benefit the target population, usually poor, rural consumers (see Chapter 20). At the central level, the largest energy-related subsidies are for LPG and kerosene. At the state level, the largest subsidies are in the power sector.

Average residential electricity tariffs in India, at 7 US cents per kWh, are about half the OECD average, excluding taxes. Industry tariffs, at 9 US cents per kWh, are slightly higher than the OECD average level. Electricity subsidies in the agriculture sector contribute to land degeneration and encourage wasteful use of water (see Box 16.6 in Chapter 16). Overall, gross electricity subsidies amounted to some $9 billion in 2005/06, according to the Ministry

---

11. Even though end-use energy prices in India are generally lower than international prices, measured in PPP they can be very high for many Indian consumers.

of Finance's *Economic Survey 2006-2007*. Subsidies provided by the SEBs cause them to incur big financial losses, harming their capacity to invest in building new generating plant and maintaining and extending the network. As a result, in many parts of India, electricity is unavailable for up to 14 hours a day. This encourages richer households and small manufacturers to use subsidised, inefficient and polluting diesel in small generators.

With the dismantling of the administered pricing mechanism in 2002, it was envisaged that the subsidies on kerosene and LPG would be phased out over four years. This has not happened. The price of kerosene sold under the public distribution system was $0.22 per litre in August 2007, less than a third of the price of kerosene in neighbouring Nepal.[12] The average price of kerosene on the Singapore market was $0.50 per litre in 2007. The government has also re-imposed price controls on gasoline and diesel.

In India, the difference between refinery-gate prices, based on import parity, and the selling price is borne largely by public oil companies. Because of recent increases in product prices at the refinery gate, the burden on the oil companies has been increasing. Total losses in 2005 incurred on the sale of gasoline and diesel reached $3.4 billion. In 2006, they surged to $4.6 billion. Losses for kerosene, LPG, gasoline and diesel combined were $8.9 billion in 2006. Government-issued bonds covered about a third of these losses.

Petroleum products, however, are also taxed heavily in India. The post-tax price of unleaded gasoline was $1.06 per litre in India, compared with $0.80 for the OECD average. State and federal government revenues from petroleum product taxes in India reached almost $27 billion in 2006, outweighing the subsidies and losses. Price rationalisation is desperately needed in the Indian downstream oil market. The IMF estimates that petroleum product prices would have to be adjusted by 40% to 45% on average to be fully in line with international prices, with kerosene and LPG requiring the largest adjustments (IMF, 2006). The greatest challenge if such an adjustment does take place is to ensure that households below the poverty line are not made worse off (see Chapter 20).

Energy Efficiency

Improving the efficiency of energy use in power generation and in final uses will be vital to curbing demand growth as the economy and the population grow. According to the Construction Industry Development Council of India, the rate of growth in residential and commercial property construction was 10% in 2005 over 2004. The Indian Society of Automobile Manufacturers estimates that vehicle sales grew by about 14% per year on average from 2000

---

12. Prices are from the PPAC website (www.ppac.org.in) and have been converted at an exchange rate of $1 = Rs 39.68.

to 2006. More efficient vehicles could help to curb the rapid growth in oil demand. There is also enormous scope to improve the efficiency of energy use in highly energy-intensive industries such as cement, steel and fertilizer. Many large steel, cement and aluminium plants in India are state-of-the-art, with efficiencies equivalent to those in OECD countries; but there are also very many small plants that are extremely inefficient. Financing efficiency improvements in these inefficient plants is often much more difficult than for larger ones.

Improving efficiency in the power generation sector is immensely challenging. India's coal-fired power plants are among the least efficient in the world and the construction of new thermal power stations would be a cost-effective way of improving efficiency.

There have been some positive achievements in energy efficiency in the past few years, such as the enactment of the Energy Conservation Law in 2001 and the creation of the Bureau of Energy Efficiency (BEE) in 2002. The BEE launched both the National Energy Labelling Programme and the Energy Conservation Building Code in 2006.[13] There are, however, several barriers to the adoption of the necessary measures, including inadequate institutional capacity, high transaction costs, lack of access to capital, a high private discount rate and a lack of enforcement of standards and codes. There is an urgent need to increase the staffing and resources of administrative agencies at the federal and state levels in order to implement energy-efficiency measures.

Infrastructure Investment

Rising energy demand is putting enormous strain on India's infrastructure. Ports, railways, roads and power plants are all in serious need of new investment. Public funds will not be sufficient to cover all investments required to support rapidly growing energy demand and to increase energy access. The authorities are turning to public-private partnerships as a way of bridging the funding gap (Box 15.3). To meet rising investment needs, India needs to create a transparent, predictable and consistent investment framework, to improve its regulatory framework and to speed up its legal process. Private investors have been more hesitant to enter the energy market in India compared to other sectors. Private companies have been deterred by the preferential treatment given to state-owned energy companies and the slow progress on tariff reform and other issues.

---

13. Strict implementation of these policies is assumed in the Alternative Policy Scenario (Chapter 18).

Box 15.3: **Public-Private Partnerships**

Public-private partnerships are designed to provide public services more efficiently and at a lower cost to the end user than either the government or the private sector could provide on their own.[14] In India, both the central government and the states intend to use public-private partnerships more intensively to help meet gaps in the provision of energy services. India has run fiscal deficits for decades and, increasingly, there are limitations on how much the public sector can spend. Public-private partnerships could play a key role in meeting investment needs over the *Outlook* period in the context of a transparent and stable business environment. One role for the government in these partnerships is to reduce start-up hurdles, such as delays in acquiring land and construction permits. This would lower investor risk.

Private-sector investment will be crucial for the rehabilitation of existing power plants and for other needed investments in transmission and generation. Powerlinks Transmission, a joint-venture between the private utility Tata Power Company and the state-owned Power Grid Corp, is the first public-private partnership in power transmission in India. Powerlinks is a $265 million project to build, own, operate and transfer five 400-kV lines and one 220-kV transmission line extending over 1 200 km from West Bengal to Delhi, with a capacity of about 3 000 MW. The system became operational in early 2007. It brings power from the Tala hydro plant in Bhutan to the north of India. Power Grid managed the consent and approval processes involved with laying the lines, but Powerlinks will maintain them.

Energy Access

While economic growth has reduced poverty levels in India, we estimate that there are still some 412 million people without access to electricity. The number of people in India relying on fuelwood, dung and agricultural residues for cooking is estimated to be about 668 million. The heavy dependence on these fuels for cooking has serious consequences for health: women and children are the most vulnerable. These issues are taken up in greater detail in Chapter 20.

Environment

India faces serious energy-related environmental damage. Congestion and pollution from motor vehicles is an increasing threat to health in all Indian cities. Over half of Indian cities have levels of particulate matter (PM_{10}) which are more

14. See OECD (2007) for guidelines for public-private partnerships.

than one-and-a-half times the Indian standard of 0.1 to 0.5 microgrammes per cubic metre. Land degradation, resulting from opencast coal mining and over-extraction of water for mining purposes, is also a major concern.

India has air quality standards prescribed for various pollutants. They vary by location but are set within a legal framework under yhe Prevention and Control of Air Pollution Act, 1981, which extends to the whole of India. The Indian government reports that Delhi, Mumbai, Kolkata, Chennai, Bengaluru, Hyderabad, Ahmedabad, Surat, Kanpur, Agra, Sholapur and Lucknow are India's most polluted cities. Greater policy efforts are needed, like expanding the use of public transport, introducing fuel economy standards and accelerating the uptake of cleaner vehicle technologies. Progress in reducing local pollution has been made in some large cities, notably Delhi, where all public transport vehicles are required to be powered by compressed natural gas (CNG).[15]

India acceded to the Kyoto Protocol in 2002 and the government is becoming more active in global climate change negotiations. Although it does not have greenhouse gas emissions commitments, it has taken active steps to address climate change, notably encouraging projects under the clean development mechanism (CDM), which play an important role in curbing global emissions. The Energy and Resources Institute of India has been selected to carry out the National Strategy Study on CDM in India sponsored by the World Bank. This initiative focuses on the following themes: strategic overview of CDM opportunities for India and international demand for emission offsets; identification of CDM projects for key sectors; key institutional, legal, financial, and regulatory prerequisites to facilitate CDM project development and implementation; human and institutional capacity building to identify, develop, implement and process CDM projects in India; and capacity to exploit global opportunities.

CDM activity in India is second only to that of China. Expected emissions reductions from proposed CDM projects in India amount to some 54 Mt of CO_2-equivalent per year during 2008-2012.[16] Slightly under half of these expected reductions are from projects which have already been officially approved by the CDM Executive Board. Energy-related projects account for almost 75% of the total savings. These projects focus mainly on renewable energy (20 Mt CO_2), energy efficiency (12 Mt) and fuel-switching (7 Mt). Most of the projects are being developed by Indian companies. The main buyers of credits worldwide are industrial companies and power generators, both in the European Union, where they are covered by the EU Emissions Trading Scheme, and in Japan, which has a voluntary trading system.

15. Air pollution and CO_2 emission trends in India are discussed in Chapter 16.
16. Based on data from the Joint Implementation Pipeline of the United Nations Environment Programme, Risø Centre on Energy, Climate and Sustainable Development (available on line at www.uneprisoe.org).

CHAPTER 16

REFERENCE SCENARIO DEMAND PROJECTIONS

HIGHLIGHTS

- Primary energy demand in India more than doubles by 2030 in the Reference Scenario, driven largely by GDP which is assumed to expand at an average annual rate of 6.3%. Coal remains the most important fuel, but oil demand also grows fast, increasing two-and-a-half times by 2030. Energy intensity declines progressively thanks to efficiency improvements and a continuation of the shift to services and less energy-intensive industry. A reduction in the share of fuelwood and dung in residential energy use also contributes. Power generation accounts for much of the increase in demand.

- The growth rate in industrial energy demand is expected to accelerate to 4.7% per year in 2005-2015, with surging demand for steel, cement and other materials for infrastructure development. It then slows down to 3.7% in 2015-2030, as end-use efficiency improves. Energy demand in the iron and steel sector grows by 5.9% in 2005-2030. Coal, mainly for steel production, and electricity remain the dominant fuels for industrial use over the *Outlook* period.

- Energy demand for transport will see rapid growth in the next two-and-a-half decades. It is projected to grow by 6.1% per year, reaching 162 Mtoe in 2030 as the vehicle stock expands rapidly with rising economic activity and household incomes. As more people can afford passenger cars, ownership of two- and three-wheelers begins to plateau towards the end of the projection period. But they still account for over 50% of the total vehicle stock in 2030.

- Residential energy demand grows by 1.6% per year over 2005-2030. The share in residential energy use of biomass, including fuelwood, dung and agricultural waste, falls from 79% in 2005 to 59% in 2030. This decline, however, masks a wide disparity between rural and urban households. Biomass consumption falls by 0.6% per year in urban households, but still grows slightly in rural households, by 0.5% per year.

- NO_x emissions, mainly from road vehicles and the power sector, are projected to rise sharply in the Reference Scenario. SO_2 emissions are set to rise even faster. India becomes the world's third-largest CO_2 emitter by 2015. It ranked fifth in 2005. Two-thirds of India's emissions come from burning coal, mainly in power stations. This share will increase slightly, to 69%, by 2030. Per-capita CO_2 emissions, though doubling over the *Outlook* period, are in 2030 still well below those in the OECD today.

Key Assumptions

The Reference Scenario takes account of government policies and measures that were enacted or adopted by mid-2007. However, not all of these policies are assumed to be fully implemented in the Reference Scenario. Lack of co-ordination among government departments and over-ambitious targets have resulted in a poor track record of policy implementation.[1] Full implementation of these policies is considered in the Alternative Policy Scenario, along with implementation of other policies which are now in contemplation or seem likely to be adopted.

The Reference Scenario projections assume that India's gross domestic product (GDP) will grow on average by 7.2% per year from 2005 to 2015 (Table 16.1). Growth is assumed to slow thereafter, bringing down the average for the entire *Outlook* period to 6.3% per year.[2] In the short and medium term, both infrastructure investments and continued market reforms, particularly in the power sector, are expected to support faster growth. The share of agriculture in GDP is assumed to decline by 5 percentage points over the *Outlook* period, while the share of the services sector rises by 4 percentage points. The output gains from labour migration from agriculture to services and, to a lesser extent, to industry, are expected to continue to contribute to economic growth.

Table 16.1: **GDP and Population Growth Rates in India in the Reference Scenario** (average annual rate of change)

	1980-2005	1990-2005	2005-2015	2015-2030	2005-2030
GDP	5.9%	6.0%	7.2%	5.8%	6.3%
Population	1.9%	1.7%	1.4%	1.0%	1.1%
GDP per capita	4.0%	4.2%	5.7%	4.7%	5.1%

India's rate of population growth is declining, from some 2.1% per year in the 1980s to 1.7% per year from 1990 to 2005. This *Outlook* assumes that the population, which stands at 1.1 billion, will increase by 1.1% per year on average to 2030, reaching 1.45 billion.[3] India's population is growing faster than China's. According to the United Nations Population Division, India is expected to become the most populous country in the world in 2031. Over 70% of India's population lives in rural areas today; this share is expected to drop to 59% by 2030.

---

1. To address this, Prime Minister Singh created the Energy Coordination Committee in 2005 (see Chapter 15).
2. The High Growth Scenario (Chapter 19) explores the impact on energy demand of even higher growth than is assumed here.
3. Population assumptions are based on the United Nations' report, *World Population Prospects: The 2006 Revision* (UNPD, 2007).

Most energy prices in India are controlled by the government and thus do not move in line with international prices. Electricity and gas prices are particularly heavily subsidised (see Chapter 15). The government had planned to phase out the subsidies on LPG and kerosene but with recent high oil prices, subsidy reform has stalled. The Reference Scenario projections assume that subsidies initially remain in place, leading to distortions in inter-fuel competition and energy use, but are gradually reduced over the second half of the projection period.

Primary Energy Demand

Primary energy demand in India is projected to increase from 537 Mtoe in 2005 to 770 Mtoe in 2015 and to 1 299 Mtoe in 2030 (Table 16.2). Demand grew by 3.5% per year in 1990-2005. Energy demand growth is somewhat faster in 2005-2015 at 3.7% per annum, slowing again to 3.5% in 2015-2030. As GDP growth is faster over the *Outlook* period, intensity improves more quickly than in the past. In 2025, India's energy demand passes that of the entire OECD Pacific region; it equals 60% today. By 2030, India is the third-largest energy consumer in the world, after China and the United States; today, it ranks fourth (Figure 16.1).

Table 16.2: **Indian Primary Energy Demand in the Reference Scenario** (Mtoe)

	1990	2000	2005	2015	2030	2005-2030*
Coal	106	164	208	330	620	4.5%
Oil	63	114	129	188	328	3.8%
Gas	10	21	29	48	93	4.8%
Nuclear	2	4	5	16	33	8.3%
Hydro	6	6	9	13	22	3.9%
Biomass	133	149	158	171	194	0.8%
Other renewables	0	0	1	4	9	11.7%
Total	**320**	**459**	**537**	**770**	**1 299**	**3.6%**
Total excluding biomass	*186*	*311*	*379*	*599*	*1 105*	*4.4%*

* Average annual rate of growth.

16

Coal remains the dominant fuel in India's energy mix over the *Outlook* period. Its share increases from 39% in 2005 to 48% in 2030, by which time almost three-quarters is used in power generation. Demand for oil, mostly for transport, increases by two-and-a-half times, but its share rises only one percentage point, from 24% of total primary energy demand in 2005 to 25%

in 2030. Natural gas is the fastest growing of the fossil fuels, more than tripling by 2030, when its share of primary demand reaches 7%. Although demand for biomass continues to rise, its share in primary energy demand drops sharply from 29% in 2005 to 15% in 2030 – mostly as a result of fuel switching in the residential sector. Other renewables, mostly wind power, grow at a rate of nearly 12% per year, albeit from a low base. Nuclear and hydropower supplies grow in absolute terms, but they make only a minor contribution to primary energy demand in 2030: 3% in the case of nuclear and 2% for hydropower.

Figure 16.1: **Primary Energy Demand in Selected Countries in the Reference Scenario**

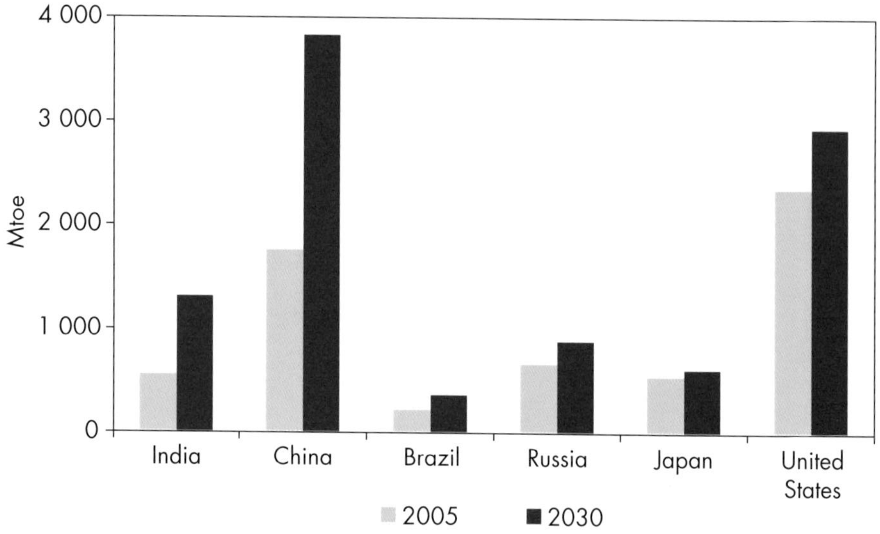

Fuels used for electricity production account for a growing share of primary energy demand over the *Outlook* period. Their use grows by 4.6% per year, boosting their share of total energy demand from 36% in 2005 to 45% in 2030. In India today, almost a third of electricity production is lost or not paid for. Losses are expected to decline over the *Outlook* period.

Final Energy Demand

Total final energy demand grew by 2.3% per year in 1990-2005, reaching 356 Mtoe. Demand is projected to accelerate to 3.3% per year over the *Outlook* period and will be 804 Mtoe in 2030. The share of transport in final energy demand in 2005-2030 grows from just some 10% today to 20% (Figure 16.2). The transport sector dominates the growth in demand for oil.

Energy demand in the residential sector grows at a steady 1.6% per year, while industrial energy demand growth accelerates from 2.4% in 1990-2005 to 4.1% in 2005-2030. Electricity use grows at 6.1% per year, resulting in a more than four-fold increase by 2030. Oil increases by 4.1% annually from 2005 to 2030. Gas grows also by 4.1%, but its share of final demand in 2030 remains small. Coal demand rises somewhat more quickly at 4.7% per year. The use of biomass increases very slowly, by 0.5% per year. Nearly all of this increase takes place in the rural residential sector. Penetration of biofuels and industrial co-generation is low in the Reference Scenario.

Figure 16.2: **Sectoral Shares in Final Energy Demand in India in the Reference Scenario**

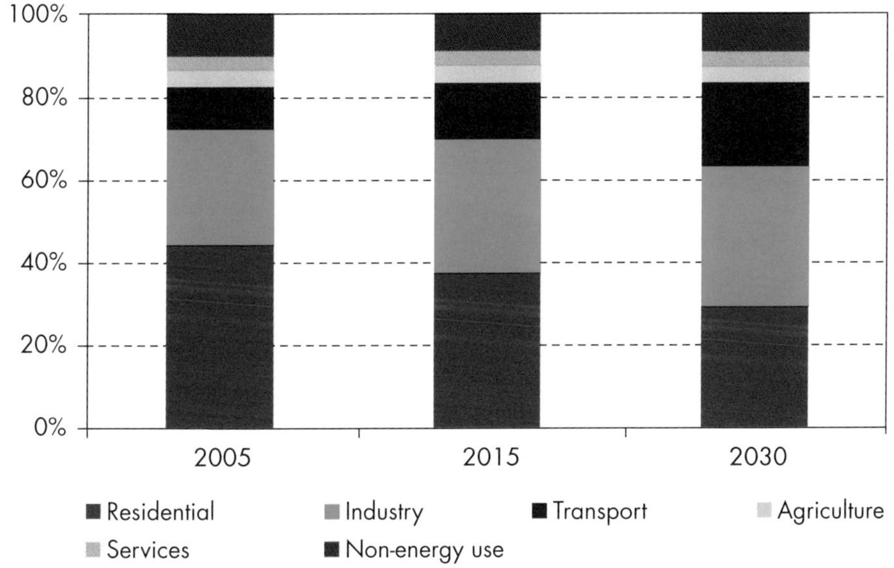

Industry Sector

Energy demand in the industrial sector accounted for 28% of final energy demand in 2005. Growth in industrial energy demand is projected to accelerate to 4.7% per year in 2005-2015 and then moderate to 3.7% per year in 2015-2030, as end-use efficiency improves. Industry's share of total final energy demand edges up to 34% in 2030. Energy demand in the iron and steel, chemical and petrochemicals, non-metallic and other minerals, food, paper and textile industries together currently represents over half of total industrial energy demand. These sectors are expected to remain the main drivers of industrial energy demand over the projection period. Coal, mainly for steel and cement production, and electricity are expected to remain the main fuels used in industry over the *Outlook* period (Figure 16.3).

16

Coal accounts for 41% of industrial energy demand in 2030, up from 30% in 2005. Electricity's share rises from 18% to 31% mainly because of an expansion of lighter manufacturing production as well as to increased penetration of large-scale, more efficient electric arc furnace technology in the iron and steel industry. The share of natural gas falls from 5% to 4%, although demand increases in absolute terms along with increased availability of gas supplies.[4]

Figure 16.3: **Industrial Energy Demand by Fuel in India in the Reference Scenario**

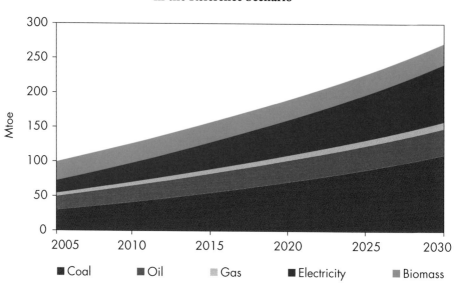

The efficiency of many processes in the Indian cement and steel industries has improved over the past 15 years, helping to lower the country's overall energy intensity. In contrast, electricity intensity in industry as a whole, which fell in the early 1990s, began to rise towards the end of the decade. Much of Indian industrial output is derived from small-scale, often village-based, enterprises, fuelled by inefficient motors and equipment, where it has been difficult to implement efficiency improvements. There is enormous scope to improve efficiency.[5]

4. Demand for gas as a petrochemical feedstock (not included in industrial demand) also increases in absolute terms. In previous editions of the *Outlook*, petrochemical feedstocks were included in energy use in the industry sector. They are now included in non-energy use.
5. The impact of government policies to enhance industrial energy efficiency is discussed in Chapter 18.

Iron and Steel Industry

India's iron and steel industry is expected to continue to boom. Energy demand in this sector is projected to make up nearly 28% of total industry energy demand in 2030, up from less than 20% in 2005 and 15% in 1990. Total energy demand in the sector grows by 5.9% per year in 2005-2030, as India ramps up production of steel-based goods for the domestic market. Growth in India's automobile industry (currently at almost 20% per year) and in the housing and white goods sectors (showing double-digit growth) is driving up steel demand. The *Outlook* projects that rising vehicle demand and growth in appliance and building stocks will continue to drive growth in steel demand in 2005-2030. Annual crude steel consumption per capita in India is extremely low by international standards; in 2005 it was about 38 kilogrammes per capita, some 20% of the world average (World Coal Institute, 2007). Coal demand in the iron and steel sector grows by 5.2% a year, gas demand by nearly 6% per year and electricity demand by nearly 8% per year over the *Outlook* period. About 44% of industrial coal demand and a quarter of industrial electricity demand are for iron and steel production at present. The share falls to 41% for coal but rises to 35% for electricity in 2030.

India is currently the world's seventh-largest steel producer. In 2006, production of finished steel amounted to 44 million tonnes (World Coal Institute, 2007). More than half of India's steel is produced in basic oxygen furnaces. Electric arc furnaces (EAF) account for about 45% and open hearth furnaces for the rest. The energy intensity of Indian steel production in both integrated plants and mini-mills, measured as energy use per tonne of steel produced, is high by international standards. This is because of the poor quality of local coking coal and iron ore supplies and outdated technologies and energy management practices. Energy intensity in electric arc furnaces is particularly high, largely as a result of their small average size in India. All the integrated steel mills in India have captive power plants to guarantee supply, but mini-mills are generally too small for this option to be economically viable.

Most of India's proven reserves of coking coal, which amount to some 17 billion tonnes, are often not suitable for steel production, so the steel industry currently has to import almost 50% of its coking coal needs. The Ministry of Steel expects this share to rise to over 85% by 2020. In order to reduce the dependence on imported coal and to increase energy efficiency, new steel capacity is likely to be in the form of blast furnaces and direct-reduced electric arc furnaces. In 2006, India was the world's largest producer of direct-reduced iron (DRI), also known as sponge iron, with 150 DRI plants producing nearly 15 million tonnes, or almost 25% of total world production (Midrex, 2006). Indian companies have favoured the DRI process, because it is smaller-scale and less capital-intensive than other technologies. About 60% of the current production comes from small-scale industry. Some 225 coal-

16

based DRI plants are at various stages of commissioning and construction and 77 existing plants are expanding production (Joint Plant Committee, 2005/06). The high use of coal-based DRI-EAF carries a heavy environmental burden, as CO_2 emissions per tonne of crude steel are about 2 500 kilogrammes, compared with about 400 kilogrammes for EAFs using scrap (IEA, 2007). Crude steel production through scrap-based EAF is limited in India by low domestic scrap availability and the high cost of imports.

Chemicals and Petrochemicals

The chemicals and petrochemicals sector is India's third-largest industrial energy consumer, accounting for 9% of total industrial consumption in 2005. Electricity accounts for about 46% of energy use, oil for 40% and coal for the remainder. The sector's energy needs are projected to grow by 4% per year in 2005-2015, slowing to 2.8% per year from 2015 to 2030. Use of coal is rapidly phased out over the *Outlook* period, its share plummeting from 14% in 2005 to 2% in 2030. Electricity gains market share, its demand growing by 4.8% per year. Fertilizer and chlor-alkali producers are the leading energy consumers in this sector.

India is the world's second-largest producer, after China, of nitrogenous fertilizer, which is made from ammonia. India produced 12.8 million tonnes of ammonia in 2005, with average energy consumption of 0.91 toe per million tonnes of ammonia, down from 2.3 toe in 1960.[6] The most efficient plants in India produce ammonia with less energy intensity than the average for the world's most efficient plants, comparing, in both cases, those ranking in the top 25% for efficiency. Today, there are 57 large-sized fertilizer plants and about 64 medium and small-scale units in operation. Coal and naphtha are the primary feedstocks but there is increasing interest in switching to natural gas (Box 16.1).

Box 16.1: **Feedstock for India's Fertilizer Industry**

The government has in the past heavily subsidised fertilizer use in India, because of its importance in maintaining food self-sufficiency. The recent surge in oil prices, however, has increased the financial burden on the government, prompting it to reduce provision for payments to fertilizer producers in the 2007/08 budget. But fertilizer prices have not been allowed to increase and now the producers are facing large losses. The share of energy in the total cost of ammonia production in India is currently about

6. See Karangle, Rashtriya Chemicals & Fertilizers (2007) for more details.

80%. Past subsidies eliminated incentives to invest in more efficient technologies or in research and development in order to cut costs. The producers are now making efforts to reduce costs, particularly those related to energy use, but these efforts are limited by controls on their product prices.

Many fertilizer companies are switching to gas as feedstock. This move will save energy as converting gas into fertilizer is considerably less energy-intensive than converting other feedstocks. Gas availability to meet expected demand growth and gas pricing are, however, matters of concern (see Spotlight in Chapter 15). LNG imports and indigenous output from recently discovered fields are expected to contribute.

Other Industries

The non-metallic and other minerals, food, paper and textile industries account for about a quarter of energy demand in the industry sector. Coal and oil meet most demand, but electricity use is rising. The Indian aluminium industry is poised for expansion and is a large consumer of electricity. The textile industries, comprised largely of small enterprises, rely heavily on coal, used in boilers for process heat. Coal and electricity are the two major fuels used in the pulp and paper industry.

Cement production in India increased from 107 million tonnes in 2004 to 134 million tonnes in 2005, or 9.3%, driven by infrastructure development and the country's housing boom (TERI, 2007). India is the world's second-largest cement producer after China. Coal is the main fuel. Annual per-capita consumption of cement is around 100 kg, much lower than the global average of 270 kg. India has 128 large cement plants, with an estimated combined capacity of 152 Mt a year, and over 300 mini-plants, with a total capacity of 11 Mt. Clinker production is the most energy-intensive step in the production of cement and it can be produced through either a wet or dry process. The latter is much less energy-intensive. Indian plants using the wet process have been phased out over the past several decades and, today, according to The Energy and Resources Institute of India, over 90% of cement is produced using the dry process.

Energy demand in these sectors as a whole is projected to grow by 4.8% per year from 2005 to 2015 in the Reference Scenario, with rapid economic growth and infrastructure development. Demand for cement to build highways and railway infrastructure will rise particularly quickly. In 2015-2030, energy demand will slow to 3.5% per year, as efficiency improves, yielding an average increase in demand of 4% per year over the entire *Outlook* period. Demand for electricity grows most rapidly, as it replaces oil currently used in inefficient pumps and motors. Coal use expands by 4.3% per year from 2005 to 2030, by which time it accounts for 53% of total energy demand in these sectors, up from 50% in 2005.

16

Infrastructure construction is driving an increase in demand in other industries for equipment, machinery, paints and related products. Almost 60% of the energy demand in these industries is met by biomass, used mostly in inefficient boilers and kilns. But this share is expected to fall sharply, as coal, gas and electricity use rises. In the Reference Scenario, it is assumed that inefficient plants are phased out over the projection period. As a result, energy demand in these industries grows at a slower pace than in industry as a whole, at 3.4% per year over the *Outlook* period.

Transport Sector

India's energy demand for transport increased by 1.9% per annum in 2000-2005, well below the rate of growth of both final energy use as a whole and GDP (Box 16.2) Our projections show a different picture, with transport demand projected to grow by a brisk 6.1% per annum over the projection period, as the vehicle stock expands rapidly with rising economic activity and household incomes. Demand almost doubles by 2015 and more than quadruples by 2030, reaching 162 Mtoe. The share of transport in final energy demand in India doubles over the *Outlook* period, increasing from 10% in 2005 to 20% in 2030. India currently accounts for only 2% of global transport energy demand. This share is projected to nearly triple over the *Outlook* period.

Box 16.2: **Recent Slow Growth in Transport Fuel Demand**

According to the latest official Indian data, energy consumption in transport grew by only 1.9% per year in 2000-2005. Demand for diesel, which makes up almost 70% of the oil used in Indian road transport, fell at the rate of 0.7% per year, while gasoline consumption grew by 5.5% per year. The modest overall increase in transport fuel use stands in stark contrast to 14% per annum growth in vehicle ownership over the same period. There are various possible explanations for the surprisingly slow increase in fuel use. Improved efficiency of new cars and trucks, higher international oil prices, increased load factors and switching to compressed natural gas (CNG) for public transport in some major cities may partly explain the drop in diesel consumption. Another explanation specific to India is the illegal blending of kerosene with diesel. This long-standing practice is encouraged by the large subsidy on kerosene (mainly used for cooking and lighting), making kerosene much cheaper than diesel, which is heavily taxed. Diesel prices increased sharply from Rs 11.84 ($0.27) per litre in 1999 to Rs 32.83 ($0.74) per litre in 2005, exacerbating the problem. The application of kerosene does not show up in transport statistics, as it is purchased for household purposes. Recent enforcement of overloading restrictions on trucks may have cut load factors, thereby increasing consumption and offsetting, to some extent, the depressive impact of kerosene adulteration on diesel demand.

The transport sector currently consumes 27% of total primary oil demand in India and this will increase to 47% by 2030. Oil contributes 95% of the total increase in transport energy use between 2005 and 2030 and, unsurprisingly, the lion's share of transport energy demand in 2030 is met by oil. Natural gas for CNG vehicles, electricity and biofuels make up the rest. Road vehicles account for 86% of total transport energy demand in 2030, aviation (which sees continued strong growth) for 9% and railways (which see slower growth than other sectors) for most of the rest (Figure 16.4). These shares remain fairly constant over the *Outlook* period, as does the proportionate demand for each transport fuel and mode.

Figure 16.4: **India's Transport Energy Demand by Mode in the Reference Scenario**

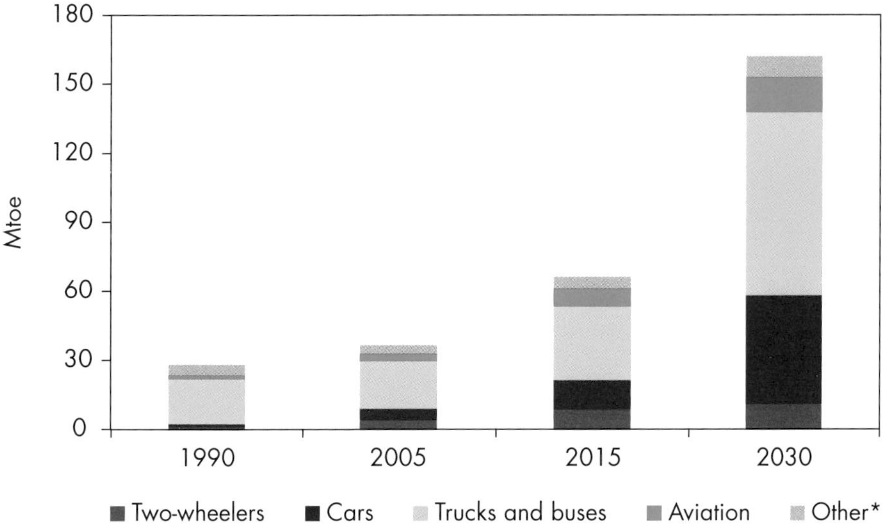

* Refers to rail, pipeline and navigation.

Road Transport

The number of vehicles on the road is the principal determinant of fuel demand for transport. The total vehicle stock in India increased from 19 million in 1990 to 68 million in 2004.[7] We project it to reach 295 million by 2030, overtaking that of the United States soon after 2025. Annual sales of new light-duty vehicles (LDVs), which reached 1.2 million in 2005, are projected to soar to 13.3 million

7. Data provided to the IEA by The Energy and Resources Institute of India (TERI).

in 2030. Strong vehicle growth will continue through the *Outlook* period, at 5.7% per annum, faster than the growth rate of GDP/capita, at 5.1%. The fleet of LDVs will increase faster than any other category of transport vehicles, from 11 million in 2005 to 115 million by 2030 – an annual average rate of growth of almost 10% per annum. Excluding two- and three-wheelers, there are currently 13 vehicles per 1 000 people in India. This ratio grows to 93 by 2030. Despite this seven-fold increase, vehicle ownership in 2030 is still only 15% that of Japan today (600 vehicles per 1 000 people).

Two-wheelers[8] make up over 80% of the current vehicle stock, yet they consume around 15% of road-transport fuels. The recent shift from two-stroke to four-stroke engines for these vehicles has greatly increased efficiency and reduced air pollution. Two-wheelers are the first step on the ladder to increased personal mobility, because they are cheaper than cars and are well suited to congested cities with poor public transport services. Towards the end of the projection period, ownership of two-wheelers begins to plateau, as more people purchase passenger cars. Two-wheelers still account for over 50% of the total vehicle stock in 2030 (Figure 16.5).

Figure 16.5: **India's Vehicle Stock in the Reference Scenario**

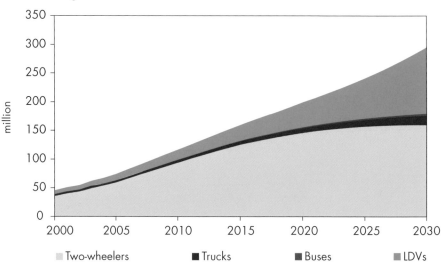

Although the Indian vehicle stock is dominated by two-wheelers, which use motor gasoline, almost 70% of the oil used in road transport is diesel. This is due to the much higher per-vehicle consumption of heavy vehicles (mainly trucks and buses) and the increasing percentage of LDVs in India that run on diesel. The share of diesel is not expected to change significantly over the *Outlook* period, because the decrease in two-wheelers and the increasing share of diesel passenger cars is partly offset by the increase in buses and trucks for freight.

8. Two-wheelers refer to two- and three-wheel vehicles.

Box 16.3: **Upside Potential of Transport Oil Demand in India**

Our Reference Scenario projections for vehicle ownership and, therefore, oil demand for road transport in India may prove to be conservative. Experience around the world shows that vehicle ownership takes off when per-capita GDP, expressed in PPP terms, reaches a level of between $3 000 and $10 000, as a large portion of the population can then afford to own a vehicle (ADB, 2006).[9] India has recently passed the $3 000 tipping point and has seen a corresponding increase in vehicle ownership rates.

As the vehicle stock is expanding rapidly, small changes in the projected rate of growth have a large impact on transport oil demand by the end of the projection period. While we have projected growth in the vehicle stock (excluding two-wheelers) to increase on average by 9.4% per year over the *Outlook* period – a rate well above the prevailing GDP growth rate assumption – only a slight increase in this rate of growth (which could result from faster GDP growth than assumed in the Reference Scenario) would push up road fuel consumption significantly, as highlighted in our High Growth Scenario (see Chapter 19).

However, even if GDP and household incomes were to grow faster, infrastructure bottlenecks might constrain vehicle ownership rates and fuel demand. Much depends on public spending on highways and measures to tackle traffic congestion. Efforts are under way to improve the road network, such as the Golden Quadrilateral Highway project (see Chapter 14).

The absence in India of mandatory vehicle fuel efficiency standards[10], such as those in OECD countries, China and many other developing countries, suggests that the fuel efficiency of vehicles on the road in India will lag that of the OECD and China.[11] On the other hand, the large number of partnerships in India between local and foreign vehicle manufactures does mean that more efficient vehicle technology is being introduced into the country. In addition, India has introduced mandatory standards for pollutant emissions comparable to those adopted in the European Union, which has probably had the effect of accelerating the introduction of more fuel-efficient vehicles. Indian emission standards on two-wheelers are stricter than EU standards, but four-wheel vehicle standards lag those in Europe (Table 16.3).

16

9. For example, China had 3 cars per 1 000 people in 1994, when per-capita GDP reached the $3 000 mark. By 2004, GDP per capita had increased to over $6 000, while car ownership had increased to 13 cars per 1 000 people.
10. The introduction of fuel-efficiency standards are analysed in the Alternative Policy Scenario (see Chapter 18).
11. Recent trends are unclear, as reliable data do not exist for India.

Table 16.3: **Four-Wheel Vehicle Emission Standards in India**

European standard	European introduction year	Indian standard	Indian introduction year	Coverage in India
EURO I	1992	India 2000	2000	Nationwide
EURO II	1995	Stage II	2001	4 Cities*
			2003	11 Cities**
			2005	Nationwide
EURO III	1999	Stage III	2005	11 Cities**
			2010	Nationwide
EURO IV	2005	Stage IV	2010	11 Cities**
			To be decided	Nationwide

* Delhi, Mumbai, Kolkata, Chennai.
** Delhi, Mumbai, Kolkata, Chennai, Bengaluru, Hyderabad, Ahmedabad, Pune, Surat, Kanpur and Agra.
Source: Ministry of Petroleum and Natural Gas (2003).

Residential Sector

Energy consumption in the residential sector grew on average by 1.6% per year in 1990-2005 and is projected to maintain this growth rate from 2005 to 2030.[12] Its share of total final consumption will decrease from 44% in 2005 to 29% in 2030. Higher incomes and urbanisation progressively reduce reliance on traditional biomass, including fuel wood, dung and agricultural waste. These resources dominate residential energy consumption today, accounting for 79% of residential energy demand. That share drops to 59% in 2030. They are replaced by more efficient fuels – liquefied petroleum gas (LPG), kerosene, gas and electricity (Figure 16.6).[13] Biomass use will nonetheless remain the primary fuel in rural households, with associated damage to the health of women and children from indoor air pollution (see Chapter 20).

An aggregate analysis of household energy consumption in India masks very wide differences in the consumption pattern of rural and urban households.

12. The growth rate of residential energy demand, excluding biomass, is 4.3% per year from 2005 to 2030. The use of traditional biomass is very inefficient so that its replacement offsets the growth in energy demand as incomes rise.
13. Biogas is also cleaner and more efficient, but its use today is limited. India's National Biogas and Manure Management Programme is discussed in Chapter 18.

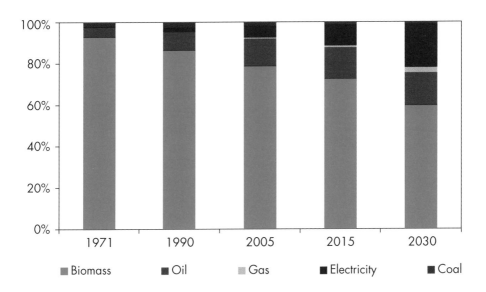

Figure 16.6: **Residential Fuel Mix in India in the Reference Scenario**

Box 16.4: **Rural and Urban Household Energy Demand Projections**

The energy demand model for the residential sector in India has been expanded for this year's *WEO* in order to evaluate rural/urban differences. A new database was created, with the rural/urban breakdown of energy demand, based on historical data from the National Sample Survey Organisation (NSSO). As a result of our bottom-up analysis, estimated aggregate biomass use in the residential sector has been revised downwards. Energy consumption for both rural and urban areas is calculated econometrically for each fuel as a function of GDP per capita, the urbanisation rate, the related fuel price and past consumption levels. The residential sector module covers five end uses by fuel and area: space heating, water heating, cooking, lighting and appliance use. Fuel demand is projected per household. Total demand for household consumption is derived from two components. An "existing stock" component bases energy consumption on historical shares for each fuel, while a portion of demand is allocated to "new stock", where fuel shares are a function both of relative prices and of existing shares of each fuel.

16

Rural households depend on biomass for almost 85% of their cooking needs; LPG meets 56% of this need in urban households (NSSO, 2007). Urban households, which make up less than 30% of India's population, account for

75% of India's residential demand for LPG. Rural households account for 92% of India's residential use of biomass. While rural households make up over 70% of the population, they account for only 42% of the residential demand for oil, gas and electricity.

Biomass consumption in rural households grew at an average annual rate of 1.1% in 1990-2005, but growth is projected to slow to 0.5% per year in 2005-2030, as per-capita incomes expand and the availability of LPG increases. It will not decline, however, because, except for some local scarcity, biomass resources are readily available and are often preferred. Even as per-capita incomes rise, the use of biomass could actually increase, as households cook more meals per day. Moreover, although, in 2030, there will be more rural households relying on LPG as their primary fuel for cooking than today, biomass will still be used as a secondary fuel. Kerosene is used primarily for lighting in rural households (Box 16.5), and demand is expected to decline by 1.4% over the *Outlook* period. LPG consumption will increase by 5% per year, but in volumetric terms it will still be less than half of total LPG demand in urban households in 2030 (Table 16.4).

Box 16.5: **Kerosene Use in Rural Areas of India**

Most rural households in India depend on kerosene lamps for lighting. Kerosene prices are controlled by the government and are heavily subsidised. About 90% of rural kerosene is distributed through a public distribution system (PDS), comprising state and district level officials, wholesalers and retailers (fair-price shops). The Ministry of Petroleum and Natural Gas fixes a quota for each state, according to historical patterns of supply (rather than actual demand or relative poverty levels). A blue dye is added to the kerosene supplied by the PDS to discourage its misuse, for example, for transport. Private operators are free, without constraint, to import and sell kerosene at market prices.

In rural Rajasthan, an estimated 80% of households use kerosene for lighting (Rehman *et al.*, 2005). Even among households reporting use of kerosene for cooking, in most cases it is used simply to ignite a biomass-fuelled stove. Most meals require baking and not direct heat, so people prefer to use traditional mud stoves. Cooking with kerosene also contaminates food, deterring households from using it. Nevertheless, the current kerosene subsidy scheme is linked to the use of LPG. Subsidised kerosene supply has been limited where households enjoy the use of LPG, primarily for cooking: households with a single LPG cylinder are entitled to half the normal quota of kerosene, while those with two cylinders are not allowed to buy any kerosene through the PDS.

Biomass use in urban households[14], by contrast, declined on average by 0.5% per year between 1990 and 2005, reflecting higher incomes, the higher cost of fuelwood in urban areas and the greater availability of kerosene and LPG. Demand is projected to fall by 15% over the *Outlook* period, and the share of biomass in urban household energy use will be a mere 12% in 2030. Switching from traditional fuels to kerosene, LPG and electricity occurs at a much more rapid pace in urban areas, where households have both more choice and more cash.

Table 16.4: **Urban and Rural Household Energy Consumption in India in the Reference Scenario** (Mtoe)

	1990	2005	2015	2030	2005-2030*
Urban					
Biomass	10.3	9.6	9.2	8.2	–0.6%
Kerosene	3.5	3.3	2.5	1.6	–2.8%
LPG	1.9	8.3	14.1	21.4	3.8%
Gas	neg.	0.6	1.4	5.5	9.1%
Electricity	1.6	5.4	11.0	28.8	6.9%
Coal	1.6	1.4	1.2	1.0	–1.2%
Rural					
Biomass	96.6	114.0	122.3	130.7	0.5%
Kerosene	5.3	6.5	5.7	4.6	–1.4%
LPG	0.2	2.8	5.4	9.7	5.0%
Gas	–	–	–	–	–
Electricity	1.1	3.5	7.1	19.9	7.2%
Coal	1.4	1.4	1.2	1.2	-0.6%
Total					
Biomass	106.9	123.6	131.5	138.9	0.5%
Kerosene	8.8	9.8	8.2	6.2	–1.8%
LPG	2.1	11.1	19.5	31.0	4.2%
Gas	neg.	0.6	1.4	5.5	9.1%
Electricity	2.8	8.9	18.0	48.7	7.0%
Coal	3.0	2.8	2.4	2.2	–0.9%

* Average annual rate of growth.
neg. = negligible.
Note: Figures do not include solar thermal.
Source: Historical figures are IEA analysis based on NSSO, 2007.

14. Predominantly fuelwood as urban households use very little dung and residues for cooking.

Residential consumption of natural gas is small in India and is limited to major cities. We project it to grow by an average 9.1% per year in 2005-2030 (several cities, including New Delhi, are expanding distribution networks to supply apartment complexes), but it will still account for only 2% of total residential energy use in 2030 – all of it in urban areas. The share of coal in total residential energy is projected to fall from 1.8% in 2005 to 0.9% in 2030, with demand falling slightly in absolute terms. Most of residential coal use will remain concentrated in areas close to mines.

Like kerosene, LPG is subsidised in India. All subsidised LPG is distributed by the state oil companies. The choice between the use of fuelwood or LPG for cooking depends on income, but also differs between rural and urban areas. LPG is popular for its efficiency, cleanliness and safety, relative to other fuels (IEA, 2006), and in 2005, LPG was the main fuel used for cooking in 56% of urban households. But this was the case in only 8% of rural households, mainly the richest (Figure 16.7). The high initial cost of the stove and the deposit on the cylinder, as well as poor distribution networks in rural areas, hold back more widespread use of LPG, even though LPG stoves are much more energy-efficient than alternatives (See Box 20.1 in Chapter 20). The government acknowledges that an expansion of supplies of LPG to rural households is needed to improve living standards, but oil companies have complained that distribution to remote villages is logistically difficult and is not profitable. There is no specific federal plan in place for expanding LPG availability to India's rural poor.

Figure 16.7: **Fuelwood and LPG Use for Cooking in India by Income Class, 2005**

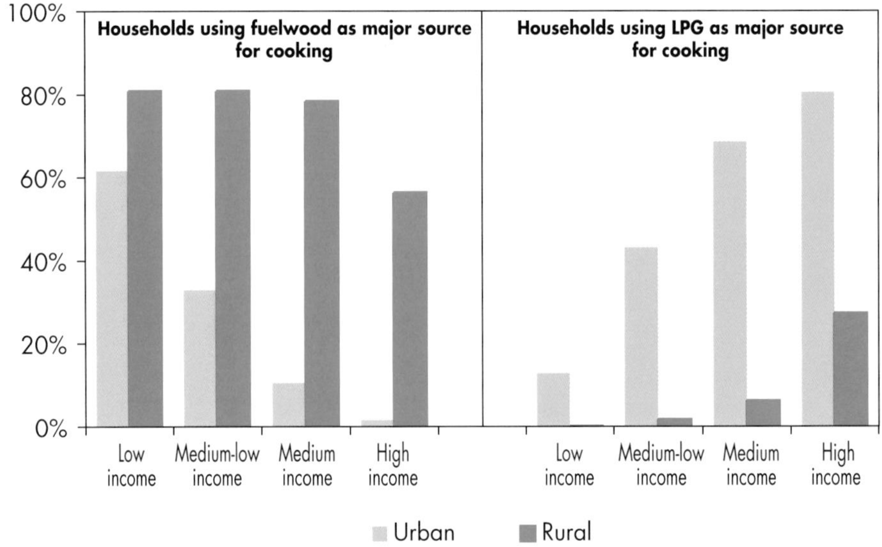

Source: NSSO (2007).

Kerosene and LPG supply 83% of the urban household energy mix for cooking in 2030 (Figure 16.8), but only 32% of the rural household mix. The fuel shares in Figure 16.8 are based on the primary fuel used in households for cooking although, in practice, households use a combination of fuels for cooking, depending on availability, the season of the year and income. This is particularly true of those in rural areas. Even if households are willing to pay higher prices for cleaner and more efficient fuels, such fuels are often unavailable or the quality of service is poor. Higher per-capita incomes alone will not lead to switching to LPG in rural areas. We project that, although rural LPG use will grow, only 15% of rural households will be using LPG as their primary fuel for cooking in 2015, compared with 66% in urban areas. Indeed, the share of urban households in total LPG consumption for cooking is set to grow – from 56% today to about 75% by 2030.

Figure 16.8: **Fuel Shares in Household Energy Consumption for Cooking in India by Area in the Reference Scenario**

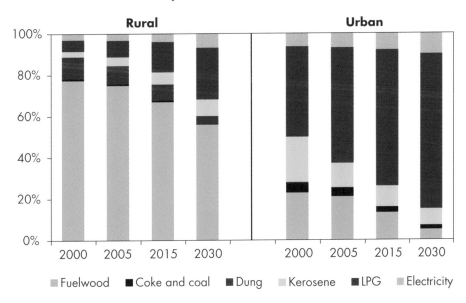

Sources: NSSO (2007) and IEA analysis.

Electricity use by Indian households is growing rapidly, even in rural areas, though most of the increase in demand in absolute terms is in towns and cities. Lighting accounts for about 70% of electricity use in the residential sector. Refrigeration and air conditioning account for almost all the rest. Use of electricity for cooking is limited. Most high-income urban households have a backup diesel generator which runs at least a few hours every day, because grid-based electricity supplies are very unreliable. There is an enormous disparity in access to electricity between rural and urban areas, which is reflected in the fuel mix for lighting. In 2005, the share of lighting met by electricity in urban households was about 90%, but it was only around 50% in rural households.[15] By 2030, all lighting is projected to be met by electricity in urban areas while in rural areas about 7% of households will still rely on kerosene as their primary fuel for lighting.

Appliance ownership per household in India, even among the richest households, is much lower than the OECD average (Figure 16.9). Appliance ownership is expected to grow as incomes rise. For example, between 3 and 4 million refrigerators are sold in India annually at present: sales are expected to nearly triple by 2020 (LBNL, 2005). The consequent rise in electricity demand will be relatively high, since appliance efficiencies are currently low in India by international standards, though they are expected to improve over the *Outlook* period, in spite of low electricity prices. Further efficiency improvements could dampen the increase in electricity demand significantly (see Chapter 18).

Figure 16.9: **Appliance Ownership in India Compared with the OECD, 2004**

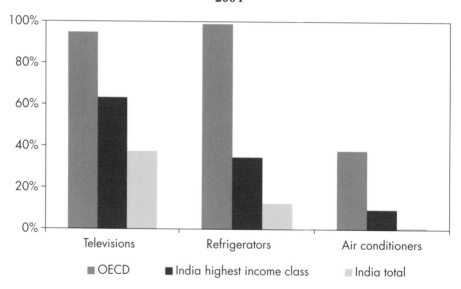

Note: The highest income class in India includes urban residents with monthly expenditure greater than Rs 1 539 and rural residents with expenditure greater than Rs 862.
Sources: National Sample Survey Organisation 2007 and IEA analysis.

15. See Chapter 20 for a discussion of access to electricity.

Other Sectors

The services sector accounted for only 3% of total final energy consumption in 2005, well below its share of economic output. Energy demand for services increased at an average annual rate of 1.7% from 1990 to 2005. It is projected to increase over the *Outlook* period by 3.8% per year. As a result, this sector's share of total final energy consumption will rise to 4%. Biomass meets about half of the sector's energy needs today, mostly in community centres, schools and hospitals, but by 2030 electricity will account for 67% of the total, growing by 7.9% per year. Most of this electricity will be used in hotels, commercial establishments, residential complexes and shopping malls. The expected growth in demand for personal computers will also drive up energy consumption in this sector over the projection period. The number of computers per 100 people in India is very low – 1.2 in 2004 compared with 4.1 in China, 4.5 in Philippines and 19.2 in Malaysia (Figure 16.10). Diesel-generated power is generally more expensive than grid-based supply, even though state electricity boards' tariffs to business carry a disproportionate share of system costs. Despite this, diesel generators are widely used by businesses to cope with India's chronic power shortages. Computer services companies are normally obliged to maintain their own generators.

Energy use in agriculture represents 4% of final energy demand. Consumption grew rapidly from 1990 to 2005, averaging 6.5% per year. It is projected to slow to 3.2% per year over the *Outlook* period. Electricity, used mainly for irrigation pumps (Box 16.6) will account for most of this growth, and oil, mainly for tractors and other machinery, for the rest.

Figure 16.10: **Prevalence of Personal Computers**

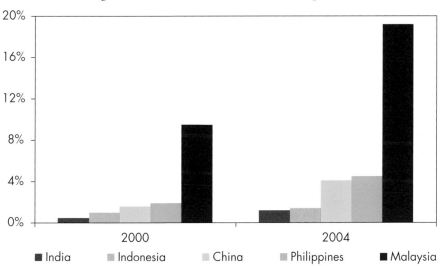

Source: www.unstats.un.org.

Box 16.6: **Energy and Water Use in the Agricultural Sector**

Energy in the agricultural sector is mainly used for land preparation and irrigation. Irrigation is vital to cultivation in areas receiving inadequate rainfall. In 2001, more than 40% of the total gross cropped area was irrigated. Increased mechanisation is the main driver for the growth in electricity demand in the sector, but the use of mechanical power in India is still far below the level in developed countries. The growth in groundwater irrigation has been massive, as reflected in the increase in the number of electrical pump sets, from 1.6 million in 1992 to 11 million in 1997 (Central Electricity Authority, 2006), and in diesel pump sets, from 1.5 million to 6.5 million, over the same period. There is considerable scope for improving the efficiency of Indian pump sets.

Electricity to farmers is subsidised or, even, free. This leads to inefficient irrigation practices and considerable waste of groundwater, which in turn has led to rapid water depletion in many regions. The required water depth for paddy field cultivation is about 800-1 000 mm, whereas farmers in India use as much as 2 000-2 500 mm. India's total water withdrawal for agriculture averaged 558 billion cubic metres per year from 1998 to 2002, the highest in the world (UNFAO Aquastat database).

Energy-Related Emissions

Pollutant Emissions[16]

India suffers from high levels of airborne pollution, largely caused by the burning of fossil fuels in power stations, factories and vehicles. The main pollutants are sulphur dioxide (SO_2), nitrogen oxides (NO_x) and fine particulate matter ($PM_{2.5}$).[17] India has set air-quality standards for various pollutants, but more will have to be done to achieve substantial emissions reductions.

Total sulphur dioxide emissions in India reached almost 7 million tonnes in 2005, 3 million tonnes more than the level of 1990. Coal is the largest source, although the sulphur content of Indian coal is relatively low, ranging between 0.2% and 0.7% (Menon-Choudhari *et al.*, 2005). Nearly two-thirds of total SO_2 emissions now come from the power sector, reflecting its heavy reliance on coal. The other main sources are oil-fired power plants, steel mills, cement plants and fertilizer factories. Emissions from Indian power plants have increased as coal consumption for electricity production has risen. Between 1990 and 2005, they increased at an average rate of 6.5% per year, pushing up the share of the power sector in total SO_2 emissions from 47% to 66%. The rate of increase in emissions

16 .The projections in this section are based on analysis carried out by the International Institute for Applied Systems Analysis (IIASA) on behalf of the IEA.

17. Fine particulate matter is particulate matter that is 2.5 microns in diameter and less. It is also known as $PM_{2.5}$ or respirable particles because it penetrates the respiratory system further than larger particles.

from other sources – mainly industry – was much lower. Emissions from transportation have declined, largely because of more stringent fuel-quality standards.[18]

In the absence of more stringent government measures, such as requirements to instal flue-gas scrubbers and coal washing, SO_2 emissions are set to rise further. The power sector will remain the main emitter. Total emissions rise to 16.5 Mt in 2030 in the Reference Scenario (Table 16.5).

Emissions of NO_x come mainly from vehicles and the power sector. They are the cause of urban smog. NO_x concentrations have increased slightly in recent years, but are still well below the national standard for residential areas. They are projected to rise sharply in the Reference Scenario and NO_x pollution will become an increasingly serious problem in the coming decades unless new control measures are introduced. In the Reference Scenario, total NO_x emissions increase from 4 Mt in 2005 to 8.5 Mt in 2030. Most of the increase will come from the transport sector, with adverse implications for air quality in urban areas, followed by the power sector. By contrast, emissions of particulate matter, which come mainly from biomass burning by households, are projected to decline, falling from 4.7 Mt in 2005 to 4.2 Mt in 2030.

Table 16.5: **Local Air Pollutant Emissions in India in the Reference Scenario** (kilotonnes)

	1990	2005	2015	2030	2005-2030*
SO_2	3 668	6 699	9 759	16 546	3.7%
NO_x	2 791	4 109	5 165	8 528	3.0%
$PM_{2.5}$	4 206	4 681	4 469	4 192	–0.4%

* Average annual rate of growth.

CO₂ Emissions

India was the fifth-largest emitter of energy-related carbon dioxide in 2005, releasing 1.1 Gt into the atmosphere or 4% of the world total. India becomes the third-largest CO_2 emitter in 2015 in the Reference Scenario, when its emissions rise by almost 60%. Emissions in the Reference Scenario are projected to rise to 3.3 Gt by 2030, an average rate of increase of 4.3% per year. By the end of the projection period, India accounts for 8% of global emissions. It is likely to pass Japan as the fourth-largest emitter well before 2010 and Russia, currently the third-largest CO_2 emitter, in 2015. Two-thirds of India's

18. Indian refineries reduced the sulphur content of diesel for cars in the four largest cities to 0.25% in 2000 and to 0.05% in 2001 (Garg *et al.*, 2006).

emissions come from burning coal, mainly in power stations. This share will increase slightly, to 69%, by 2030.[19]

Measured on a per-capita basis, India's CO_2 emissions are very low at just over 1 tonne in 2005, compared with 11 tonnes in the OECD (Figure 16.11). They are about half those of developing countries on average. By 2030, per-capita emissions are projected to double, but they will still be well below those of the OECD.

Figure 16.11: **Per-Capita Energy-Related CO_2 Emissions in India, Compared with Developing Countries and the OECD in the Reference Scenario**

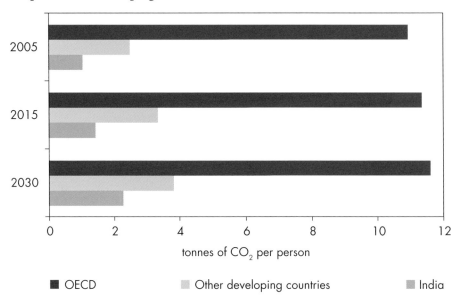

Because power station efficiency is low by international standards, India's power sector is one of the most CO_2-intensive in the world. Power stations emitted, on average, 943 grammes of CO_2 per kWh of electricity produced in 2005 – more than 50% higher than the average for the world. While this is slightly lower than the corresponding figure for China, where coal is even more dominant, it is higher than in other countries that rely heavily on coal for power generation, such as South Africa, Australia, Indonesia and the United States. Total emissions of CO_2 from power plants in 2005 were 659 Mt, nearly 60% of total CO_2 emissions in India.

19. Projections by TERI and by the Indian government (some scenarios) show higher demand for fossil fuels, especially coal, and therefore higher CO_2 emissions (TERI, 2006 and Planning Commission, 2006).

Despite improvements in thermal efficiency, in the Reference Scenario the power sector continues to be responsible for most of the increase in CO_2 emissions to 2030 and its share in total emissions remains broadly constant (Figure 16.12). This is because of fast growing demand for electricity and because the share of coal in the electricity mix is projected to remain high. Power sector emissions fall dramatically in the Alternative Policy Scenario (Chapter 18).

Figure 16.12: **Increase in India's CO_2 Emissions by Sector in the Reference Scenario**

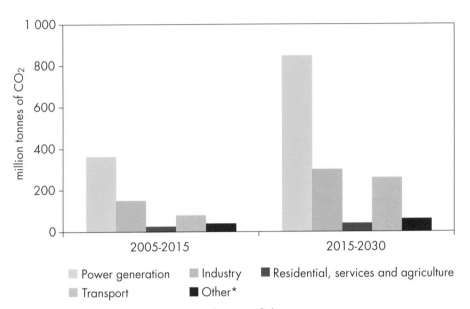

* Includes other energy sector, non-energy use and non-specified energy use.

The lack of non-fossil alternative fuels means that CO_2 emissions trends in the transport sector closely follow energy demand. The transport sector is responsible for 8% of India's CO_2 emissions today. This share grows with rapidly rising transport demand, particularly after 2015 as vehicle ownership increases, to 13% in 2030. Transport's share of emissions in India at the end of the *Outlook* period is, nonetheless, much lower than that in developed countries today. For example, the share of transport in total CO_2 emissions in 2005 was 31% in the United States and 24% in the European Union.

16

CHAPTER 17

REFERENCE SCENARIO SUPPLY PROJECTIONS

HIGHLIGHTS

- India's proven reserves of oil are limited. About 80% of current production is estimated to be from fields which have passed their peak. India overtakes Japan to become the world's third-largest net oil importer, after the United States and China in the first half of the 2020s. India's role as a major oil refiner grows, assuming the necessary investments are forthcoming. Distillation capacity reaches 8.1 mb/d in 2030 in the Reference Scenario.

- Recent discoveries are expected to boost gas production. Nevertheless, it is projected to peak between 2020 and 2030, falling to 50 bcm by the end of the projection period. Further pricing reform will determine whether the requisite supply infrastructure is built in a timely manner.

- Coal production increases from 262 Mtce in 2005 to 637 Mtce in 2030, but demand rises even faster, to 886 Mtce. Because of the low quality of domestic coking coal, about 105 Mtce is imported in 2030. Steam coal imports reach 139 Mtce, as power generation demand is largely concentrated in coastal regions where domestic resources are scarce. Total coal imports in India in 2030 are more than 10% above the coal imports into the entire European Union.

- Total electricity generation increases from 699 TWh in 2005 to 2 774 TWh in 2030, an average increase of 5.7% per year. Per-capita electricity generation rises to more than 1 900 kWh, compared with 8 870 kWh in OECD countries today. Coal-fired power generation remains the backbone of India's electricity sector, because it is the cheapest way to produce electricity. Its share is projected to increase from 69% to 71%.

- Over the projection period, the average efficiency of coal-fired power generation is projected to improve considerably, as new plants will be larger and more efficient and as supercritical units are built. On average, efficiency is expected to increase from 27% now to 38% in 2030 - slightly above the current level of efficiency in the OECD.

- India needs to invest about $1.25 trillion in energy infrastructure in the period 2006-2030 to meet demand in the Reference Scenario. Three-quarters of this investment, almost $1 trillion, is in power infrastructure. Attracting investment in a timely manner will be crucial for sustaining economic growth. Power-sector reforms are on the right path but reform implementation needs to be strengthened. For the sizeable investments that India will need over the next two-and-a-half decades, improving the investment conditions in the sector and moving towards a transparent, predictable and consistent power-sector framework based on market principles will remain of paramount importance.

Oil Supply

Resources and Reserves

India's proven reserves of oil amounted to 5.6 billion barrels at the end of 2006, equal to 0.4% of world reserves.[1] Official government data put "proved and indicated" reserves on 1 April 2006 at 756 million tonnes or about 5.4 billion barrels. Reserves are almost equally shared between onshore and offshore (Table 17.1). But the average size of the offshore fields is much bigger. Onshore reserves are fragmented in small to medium-size fields. Despite the uncertainty related to the level of production so far from a few fields, we estimate that more than 50% of the ultimately recoverable reserves from identified fields have been produced, excluding the volumes in fields yet to be discovered.

Table 17.1: **India's Oil Reserves, End-2005**

	Onshore	Offshore	On/Offshore	Total
Number of fields	242	91	3	336
Proven and probable reserves (million barrels)	2 650	2 525	180	5 355
Cumulative production to date (million barrels)	2 603	3 414	167	6 184

Sources: IHS Energy databases; IEA estimates.

The reserves are mainly located in five sedimentary basins: Mumbai (38%), Cambay (20%) and Barmer (15%) in the north-west, close to the border with Pakistan, Assam shelf (18%) in the north-east and Krishna-Godavari (7%) in the south. The Assam shelf, Barmer and Cambay basins are almost entirely onshore while the Mumbai and the Krishna-Godavari basins are exclusively offshore. Since 1997, 97 oilfields have been discovered, mainly located in the Assam shelf, Cambay and Barmer basins. These fields represent almost 30% of all fields ever discovered in India, highlighting the effort deployed in exploration under the new upstream licensing regime (see Oil Production below). However, the 1 320 million barrels of reserves that were added represent only 13% of the oil discovered before 1997, as recent discoveries have generally been much smaller. We estimate that about 43% of the undiscovered volumes of oil that were estimated to exist in 1995 by the United States Geological Survey in 2000 have already been discovered.

---

1. *Oil and Gas Journal,* 18 December 2006.

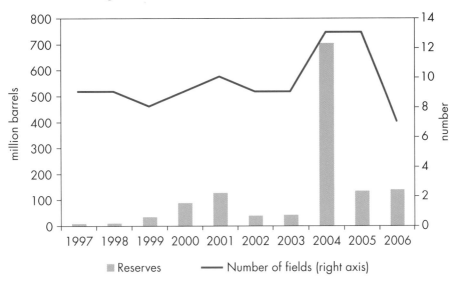

Figure 17.1: **Oil Discoveries in India Since 1997**

Reserves ▪ ——— Number of fields (right axis)

Sources: IHS Energy databases; IEA estimates.

Oil Production[2]

Oil production in India amounted to 793 thousand barrels per day in 2006, of which 687 kb/d were crude oil and 106 kb/d natural gas liquids (NGLs). About three-quarters of total production came from nine offshore fields, with 85 onshore fields making up the rest (Table 17.2). A few major fields discovered in the 1970s and 1980s account for the bulk of India's oil output. The top five producing fields contribute half (Table 17.3). Most producing fields are now in decline, including the nine offshore fields which account for 60% of total production. We estimate that 80% of current production comes from fields which have passed their peak. Despite some significant discoveries since the New Exploration Licensing Policy (NELP) has been implemented and the fact that obviously only a portion of the sedimentary basins has been actively explored, India can be regarded as a mature oil-producing country. Reforms undertaken by the Indian authorities since the beginning of the 1990s, one of which is the NELP, have had a positive impact on domestic production by allowing the private sector partly to compensate for the declining contribution of the national companies.

In the Reference Scenario, India's total oil output is projected to increase from 793 kb/d in 2006 to 870 kb/d in 2010 and then fall back to about 520 kb/d in 2030. Crude production is projected to peak at around 750 kb/d in 2010

2. Crude oil, natural gas liquids (NGLs) and condensates.

and decline to just under 400 kb/d in 2030. A number of new projects, including Mangala, Bhagyam, D6-MA-1, Aishwariya, GS-29-1 and Vijaya, and presumably fields yet to be discovered only partly compensate for the decline of existing mature fields. NGLs production increases by 25% between now and 2020, helping to slow the decline of overall oil production.

Table 17.2: **India's Crude Oil Production, 2006**

	Onshore	Offshore	On/offshore	Unidentified *	Total
Number of fields	85	9	1	41	136
Of which beyond peak	56	9	1	-	66
Production (kb/d)	157	414	35	80	687
Share in production (%)	23	60	5	12	100

* Fields for which production is not accurately reported.
Sources: IHS Energy databases; IEA estimates.

Table 17.3: **India's Oil Production in the Reference Scenario** (kb/d)

	2006	2015	2030
Total crude oil	**687**	**622**	**394**
Top 12 existing producing fields	407	199	73
Mumbai High	*218*	*122*	*46*
Ravva	*42*	*8*	*1*
Neelam	*17*	*3*	*0*
Heera South	*31*	*9*	*0*
Heera	*29*	*17*	*7*
Gandhar	*25*	*13*	*5*
Lakwa	*12*	*9*	*5*
Kadi North	*9*	*6*	*3*
Panna	*8*	*1*	*0*
Kalol	*6*	*5*	*4*
Santhal	*6*	*5*	*3*
Hapjan	*4*	*1*	*0*
Fields awaiting development	–	142	107
Other currently producing fields, reserve additions and discoveries	280	282	215
NGLs	**106**	**109**	**123**
Total	**793**	**730**	**517**

Sources: IHS Energy databases; IEA analysis.

Our oil-production projections are derived from a bottom-up assessment of the 12 largest producing fields in 2006 and new oilfield developments in the coming years, as well as from a top-down analysis of longer-term development prospects. The delay between the date of any discovery and the start of production is assumed to be eight years, in line with the average delay since the 1990s. Inevitably, these projections are subject to considerable uncertainty, notably with respect to the rate at which discovered fields can be brought into production and to decline rates. About 100 small fields discovered in the 1980s and 1990s, each with proven and probable reserves of less than 10 million barrels, are assumed to be brought into production before 2012. This represents a major challenge, as the highest number of new field start-ups in a single year previously was 28 in 1999. We assume that an average of 25 fields annually will be brought into production between 2007 and 2013, though it is not clear that the investment needed will be forthcoming under the existing fiscal regime. Moreover, the increase in demand for drilling rigs implied by our projections would be likely to drive up costs. Only about 10% of the 880 wells agreed under licenses awarded in the first six rounds of the NELP introduced in 1997[3] have been drilled so far, partly because of a lack of available drilling rigs.[4] Delays in drilling would result in a smaller contribution of those fields before 2015, but higher between 2020 and 2030. Indian oil production would still peak in the near future.

Oil Refining

India has 19 refineries with total installed refining capacity of 2.9 mb/d. The state-owned Indian Oil Corporation (IOC) owns ten refineries directly and another one through a subsidiary. Six refineries are owned by other public companies. The private firms, Reliance and Essar Oil, commissioned two new refineries at Jamnagar in 2000 and Vadinar in 2006. The refining sector was opened up to private investment in 1996 with Mangalore Refinery and Petrochemicals (MRPL) commissioned as a joint-venture refinery with private actors, although it was later purchased by the Oil and Natural Gas Corporation (ONGC). Atmospheric distillation capacity doubled to 2 mb/d between 1993 and 2000, with the commissioning of five refineries, including Reliance's 580 kb/d refinery in Jamnagar – the third-largest in the world. Capacity expansions at existing refineries added almost 700 kb/d to distillation units between 2000 and 2006. The complexity of the refining sector has increased markedly since the end of the 1990s thanks to capacity additions and

3. See Box 15.2 in Chapter 15 for an overview of the NELP.
4. The Ministry of Petroleum and Natural Gas recently accepted the proposal of the Directorate General of Hydrocarbons to merge the first two phases of the production-sharing contracts signed under the NELP-III and IV licensing rounds to at least partly remove the threat of withdrawing licences because of delays in drilling.

expansions of cracking units. Between 1997 and 2006, visbreaker capacity increased from 65 to 130 kb/d, coking capacity from 30 to 215 kb/d, catalytic cracking capacity from 150 to 470 kb/d and catalytic hydrocracking capacity from 25 to 310 kb/d.

Distillation capacity is projected to almost double by 2014 to 5.2 mb/d as a result of capacity expansions and major new greenfield refineries (Figure 17.2): Vadinar in 2007 (210 kb/d) and 2008 (110 kb/d), Jamnagar (580 kb/d) in 2009, Bina (120 kb/d) in 2010, Paradeep (300 kb/d), Barmer (150 kb/d) in 2013 and Bathinda (180 kb/d) and Cuddalore (120 kb/d) in 2014. However, continuation of the increasing delays and significant budget overruns being experienced by major projects around the world could cause delays or cancellations. Although detailed conversion capacities are not fully reported for those projects and their refining configuration may be subject to modification, we expect the complexity of the refining sector to continue to increase.

Figure 17.2: **Distillation Capacity in India**

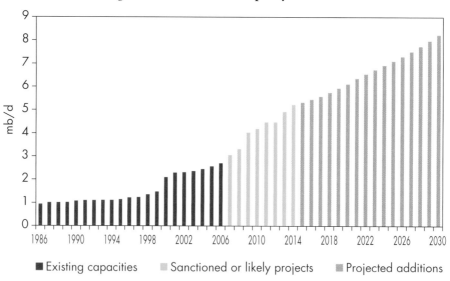

Sources: IEA databases and estimates.

Beyond 2014, we expect refining capacity to grow in line with domestic oil consumption. Distillation capacity reaches 8.1 mb/d in 2030 – about two-and-a-half times the current level. India would therefore remain an export refining hub over the projection period. India is geographically well placed, close to both Middle Eastern crude oil supplies and to rapidly expanding refined products markets in Asia and the Middle East. Lower costs compared to other countries contribute to the competitiveness of Indian refining. Nonetheless, maintaining export capacity beyond that which is already planned will require another five large refineries like the Jamnagar plant to be built – one every three years. The attractiveness of future investment in the sector will hinge partly on custom duties, which currently favour crude imports over products imports, and domestic pricing policies.

Oil Trade

India currently imports about 100 million tonnes, or 2 mb/d, of crude oil – mostly from the Middle East (67%) and West Africa (21%). Saudi Arabia is the largest supplier, accounting for 25% of India's crude oil imports (Figure 17.3). Although India is a large net oil importer, it has recently become a net exporter of refined products, thanks to the rapid expansion of refining capacity in recent years. For the fiscal year 2004/05, India was a net exporter of diesel (125 kb/d), gasoline (60 kb/d), aviation fuels (52 kb/d), heavy fuel oil (18 kb/d) and naphtha (16 kb/d). India is still a net importer of liquefied petroleum gas (73 kb/d) and kerosene (4 kb/d), most of which is used in the residential sector.

Figure 17.3: **Crude Oil Imports by Origin, Fiscal Year 2004/05**

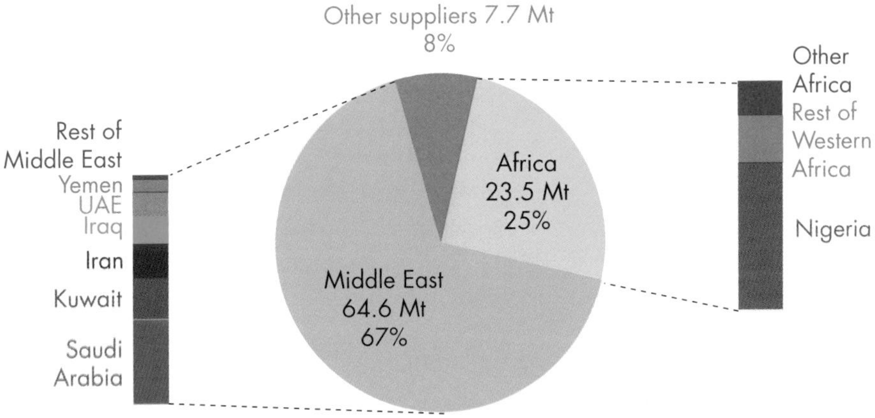

Source: Planning Commission (2006).

In the Reference Scenario, net oil imports in total are projected to increase steadily to 2.3 mb/d in 2010, 3 mb/d in 2015 and 6 mb/d in 2030 (Figure 17.4). Gross oil imports are projected to be even higher, reaching 7.6 mb/d in 2030. Net product exports reach close to 1.6 mb/d by 2015 and then stabilise. India's overall dependence on imports net of exports rises from less than 70% today to around 90% by the end of the projection period.

For crude oil alone, India's import dependence reaches 94%. In volumetric terms, India overtakes Japan to become the third-largest oil importer, behind the United States and China, in 2024.

Figure 17.4: **India's Oil Balance in the Reference Scenario**

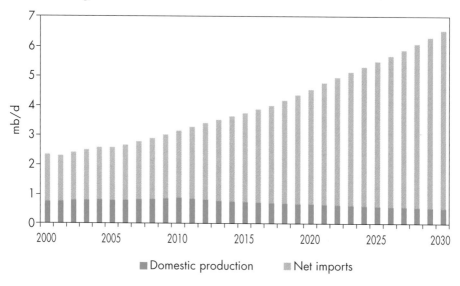

■ Domestic production   ▨ Net imports

Box 17.1: **India's Emergency Oil Stocks**

To mitigate the risks of short-term supply disruptions, the Indian government decided in early 2004 to build a strategic stockpile of crude oil. For that, the Indian Strategic Petroleum Reserves Ltd (ISPRL) was established, which is now under the control of the Oil Industry Development Board (OIDB), the state-controlled organisation that manages loans and grants to the oil industry.

The emergency oil stocks will supplement crude oil and petroleum products stocks already held by Indian oil refiners to meet their own operational needs. The stockpile will eventually reach 15 Mt or about 110 million barrels in 2016, to be accumulated in three phases of 5 Mt each. Each phase will cover 14 days of domestic consumption or 19 days of net oil import, at 2006 levels. The oil will be held in rock caverns with interconnected galleries. The first phase includes three locations: Padur (2.5 Mt) and Mangalore (1.5 Mt) on the west coast and Visakhapatnam (1 Mt) on the east coast, and will be finished by 2012. The total cost of this first phase of the project was estimated in 2006 at $2.7 billion, including the construction of the storage facilities ($600 million) and the buying of the oil (at $55 per barrel). The government has set aside funds for this first phase. The following phases could be financed through public-private partnerships. The construction of the first phase started in the second half of 2007.

The oil industry, including government-owned companies, is currently under no obligation to hold emergency stocks or to release oil in the event of an emergency. A decision has not yet been taken, but it is understood that the government will use the emergency stocks only in the event of a supply disruption that affects India. International co-ordination would render the use of these stocks more effective in the event of a supply disruption.

Natural Gas Supply

Resources and Reserves

India's proven reserves of natural gas amounted to 1 101 bcm at the end of 2006, equal to 0.6% of world reserves (Cedigaz, 2007). According to official Indian data, "proved and indicated" reserves amounted to 1 075 bcm on 1 April 2006. IHS data show that proven and probable reserves in discovered fields yet to be produced amounted to about 1 500 bcm in 2005. About two-thirds of remaining gas reserves are non-associated gas (Table 17.4). Most reserves are located offshore (88% of non-associated gas and 56% of all gas reserves). The breakdown of the reserves by field shows that reserves of associated gas fields currently producing are about 49% depleted. Although 123 associated gas fields are still awaiting appraisal or development, they are not expected to add more than 100 bcm to reserves. The potential for boosting non-associated gas production is greater, as the 29 fields currently in production are only 30% depleted and the 154 discovered fields not yet producing contain almost 900 bcm of proven and probable reserves. The prospects for Indian gas production hinge, therefore, on the development of non-associated gas fields.

The Krishna-Godavari sedimentary basin on the east coast holds just over half of India's proven and probable reserves and the Mumbai basin on the west coast another 23%. The Cambay basin and the Assam shelf together account for 16%. The Krishna-Godavari basin is mostly located offshore; it holds more than 86% of India's offshore non-associated gas reserves and 62% of all offshore reserves. The fields of the Mumbai basin are all offshore, with more than 80% of the gas associated with oil. This basin represents about 50% of all associated gas reserves. However, this basin is mature as this is, with the Assam shelf, one of the two long-standing producing regions of India, thanks to two major oil and gas fields, Mumbai High and Bassein, both of which were discovered in the mid-1970s.

17

Figure 17.5: Main Oil and Gas Infrastructure in India

—— Existing oil pipeline	⟶ Tanker terminal
- - - Under constr./planned/proposed oil pipeline	✸ Existing LNG import terminal
—— Existing product pipeline	✿ Under const./planned LNG import terminal
—— Existing gas pipeline	☆ Speculative LNG import terminal
- - - Under constr./planned/proposed gas pipeline	● Oilfield
⫼ Refinery	● Gas field
⫼ Refinery under construction	● Oil and gas field

The boundaries and names shown and the designations used on maps included in this publication do not imply official endorsement or acceptance by the IEA.

Sources: The Petroleum Economist Ltd; IEA analysis.

In the last ten years, 90 non-associated gas fields and 50 associated gas fields have been discovered, adding 922 bcm to proven and probable reserves (Figure 17.6). This is almost equivalent to the volume of all gas discoveries before 1997 (1 040 bcm). The vast majority of reserves additions since 1997

Table 17.4: **Natural Gas Reserves in India, End-2005**

	Onshore	Offshore	On/Offshore	Total
Total gas				
Number of fields	261	167	4	432
Proven and probable reserves (bcm)	312	1 171	54	1 537
Cumulative production to date (bcm)	131	274	24	429
Non-associated gas				
Number of fields	98	83	1	182
Proven and probable reserves (bcm)	117	849	1	967
Cumulative production to date (bcm)	18	17	0	35
Associated gas				
Number of fields	163	84	3	250
Proven and probable reserves (bcm)	195	322	53	570
Cumulative production to date (bcm)	113	257	24	394

Sources: IHS Energy databases; IEA estimates.

Figure 17.6: **Natural Gas Discoveries in India since 1997**

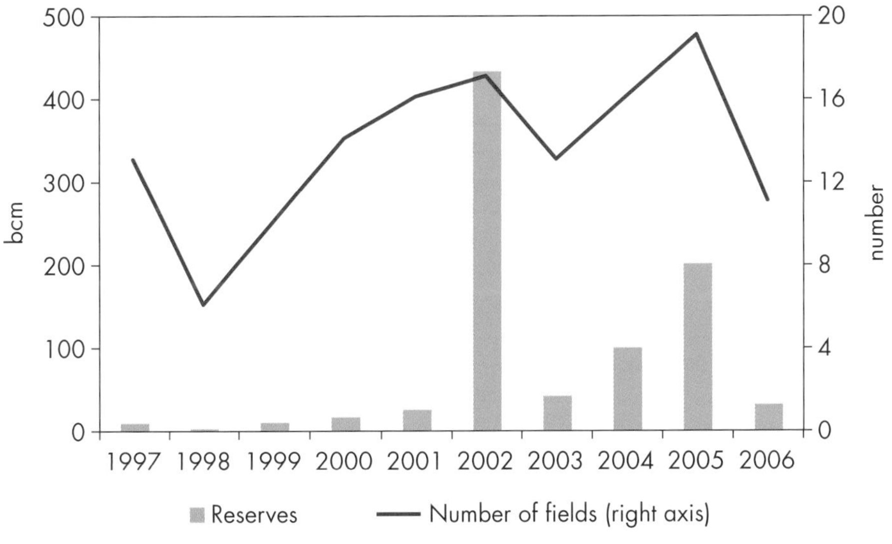

Sources: IHS Energy databases; IEA estimates.

come from offshore non-associated gas fields in the Krishna-Godavari basin, which make up about half of the fields discovered since 2002. The biggest fields are in the prolific KG-D6 and KG-DWN blocks. These discoveries, which resulted directly from the NELP, have fundamentally changed the prospects for gas supply in India.

India has estimated coal-bed methane (CBM) resources of over 1 400 billion cubic metres (Planning Commission, 2006) and underground coal gasification (UCG) could facilitate exploitation of the substantial deep coal resources that are currently uneconomic to mine, including those lying deeper than 1 200 metres. Some commercial production of CBM is under way.

Production

In the Reference Scenario, India's natural gas production is projected to reach 45 bcm in 2015 and 51 bcm in 2030 (Table 17.5). Production from the mature fields selected in our analysis is projected to decrease from 24.78 bcm in 2006 to 3.53 bcm in 2030. Indian domestic gas production has the potential to reach more than 50 bcm in 2020, thanks to the major discoveries over the

Table 17.5: **India's Gas Production by Field in the Reference Scenario** (bcm)

	2005	2015	2030
2005 Top 12 producing fields	**24.78**	**11.59**	**3.53**
Bassein	8.42	3.60	0.00
Mumbai High	5.34	2.79	1.05
Tapti South	2.34	0.70	0.12
Gandhar	1.82	0.90	0.32
Nahorkatiya	1.48	1.41	1.31
Hazira	1.18	0.44	0.10
Ravva	0.87	0.63	0.39
Pasarlapudi	0.67	0.26	0.04
Tarapur (B-55)	0.77	0.35	0.06
Neelam	0.65	0.11	0.01
Heera	0.64	0.35	0.14
Panna	0.60	0.05	0.00
Other fields and developments			
Fields awaiting development	–	28.48	30.29
Other currently producing fields and new discoveries	4.02	4.82	16.85
Total	**28.80**	**44.89**	**50.67**

Sources: IHS Energy databases; IEA analysis.

last seven years in the Krishna-Godavari basin. However, we expect production to reach a peak between 2020 and 2030 and then to drop to 51 bcm at the end of the projection period.

The outlook for India's gas production faces uncertainties similar to that for oil. About 100 fields are already being appraised or developed, and they will undoubtedly boost output in the next decade. But another 150 fields, discovered in the last thirty years, are still awaiting appraisal. How quickly this happens will depend on the business confidence and availability of both capital and drilling rigs. We assume that the delay between the date of any discovery and the start of production averages 8 years for larger fields, compared with about 10 years for offshore projects in the recent past. The New Exploration Licensing Policy (NELP) is expected to help speed up development. For example, there are more than 20 Dhirubai fields in the Krishna-Godavari basin, all of which were discovered since 2002, which are expected to start producing as early as 2008 and gradually increase production up to 2015. However, constraints on drilling activity are likely to limit the number of fields that can be brought on stream in the near term. We assume that a maximum of 17 fields can be brought into production on average each year between 2007 and 2021, resulting in delays between discovery and production of more than 8 years for a number of small fields.

There are other uncertainties surrounding gas production prospects. As production shifts from the mature producing fields in the Mumbai basin on the west coast to the relatively under-developed offshore Krishna-Godavari basin on the east coast, substantial investment in transmission and distribution infrastructure will be needed, especially in the south-east. Demand is currently concentrated in the centre and the north. It is unclear whether the business climate is interesting enough to attract all this investment, much of which is expected to come from the private sector. In addition, although the NELP provides for gas to be sold at market-related prices, it is not certain that end users – notably power stations and fertilizer plants – will be willing or able to pay (see Spotlight in Chapter 15).

Gas Imports

Gas imports are projected to double between 2005 and 2010, reaching about 12 bcm. Imports are then projected to stabilise before quadrupling between 2020 and 2030 as demand continues to grow and production peaks, reaching 61 bcm at the end of the projection period (Figure 17.7). Liquefied natural gas import capacity, at two existing plants in Hazira and Dahej, currently amounts to 10.2 bcm/year (Table 17.6). If another two plants currently under construction, or planned, come to fruition, capacity would reach 24 bcm by the middle of the next decade. This implies that, in the Reference Scenario, India will have excess import capacity of about 10 bcm on average over the

17

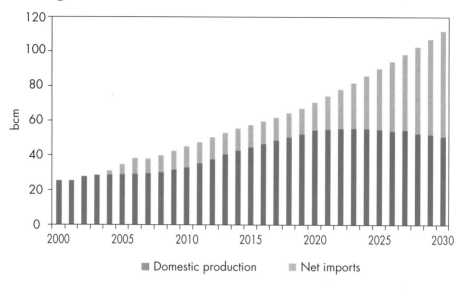

Figure 17.7: **India's Natural Gas Balance in the Reference Scenario**

■ Domestic production   ■ Net imports

Table 17.6: **Existing and Planned LNG Regasification Terminals in India**

LNG terminal	Main operator	Status	Start-up	Initial capacity (bcm)	Initial capacity (Mt)
Hazira	Shell-Total	Operating	April 2005	3.4	2.5
Dahej I	Petronet LNG	Operating	January 2004	6.8	5
Dahej II	Petronet LNG	Under construction	2009	6.8	5
Dabhol / Ratnagiri	Petronet LNG	Under construction	2007-2008	3.4	2.5
Kochi	Petronet LNG	Construction expected to start in 2007	2011	3.4 (possible extension to 6.8)	2.5 (possible extension to 5)
Ennore	IOC - Petronet LNG	Under study	2010	3.4	2.5
Mangalore	HPCL-Petronet LNG-MRPL	Proposed	2012	6.8	5
Total				**34.0**	**25.0**

Sources: Company reports and IEA estimates.

following decade. But major new capacity additions would be needed after 2025. By 2030, the gap between currently planned capacity and demand would reach 20 bcm, requiring the equivalent of four additional 5-bcm terminals. However, these projections depend on the availability of supplies from LNG-exporting countries and the competitivity of gas against other fuels in the Indian market.

Protracted negotiations over LNG imports have led to delays in finalising a number of planned projects. Price has been the main stumbling block, as increasingly expensive LNG struggles to compete against cheap coal and domestic gas in the Indian market. LNG is nonetheless the most competitive way of filling the gap between rising demand and indigenous production. It is very unlikely that alternative supplies through pipelines will reach the Indian market for many years. Several pipeline projects have been mooted, including those from Iran, Myanmar, Turkmenistan and Oman. The Iran-Pakistan-India pipeline, which would be more than 2 660 km-long and have a capacity of 60 mcm/day (of which half would be allocated to India), has been under discussion since 1994. Geopolitical factors, transit charges and pricing are still complicating negotiations. We do not take any of these pipeline projects into account in the Reference Scenario.

Coal Supply

Resources and Reserves

India has the world's fourth-largest hard coal resources, after Russia, China and the United States (BGR, 2007). India's total hard coal resources, to a depth of 1 200 metres, are estimated at 255 billion tonnes (Table 17.7). Some 60% of these resources lie within 300 metres of the surface, making them potentially exploitable by surface mining techniques. Proven reserves of hard coal total

Table 17.7: **Coal Resources and Reserves in India**

	Billion tonnes	Notes
Hard coal resources	255.2	to a depth of 1 200 metres - 60% lie within 300 metres of the surface
Proven hard coal reserves	97.9	17% of these reserves are classified as coking coal - located mainly in Jharkhand
Recoverable hard coal reserves	34.7	assuming 35.4% recovery, after accounting for unmineable reserves and mining methods
Lignite resources	34.8	mainly in Tamil Nadu and Puducherry

Sources: Geological Survey of India (2007); Planning Commission (2006); Ministry of Coal (2005).

98 billion tonnes, principally in Jharkhand, Orissa, Chhattisgarh and West Bengal (Figure 17.8). State-owned Coal India holds 73% of proven reserves (Ministry of Coal, 2005). Recoverable coal reserves are some 35 billion tonnes, giving a reserve-to-production ratio of over 80 years at current production rates.

Indian coal is typically of poor quality, with an average heating value of about 4 500 kcal/kg, compared to over 6 000 kcal/kg for most internationally traded coals, and high moisture, especially during the monsoon season. It is high in ash, typically 30% to 50% (Coal Industry Advisory Board, 2002), but low in sulphur and very little is suitable for iron and steel making. Even with washing, the ash content remains around 30%, leading to inefficient power generation and relatively high transport costs.

Average coal quality has fallen in India with the depletion of better deposits. Power stations often now rely on poor grades that fall below specification and compromise performance. Blending with better quality imported coals and coal washing can improve performance. Total annual capacity at coal washeries is now around 100 Mt per year (Ministry of Coal, 2007b), but Coal India is encouraging other companies to invest in coal washing at its sites and now insists that all large new mines include coal washing.

Coal Production

Total hard coal production in 2005 was 252 Mtce (403 Mt)[5], with a further 10 Mtce (30 Mt) of lignite. Coal India accounts for 86% of total coal supply. Other state-owned coal mining companies, Singareni Collieries, Neyveli Lignite Corp. and Gujarat Mineral Development Corp. produced 9% (and 98% of lignite or brown coal). Captive mining companies, whose sales are restricted to particular customers in the power, steel and cement sectors, account for the remainder (Table 17.8). In the Reference Scenario, total coal production is projected to rise to 354 Mtce (580 Mt) in 2015 and to 637 Mtce (1 059 Mt) in 2030.

The government Expert Committee on Coal Sector Reforms supports a production target for the 11th Five-Year Plan of 680 Mt in 2011/12, with production then rising to 1 100 Mt by the end of the 12th Five-Year Plan (2016/17) and to 1 900 Mt by 2031/32 (Ministry of Coal, 2005 and 2007b). This represents an average production growth rate of 6.1% per year, compared with the 5.6% per year achieved during the 10th Five-Year Plan. About a third of the coal targeted for production in the 11th Five-Year Plan is from mines

5. The following net calorific values of Indian coal production have been used: 5 800 kcal/ kg for coking coal, 4 410 kcal/ kg for steam coal and 2 280 kcal/ kg for lignite.

Figure 17.8: Major Coal Fields and Mining Centres in India

Legend:
- Major coal field
- Main coal-fired power plant
- Main steel plant
- Coal-importing port

Major coal fields
1. Raniganj
2. Jharia
3. East Bokaro & West Bokaro
4. Singrauli
5. Pench-Kanhan, Tawa Valley
6. Talcher
7. Chanda-Wardha
8. Godavari Valley

The boundaries and names shown and the designations used on maps included in this publication do not imply official endorsement or acceptance by the IEA.

Source: CIAB, 2002.

Table 17.8: **Coal Production in India by Company, 2005/06**
(million tonnes)

Company	Subsidiary	Output	
		Coal	Lignite
Coal India	Eastern Coalfields	31.11	
	Bharat Coking Coalfield	23.31	
	Central Coalfield	40.51	
	Northern Coalfields	51.52	
	Western Coalfields	43.20	
	South-eastern Coalfields	83.02	
	Mahanadi Coalfields	69.60	
	North-eastern Coalfield	1.10	
Singareni Collieries		36.14	
Neyveli Lignite Corp.			20.44
Gujarat Mineral Development Corp.			8.94
Captive Mining Companies, including Damodar Valley Corp., Jindal Steel & Power and Tata Iron & Steel Co.		27.51	0.69
Total		**407.02**	**30.07**

Note: Data in the table are for fiscal year 2005/06 and so do not match exactly IEA statistics (see Box 15.1 in Chapter 15).
Source: Ministry of Coal (2007a).

which have yet to be developed (Ministry of Coal, 2007b). Procedures for obtaining the required approvals and permits result in long delays, so there is considerable uncertainty over the coal sector's ability to meet this target. Production growth is also limited by poor productivity, transport capacity limitations and lack of investment.

In contrast to China, coal in India is found mostly at relatively shallow depths. Some 85% of production comes from over 170 opencast mines. By international standards, the equipment employed is not the most productive, being of relatively small size (Box 17.2). There are around 390 underground mines, typically labour-intensive, bord and pillar operations with less mechanisation than would be expected, given India's position as the world's third-largest hard coal producer. India has so far been unsuccessful in adopting the more productive longwall mining technology. This will be necessary to economically extract coal reserves below 300 metres.

The economics of coal production in India depend critically on the cost of inland transport and pricing. Coal deposits are located mainly in the east, while demand centres are mostly in the north, south and west. Almost half of Indian coal production is dispatched by rail, accounting for 45% of total rail freight

Coal India is the world's largest coal mining company by manpower and output. It had almost 470 000 employees at the end of 2005 and produced 343 million tonnes of coal in the year to 31 March 2007 (Ministry of Coal, 2007a and 2007b). Productivity at Indian mines can be as low as 150 tonnes per man-year, largely because of overmanning, poor working methods and low equipment capacity. For example, at opencast mines, some 170-tonne trucks are used but many are much smaller. At the most efficient mines elsewhere, trucks with up to 380 tonnes capacity are employed. A voluntary retirement scheme, closure of uneconomic mines, prioritised investment and improved equipment utilisation rates have started to contribute to higher productivity of up to 2 650 tonnes per man-year (IndiaCore, 2006). Average productivity is around 700 tonnes (Figure 17.9).

Around the world, coal mining productivity will continue to improve. Greater automation, increasingly powerful drives, remote control, more sophisticated monitoring and predictive diagnostics will enable equipment to operate more reliably at higher utilisation rates. Greater recovery rates can be expected from larger equipment and advanced methods, such as sub-level caving whereby coal from the falling roofs of thick seams can be recovered as they collapse behind an advancing longwall face. Importantly, productivity improvements bring health and safety benefits. Fewer mine workers are needed to operate modern machinery. Safe working practices also enhance productivity, by reducing down time. In India further reforms will be needed to bring in more private investment and modernise coal mining.

(Ministry of Finance, 2007). Long-distance transport by rail adds considerably to the delivered cost, particularly if the coal is not washed first. Coal transport and handling costs can double or triple the pit-head price when delivered to states distant from the coalfields. The capacity and reliability of Indian Railways needs to be upgraded to achieve higher speeds and new investment is needed in dedicated freight tracks to meet rising coal demand.

The price and distribution of coal was, in principle, deregulated in 2000, under the Colliery Control Order. However, the coal market is still not competitive in practice. Coal India is the price-setter and Indian Railways, the national state-owned rail company, cross-subsidises passenger traffic through high coal freight rates.[6] A system of "coal linkages" still operates, whereby long-term coal

6. The *Economic Survey 2006-2007* by the Ministry of Finance reports that approximately two-thirds of Indian Railways revenues are from freight haulage.

Figure 17.9: **Coal-Mining Productivity in Australia, China, India and the United States**

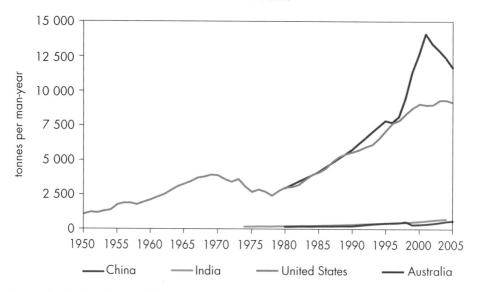

Sources: For the United States, US Department of Labor, Mine Safety and Health Administration (2007); for Australia, Barlow Jonker (2006); for India, Ministry of Coal (2007b); for China, IEA estimates.

production and transport is allocated to core sector consumers, including the power, steel, cement and sponge-iron sectors (Ministry of Coal, 2007a). In the absence of a competitive coal market, other users have struggled to obtain coal supplies, despite being willing to pay high prices. A more transparent and competitive coal market, with prices determined by quality and the timeliness of delivery, would encourage investment in coal mines and the supply infrastructure.

Coal India's pit-head prices typically range from $10 to $35 per tonne (Coal India, 2007). These prices cannot easily be compared with internationally traded coal because the energy content varies widely, using an archaic grading system. The poorest grades (F and G) have high ash and moisture and less than half the energy content of imported coal, often making these grades an expensive source of energy. A typical power station coal (Grade E) is equivalent to an imported coal costing $31-$39 per tonne before adding any inland transport costs. The Integrated Energy Policy[7] recommends moving to prices based on energy content (Planning Commission, 2006).

Domestic coal has always been competitive for mine-mouth users, while imported coal is generally competitive for users adjacent to a port, particularly so in locations distant from the coalfields, like the western and

7. See Chapter 15.

southern coasts. At the high coal prices seen since the end of 2003, Indian coal is competitive at most locations. However, for some Indian consumers, imported coal is the only option since their quality requirements cannot be met by local producers.

Coal Imports

India imported 36 Mtce of coal in 2005, covering 12% of demand. Steam and coking coal imports were 18 Mtce each. Coal imports have grown strongly over the past two decades. In 1990, India imported about 5.9 Mtce (5.1 Mt), but imports have surged and, by April 2007, steam coal imports reached an all-time monthly high of 3.3 Mt (McCloskey, 2007). Steam coal imports are projected to rise further to 52 Mtce in 2015 and 139 Mtce in 2030; coking coal imports are projected to rise more slowly to 45 Mtce in 2015 and 105 Mtce in 2030 (Figure 17.10). Overall the share of imports in Indian primary coal demand increases from 12% in 2005 to 28% in 2030.

There is inevitably considerable uncertainty surrounding coal import prospects in India though it is certain that the country will continue to rely on imported coal for quality reasons in the steel sector and for economic reasons at power plants distant from mines but close to ports. Because of the poor quality of indigenous coal for steel production, coking coal imports are expected to grow. Imports of steam coal for power generation are also set to increase, as indigenous production lags demand and may not meet the specifications required for more efficient operation of coal-fired power plants. Port capacity has grown to meet rising imports and domestic shipments, from 8 Mt in 1996/97 (Government of India, 2002) to around 70 Mt in 2007. Future imports are unlikely to be constrained by a lack of port capacity.

Lower coal import tariffs have encouraged increasing quantities of steam coal imports from Indonesia for power generation,[8] which are blended with local coal to reduce ash and enhance energy content. Smaller quantities come from China and Australia. Australia is the principal source of coking coal imports into India, accounting for 81% of imports in 2005, and it will continue to dominate in the future. Imports of Chinese coking coal are likely to decline as China restricts its exports to meet domestic demand. To secure future imports of steam and coking coal, the Indian government has tasked Coal India to invest in overseas mines, extending a practice established by the Indian steel industry. Investments have been announced in Australia and Indonesia (McCloskey, 2007), with interest also being shown in Mozambique, Zimbabwe and Kazakhstan (IndiaCore, 2006).

17

8. These amounted to 13.4 million tonnes in 2005.

Figure 17.10: **India's Coal Production and Imports in the Reference Scenario, 1990-2030**

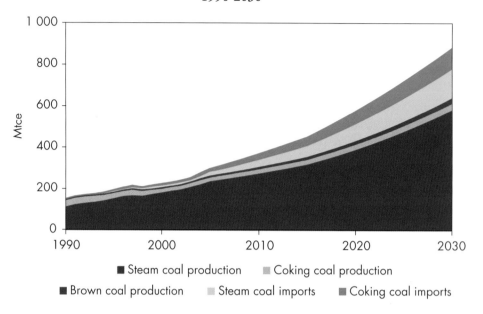

- Steam coal production
- Coking coal production
- Brown coal production
- Steam coal imports
- Coking coal imports

Power Generation

Overview of the Power Sector

India has the fifth-largest installed power-generating capacity in the world, with 146 GW in 2005, including utilities and industrial autoproducers of electricity. Total capacity owned by utilities was about 125 GW.[9] Private power producers selling their electricity to the State Electricity Boards (SEBs) have been allowed since the electricity-market reforms were launched in 1991. They owned some 15 GW in 2005, of which independent power producers (IPPs) held about 6 GW. There are also several industrial producers that generate electricity for their own use, with a total capacity of around 21 GW in 2005.

In the public domain, electricity generation falls under the responsibility of the states and the central government. The State Electricity Boards (SEBs) owned about 17 GW at the beginning of 2006 (CEA, 2007). The SEBs are also responsible for buying electricity from other companies and selling it, accounting for 95% of retail electricity sales. The central government owns some 40 GW through its companies, the largest being the Nuclear Power Corporation of India Ltd. (NPCIL), NTPC (formerly known as the National Thermal Power Corporation) and the National Hydro Power Corporation (NHPC). NTPC is the single largest company, with an installed capacity of about 26 GW and annual generation exceeding 180 TWh. It is majority-

9. Utilities include all power producers that sell power to the grid.

owned by the central government. There are also a number of other companies, electricity departments, power corporations and management boards separately or jointly owned by the states and the central government. Power Grid Corporation of India Ltd. (PGCIL) integrates India's five regional grids into a national grid, though inter-regional capacity is still limited.

Power Generation Mix

Total electricity generation reached 699 TWh in 2005, a little higher than in Canada or Germany. In the Reference Scenario, it rises to 2 774 TWh in 2030, an average annual increase of 5.7%, compared with 6.1% per year between 1990 and 2005 but only 4.5% per year over the past five years. In the period 2005-2015, electricity generation is projected to grow by 6.6% per year. Per-capita electricity generation, at 639 kWh in 2005, is one of the lowest in the world – over four times lower than the world average and 14 times lower than the average in the OECD (8 870 kWh). It is comparable to that of Vietnam and Mozambique. While power generation is projected to grow to more than 1 900 kWh per capita in 2030, it will remain very low by OECD standards.

Coal is the dominant fuel in India's electricity generation, accounting for over two thirds of total electricity produced. This share has often exceeded 70% over the past fifteen years. India's heavy dependence on coal will continue into the future (Figure 17.11). In the Reference Scenario, the share of coal increases to 71% by 2030.

Figure 17.11: **Changes in India's Electricity Generation Mix in the Reference Scenario**

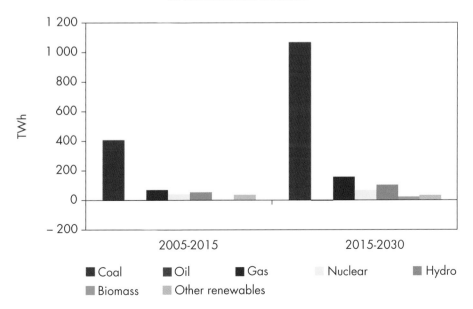

17

India's coal-fired power plants are among the least efficient in the world. The average conversion efficiency has been fluctuating between 27% and 30%, compared with 37% on average in the OECD.[10] The Central Electricity Authority, an advisory body to the government that monitors developments in the country's power sector, estimates that the average efficiency is about 15% less than the design efficiency (CEA, 2006). The poor quality of the coal available and inadequate maintenance of power plants contribute to the low performance. Improved performance could help the country's chronic electricity supply shortage.[11] Over the projection period, the efficiency of coal-fired power generation is projected to improve considerably, as the new plants will be larger and more efficient, and more supercritical units will be built (Box 17.3). On average, efficiency is expected to increase from 27% now to 38% in 2030. It will nonetheless remain below the 42% average efficiency expected to be attained in the OECD in 2030. No integrated gasification combined-cycle (IGCC) plants nor plants with CO_2 capture and storage (CCS) facilities (Box 17.4) are expected to be built before 2030 in the Reference Scenario.

Box 17.3: **Coal-Fired Power Plant Technology in India**

All of India's operating coal-fired power stations use subcritical technology. In the short run, most new coal-fired power plants are expected to continue to be based on subcritical technology, although new units are getting larger and more efficient. Planned capacity additions in the 11th Plan include units ranging mainly from 210 MW to 800 MW, with a few small units between 100 MW and 125 MW. Half of the planned capacity is based on 500 MW units. Six supercritical coal-fired units, with a capacity of 660 MW each, were included in the 10th Plan. However, none of these units was built within the expected time frame and the first one is expected to come on line in 2007/08. The 11th Plan includes twelve supercritical units, with a combined capacity of 8 GW.

The main supplier of coal-fired power plants in India is Bharat Heavy Electricals Ltd. (BHEL) and it is likely to maintain its dominant position in the future. Manufacturers from industrialised countries are more prominent in the provision of gas turbines and hydro plants. The 11th Five-Year Plan calls for BHEL's manufacturing capacity to expand from 6 000 MW a year now to around 10 000 MW. BHEL's R&D expenditure is around 1% of sales, while internationally this ratio is between 1.8 and 6% (Ministry of Power, 2007). Many uncertainties exist as to the rate at which BHEL will be

10. IEA statistics indicate that the average gross conversion efficiency of coal-fired generation was 27% in 2005, while India's Central Electricity Authority places this efficiency at 31%.
11. For power deficit, see the Investment section.

Box 17.3: **Coal-Fired Power Plant Technology in India** *(Continued)*

able to expand its manufacturing capacity and when it will be in a position to produce far more efficient power plants, notably supercritical ones. In any case, with increasing demand for coal-fired power stations, it is likely that more plant purchases will have to be made from other manufacturers. Tata Power has selected Doosan Heavy Industries of Korea as supplier of five boilers for the 4 GW Mundra project, one of the largest plants ever in India. Competition between manufacturers is likely to encourage innovation.

Besides work on supercritical power plants, research efforts in India also focus on fluidised bed-based IGCC. Research is carried out mainly by BHEL and NTPC, but private participation may prove necessary. Efforts are under way to develop a 125-MW IGCC unit. Its efficiency is expected to reach 39.5%, well below that of the IGCC plants that are now on order in OECD countries.

In 2006, the Ministry of Power launched an initiative to develop large coal-based plants, known as ultra-mega power projects. Each of these plants will have a minimum capacity of 4 GW. The intention is to promote the construction of large supercritical units (800 MW each). The selection of the projects is based on competitive bidding and both coastal and pit-head projects can be considered. To streamline these projects, the government set up project companies to obtain the necessary clearances before offering the project to bidders and to allocate mining blocks to the pit-head projects.

Gas-fired generation accounted for 9% of total generation in 2005. This share has risen somewhat over the past decade as gas production increased. The power sector faces gas supply shortages both because the government favours allocation of gas supplies to the fertilizer industry (a non-energy use of gas) and because adequate supplies at the agreed price have not been forthcoming.[12] Many gas-fired power plants still have to run on naphtha as a substitute or remain idle because naphtha is too expensive to use. CEA estimates that around 7 TWh of generation was lost in 2005 because of a lack of gas. Out of the 62 TWh of gas-based electricity produced in 2005, about a quarter came from industrial autoproducers.[13] Those autoproducers with access to gas in recent years have elected to build gas-fired plants and gas now accounts for about 20% of their total production. Future growth in gas-based generation, both in utilities and autoproducers, crucially depends on the availability of gas – how fast domestic production will expand and how much imported gas will

12. Gas pricing is discussed in the Spotlight in Chapter 15.
13. Industrial autoproducers produce electricity on site mainly for their own use. They are also referred to as captive power producers.

India has joined a number of international efforts to speed up the development and dissemination of CCS technologies, including the Carbon Sequestration Leadership Forum, the Government Steering Committee for the US FutureGen project, the US Big Sky CCS partnership and the Asia-Pacific Partnership on Clean Development and Climate. However, India has adopted a reserved position towards the assessment of CO_2 storage potential in India or building a zero-emissions fossil-fuel power plant, because of the higher cost and technical uncertainties associated with CCS technologies.

The Indian CO_2 Sequestration Applied Research network was launched in 2007 to develop a framework for activities and policy studies. CO_2-EOR scoping studies are being carried out in the Ankleshwar oilfield which is mature, where acid gas from the Hazira processing plant could be injected. The reservoir properties indicate that the project would be feasible (Malti, 2007).

Estimates for the geological storage potential of India are in the range of 500 to 1 000 Gt of CO_2, including onshore and offshore deep saline aquifers (300 to 400 Gt), basalt formation traps (200 to 400 Gt), unmineable coal seams (5 Gt) and depleted oil and gas reservoirs (5 to 10 Gt) (Singh *et al.,* 2006). One of the largest potential areas for CO_2 storage is the basalt rock region Deccan Volcanic Province (DVP) in the north-west of India. Storage volumes are in the range of 300 Gt of CO_2 (Sonde, 2006). The second important area is the Indo-Gangetic foreland (Friedmann, 2006). The Ganga Eocene-Miocene Murree-Siwalik formations are fluvial sandstones that, as saline aquifers, have good storage potential. Their high salinity and depth prevent them from being economical for surface use. The existence of important CO_2 sources close to the potential storage site makes it a good candidate for a pilot CCS project.

Early opportunities for CCS in India, matching sources and sinks, have been analysed by Beck *et al.* (2007) using the IEA greenhouse gas methodology. A preliminary analysis indicates a potential for disposal of 5 Mt/year within 20 kilometres of large CO_2 sources, storing the CO_2 in depleted oil and gas fields or using it for enhanced oil recovery. Saline aquifers could absorb a further 40 Mt per year. Over 30 large-scale sources could be considered for early trials.

be available – and on its pricing. In the Reference Scenario, total gas-based electricity generation is projected to increase by 6.4% per annum to 2030 and its share in electricity generation is projected to reach 11%.

Oil plays a minor role in electricity generation, accounting for just over 4% of total output in 2005. Diesel is the main oil product used in this application. In practice, diesel use for power is higher than reported, as the statistics do not include the fuel used in the stand-by generators which are in widespread use in buildings in India to cope with power cuts. Some fuel oil is used in power stations and small amounts of naphtha are also used where gas is not available. Over the projection period, oil-based electricity generation is expected to remain roughly at current levels, as switching to gas progresses. The share of oil is projected to fall to 1% of total generation in 2030.

Nuclear power accounted for 2.5% of total electricity generation in 2005, when installed nuclear power capacity was 3 GW. This rose to 3.6 GW in 2006, with the connection to the grid of Tarapur-3. One unit at Kaiga was connected to the grid in April 2007 and three more units are expected to be connected to the grid by the end of 2007. Three additional units, of which one is a fast-breeder reactor, are under construction. The Indian government's nuclear power generation programme is ambitious (Box 17.5). The current target is to raise nuclear power generation capacity to 20 GW by 2020 and to 40 GW by 2030. Earlier targets, such as the target set in the 1984 Nuclear Power Profile of 10 GW by 2000, have not been met (IEA, 2006). Installed capacity in 2000 was only a quarter of that target. The programme seems to have accelerated now.

Box 17.5: **India's Nuclear Power Generation Programme**

India's nuclear power generation programme started with the construction of two boiling water reactors (BWRs) at Tarapur in the 1960s. India was the first developing country to have nuclear power plants. The BWR units were built by the General Electric Company and Bechtel. Atomic Energy of Canada, Ltd. built two pressurised heavy water reactors (PHWRs) in Rajasthan. India subsequently developed its own technology, based on PHWR technology.

India has modest uranium resources, about 1.4% of the world's reasonably assured and inferred resources (NEA/IAEA, 2006). It has vast thorium resources. On the basis of these resources and in an attempt to overcome its relative isolation in international trade (as it has not joined the Non-Proliferation Treaty), India has drawn up a three-stage nuclear programme. The first stage involves the development of mainly domestic-built PHWRs, although two Russian VVERs are also under construction. It also has plans to build other light water reactors (LWRs), depending on access to international markets.

17

The second stage envisages the development of fast breeder reactors (FBRs) coupled with reprocessing plants and fuel fabrication plants using plutonium. This stage has begun with the construction of a 500-MW FBR at Kalpakkam. The FBR is based on Indian technology and could be completed by 2012. The government envisages building four additional FBRs by 2020.

The third stage will be based on the thorium-uranium-233 cycle. The Bhabha Atomic Research Centre (BARC) has designed a 300-MW advanced heavy water reactor (AHWR). Construction of a demonstration project could start at the end of 2007. India is also a partner in the ITER project to develop a fusion reactor.

India's nuclear power plants had a very low capacity factor until the 1990s, but this has steadily improved to reach 87% in 2002, comparable to OECD levels. It fell back to 65% in 2005, because nuclear fuel supply was constrained. A recent agreement with the United States is expected to improve the supply of nuclear fuel and access to technology – though the agreement is not yet ratified. NPCIL, the owner of India's nuclear power stations, is responsible for the construction of new nuclear power plants. The fast breeder reactor is being developed by Bharatiya Nabhikiya Vidyut Nigam Limited (BHAVINI), a company owned by the government of India, under the control of the Department of Atomic Energy.

In the Reference Scenario, India's nuclear power capacity is projected to rise to 8 GW by 2015 and to 17 GW by 2030, well below the level targeted by the government. This reflects the difficulties India has experienced in building nuclear power plants, because of high construction costs and because of its exclusion from international trade in nuclear power plants and materials. These challenges could persist. Electricity generation from nuclear power is projected to grow from 17 TWh in 2005 to 128 TWh in 2030, assuming an 85% capacity factor. The share of nuclear power in electricity generation is projected to rise to 5% by 2030.

Renewable energy sources accounted for 15% of total electricity generation in 2005, mostly hydropower. This share is projected to fall to 13% in 2030 because of a falling share of hydropower. Increases in biomass, wind and solar power compensate for some of this decrease.

To meet projected electricity demand, India's power generating capacity in total will need to increase from 146 GW now to 255 GW in 2015 and 522 GW in 2030. Total capacity additions between 2006 and 2030 are projected to amount to 410 GW, including the replacement of some older power plants, mainly coal-fired. This is about the same as the current installed capacity of Japan, Korea and Australia. In the period 2006-2015, India is projected to build 113 GW of capacity. More than half of this capacity is projected to be coal-fired (Figure 17.12). These new coal-fired power plants alone would increase India's CO_2 emissions by about 0.3 Gt above current levels by 2015 and by 1.1 Gt between now and 2030. About 15 GW of coal-fired capacity was under construction at the beginning of 2007.

Figure 17.12: **India's Capacity Additions by Fuel**

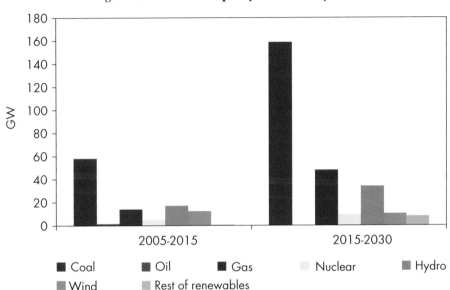

Power Generation Economics

The most economic fuel for electricity generation in India is coal, though nuclear power can compete at higher coal prices (Figure 17.13). The generating cost of CCGTs is largely dependent on gas prices. A range of $4.5 to $8 per MBtu has been used; the current price for power producers is lower than this but are expected to be higher in the future. CCGT construction costs are assumed to move in line with international trends.[14] Private investors may nonetheless still favour gas over coal, because of its flexibility and low initial cost. With generating costs ranging between US cents 4.7 and 7.7 per kWh, gas can be competitive for on-site electricity production in industry. The current electricity price for industry is about US cents 9 per kWh. Gas costs are reduced where the plants earn CDM credits.

14. Construction costs of CCGT projects in India in the 1990s have been higher than in OECD countries.

Figure 17.13: **Electricity Generating Costs in India**

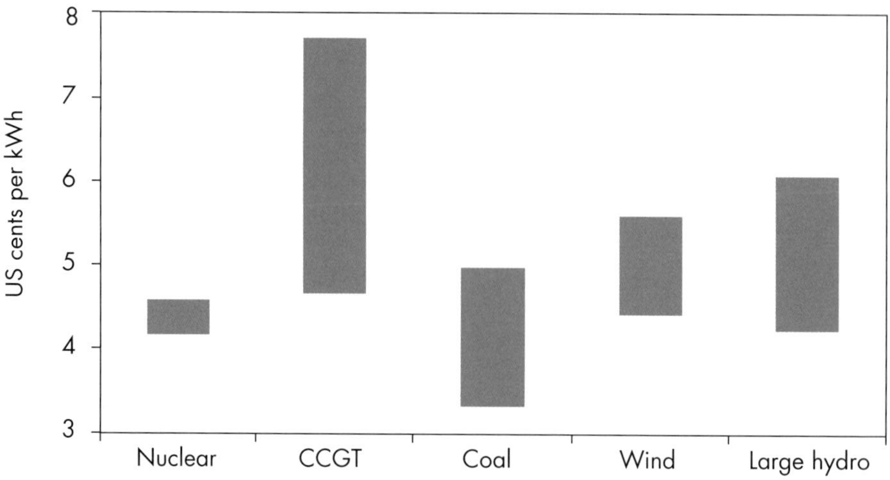

Source: IEA analysis.

Renewable Energy

Renewable energy supplied nearly a third of India's energy needs in 2005. Most of this was traditional biomass. Hydropower was the second-largest source, while wind power is emerging as a relatively important source of electricity.[15] Total renewable energy demand is projected to rise to 225 Mtoe in 2030.

India consumed a total of 158 Mtoe of biomass in 2005, most of it by rural households. At least three-quarters of rural households (668 million people) use traditional biomass fuels, fuelwood, animal dung or agricultural residues, for cooking and heating (see Chapter 20). In the Reference Scenario, biomass use is projected to continue to grow, but more slowly than in the past. It reaches 171 Mtoe in 2015 and 194 Mtoe in 2030. While demand for traditional biomass is expected to increase only marginally, the use of biomass in power generation and in biofuel production is projected to increase more quickly. The potential installed capacity for biomass power generation is about 20 GW. Current installed capacity is 0.3 GW. It reaches 4.5 GW in 2030 in the Reference Scenario.

In addition to power generation, biomass is also used in thermal gasifiers. The current installed capacity of thermal gasifiers is 87 MW. In addition to the direct use of biomass solids, biogas technology is primarily used in India in thermal applications. The biogas plants vary in capacity from 2 to 10 m³. The total number of biogas plants installed in India now is 3.9 million.

15. For additional information, see IEA (2008, forthcoming).

India started using ethanol recently. India has a large sugar industry indicating a large potential for ethanol production. Biodiesel potential based on jatropha is also large (see Chapter 18 for a more detailed discussion). Total biofuel consumption is projected to increase to 1.9 Mtoe in 2030.

Hydropower is India's second-largest source of electricity, with a share of 14% in total electricity generation at present. That share has declined considerably over time, from over 40% in 1971 to 25% in 1990 and 13% in 2000, with a marginal increase in recent years. Between 1990 and 2005, India added 12 GW of hydropower, but coal-fired power plant additions were more than three times higher. The development of hydropower faced severe constraints during that period, with strong environmental opposition and financing difficulties.

Installed hydropower capacity reached 34 GW in 2005, within an estimated total economic potential of about 150 GW. In order to speed up the development of hydropower, the government launched the 50 000 MW hydro initiative in 2003. Nearly all the projects needed to realise this target have been identified. In the Reference Scenario, these plants are expected to be completed by 2030, bringing total hydropower capacity to 85 GW. The Indian government envisages that these plants will become operational during the 12th Five-Year Plan (2012-2017).

Environmental and social concerns have, in the past, greatly extended the time scale of a few high-visibility hydropower projects, causing cost overruns and poor economic returns. Resettlement has been the major issue. Public suspicions are in any case high, since proposals for large hydropower projects are often linked in the public mind to environmental disasters in Central Asia (Naryn and Amu Rivers). Generally, large hydropower projects in remote mountainous regions with low population density should face fewer obstacles. Negotiations for compensation can be facilitated by allocating to the dispossessed a share of project revenues over the life of the project. Suitable compensation models need to be developed. Appropriate action along these lines would help revive interest in hydropower by international financing organisations and contribute to future growth in hydropower.[16]

Among other sources of renewable energy, wind power is the most prominent. India had the fourth-largest windpower installed capacity in the world in 2006, exceeding 6 GW. Total onshore potential is estimated at 45 GW, although this potential could increase with further assessment of the resource in the future. India is home to one of the world's leading wind turbine manufacturers, Suzlon. Wind power capacity is expected to rise to 27 GW by 2030 and to account for 2.5% of total electricity generation.

17

16. See IEA (2006), Box 6.1.

India receives abundant solar radiation, indicating a very large potential for solar energy use. Photovoltaic (PV) systems are being promoted primarily for rural and remote applications, but their use is limited. The decentralised systems, used mainly in rural areas, are solar power plants with minigrids, solar home systems, solar lanterns and solar streetlights. Photovoltaics capacity rises to 4 GW in the Reference Scenario, but its share in total generation remains minimal.

Currently, about 1.9 million m² of solar water heaters are used in buildings and in industry. This is a small fraction of the total potential, estimated at 140 million m². Solar water heaters can be cost-effective, with a payback period of around 3 years (see Chapter 18). Energy use from solar water heaters remains marginal in the Reference Scenario.

India has multiple strategies to promote renewable energy, at the state and central government level. A number of states have set targets and have provided incentives. Renewable energy is seen as an important element in rural electrification and is promoted by the Ministry of New and Renewable Energy through various programmes.

Investment

India will need to invest $1.25 trillion in energy infrastructure in the period 2006-2030. More than three-quarters of this investment will be in power infrastructure (Figure 17.14). Attracting investment in a timely manner will be essential if economic growth is to be sustained.

Figure 17.14: **India's Investment in Energy Infrastructure, 2006-2030**

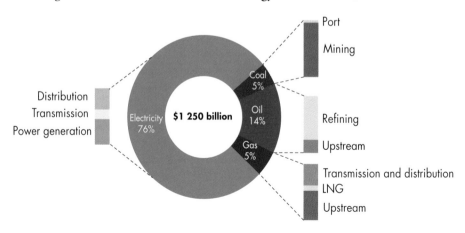

Note: Investment in biofuels is negligible and is included in oil.

Oil and Gas

The Reference Scenario implies a need for India to invest $233 billion in the oil and gas sectors. More than three-quarters of the $169 billion oil investment will be absorbed by the refining sector which will be partly export-oriented. The question remains whether those investments will actually happen on subsidised local Indian markets. More than 90% of the $63 billion gas investments will be mainly oriented towards developing the upstream capacities as well as transport and distribution infrastructures. Even more than in the case of the refining sector, upstream gas investments, largely influenced by international markets such as drilling equipment and skilled manpower, require a consistent and stable pricing framework. Investments in LNG regasification plants will only represent a minor share, amounting to $6 billion.

Coal

The Reference Scenario projections call for India to invest $57 billion in its coal sector over 2006-2030. Most of this investment – $54 billion – will be in mining and the remainder in development of port infrastructure. Attracting this investment will pose many challenges: the sector is currently unprofitable and lacks commercial discipline. Unpaid dues from SEBs mean that Coal India and Singareni Collieries have difficulty in raising capital. Rail bottlenecks and poor infrastructure create considerable inefficiencies. The absence of long-term, legally-binding fuel supply and transport agreements increases the risks of long-term investment in the coal sector and results in a lack of accountability when problems arise.

Currently, the implementation of mining projects in India is highly bureaucratic and, in the absence of competitive pressures, does not result in the most efficient use of capital and other resources. The government needs to ensure that Coal India Ltd is not allowed to abuse its powerful, monopoly position and that trade unions understand the need for reform, which ultimately must mean breaking the company up to foster competition in the coal sector. A clear division needs to be drawn between those activities best undertaken by commercial enterprises and those best managed by the state, such as regulation and community development; and the structure and organisation of the industry need to be revised accordingly. Regulators must then be left free to pursue their defined responsibilities, free from political interference. Our projections assume that market-based solutions do succeed with the result that private investors do respond to the burgeoning coal market and provide India with a secure and competitive source of energy for power generation.

Electricity

The Reference Scenario projections imply a need for India to invest $956 billion in new power infrastructure over the period to 2030. Investment

17

in power generation alone is estimated at $435 billion. Transmission networks will need $164 billion and distribution networks another $357 billion. The underlying assumption of the Reference Scenario is that investment will be available and that this power infrastructure will be built in a timely fashion, but many challenges remain. These are discussed in the section below.

Focus on Investment Challenges in India's Power Sector

Chronic underinvestment in India's power sector has been a major constraint to the country's development. Nearly 40% of the country's population still does not have access to electricity.[17] Demand for electricity from those who have a supply is growing rapidly, but part of this demand is not being met. The Indian government estimates that the current power deficit stands at about 9%, reaching 14% for peak power, with both rates deteriorating (Ministry of Power, 2007). Many experts consider that these figures are largely underestimated. The quality of power is a major concern, both for industrial and private consumers. Power cuts, unstable voltage and low or high supply frequency are commonplace.

The capacity addition targets set in the five-year plans have generally not been met and performance has deteriorated over the past three plans. Performance in the latest five-year plan period, which ran until March 2007, was the worst ever. Less than half of the capacity envisaged by the government was built. Insufficient investment resulted in electricity generation increasing at a rate well below the rate of growth in GDP for five consecutive years between 2001 and 2006, a situation never seen in the past and one that is not sustainable.

The pace of capacity additions stagnated in the 1990s (Figure 17.15). Until the early 1990s, the power sector had received between 15% and 20% of the total central government budget. This share declined after economic reforms were introduced in 1991, in the expectation that part of the required investment would come from the private sector (IEA, 2003). But many of the projects proposed have not proceeded, in large part because of an inadequate legal and commercial framework, involving lack of law and contract enforcement and delays in obtaining regulatory approvals. More than fifteen years after the reforms were initiated, only 6 GW of IPP plants have been put into operation. Over the same period, total installed capacity increased by more than 60 GW.

There have, however, been some encouraging signs recently. Investment has been on an upward trend since 2003 and is coming from both the public and private sectors. Utilities had over 30 GW of capacity under construction in 2006, higher than the 20 GW added in 2001-2005. Timely completion of plants has been a problem in the past, but if problems are overcome, India's installed capacity

17. This is an IEA estimate based on National Census data for India and information obtained from TERI.

could increase by nearly a third between 2005 and 2010. Even this could still fall short of the target in the 11th Five-Year Plan (April 2007 through March 2012) which foresees capacity additions of 69 GW, much higher than the unmet target of 41 GW set in the 10th Five-Year Plan. Most power plants under construction or expected to be built under the 11th Five-Year Plan are coal-fired or based on hydropower.

Figure 17.15: **India's Power-Generating Capacity Increases**

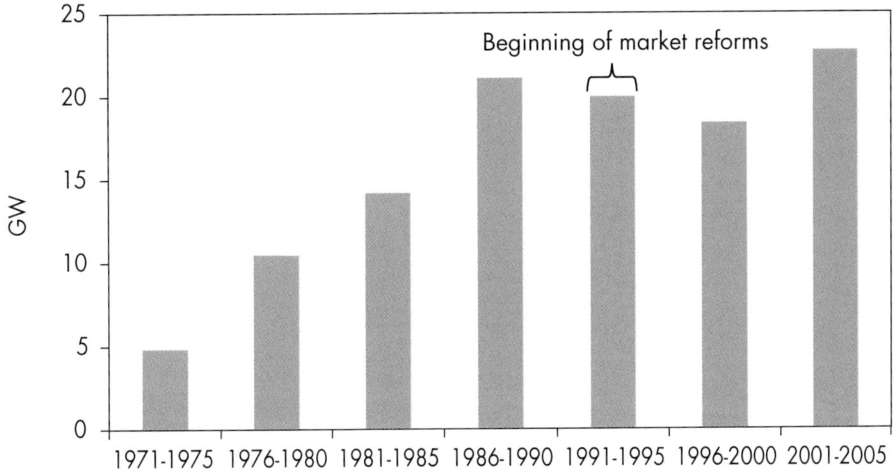

Sources: Platt's World Electric Power Plants Database (2006); Ministry of Power (2007).

Not content with the results, the government provided for a major overhaul of the electricity industry in the Electricity Act of 2003.[18] In accordance with the requirements of the act, the government announced in 2005, after consultation with the states and industry, a National Electricity Policy aimed at further accelerating power-sector development. It emphasises the need for tariffs to provide for full cost recovery. It also recognises the need for reform with respect to cross-subsidisation.

Public-Sector Investment

Most power infrastructure projects in India are publicly financed either directly by the government or by government-backed loans. The financial health of most of India's State Electricity Boards has been deteriorating because of high operating costs, pricing policies that keep tariffs to most customers below the cost of supply, and failure to collect revenues for much of the electricity consumed. Insufficient revenues drive up debt and force the SEBs to seek other means to finance investment.

18. See Chapter 15 for more detail.

In 2005, revenues from electricity sales were about 85% of costs. This share has ranged between 76% and 86% since 1990. The gap appears to have closed somewhat in recent years. About 40% of revenues comes from subsidies, largely required to support electricity sales to farmers. Total subsidies reached $9 billion in 2005.

The SEBs' rate of return on capital has been negative, ranging from –28% to – 44% since 1990 (Figure 17.16).[19] Their required rate of return is theoretically 3%. In the OECD, the rate of return on investment by power companies typically ranges between 5% and 12%, the higher end applying in competitive markets and the lower end to public entities. Some merchant projects in OECD countries have had returns of the order of 15% to 17%. The insolvency of the Indian SEBs is a problem not only for public-sector investment. Its underlying causes are also an impediment to private investment, which relies on cash flows from properly enforced payment of charges.

Figure 17.16: **Comparison of Returns on Investment, India and OECD**

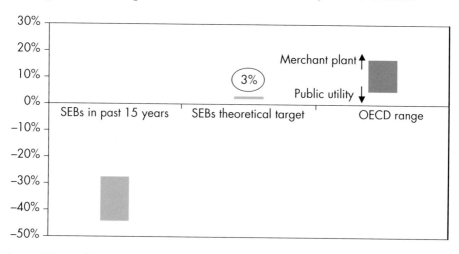

Sources: Ministry of Finance (2007); Government of India (2002); IEA analysis.

There are wide variations in economic and operational performance across states. The Ministry of Power has introduced a system of rating the state power sector in order to monitor and, hopefully, improve performance. Each of the 28 states is given marks out of 100, based on a number of parameters. The main parameters measure progress in relation to state government performance in implementing reforms and in rural electrification, the regulatory process

19. A handful of SEBs such as those in Orissa, Goa and Chhattisgarh have managed positive results in recent years.

(mainly related to tariff policy), business risk (the performance of power plants and networks, distribution reforms and management of technical and commercial losses), financial risk (cost recovery) and in attaining overall commercial viability. Negative marks can also be assigned (for example if the rate of losses is deteriorating).

Table 17.9: **Performance Ratings of the State Power Sector, Selected States**

State	Score	Rank
Andhra Pradesh	55.81	1
Gujarat	54.46	2
Delhi	50.87	3
Karnataka	46.92	4
West Bengal	46.24	5
Maharashtra	35.41	8
Uttar Pradesh	24.38	18
Madhya Pradesh	21.97	20
Jharkhand	4	24
Bihar	–3.06	27

Source: CRISIL and ICRA (2006).

The levels of transmission and distribution (T&D) losses and own use of electricity in power stations in India, as a share of total electricity generation, are among the highest in the world. They compare with those in several countries in Sub-Saharan Africa and some countries in Eastern Europe. The combined rate was 32% in 2005 across India, but there is wide variation between states, with the loss rate in some states exceeding 50% (Figure 17.17).[20] The average rate in the OECD is 14% (about 2% for transmission losses, 5% for distribution losses and 7% for own use). Losses arise from insufficient investment in, and poor maintenance of, networks (they are estimated at around 17%) and from theft (the remainder, *i.e.* about 15% in 2005). India is now making efforts to monitor and reduce losses. High losses (electricity generated but not paid for) render the sector financially unviable and detract from investment.

Private Investment

Between 1991 and 2005, total investment in electricity-sector projects involving the private sector amounted to $20 billion.[21] Figure 17.18 shows that

20. The 32% rate has been calculated using IEA statistics.
21. Investment has been calculated using only projects in operation or under construction.

Figure 17.17: **Electricity Losses by State (%)**

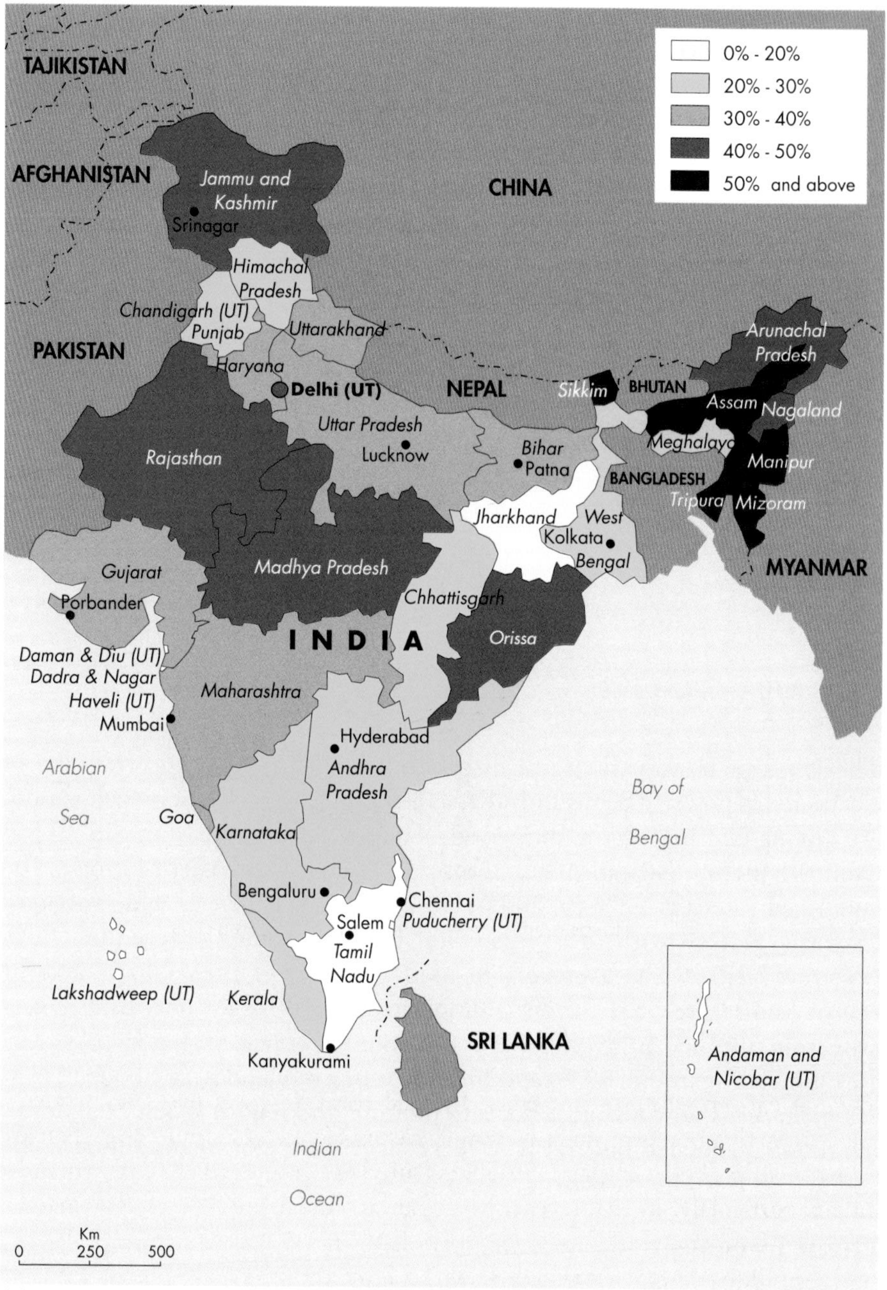

Legend:
- 0% - 20%
- 20% - 30%
- 30% - 40%
- 40% - 50%
- 50% and above

The boundaries and names shown and the designations used on maps included in this publication do not imply official endorsement or acceptance by the IEA.

Source: IEA analysis based on CEA (2007).

in 1991 – right when the market opened – investment in India's power sector topped $840 million. At that time, this was 62% of total investment going into electricity projects in middle- and low-income countries. Between 1992 and 2005 this share fluctuated between almost 0% and 13%, with the exception of 2004, when investment surged again and the share reached 29%. That year, total Indian private-sector electricity investment reached $4 billion, a sign that the private sector evaluation of the conditions for investment had changed. It is too early to say if this trend will be sustained: there was a sharp fall in 2005. The government remains fully alert to the need to get the conditions right.

Figure 17.18: **Private Investment in India's Electricity Sector, 1991- 2005**

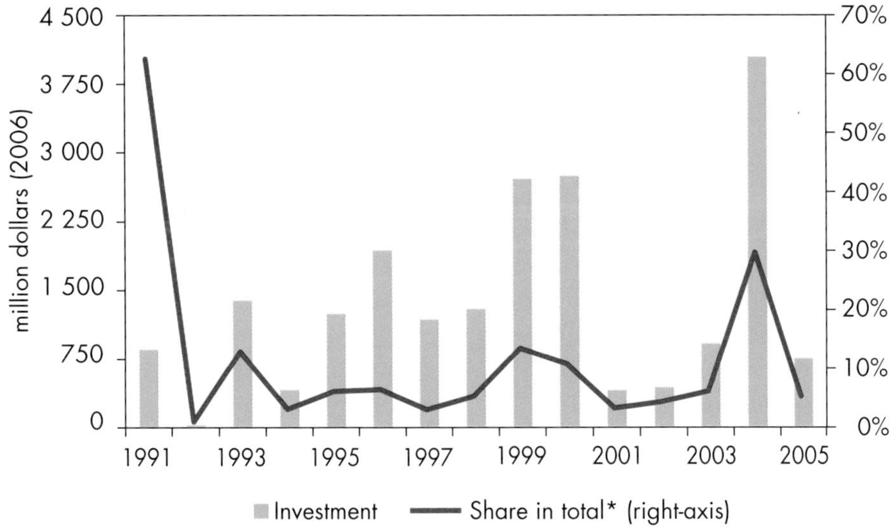

* Total in low- and middle-income countries as defined by the World Bank, including the transition economies.
Source: IEA analysis based on the World Bank's Private Participation in Infrastructure database.

Most private investment has been directed towards greenfield (new) power stations, with a few projects involving transmission, connections and changes of ownership. The companies involved in these projects have been mainly Indian, particularly in recent years. Companies outside India have come mainly from the United States, but companies from the United Kingdom, Japan, Switzerland, Norway, China and Malaysia have also been involved.

Many private projects have faced delays and many have been criticised as being too costly. Some projects have faced fuel supply problems once they became operational. For example, as noted earlier, some gas-fired power plants have had to rely on expensive naphtha, because gas supply was short, while others were shut down. Despite past problems with gas supply and the many uncertainties surrounding future gas availability and pricing in India, many of the recently announced power-generation projects are based on natural gas.

Since the beginning of reforms, the increase in IPP capacity has been lower than the increase in captive power (power plants used by the industry for its own needs). The increase in captive power was prompted by irregular and insufficient public electricity supply and by high tariffs, and was facilitated by certain provisions in the Electricity Act 2003. As a result, many industries now use their own power plants for processes or for in-house power consumption, using the grid only as backup. The government has plans to tap the excess power of these plants. A total of 1 100 MW was offered by various industries in 2005, but this level may increase in future, depending on tariffs and technical issues (CEA, 2005). Open access – a provision of the Electricity Act 2003, which allows industrial autoproducers to sell excess power to the grid – can be one way to improve national supplies. In practice, however, there is a lot of resistance by the SEBs who fear that they will lose those customers who actually pay their bill (notably large industries). It is anticipated that full implementation of these measures could stimulate some investment, although the level is likely to be rather small compared to the country's total needs (Desai, 2004).

Power-Sector Finance

Financing for public-sector power projects in India comes mainly from the federal government budget, in the form of equity or loans coming mainly from the Power Finance Corporation (PFC), which operates under the Ministry of Power. PFC provided about $2.5 billion in loans in 2005. The Rural Electrification Corporation finances projects in rural areas. Multilateral lending agencies, such as the World Bank and the Asian Development Bank, also lend money to India's power sector.

The unprofitability of India's power sector generally remains the major obstacle to attract private investment. Attracting private-sector investment in the generation sector is not exclusively a question of financial performance. Private generation projects have suffered from other factors. Foremost is the need to ensure reliable and sufficient fuel supplies. Reforms in other fuel markets will also need to move in line with reforms in the electricity sector. A concerted reform effort in the coal sector is needed to ensure delivery of the quantity and quality of coal required by a modern power sector. In the gas sector, the new transmission policy that allows private investment, domestic gas finds made by private companies, and increasing LNG import capacity should contribute to addressing the problem. Other obstacles in attracting private investment into generation are land acquisition and cumbersome procedures to obtain statutory approvals. These have often contributed to substantial implementation delays and escalating project costs, thus increasing generation costs. The government is addressing these concerns in its ultra-mega power project policy under which the government takes responsibility for statutory clearances, land acquisition and land preparation. If this policy proves a successful model, it could be considered to also extend it to smaller-sized projects.

To ensure adequate financing in the transmission or the generation businesses, the distribution business has to be addressed first. Other measures, such as government guarantees for off-take and commercial risks, can work only for a limited period and can be quite harmful in the long term, partly because of the take-or-pay contracts with the IPPs.

Power-sector reforms seem to be on the right path but reform implementation needs to be strengthened. Also, the widely varying reform progress between states needs to be addressed more pointedly. For the sizeable investments that India will need over the next two-and-a-half decades, improving the investment conditions in the sector and moving continuously towards a transparent, predictable and consistent power-sector framework based on market principles and financially profitable will remain of paramount importance.

17

CHAPTER 18

ALTERNATIVE POLICY SCENARIO PROJECTIONS

HIGHLIGHTS

- New policies that the Indian government is considering could result in significant energy savings. In the Alternative Policy Scenario, in which these policies are assumed to be fully implemented, primary energy demand is 17% lower than in the Reference Scenario in 2030.

- Coal savings are the greatest in both absolute and percentage terms. Most of the coal saved arises from reduced requirements for power generation. Lower electricity-demand growth, higher power-generation efficiency and fuel-switching explain this trend. More efficient production of iron and steel and cement, combined with higher efficiency in less energy-intensive industries, contributes to the savings.

- In the Alternative Policy Scenario oil imports are 1.1 mb/d lower in 2030 than in the Reference Scenario, but oil import dependence remains high at 90%. Gas imports fall by 4.8 bcm. Most remarkably, coal imports fall by 97 Mtce, mostly thanks to lower demand for steam coal for power generation.

- Lower energy demand in the power and transport sectors reduces SO_2 emissions by 27% and NO_x emissions by 23% in 2030, compared with the Reference Scenario. Lower overall energy consumption, combined with a larger share of less carbon-intensive fuels in the primary energy mix, yields savings of 27% in carbon-dioxide emissions by 2030. Energy-efficiency improvements on both the demand and supply sides account for most of the savings.

- Implementation of the policies considered in the Alternative Policy Scenario reduces cumulative investment by $78 billion over the *Outlook* period, with total supply-side investment savings of $132 billion being offset by increased investment on the demand side of $54 billion. These policies are cost-effective, though the payback periods for those investing in efficient appliances in India are longer than in many other countries because of price subsidies. Electricity-tariff reform, combined with improvements in bill collection, would increase the cost-effectiveness of the policies.

- Because of lower demand, India's cumulative energy-import bill is much lower in the Alternative Policy Scenario. The cost of India's coal imports is $73 billion less over 2006-2030 than in the Reference Scenario. The oil-import bill is reduced by about $250 billion, and the gas bill by $7 billion.

Background and Assumptions

Like China, India is on an unsustainable energy path. In the Reference Scenario, which takes account of only those policies already enacted or in place, oil, gas and coal imports increase substantially and local pollution and CO_2 emissions worsen alarmingly. Rapid growth in appliance ownership and in the building stock puts more pressure on India's already weak electricity infrastructure. Despite some improvement, the inefficient use of biomass for cooking and heating in rural households continues to cause too many premature deaths from indoor air pollution and drudgery for women and girls. While there have been improvements in energy efficiency in recent years, many industries – particularly small enterprises – continue to rely mainly on technologies and equipment that are far below the best.

However, there are many energy-policy actions that the Indian government is considering that could alter significantly these worrying trends. These policies are assumed to be adopted and fully implemented in the Alternative Policy Scenario. In addition, some of the policies that have already been adopted and which are, therefore, incorporated into the Reference Scenario are assumed to be implemented and enforced more rigorously. We have analysed some 80 policies and measures for India covering all energy sectors, ranging from efficiency improvements in the residential and services sectors to new technologies in the power-generation sector.[1] They reflect the proposals under discussion in India in the current energy-policy debate. These policies result in a reduction of dependence on coal and oil and in the faster development and deployment of more efficient and cleaner energy technologies.

The Report of the Expert Committee which formulated the Integrated Energy Policy in 2006 points out that reliable and stable energy supplies will be crucial to sustaining high economic growth. There is a broad consensus among policy makers that India's current energy system is far from sustainable. The current administration has stepped up efforts to move towards a more efficient and environment-friendly energy mix. Nonetheless, India will need strong political commitment and effective public communication to put these policies and measures in place.

The rate of economic growth and structure of GDP in India are assumed to follow the same trajectory as in the Reference Scenario. International oil and gas prices are also the same as in the Reference Scenario. The international price of coal is lower. Domestic prices are assumed to follow international energy prices, and subsidies are assumed to be reduced progressively over the *Outlook* period.

---

1. A full list of the policies and measures considered for India in the Alternative Policy Scenario can be found at www.worldenergyoutlook.org.

Key Results

Energy Demand

Primary energy demand in the Alternative Policy Scenario is reduced by about 17% in 2030 relative to the Reference Scenario (Table 18.1). The pace of energy demand growth in 2005-2030 is reduced to 2.8% per year, against 3.6% in the Reference Scenario. Changes in the capital stock occur slowly, so the average efficiency of power plants, industrial processes and household appliances approaches that of OECD countries today only by the end of the *Outlook* period. Between 2005 and 2030, energy intensity improves rapidly, at 3.3% per year, in the Alternative Policy Scenario, compared with 2.6% in the Reference Scenario. More efficient industrial processes and power plants, an increase in the use of energy-efficient appliances and equipment, and a reduction in the inefficient use of biomass for cooking and heating all contribute to this improvement.

Table 18.1: **India's Primary Energy Demand in the Alternative Policy Scenario**
(Mtoe)

	2005	2015	2030	2005-2030*	Difference from the Reference Scenario in 2030 Mtoe	%
Coal	208	289	411	2.8%	−209	−33.7
Oil	129	173	272	3.0%	−56	−17.1
Gas	29	47	89	4.6%	−4	−4.3
Nuclear	5	19	47	9.9%	14	41.9
Hydro	9	17	32	5.3%	9	42.3
Biomass	158	168	211	1.2%	17	8.5
Other renewables	1	6	21	15.8%	12	145.5
Total	**537**	**719**	**1 082**	**2.8%**	**−217**	**−16.7**

* Average annual rate of growth.

Demand for coal falls the most in both absolute and percentage terms, with coal savings reaching 209 Mtoe in 2030 (Figure 18.1). The majority of the coal saved – over 70% – comes about as a result of the lower requirement for power generation. Lower electricity-demand growth, higher power-generation efficiency and fuel-switching explain this trend. More efficient production of iron and steel and cement, combined with higher efficiency in less energy-intensive industries like textiles and ceramics, brings additional coal savings. Coal demand grows much slower in the Alternative Policy Scenario, by 2.8% per year on average, compared with 4.5% in the Reference Scenario.

Figure 18.1: **India's Energy Demand in the Reference and Alternative Policy Scenarios**

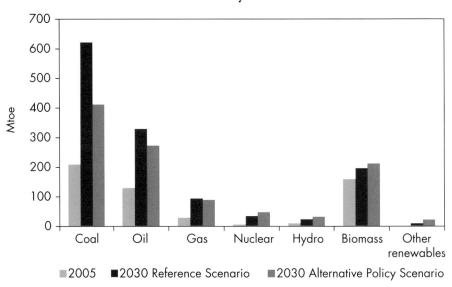

Oil savings are also significant, at 1.1 mb/d, or 17%, in 2030 in the Alternative Policy Scenario. Oil demand grows on average by 3% per year, reaching 5.4 mb/d in 2030. More than two-thirds of the oil savings come from the transport sector, thanks to the faster introduction of biofuels and compressed natural gas (CNG) and the increased fuel efficiency of new vehicles. Oil demand savings accelerate in 2015-2030, reflecting the demand effect of a modal shift to public transport that reduces the demand for two-wheelers and cars.

Natural gas demand continues to grow steadily over the *Outlook* period in the Alternative Policy Scenario, by 4.6% per year – 0.2 percentage points less than in the Reference Scenario. Gas consumption reaches 107 billion cubic metres in 2030, about 5 bcm less than in the Reference Scenario. But the share of gas in India's primary energy mix is slightly higher, rising from 5% in 2005 to 8% by 2030, compared with 7% in the Reference Scenario.

Demand for non-fossil fuel energy sources is 20% higher in the Alternative Policy Scenario. Hydropower, especially, grows faster and is 42% higher in 2030. Growth runs at 5.3% per year over the period 2005-2030, compared with 3.9% in the Reference Scenario. Biomass accounts for 19% of primary energy demand in the Alternative Policy Scenario in 2030, compared with 15% in the Reference Scenario. The share is greater because the dampening effect of fuel-switching from traditional biomass in the residential sector is outweighed by increased use of biomass in combined heat and power (CHP) plants in industry and for the production of biofuels for transport. Demand for other renewables – mostly wind and solar power – increases considerably and their share in primary energy demand in 2030 triples, compared with the Reference Scenario. The share of nuclear power in the primary energy mix gains two percentage points, reaching 4% in 2030.

Final energy consumption is also lower than in the Reference Scenario, but it falls slightly less in percentage terms than primary demand. This is because of efficiency improvements in the power generation sector and substantial reductions in transmission and distribution losses. Savings in final demand are still significant – electricity demand is 3% lower in 2015 and 10% lower in 2030 compared with the Reference Scenario. More efficient electrical appliances, air conditioning and lighting in the residential and services sectors contribute two-thirds of the savings. The other one-third comes mainly from improvements in the efficiency of industrial processes. Coal use in industry is reduced by over a quarter in the Alternative Policy Scenario by 2030.

Implications for Energy Markets and Supply Security

Energy-supply security is one of the main concerns voiced in India's Integrated Energy Policy. There is emphasis on the importance of increasing the diversity of India's fuel mix. Energy security is improved in the Alternative Policy Scenario, largely thanks to lower import dependence on fossil fuels, particularly coal, and to a much higher share of renewables in the power generation fuel mix and in transport.

In the Alternative Policy Scenario, net oil import dependence is 90% in 2030, two percentage points lower than in the Reference Scenario. By 2015, oil imports are already 300 kb/d less, and in 2030, India imports 4.9 mb/d, compared with 6 mb/d in the Reference Scenario (Figure 18.2). A greater penetration of biofuels in transport in the Alternative Policy Scenario contributes to this decline in oil-import dependence. Gas imports fall by 4.8 bcm. Gas-import dependence falls by two percentage points to 53% in 2030. The decline in coal imports is even more marked, in proportionate terms. Dependence on imported coal in 2030 is 25% in the Alternative Policy Scenario, compared to 28% in the Reference Scenario, with actual imports 97 Mtce lower.

India would have to import far less coal in the Alternative Policy Scenario because of the large reduction in coal demand in the power generation sector. Three factors contribute to this trend: fuel-switching to renewables and nuclear power; lower electricity demand from more efficient end-use electrical appliances, motors and other equipment; and lower demand from more efficient coal-fired power plants. Compared with the Reference Scenario, steam coal imports are 49 Mtce less in 2030 in the Alternative Policy Scenario. The decline in coking coal imports, owed to efficiency improvements in the production of iron and steel, accounts for 48 Mtce by 2030.

Figure 18.2: **India's Fossil Fuel Imports in the Reference and Alternative Policy Scenarios in 2030**

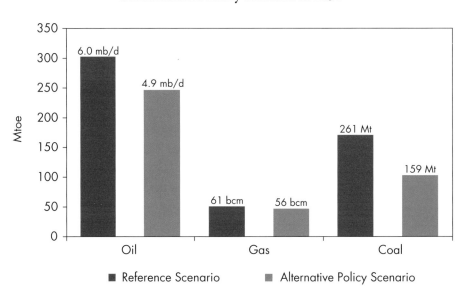

Environmental Implications

Local Pollution

Emissions of local pollutants are substantially lower in the Alternative Policy Scenario (Figure 18.3). The largest decrease is in SO_2 emissions, which are 27% lower in 2030, compared with the Reference Scenario. Emissions fall faster than coal use – the main source of SO_2 – as a result of equipping more plants with sulphur scrubbers. Emissions fall also because of greater use of ultra-low-sulphur diesel in transport. Total SO_2 emissions reach 12 million tonnes in 2030, compared with 16.5 million tonnes in the Reference Scenario. NO_x emissions are 23% lower. Most of the decrease comes from the power and transport sectors. $PM_{2.5}$ emissions fall slightly faster in the Alternative Policy Scenario than in the Reference Scenario, mainly in the residential and power sectors.

Figure 18.3: **Local Air Pollutant Emissions in India in the Reference and Alternative Policy Scenarios**

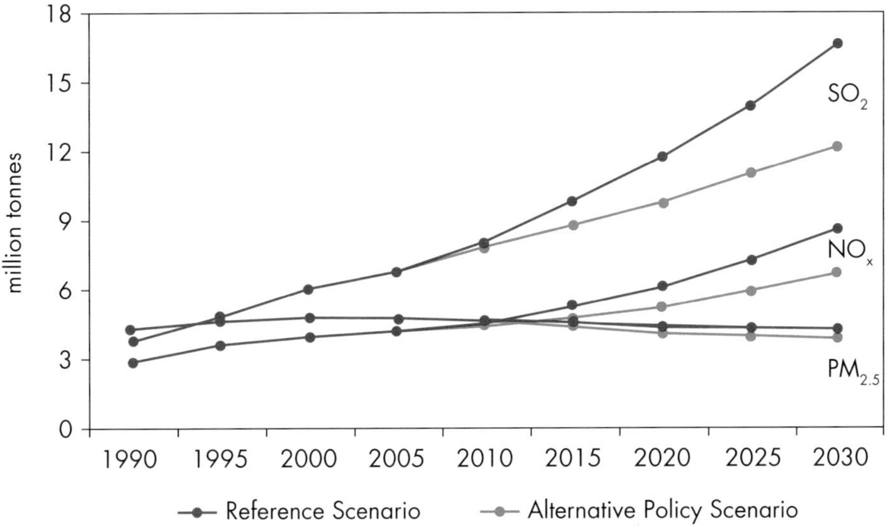

Energy-Related CO_2 Emissions

The policies and measures analysed in the Alternative Policy Scenario significantly curb the growth of India's energy-related CO_2 emissions. Lower overall energy consumption, combined with a larger share of less carbon-intensive fuels in the primary energy mix, yields savings of 27% in emissions by 2030, compared with the Reference Scenario. The savings in emissions are already marked by 2015, when they reach 11%. Nevertheless, emissions continue to rise, reaching 1.6 Gt in 2015 and 2.4 Gt in 2030, compared with 1.1 Gt in 2005. Per-capita emissions, which are 2.3 tonnes in 2030 in the Reference Scenario, fall to 1.7 tonnes.

The largest single contribution to emission savings comes from greater energy efficiency. Overall, energy-efficiency improvements on both the demand and supply sides and fuel-switching in power generation account for 72% of the total savings (Figure 18.4). Renewable energy, notably hydropower in the power sector and biofuels in transport, contributes 24% to the overall reduction in emissions. Nuclear power contributes 5%.

The biggest reductions are in the power sector, where emissions are lower by a third, as a result of policies to encourage low-carbon generation and to use coal more efficiently, and because demand for electricity is lower. The CO_2 intensity of electricity generation declines even further, compared with the Reference Scenario (Figure 18.5). At 550 g/kWh, it is 18% lower in 2030 and 42% lower than in 2005. Transport emissions fall by 22%, reflecting greater vehicle efficiency and greater use of alternative fuels. Emissions in the residential, services and agriculture sectors are 13% lower in 2030, compared with the Reference Scenario.

18

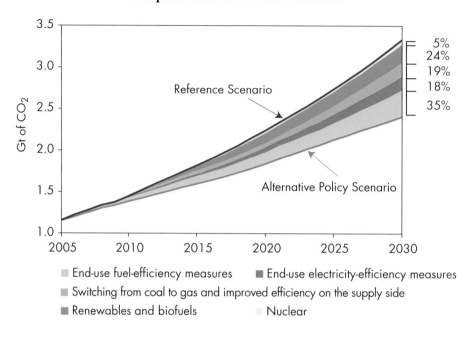

Figure 18.4: **India's CO$_2$ Emissions in the Alternative Policy Scenario Compared with the Reference Scenario**

Legend:
- ■ End-use fuel-efficiency measures
- ■ End-use electricity-efficiency measures
- ■ Switching from coal to gas and improved efficiency on the supply side
- ■ Renewables and biofuels
- ■ Nuclear

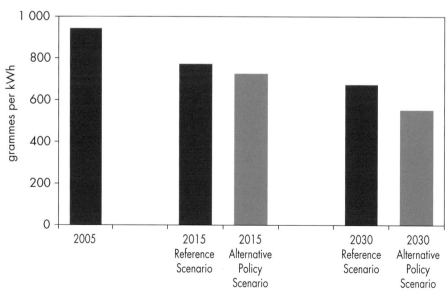

Figure 18.5: **CO$_2$ Intensity of India's Electricity Generation in the Reference and Alternative Policy Scenarios**

Box 18.1: **The Impacts of Climate Change in India**

The UNFCCC Intergovernmental Panel on Climate Change predicts an increase in the future in the magnitude of extreme climate events, including droughts, floods and cyclones. India, with its low lying coastline and economy tied to agriculture, is vulnerable to the impacts of these events. One-quarter of the Indian population live along the country's coasts and climate change can impact coastal areas through an increase in mean sea level and through increased frequency and intensity of coastal surges and storms. In India at large, water resources will be greatly altered by climate change, affecting fresh water drinking supplies, rain-fed agriculture, groundwater supply, forestry and biodiversity. Increased variability in summer monsoon precipitation will have an adverse impact on the large part of the rural population which depends on rain-fed agriculture for its livelihood. Irreversible damage could be caused to some forestry resources, rendering several species extinct and affecting markets, water supply and energy production. The main impacts on industry, energy and transport are expected to relate largely to transport and distribution systems, machinery, power plants and water and wastewater systems. Temperature increases in India can affect the stability and strength of building materials, while rainfall increases can cause water-logging and erosion, leading to increased maintenance costs. Changes at sea level can cause land erosion and flooding which would increase infrastructure maintenance costs. Climate change could increase the incidence of malaria in areas in India that are already malaria-prone and introduce malaria into new areas.[2]

The potentially serious effects of climate change suggest a strong need to introduce a strategy in India to respond to the prospect of future climate variability and change, including extreme climatic events, and deal with the adverse consequences. The Energy and Resources Institute in India has prepared such a strategy in co-operation with the World Bank. The policy recommendations include enhancing the role of local authorities in reducing the vulnerability of infrastructure to droughts and floods; developing non-farm opportunities to diversify incomes in rural households; converting rain-fed land to irrigated land and expanding credit and insurance networks in rural areas. (TERI, 2007).

18

---

2. See the report by UK Department of Environment, Food and Rural Affairs and the Indian Ministry of Environment and Forests (available at www.defra.gov.uk) for more information.

Results by Sector
Power Generation
Policy Assumptions and Effects

Over a third of India's total energy use and about half of its fossil-energy use is for power generation. Under current policies, India's power sector will remain heavily dependent on fossil fuels. There is considerable scope to save electricity and to make the electricity sector more efficient by switching to cleaner technologies and relying less on fossil fuels. This will bring economic benefits, cut pollution and CO_2 emissions and reduce the burden of investment in power infrastructure, helping to liberate funds for more investment in rural electrification. The main policies included in the Alternative Policy Scenario are described in Table 18.2. For the most part, they involve more rigorous implementation of current policies and the strengthening of the current framework.

Table 18.2: **Key Policies in India's Power-Generation Sector in the Alternative Policy Scenario**

Policy/measure	Assumption
Integrated Energy Policy recommendation to increase coal plant efficiency from 30.5% to 39%	Two percentage points higher efficiency for new plant compared to Reference Scenario
Development of IGCC programme	More R&D, IGCC becomes available in 2020
Renovation of electricity networks, Accelerated Power Development and Reform Programme (APDRP)	Six percentage point decline in losses compared to Reference Scenario in 2030
R&M (renovation and modernisation) programme of power stations	One percentage point efficiency improvement of existing coal-fired power stations
Greater use of hydropower	Approaches full economic potential by 2030
New and Renewable Energy Policy Statement 2005 - Draft II, Rural Electricity Supply Technology (REST) Mission, Remote Village Electrification Programme (RVE)	Faster deployment of renewable energy technologies through incentives
Expand use of nuclear	24 GW by 2030

Summary of Results

Total electricity generation in the Alternative Policy Scenario reaches 2 305 TWh in 2030, 17% lower than in the Reference Scenario. Installed capacity is 45 GW lower. Measures to cut final electricity demand and to increase the efficiency of generation reduce the need for energy inputs. The fuel mix in generation changes (Figure 18.6). Coal remains the dominant fuel, but its share is much lower than in the Reference Scenario and compared with today. In the Reference Scenario, this share rises from 69% in 2005 to 71% in 2030. In the Alternative Policy Scenario it drops to 55%. This results mainly from policies to support renewable energy and nuclear power.

Figure 18.6: **India's Power Generation Fuel Mix in the Reference and Alternative Policy Scenarios**

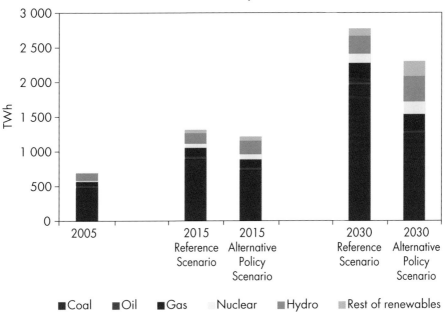

The average thermal efficiency of coal-fired generation increases from 27% in 2005 to 37% in 2030 in the Alternative Policy Scenario (Figure 18.7). The gap in efficiency between India and the OECD narrows, but in 2030 efficiency is still lower in India – by about four percentage points – for several reasons:

n The very low efficiency of existing power stations, most of which will still be operating in 2030 (Box 18.2). Their efficiency is assumed to improve through the implementation of renovation programmes, but the effect is marginal.

n The efficiency of the power plants expected to be built over the next ten years will be low by international standards on average (see Chapter 17). The lock-in effect of these plants will hinder overall efficiency improvements. The plants built after 2015 are also expected, on average, to be less efficient than those built in the OECD, despite some being built to world-class standards, as a result of greater private-sector participation.

n The quality of coal used in Indian power stations will continue to depress efficiency, though increased use of imported coal and coal washing could moderate this effect.

Box 18.2: Performance of India's Coal-Fired Power Plants

Historically, Indian coal-fired power stations have suffered from low plant load factors, low unit availability and low unit efficiencies relative to their counterparts in OECD countries. Even within the coal-fired sector, considerable differences in performance are evident between some of the units run by the State Electricity Boards (SEBs) and those run by the private sector and NTPC. NTPC's power plants operate at higher capacity factors and their performance deteriorates less rapidly, demonstrating that good operational practice and maintenance are essential for improved efficiency. The performance of the sub-200 MW units, which account for 20% of coal-fired generating capacity, has been identified by the Ministry of Power as particularly poor. The poor performance of the coal-fired power stations can be attributed largely to:

■ Lower-quality coal supply to the plants, relative to that specified during design, particularly involving higher ash content.

■ Design and manufacturing deficiencies, compounded by inadequate operation and maintenance regimes resulting in prolonged and repetitive forced outages; together with undue delay in implementing the renovation and modernisation programme for the ageing fleet.

■ Inadequate and untimely availability of spare parts, especially for the ageing stock of imported equipment.

■ Lack of properly trained manpower for the operation and maintenance of the plant.

The operators' lack of cash, caused by poor financial results, lies behind or aggravates these problems. The general performance of coal-fired units could be improved by enforcing a stricter coal quality control regime, in co-operation with the coal suppliers and the government; expanding the use of coal washing to lower the average ash content; introducing circulating fluidised bed combustion (CFBC) which can handle a wide variation of ash content, volatile matter and moisture content; promoting coal-blending wherever required in addition to coal quality control; mandating a timely renovation and modernisation regime of ageing coal plants, particularly those belonging to the SEBs to improve the average operating efficiency; offering regular training of the plant personnel with improved operation practices, particularly for the poorly performing SEB plants.[3]

In the longer term, consideration will need to be given to widespread replacement of older (*i.e.* more than 25 years) and smaller units, with larger state-of-the-art supercritical and ultra-supercritical units, as are currently pursued to a limited extent through the ultra-mega power projects.[4] The lessons learned from the operation of these large supercritical units under Indian conditions and using Indian coal should be rapidly disseminated to other utilities to promote wider uptake of advanced technologies.

India plans to develop its own supercritical plant technology as well as integrated gasification combined-cycle (IGCC) technology, both of which could contribute towards much greater efficiency. No CO_2 capture and storage (CCS) plants are assumed to be built in the Alternative Policy Scenario. However, there are several initiatives to develop this technology in India (Box 17.4 in Chapter 17).

Figure 18.7: **Average Coal-Fired Power Plant Efficiency in India Compared with the OECD in the Alternative Policy Scenario**

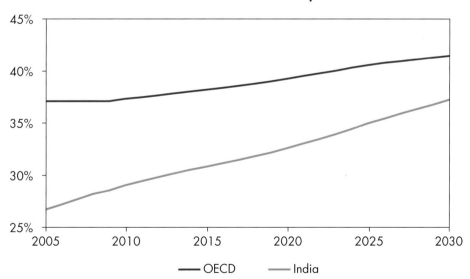

3. See IEA (2008, forthcoming).
4. Ultra-mega power projects are large-scale projects for electricity generation planned by the government of India, see Chapter 17.

Gas-fired power generation accounts for 11% of total generation in 2030 in both scenarios, but in the Alternative Policy Scenario is lower in absolute terms by 16%. This is because there is substitution of biomass for gas, notably in on-site production of electricity in industry.

Nuclear power capacity in the Alternative Policy Scenario reaches 24 GW in 2030, compared with 17 GW in the Reference Scenario. The share of nuclear power in total electricity generation rises from 2% in 2005 to 8% in 2030, compared with 5% in the Reference Scenario. The level of installed nuclear capacity in 2030 is well below the 40 GW targeted by the government, reflecting doubts about the speed at which India can increase its nuclear power capacity. How the recently announced USA-India nuclear co-operation agreement will unfold is a major uncertainty. India plans to build light water and fast breeder reactors.

Because of limited domestic uranium, but plentiful domestic thorium resources, India is developing fast breeder reactor (FBR) technology, but it is very uncertain whether India will be able to deploy fast breeders on a large scale before 2030. A number of countries in the OECD have experimented with the development of this technology, originally in the belief that uranium would be scarce and expensive. In practice, uranium supplies have been plentiful and FBRs are now considered as possibly being ripe for commercial deployment around 2050. While early prototypes suffered from a number of problems, they also demonstrated that large-scale FBRs are a practical proposition. The small number actually brought into operation achieved better availability, in some years, than some more established thermal reactor designs. Outside the OECD, Russia has the greatest experience with FBRs, including the BN600 reactor in Beloyarsk, which has operated relatively successfully for many years.

The FBR fuel cycle involves plutonium, which is usually extracted from spent nuclear fuel through reprocessing. Breeder reactors are expected to produce more plutonium than they actually consume (reprocessed from spent fuel and recycled into fresh fuel). The reprocessing of spent fuel and separation out of plutonium, however, raise proliferation concerns which have to be addressed through strict adherence to the Non-Proliferation Treaty (NPT), its additional protocol and any further safeguard agreements which may be entered into under this treaty. Another possible approach to proliferation risks includes putting sensitive parts of the fuel cycle under international control, under a new, multilateral framework for the nuclear fuel cycle.

India's technical hydropower potential is among the highest in the world (WEC, 2007). Estimated at 660 TWh annually, it is the second-largest potential in Asia, after China, and the sixth-largest in the world. In the early

1970s hydropower accounted for around 40% of India's electricity generation but, with very few plants built in the 1980s and 1990s, that share had fallen to 14% by 2005. India is now starting to refocus on hydropower and, in the Alternative Policy Scenario, it is assumed that India develops hydropower plants very quickly, exceeding the current 50 000 MW Hydro Initiative target. Installed hydropower capacity reaches 120 GW in 2030, approaching its full economic potential. This brings the share of hydropower in total electricity generation in 2030 to 16%, compared with 9% in the Reference Scenario.

Biomass and other renewables also grow considerably, contributing 9% of electricity generation in 2030. Biomass-based power rises to 12 GW in 2030, compared with 4.5 GW in the Reference Scenario in 2030 and less than 1 GW in 2005. Wind power capacity reaches 48 GW in 2030, including 6 GW of offshore wind. Solar-powered capacity expands considerably after 2020, reaching 9 GW in 2030.

Industry

Policy Assumptions and Effects

The stricter enforcement of recently enacted policies and the implementation of polices under consideration result in energy efficiency improvements in the industry sector in the Alternative Policy Scenario (Table 18.3). A relatively large share of industrial output comes from small-scale operations, with less efficient technology, often located in inner-city slums. In the Alternative Policy Scenario, inefficient kilns and boilers in the textiles and ceramics industries are assumed to be replaced or retrofitted sooner than in the Reference Scenario. Producers in the pulp and paper, textiles, chemical and fertilizer industries employ more combined heat and power (CHP).

The Steel Authority of India is encouraging several policies and measures which would improve operational efficiency and technologies to reduce energy consumption, such as improved casting techniques, the introduction of coal dust in blast furnaces, the use of sponge in blast furnaces, the recovery of waste heat from furnace-flue gases and the use of energy-saving equipment. The Energy Conservation Act 2001 provides for mandatory auditing in 15 energy-intensive industries. Audits are important because they reveal the potential for energy savings if the recommended energy efficiency measures are implemented. Energy conservation resulting from these audits is expected to be much greater in the Alternative Policy Scenario. Average efficiency of motors is 15% higher, for example, and cement production sees efficiency gains of 3% per year.

18

Table 18.3: **Key Policies in India's Industry Sector in the Alternative Policy Scenario**

Policy/measure	Assumption
National Steel Policy – aims to reduce costs and improve efficiency and productivity in the iron and steel sector	Efficiency improves by 15% over Reference Scenario
Greater use of CHP	Increased use of biomass potential in CHP
Higher efficiency processes in energy-intensive industries, particularly cement	Reduction in energy intensity of cement industry of 3% per year
Energy Conservation Act 2001	Stricter enforcement; increased efficiency of motors by 15%

Summary of Results

The Alternative Policy Scenario sees savings in industrial energy use of 5% in 2015 and 14% by 2030 relative to the Reference Scenario (Table 18.4). Industry represents 17% of the total energy savings in 2030 in this scenario. Coal accounts for most of the industry savings (Figure 18.8). Some 34% of the fall in coal consumption occurs in the iron and steel industry, thanks to more efficient blast furnaces. The iron and steel sector accounts for a third of all industry savings and non-metallic and other minerals, food, paper and textiles industries account for a further 24%. Cement production, in particular, is less energy-intensive than in the Reference Scenario (Box 18.3).

Table 18.4: **India's Industrial Energy Consumption and Savings in the Alternative Policy Scenario** (Mtoe)

	2005	2015	2030	Difference from the Reference Scenario in 2030	
				Mtoe	%
Coal	29	50	79	31.3	−28.3
Oil	19	25	34	3.8	−10.2
Gas	5	7	9	0.8	−7.8
Electricity	18	38	78	5.2	−6.2
Biomass	27	30	33	−2.9	9.7
Total	**99**	**149**	**234**	**37.8**	**−13.9**

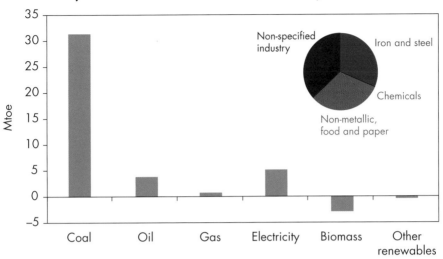

Figure 18.8: **India's Energy Savings in Industry and Shares of Savings by Sub-Sector in 2030 in the Alternative Policy Scenario**

Box 18.3: **Energy Savings in Cement Production**

In the Reference Scenario, cement production increases to 600 Mt in 2030 with a clinker/cement ratio of 0.83. Fuel use is assumed to average 3.5 gigajoules (GJ) per tonne of clinker and electricity use amounts to 110 kWh per tonne of cement produced. In the Alternative Policy Scenario, fuel use drops to 3.3 GJ/tonne of clinker and 100 kWh/tonne of cement, while the clinker/cement ratio declines to 0.7. As a result, coal use in cement production declines by 17%. The electricity savings amount to 9%.

The key factor in these savings is the reduced clinker/cement ratio. A reduction on the scale assumed here is only possible if most fly ash from coal-fired power plants is used, together with other new clinker substitutes such as steel slag. Today only part of the fly ash is of sufficient quality to be used. If the carbon content of the remaining fly ash can be reduced, all fly ash can be used in cement, as against the less-than-half assumed in the Reference Scenario. Other materials, such as steel slag, are also assumed to be used in the Alternative Policy Scenario. Reburning or electrostatic separation methods can be used to reduce their carbon content. Both methods are new and not yet widely applied worldwide.

Fuel use can be reduced by eliminating small-scale vertical kilns. However, this would involve transporting cement longer distances, as these kilns have been developed for smaller local limestone deposits where large-scale rotary kilns are not suitable. Another measure that is widely applied in Japan and China is power generation from clinker kiln waste heat. This option is not yet applied in India. This technology is attractive as it reduces the reliance on power from the grid. It can save 10 to 20 kWh/tonne of clinker.

Transport

Policy Assumptions and Effects

The growing problems of pollution and congestion in India's transport sector are widely recognised, but policy action has been limited in recent years. A summary of key policies recently enacted and under discussion is shown in Table 18.5. The 11th Five-Year Plan discusses many of the problems, including the rapid growth of car and two-wheeler fleets, a sharp decline in the share of public transport, infrastructure limitations and the need for fuel efficiency standards, without making concrete proposals for addressing them. Yet there are some success stories, notably the Supreme Court decision in 1998 mandating the conversion of all public buses and municipal vehicles in Delhi

Table 18.5: **Key Policies in India's Transport Sector in the Alternative Policy Scenario**

Measure	Description	Assumption
Fuel economy standards for LDVs	India has yet to enact fuel economy standards.	10% increase over all vehicles compared with Reference Scenario.
Vehicle emission standards	Following the European Vehicle Emission Standards (see Table 16.3 for details).	Impact on pollution and CO_2 emissions, secondary impact on fuel consumption.
Biofuels	5% ethanol blended gasoline was introduced in 9 states and 4 union territories in 2003, and was reintroduced and extended nationwide in 2006, although subject to availability.	Ethanol share in gasoline increases to 10% in 2012*. Biodiesel blending in diesel starts in 2009, increasing to 5% by 2015 and 8% share by 2018.
CNG	All commercial vehicles in Delhi, Mumbai and Kolkata run on CNG.	Doubling of CNG vehicles compared with Reference Scenario.
Public transport and infrastructure development	Construction of bus lanes and suburban and underground rail systems to ease road congestion.	5% increase in the number of buses (+200 000) compared with Reference Scenario in 2030.

* In September 2007 the Indian Agriculture Minister announced that the government would soon mandate an increase in the ethanol content in gasoline from 5% now to 10%.

to compressed natural gas (CNG), which has already led to a marked improvement in local air pollution.[5] The introduction of vehicle emission standards, based on those in the European Union, has also helped to curb fuel use and pollution. The government encourages the use of ethanol through a requirement for a 5% ethanol component in gasoline blends.

The Alternative Policy Scenario assumes faster vehicle fuel-efficiency improvements than in the Reference Scenario, resulting from increased co-operation with foreign manufactures and deployment in India of more advanced technology. Biofuels are also heavily promoted. Public transport systems are improved, particularly through an expansion of bus rapid transit systems and, in some cases, suburban rail, resulting in a 15% reduction in the use of cars and two- and three-wheelers. While there is already discussion of the possible penetration of hybrids, hydrogen vehicles and CTL in the Indian transport sector, a combination of the technological challenges, the lack of infrastructure and a lack of policy support delay their arrival until after 2030 in the Alternative Policy Scenario.

Urban rail services exist in only four Indian cities – Mumbai, Delhi, Kolkata and Chennai. Bus travel accounts for over 90% of public transport use in India (Pucher *et al.*, 2005). There are few proposals to improve bus systems. Buses could be suitable for Indian cities with large urban sprawl, but they may face space constraints. An increase in rapid-transit bus systems has been assumed in the Alternative Policy Scenario.

Summary of Results

The combination of policies considered in the Alternative Policy Scenario yields savings in India's transport energy demand of 8% by 2015 and 16% by 2030. These savings account for 12% of the total energy savings in the Alternative Policy Scenario and, in volumetric terms, are double the savings in transport energy demand in the entire OECD Pacific region. In percentage terms, they are almost double those achieved in the OECD's transport sector in the same scenario. Most of the savings come from road transport. Oil savings amount to 8 Mtoe in 2015 and 39 Mtoe in 2030, because of substitution by biofuels and CNG (Figure 18.9). Total oil use in the sector grows at a rate of 4.8% per annum in 2005-2030, compared to 6.1% per annum in the Reference Scenario. Natural gas use increases more quickly, by 11.4% per year as against 7.6% in the Reference Scenario, although this usage in the transport sector still accounts for only 11% of total Indian gas demand by 2030. A shift towards increased rail travel, both for urban passenger transportation and freight, boosts electricity consumption in transport in the Alternative Policy

18

5. A 32% reduction in carbon monoxide and a 39% reduction in sulphur-dioxide levels in 2002 compared with 1997 (Department of Environment, Government of NCT of Delhi & Delhi Pollution Control Committee, 2003).

Scenario by 28% more in 2030 than in the Reference Scenario. Biofuels, both ethanol and biodiesel, while starting from a very small base, are projected to increase by over 30% per year from 2005 to 2030, compared with 23% in the Reference Scenario.

Figure 18.9: **Road Transport Energy Use in India in the Reference and Alternative Policy Scenarios** (Mtoe)

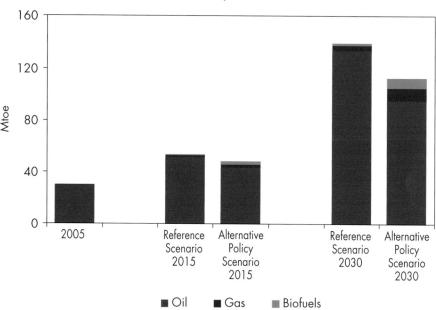

Oil savings in road transport arise from increases in efficiency, plus the faster uptake of advanced vehicle technology through local and foreign partnerships in the automobile industry (Ministry of Heavy Industries and Public Enterprises, 2006). Oil savings also result from a shift towards public transport and from greater use of CNG and biofuels. Biodiesel accounts for 69% of the biofuels in use in India by 2030 (close to the share of diesel in total road-transport fuel use), with ethanol blended into gasoline making up the rest.

There is considerable potential for expanding the production and use of biofuels in India, particularly biodiesel from jatropha. Jatropha is a plant that can grow on both good and poor or degraded soils and can produce a good number of seeds even with minimal water.[6] It is not grazed by animals and is highly pest- and disease-resistant. However, biodiesel is not currently consumed in India and there is only limited planting of jatropha. To supply the transport sector with a 5% biodiesel share by 2015, as projected in the Alternative Policy Scenario, 1.2 to 1.4 million hectares of jatropha would be needed; there are 0.4 million hectares

6. Estimates of biodiesel production range between 0.9 tonnes per hectare to 1.6 tonnes per hectare.

Delhi has encountered problems related to growing demand for mobility before the rest of the country, as it has high population density and has experienced rapid population growth. So efforts by the city to address congestion and pollution provide clues about which solutions might work in other cities. The last two decades have seen exponential growth in the vehicle fleet in Delhi, with the total number of registered vehicles recently passing 4 million, of which more than 90% are for personal use. This huge growth has resulted from increased wealth and inefficient and unreliable public transport. The increase in transport demand expected during the Commonwealth Games in 2010 is adding further impetus to efforts to tackle worsening congestion and pollution.

Several major policy initiatives have been implemented:

■ The entire public transport fleet, including buses, three-wheeler auto-rickshaws and taxis, has been converted to CNG.

■ Three lines of the Delhi metro rail system have been built.

■ Inter-state bus terminals, which facilitate inter-state transport and remove excess buses from the centre of Delhi, have been upgraded and expanded.

■ Priority lanes for buses have been introduced, together with preferential signalling for buses at intersections.

There were delays in implementing the Delhi CNG law in the first years after its passage, because of a lack of gas supply. Gas for power generation is given priority over CNG during supply disruptions. Growth in the use of CNG over the *Outlook* period will be held back not by demand, but by supply and infrastructure limitations.

planted to date (Singh, 2007). A 5% ethanol blend in gasoline was introduced in 2003 in nine states and four union territories. However as ethanol prices increased, partly because of a decline in sugar production in 2003/04, blending became temporarily optional, and fell to nearly zero. The 5% blend was re-imposed nationwide in 2006. Even though this target is mandatory, a lack of ethanol availability has meant that the 5% target has still not been met.

Residential and Services

Policy Assumptions

The tremendous increase in construction activity and the growing importance of electrical equipment and appliances are creating rapid energy-demand growth in the residential and services sectors. The Bureau of Energy Efficiency (BEE) has introduced policies which could lead to substantial reductions in energy demand over the *Outlook* period (Table 18.6). In 2007, BEE released a National Building Code for commercial and large residential buildings, with provisions for better building orientation, roof and wall insulation and the adoption of energy-efficient lighting and air-conditioning systems. BEE launched a National Energy Labelling Programme in 2006, requiring labelling for frost-free refrigerators and tubular fluorescent lamps. Labelling for other products, including direct cool refrigerators[7], air conditioners, electric motors and ceiling fans, will be introduced in a phased manner. In the Alternative

Table 18.6: **Key Policies in India's Residential and Services Sectors in the Alternative Policy Scenario**

Measure	Description	Assumption
Building codes & standards	Set minimum requirements for the energy-efficient design and construction of commercial buildings or complexes with electricity load of 500 kW or capacity of 600 kVA or more.	Greater building stock efficiency improvements.
Energy efficiency labelling	Mandatory labelling covers frost-free refrigerators and tubular fluorescent lamps. Labelling for other products will be introduced in a phased manner.	50% of all light bulbs are CFLs in 2030; average appliance efficiency is 30% higher in 2030.
Improved cookstoves (chulhas)	Installation of improved chulhas in rural and semi-urban households.	120 million improved cookstoves by 2030, scale-up of the pilot programmes.
Biogas	Promote family type biogas units for recycling of cattle dung to harness its fuel value without destroying manure value.	12 million biogas plants by 2030.
Solar devices	Construction of solar water heating systems, solar air heating/steam generating systems for community cooking.	Increased penetration of solar water heaters.

7. Direct cool refrigerators which require manual defrosting account for a large portion of overall refrigerator sales in India.

Policy Scenario, it is assumed that these standards and codes are introduced and more strictly enforced than in the Reference Scenario.

The Ministry of New and Renewable Energy (MNRE) has three main programmes to cut biomass consumption in rural areas: the Integrated Rural Energy Programme, the National Biogas and Manure Management Programme and the National Programme on Improved Chulhas.[8] This last programme currently covers 29% of the estimated potential of 120 million stoves.[9] The Biogas Programme promotes household biogas units for recycling cattle dung. The biogas digesters have the added benefit of improving sanitation in villages. In the Alternative Policy Scenario, the targeted potential for biogas plants and improved cookstoves is assumed to be met.

The MNRE is also promoting the development and application of renewable energy technologies, such as solar water heating systems for homes and commercial buildings and solar air heating/steam generating systems for community cooking and industrial applications. The overall solar potential in India is estimated to be 140 million m^2 of collector area, of which about 1.9 million m^2 has been installed.

Summary of Results

In the Alternative Policy Scenario, energy demand in the residential and services sectors is 12% lower in 2030 than in the Reference Scenario. Most of the savings come from biomass and electricity, the largest savings in percentage terms come from electricity and coal, at 15% each. The residential and services sectors account for 34% of the saving in total final energy consumption in 2030 and for 74% of the total savings in electricity. In the Alternative Policy Scenario, the implementation of the National Energy Labelling Programme, achieving substitution of the inefficient incandescent lamps by 60%-more efficient CFLs and greater penetration of efficient refrigerators, fans and air-conditioning systems, results in savings in electricity consumption of 17% in the residential sector and 20% in the services sector.

Fuel-switching away from biomass and the installation of improved cookstoves and biogas digesters in rural areas lowers biomass demand by almost 21 Mtoe, or 14%, compared with the Reference Scenario (Figure 18.10). Fossil-energy use in these sectors is reduced by 13% in 2030 as a result of more efficient buildings and the more widespread adoption of solar energy for water heating in buildings.

18

8. The National Programme on Improved Chulhas (a type of Indian cookstove) has been discontinued at the federal level and chulhas are currently subsidised at the state level or through individual targeted projects.
9. Numbers obtained from the MNRE website, http://mnes.nic.in, last accessed July 2007.

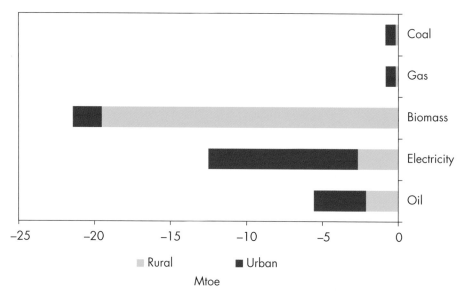

Figure 18.10: **Reduction in Final Energy Consumption in the Residential and Services Sectors in the Alternative Policy Scenario* by Fuel, 2005-2030**

* Compared with the Reference Scenario.

Cost-Effectiveness of Policies

The implementation of the policies considered in the Alternative Policy Scenario yields vastly different investment patterns, compared with the Reference Scenario. Cumulative savings of $78 billion are possible over the *Outlook* period, with total supply-side investment savings of $132 billion being offset by increased investment on the demand side of $54 billion (Figure 18.11). Supply-side savings come mainly from the power sector, where lower demand for electricity greatly reduces investment needs in power generation and networks.

The role of government is important, as investment requirements fall differently across the community. Greater investment in energy efficiency results in lower energy bills for individual consumers, but at the expense of higher initial funding. In the Alternative Policy Scenario, in the year 2030 alone each household saves $152 on average on transport fuel and $30 on running costs for domestic energy needs (mainly for lighting and cooking) (Figure 18.12). The total energy bill is lower by 23%.

Payback periods for end users in India tend to be long, as energy prices are heavily subsidised; so the time taken to recover the initial investment is artificially long. Reducing subsidies, so as to allow prices to reflect the true costs of electricity supply, would lead to shorter payback periods and encourage the

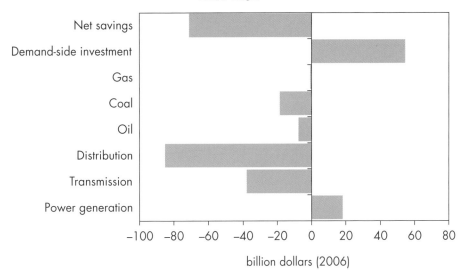

Figure 18.11: **Change in Investment in the Alternative Policy Scenario, 2006-2030**

billion dollars (2006)

Figure 18.12: **Annual Energy-Related Expenditure per Household**

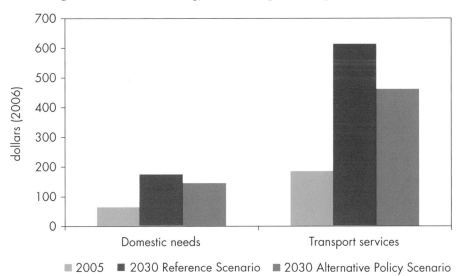

■ 2005   ■ 2030 Reference Scenario   ■ 2030 Alternative Policy Scenario

introduction of more efficient air conditioners, refrigerators and solar water heaters (Figure 18.13). The removal of electricity subsidies would see the average payback period for the household appliances analysed reduced by about a third. But the removal of subsidies would have to be accompanied by appropriate measures to protect the least well-off. Efforts to expand the use of meters and improve bill collection would also increase the cost-effectiveness of policies.

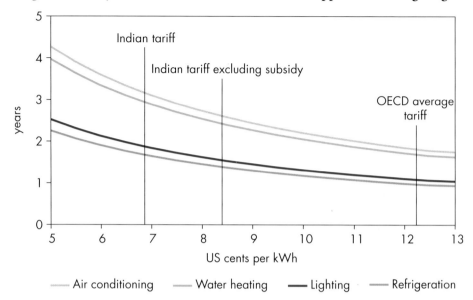

Figure 18.13: **Payback Periods for Various Household Appliances and Lighting**

The agriculture sector is a particularly striking example of the effects of subsidisation. The electricity tariff for agricultural users is very low, often less than 10% of the industrial tariff and even zero in many states. As a result, there is no incentive to invest in efficient equipment and electricity is extensively wasted. The introduction of even a small tariff would sharply diminish the payback period for an electrical pump (Figure 18.14). By contrast, electricity

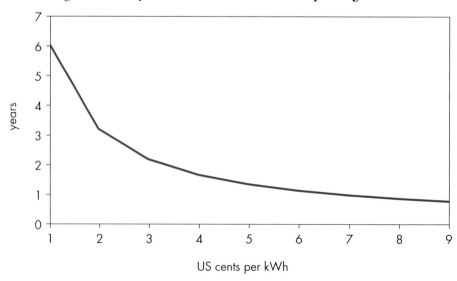

Figure 18.14: **Payback Period for Electrical Pumps in Agriculture**

tariffs in the services and industry sectors are already sufficient to produce very short payback periods. When a business chooses a more efficient light bulb or motor, it will recuperate its initially higher investment in the first year (Table 18.7). Since motors account for 70% of industrial electricity consumption, the replacement of standard motors with more efficient ones saves 42 TWh of electricity in the Alternative Policy Scenario in 2030.

Table 18.7: **Payback Periods for Lighting in the Services Sector and Motors in Industry**

	Services sector lighting		Industrial motors	
	Incandescent lamp	CFL	Standard	Efficient
Cost ($)	0.18	5.58	868	1 066
Annual electricity consumption (kWh)	131	39	60 000	57 000
Annual saving in electricity expenditure ($)		6.84		271
Payback period (years)		**0.8**		**0.73**

Because of lower demand in the Alternative Policy Scenario, India's fossil-fuel import bills are much lower. India's coal import bill will be $165 billion in 2006-2030, $73 billion less than in the Reference Scenario (Table 18.8). The oil import bill is about $28 billion lower in 2006-2015 in the Alternative Policy Scenario and $223 billion lower in 2016-2030, compared to the Reference Scenario. India's gas bill is also lower, by more than $7 billion in 2006-2030.

Table 18.8: **Cumulative Fossil-Fuel Import Bills in the Reference and Alternative Policy Scenarios** (in billion year-2006 dollars)

	Coal		Oil		Gas	
	2006 - 2015	2016- 2030	2006- 2015	2016- 2030	2006- 2015	2016- 2030
Reference Scenario	48	190	514	1 485	27.8	115
Alternative Policy Scenario	44	121	486	1 261	27.7	108
Savings*	**4**	**69**	**28**	**223**	**0.2**	**7**

* Compared with the Reference Scenario.

CHAPTER 19

HIGH GROWTH SCENARIO PROJECTIONS

HIGHLIGHTS

- India's economic outlook is very uncertain. One strong possibility is that the economy will grow faster than the rate assumed in the Reference Scenario. In the High Growth Scenario, GDP growth in India is assumed to average 7.8% per year in 2005-2030 compared with 6.3% in the Reference Scenario. GDP per capita is five times higher in 2030 in the High Growth Scenario than in 2005. Higher economic growth would require a substantial acceleration and deepening of structural, institutional and price reforms and more rapid infrastructure development.

- India's primary energy demand expands by 4.2% per year, compared to 3.6% in the Reference Scenario. Coal and oil demand accounts for most of the increase. Oil demand rises to 8.3 mb/d in 2030, 1.8 mb/d more than in the Reference Scenario, driven by a dramatic increase in the vehicle stock. Reliance on biomass is reduced, as higher incomes lead to greater use of cleaner, more efficient fuels. Electricity generation per capita rises to 2 400 kWh in 2030, slightly higher than in Turkey now.

- Net oil imports rise to 7.7 mb/d in 2030 and India becomes the third-largest oil importer, after the United States and China, soon after 2020. Because of the higher prices associated with higher demand, and greater import needs, India's cumulative fossil-fuel import bill in 2006-2030 is $1.1 trillion higher.

- Cumulative investment in energy infrastructure rises to $1.7 trillion in 2006-2030, compared with $1.25 trillion in the Reference Scenario. More than 70% of the increase is in power generation. Oil investments are $90 billion higher because of the deployment of enhanced recovery techniques. By contrast, gas investments increase by only 17%, to $74 billion, and $11 billion more is needed in coal infrastructure.

- CO_2 emissions are projected to rise to 3.9 Gt in 2030, an increase of 2.8 Gt above current levels. India's emissions exceed those of Japan before 2010 and those of Russia just before 2015. Per-capita emissions by 2030, however, are still low, at 2.7 tonnes. Greater reliance on fossil fuels results in higher emissions of SO_2 and NO_x, but emissions of particulate matter arising from burning biomass for cooking and heating decline.

- Faster economic growth accelerates the alleviation of energy poverty in the High Growth Scenario. All households in India have access to electricity by 2030, compared with an electrification rate of 96% in the Reference Scenario. But higher import bills and rising local pollution and CO_2 emissions highlight the need for strong and immediate policy action, as described in the Alternative Policy Scenario.

Background and Assumptions

There are wide-ranging views about the prospects for India's economic growth. We present here a High Growth Scenario[1] to illustrate the potential impact on energy demand and energy-related emissions of higher economic growth than that assumed in the Reference Scenario. For India's economy to grow faster, acceleration and deepening of structural, institutional and market reforms would be needed, accompanied by more rapid infrastructure development. Perhaps the greatest uncertainty surrounding the outlook for economic growth is whether or not India's government can command the necessary will and political consent to carry through the reforms needed to sustain high growth.

The critical reforms include a continued and sustained tightening of monetary policy by the Reserve Bank of India to keep inflation under control, a further reduction in fiscal deficits at the federal and state levels, and further deepening of capital markets and access to credit (Brookings Institute, 2007). Another important component is finding employment for the large and growing number of working-age people. This requires the creation of low-wage, low-value-added manufacturing jobs (unlike the current trend towards high-wage, high-value-added jobs in the services sector) and improvements in education levels. Institutional reforms include improving legal and tax administration and civil service reform. Subsidies need to be better targeted and properly financed so that commercial enterprises can generate funds for investments to meet more rapid growth in energy demand and infrastructure. Precision in the design and the use of subsidies is even more important in the High Growth Scenario, because oil and gas prices are higher. Infrastructure spending as a percentage of gross domestic product (GDP) needs to increase. India's currently weak infrastructure, particularly in the electricity and water sectors, is widely seen as one of the main obstacles holding the economy back from achieving even higher rates of growth. Investment in maintenance of the infrastructure is as important as investment in its expansion.

A key challenge for India, in this as in other scenarios, is to make growth more broad-based and inclusive. The rate of decline in poverty in India's poorer states has historically been much slower than in the richer states (Besley et al., 2006). While it is outside the scope of this analysis to determine the effect of higher economic growth on poverty, our analysis of access to electricity and clean cooking fuels confirms that higher growth would alleviate energy poverty more swiftly (see Chapter 20).

In the Reference Scenario, India's GDP grows by 7.2% per year on average in 2005-2015. It then slows towards the end of the projection period, averaging 5.8% per year in 2015-2030. In the High Growth Scenario, annual growth in 2005-2015 is 1.1 percentage points higher on average than in the Reference

---

1. The High Growth Scenario assumes the same set of policies as in the Reference Scenario.

Scenario and, in 2015-2030, 1.7 percentage points higher. This brings average growth over the entire projection period to 7.8% per year compared with 6.3% per year in the Reference Scenario. By 2030 GDP reaches a level 42% higher than in the Reference Scenario. It doubles by about 2014 in the High Growth Scenario, rather than 2015 in the Reference Scenario. The shares of the services and industry sectors in GDP increase by 2030 and the share of agriculture falls to 11%. Per-capita income increases five-fold over current levels.

Table 19.1: **Key Assumptions in the High Growth Scenario**
(average annual rate of growth)

	2005-2015	2015-2030	2005-2030
GDP	8.3	7.5	7.8
GDP per capita	6.8	6.5	6.6

Energy prices are higher in the High Growth Scenario, particularly for oil and gas, partly in response to an increase in global demand, led by China and India (see Introduction). The energy policy assumptions in the High Growth Scenario match those of the Reference Scenario, in that no new government policies and measures beyond those already enacted by mid-2007 are taken into account. Higher economic growth in India has significant consequences for energy demand and supply trends in India and in the rest of the world (see Chapter 1).

Energy Demand

In the High Growth Scenario, India's primary energy demand expands by about 180% between 2005 and 2030, an average annual increase of 4.2%, compared with 3.6% in the Reference Scenario. Demand grows slightly faster in 2015-2030, due to a burst in transport demand towards the end of the projection period. Demand in 2030 is 16% higher than in the Reference Scenario. Coal and oil demand account for most of the increase. Energy intensity falls faster in the High Growth Scenario, particularly towards the end of the *Outlook* period. By 2030, primary energy intensity is 18% lower than in the Reference Scenario.

Demand for coal in the High Growth Scenario grows by 5% per year – 0.5 percentage points faster than in the Reference Scenario. Nevertheless, in 2030 the share of coal in primary energy demand is 46%, 2 percentage points lower than that of the Reference Scenario (Table 19.2 and Figure 19.1). Oil demand rises to 4.1 mb/d in 2015 (0.4 mb/d more than in the Reference Scenario), and to 8.3 mb/d (1.8 mb/d more) in 2030. A big increase in the vehicle stock, which expands on average by 8.7% per year in 2005-2015 and 6% per year in 2015-2030, driven firstly by two-wheelers[2], then towards the end of the projection

2. Two-wheelers refers to two- and three-wheel vehicles.

period by passenger cars, is the main reason for the much stronger growth in oil demand. Natural gas demand is projected to grow briskly by 7.8% per year from 2005 to 2015, as gas replaces naphtha as a feedstock in the petrochemical industry, then to slow to 5.5% per year in 2015-2030.

Table 19.2: **India's Energy Demand in the High Growth Scenario** (Mtoe)

| | 2005 | 2015 | 2030 | 2005-2030* | Difference from the Reference Scenario in 2030 | |
					Mtoe	%
Coal	208	337	700	5.0%	79.9	12.9
Oil	129	204	416	4.8%	88.3	26.9
Gas	29	61	136	6.4%	43.2	46.7
Nuclear	5	17	40	9.2%	6.9	20.7
Hydro	9	14	24	4.1%	1.4	6.3
Biomass and waste	158	167	183	0.6%	–11.6	–6.0
Other renewables	1	5	10	12.3%	1.1	13.2
Total	**537**	**804**	**1 508**	**4.2%**	**209.2**	**16.1**

* Average annual rate of growth.

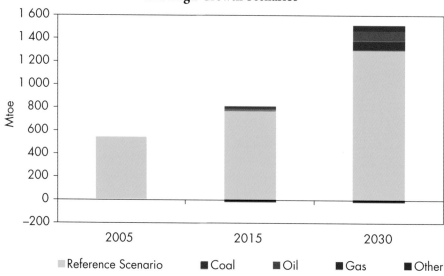

Figure 19.1: **India's Energy Demand in the Reference and High Growth Scenarios**

Note: Other includes nuclear, hydro, biomass, wind and solar power.

Biomass is the only fuel for which demand is lower in the High Growth Scenario than in the Reference Scenario, because of fuel-switching in the residential sector. Higher incomes allow households to afford more efficient, cleaner fuels for cooking and heating. The use of inefficient dung, fuel-wood and agricultural waste is much lower than in the Reference Scenario by 2030, particularly in rural households. Hydropower is 6% above its Reference Scenario level in 2030. Other renewables grow by over 12% per year on average, but they supply only 1% of primary energy supply by 2030. Nuclear power supply is 21% higher in the High Growth Scenario. With higher growth, India is assumed to speed up construction of planned reactors, under the expectation that financing for nuclear power projects will be more readily available.

Total final energy consumption in India in the High Growth Scenario grows at an average annual rate of 3.9%, compared with 3.3% in the Reference Scenario. The largest increase in demand comes from the transport sector, mostly because of the rate of increase in the vehicle stock (Box 19.1). The share of transport in total final consumption increases from 10% in 2005 to 25% in 2030 in the High Growth Scenario, 5 percentage points higher than in the Reference Scenario (Figure 19.2). In the High Growth Scenario, oil demand for transport grows by 7.6% per year over the *Outlook* period, 1.5 percentage points faster than in the Reference Scenario.

Figure 19.2: **India's Final Energy Demand by Sector in the Reference and High Growth Scenarios**

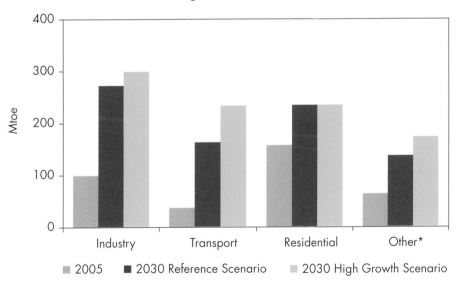

* Includes services, agriculture and non-energy use.

Box 19.1: **The Vehicle Stock in the High Growth Scenario**

The transport sector is responsible for 74% of the additional final oil demand in the High Growth Scenario. Higher incomes lead to faster growth in the total vehicle stock, which reaches 404 million in 2030 – 109 million more than in the Reference Scenario. Most of the additional vehicles are passenger cars, with only minimal stock growth for freight vehicles and buses, even though their consumption of fuel rises in line with passenger cars. Demand for two-wheelers levels out earlier than in the Reference Scenario, the rapid increase in total vehicles towards the end of the *Outlook* period being partly due to consumers leap-frogging directly to passenger cars. More vehicles on the road in the High Growth Scenario puts more strain on India's already congested roads and increases local pollution. The need to increase funding for public transport and for expanded and better roads is even more pressing than in the Reference Scenario.

Figure 19.3: **India's Vehicle Ownership and Stock in the Reference and High Growth Scenarios Compared with Selected Countries**

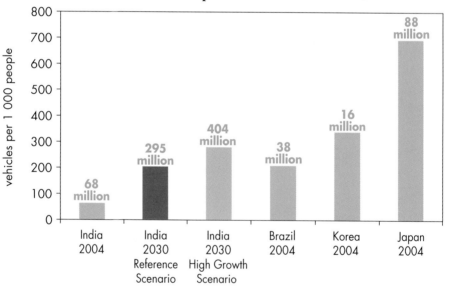

In the High Growth Scenario, industrial energy demand grows faster, by 4.5% per year over the *Outlook* period compared with 4.1% in the Reference Scenario. By 2030, it is 10% higher than in the Reference Scenario. Higher economic growth pushes up demand for steel and cement to build power plants, roads, ports and other infrastructure. Electricity demand in the industrial sector is much higher in the High Growth Scenario as the stronger economy boosts output in manufacturing, textiles and other small enterprises. The share of the residential sector in final energy demand declines to 25% in the High Growth Scenario, down from 44% in 2005. This dramatic decline is due to structural changes in the economy and changes in the fuel mix. There is a much faster penetration of liquefied petroleum gas (LPG) and electricity, and a decline in the use of fuelwood and dung (see below).

Implications for Energy Markets and Supply Security

Coal

Coal demand is 13% higher in 2030 in the High Growth Scenario, compared to the Reference Scenario. Domestic coal production provides about two-thirds of the additional demand for power generation, rising to 717 Mtce compared with 637 Mtce in the Reference Scenario, a rise in output of 4.1% per year in 2005-2030 in the High Growth Scenario, compared with 3.6% in the Reference Scenario (Figure 19.4). The logistical and economic challenges of achieving this increase in production are more pronounced than in the Reference Scenario (see Chapter 17). The negative environmental consequences would also increase, on the assumption of unchanged government policies. Hard-coal imports are 39 Mtce, or 16%, higher in the High Growth Scenario in 2030. India's dependence on imported coal is 28% in 2030 in the High Growth Scenario as well as in the Reference Scenario. But the import bill rises faster than import growth as the growing world demand puts additional pressure on international coal prices.

Figure 19.4: **India's Coal Supply in the Reference and High Growth Scenarios**

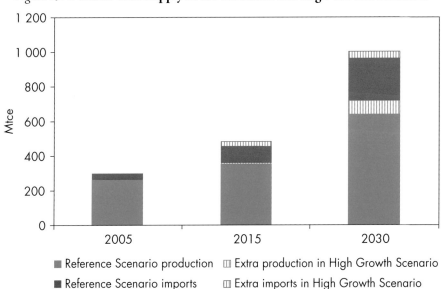

■ Reference Scenario production   ▥ Extra production in High Growth Scenario
■ Reference Scenario imports   ▥ Extra imports in High Growth Scenario

Oil and Gas

The production profile of existing and future fields in India is a major source of uncertainty. Enhanced oil recovery techniques have been applied to only 26 of the 136 fields currently producing, most of which have proven and probable reserves of more than 50 million barrels (Mb). With higher oil prices in the High Growth Scenario, enhanced recovery techniques could be profitably applied to more fields, even though these techniques are expensive for very small fields. We estimate that the application of such techniques to fields with proven and probable reserves as low as 20 Mb could add 120 thousand barrels per day (kb/d) to Indian production by 2030, two-thirds of which could come from the 12 largest existing fields (Table 19.3). It would significantly raise the recovery rate[3], which currently averages 24% at existing fields and 21% for fields where only primary recovery techniques are used. At fields where improved and enhanced recovery techniques are used, recovery rates today average close to 33%. Our analysis only covers the 12 largest producing fields. The potential for raising production from eight currently producing fields holding more than 50 Mb, which are not covered by our analysis, is not thought to be large. In practice, it is unlikely that it would be economic to deploy secondary and tertiary recovery at all fields. In addition, the application of enhanced recovery techniques in the higher price environment of the High Growth Scenario is based on the uncertain assumption that price reform in the oil and gas sector, by alleviating the burden of funding subsidies, allows the oil companies to invest more in upstream projects.

Table 19.3: **Incremental Oil Production from the Deployment of Enhanced Oil Recovery in the High Growth Scenario*** (kb/d)

	2010	2015	2030
Top 12 existing fields**	83	92	82
Identified non-producing fields	22	51	38
With reserves higher than 50 Mb (6 fields)	16	33	25
With reserves higher than 30 Mb (5 fields)	6	15	11
With reserves higher than 20 Mb (4 fields)	0	3	2
Total	**105**	**143**	**120**

* Relative to the Reference Scenario.
** See Table 17.3 for the names of the 12 largest producing fields.
Note: Production figures are based on the assumption that EOR techniques are applied to all fields, allowing recovery of an additional 10% of the oil originally in place.

3. The recovery rate is the percentage of the estimated volume of oil in a field that is actually recovered over the production life of the field.

Despite higher oil recovery rates, production still rises more slowly than demand in the High Growth Scenario, with India's dependence on oil imports growing even faster than in the Reference Scenario. Net oil imports reach 7.7 mb/d in 2030, 1.6 mb/d more than in the Reference Scenario. The rise is much more dramatic in 2015-2030 (Figure 19.5), with rapidly growing demand in the transport sector. India surpasses Japan to become the third-largest oil importer, after the United States and China, soon after 2020.

Figure 19.5: **Oil and Gas Net Imports in the Reference and High Growth Scenarios**

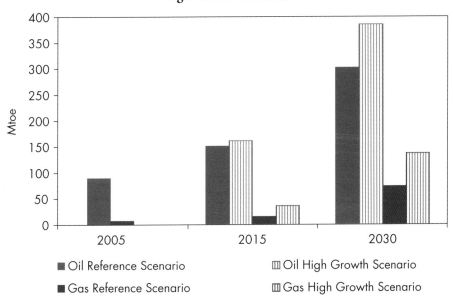

Gas production is projected to benefit slightly from the higher price environment and higher oil production. In 2030, it reaches 52.2 bcm – 1.5 bcm higher than in the Reference Scenario. An increase in production in smaller, more marginal fields is responsible for most of the increase in gas production in the High Growth Scenario. Because domestic demand rises even faster, gas imports are 51 bcm, or 83%, higher in 2030 than in the Reference Scenario. Gas import dependence reaches 68%, 13 percentage points higher. Because of recent gas discoveries in India and the expectation that similar discoveries will be made over the *Outlook* period, the projections here might underestimate the potential for gas production in India. However, any upside potential will only materialise if gas prices under future New Exploration Licensing Policy rounds are allowed to follow market trends, thus creating an environment where necessary investments in exploration and production will be forthcoming.

Electricity

Electricity generation in the High Growth Scenario reaches almost 3 500 TWh in 2030, growing at 6.6% per annum between 2005 and 2030. This is 25% more than in the Reference Scenario and almost five times the level of 2005. Installed capacity rises to 644 GW in 2030, up from 146 GW in 2005 and 522 GW in the Reference Scenario.

Per-capita electricity generation reaches 2 400 kWh, slightly higher than in Turkey now. Most of the additional demand for electricity is met from coal- and gas-fired power stations. Nuclear power and renewables capacity do not expand fast enough to match the faster growth in demand, so their shares in the power-generation fuel mix drop. A change in government policy would be needed to increase the share of these sources and reduce that of coal.

Energy Import Bills

Because of the higher prices in the High Growth Scenario, the cost of India's energy imports rises considerably by 2030. In total, India spends an additional $1.1 trillion, or almost 50%, over 2006-2030 on imported oil, coal and natural gas. Oil accounts for almost 75% of this increase. The gas import bill is over two-and-a-half times higher. The coal import bill sees the smallest increase among fossil fuels, $12.5 billion higher in 2006-2015 in the High Growth Scenario and $66.8 billion higher in 2016-2030, than in the Reference Scenario (Table 19.4).

Table 19.4: **Cumulative Fossil Fuel Import Bills in the Reference and High Growth Scenarios** (in billion year-2006 dollars)

	Coal		Oil		Gas	
	2006-2015	2016-2030	2006-2015	2016-2030	2006-2015	2016-2030
Reference Scenario	47.7	189.9	513.6	1 484.6	27.8	115.1
High Growth Scenario	60.2	256.7	568.8	2 261.9	45.3	317.3
Difference	12.5	66.8	55.2	777.3	17.4	202.2

Implications for Investment

To meet projected energy demand growth of 4.2% per year in 2006-2030, a cumulative total of $1.7 trillion (in year-2006 dollars), or some $66 billion a year, needs to be invested in India's energy-supply infrastructure in the High Growth Scenario. This is more than $400 billion, or 33% more than in the Reference Scenario. More than 70% of the increase is in power generation. For the energy sector as a whole, over three-quarters of investment will be needed for the power sector, where cumulative investment until 2030 is almost $1.3 trillion.

Total oil investment is projected to rise to $260 billion over 2006-2030, $90 billion more than in the Reference Scenario. Deploying enhanced recovery techniques to more fields results in upstream investment nearly tripling, from $39 billion to $105 billion. Investment in the refining sector, at $155 billion, is 19% higher. The additional investment requirements are due to higher domestic demand for oil products. Gas investment is projected to increase by only 17% compared with the Reference Scenario, to $74 billion. Much of the difference comes from the doubling of investment in LNG regasification terminals and from the expansion of transport and distribution infrastructure. Upstream gas investment accounts for 45% of total cumulative gas investments in 2005-2030. India needs to invest $11 billion more in coal infrastructure, almost entirely in new mines, in the High Growth Scenario. Cumulative coal investments are $68 billion in 2006-2030, or about $2.7 billion per year.

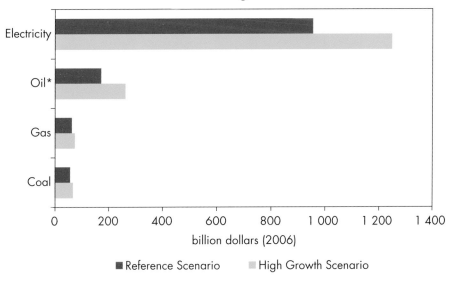

Figure 19.6: **Cumulative Investment in India's Energy-Supply Infrastructure in the Reference and High Growth Scenarios, 2006-2030**

*Investment in biofuels included in oil.

Environmental Implications

Energy-related CO_2 emissions in the High Growth Scenario are projected to rise to 1.9 Gt by 2015 and to 3.9 Gt in 2030, accounting for 9% of the world total. Emissions are 103 Mt (6%) higher than in the Reference Scenario in 2015 and 635 Mt (19%) higher in 2030. India is projected to emit more CO_2 than Japan before 2010 and more CO_2 than Russia just before 2015, becoming the world's third-largest emitter. India's per-capita emissions

19

in 2030 will still be low, at 2.7 tonnes, far below the average for OECD countries but higher than the 2.3 tonnes per capita in the Reference Scenario. Per-capita CO_2 emissions in India by 2030 will be less than the level in China, Japan and Russia today (Figure 19.7).

Figure 19.7: **Energy-Related CO_2 Emissions per Capita in the Reference and High Growth Scenarios Compared with Selected Countries**

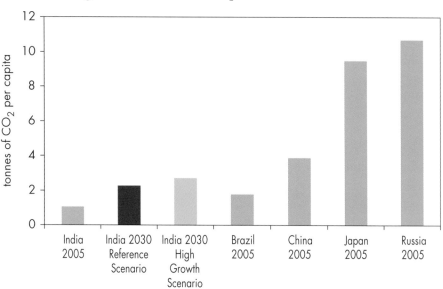

In the High Growth Scenario, greater reliance on coal, oil and gas compared with the Reference Scenario results in higher emissions of SO_2 and NO_x – on the assumption that no new government measures are introduced to control pollution. Emissions from transport rise most in percentage terms, causing air quality, especially in urban areas, to worsen. By contrast, emissions of particulate matter arising from burning biomass for cooking and heating decline more rapidly than in the Reference Scenario.

Implications for Access to Energy

Low incomes are the greatest constraint on expanding access to cleaner cooking fuels and electricity in India. In the High Growth Scenario, per-capita GDP in 2030 is 42% higher than in the Reference Scenario. This increase is assumed to be spread equally among rural and urban households. As a result, higher growth contributes substantially to alleviating energy poverty. Reliance on biomass is reduced sharply and consumption of cleaner, more efficient fuels is higher. Per-capita electricity demand rises to 2 400 kWh in 2030 in the High Growth Scenario, compared with some 1 900 kWh in the Reference Scenario.

In the High Growth Scenario all households in India have access to electricity in 2030 (see Chapter 20).

Higher incomes accelerate switching from biomass to LPG in rural and urban areas and, to a lesser extent, to natural gas in towns and cities. The share of biomass in residential energy demand declines from 79% in 2005 to 53% in the High Growth Scenario, six percentage points less than in the Reference Scenario. Most of the reduction in biomass consumption occurs in rural households, but consumption of LPG rises almost equally in both areas (Figure 19.8). Consumption of LPG grows by 5.4% per year on average and electricity demand grows by 7.6% per year over the *Outlook* period. Consumption of kerosene in 2030 is 10% lower in the High Growth Scenario, compared to the Reference Scenario, its use declining in both rural and urban areas. In urban areas, households switch mainly to LPG for cooking while, with expanded access to electricity, rural households rely less on kerosene for lighting.

Figure 19.8: **Change in India's Residential Energy Demand in the High Growth Scenario Relative to the Reference Scenario in 2030**

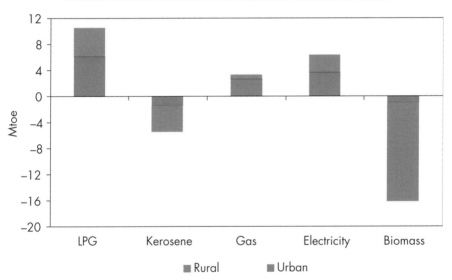

Note: Coal and renewables other than biomass are not shown as the change is negligible.

Implications for Policy

Higher economic growth would be a great benefit for India, bringing higher per-capita incomes and more funds for infrastructure investment. Access to modern energy, for example, is expected to improve much faster in the High Growth Scenario, improving the conditions of life for millions of people. On the other hand, higher economic growth would result in higher import bills and rising emissions of CO_2, NO_x and SO_2. Prompt government action could counter these worrying trends. Implementation of the policies and measures presented in the Alternative Policy Scenario would achieve substantial energy savings and emissions reductions (see Chapter 18). In practice, it will be a considerable political challenge to put India's energy system onto a more sustainable path, both in terms of establishing as a priority the need for action and winning consensus for vigorous implementation of the required measures. The possibility of stronger economic growth, however, demonstrates the urgency of taking advantage of the energy-security and environmental benefits of more rigorous policy action.

CHAPTER 20

FOCUS ON ENERGY POVERTY

HIGHLIGHTS

- Energy poverty affects many Indians and is an important issue for the Indian government. The number of households with access to electricity has risen over the past couple of decades, but access is still far from universal and the availability of modern cooking fuels and technologies is still limited, especially in rural areas. We use an energy development index, based on access to electricity and cleaner cooking fuels and on overall electricity generation per capita, to emphasise the disparity in energy poverty across India and relative to other developing countries.

- There are still some 412 million people without access to electricity in India. In all three *WEO* scenarios, the number of people without access declines, but it falls much faster in the High Growth Scenario. In that scenario, all households in India have access to electricity in 2030. In the Reference Scenario, the electrification rate in 2030 in India is 96% but nearly 60 million people in rural areas will still lack access.

- At an investment cost of $41 per person, it would cost some $17 billion to connect all those without electricity today to the central grid. But grid-based electrification is often not available to remote villages and households, because of the high cost of expanding the network. Diesel generators, mini-hydro, wind turbines, biomass gasifiers and photovoltaics, or a combination of these, could be more economic.

- The number of people relying on fuelwood and dung for cooking and heating declines from 668 million in 2005 to 395 million in 2030 in the High Growth Scenario, 77 million fewer people than in the Reference Scenario. About 22% of the population would still rely on these fuels in India in 2030, even with higher growth.

- According to the World Health Organization, the use of fuelwood and dung for cooking and heating causes over 400 000 premature deaths in India annually, mostly women and children. The concentration of particulate matter in the air in Indian households using biomass is over 2 000 microgrammes per cubic metre, compared to the US standard of 150.

- LPG and kerosene subsidies have been very ineffective in improving the welfare of the poor, particularly in rural areas. The current subsidy scheme benefits most richer households, mainly in urban areas, and has, for the most part, failed to shift fuel consumption patterns away from biomass in poor households. It is estimated that 40% of the subsidies for LPG and kerosene go to the richest 7% of the population.

Outlook for Clean Cooking Fuel and Electricity Access

Poor people in India have minimal access to clean, reliable and efficient energy sources. This is, of course, mostly a result of low incomes – India accounts for about one-third of the world's population living on less than a dollar a day. But there are other barriers to energy access in the poorest households in India, including unreliable energy service delivery, ineffective and regressive subsidies, gender discrimination in policy planning, inadequate information about the health impacts of current fuels and technologies, and administrative hurdles in getting connections.

Electricity access has improved in India – the electrification rate was 62% in 2005,[1] compared to 42% in 1991. From 499 million people with no access in 1991, the number fell to an estimated 412 million in 2005. In contrast, the number of people in India relying on fuelwood, dung and agricultural residues for cooking and heating rose, from 580 million in 1991 to 668 million in 2005.[2] India contains about one-third of all the people in developing countries who rely on inefficient, polluting fuels for cooking and heating. There are great disparities in energy access along the rural/urban divide. In rural areas, some 84% of households rely on fuelwood and dung for cooking, while only 22% do so in urban areas. Some 50% of people living in rural areas have access to electricity, compared to 90% in urban areas.[3]

When a household gains access to electricity, the normal first use is as a substitute for kerosene or biomass for lighting. To meet this basic electricity need is estimated to require 73 kWh per person per year.[4] Actual consumption will rise as per-capita incomes rise. In both the Reference and High Growth Scenarios, access to electricity and to cleaner, more efficient fuels for cooking and heating improves, but the improvements are much greater with higher economic growth (Table 20.1). In the Reference Scenario, the electrification rate in 2030 in India is 96% but nearly 60 million people in rural areas will still lack access. In the High Growth Scenario, all households in India have access to electricity in 2030.

1. This is an IEA estimate based on National Census data for India and information obtained from The Energy and Resources Institute.
2. The number of people relying on biomass has been revised since *WEO-2006*, on the basis of new data (see Box 16.4 in Chapter 16).
3. According to the 2001 Census of India, 44% of rural households and 88% of urban households had access to electricity.
4. The government's rural electrification scheme, Rajiv Gandhi Grameen Vidhyutikaran Yojana, sets minimum electricity consumption at 1 kWh per day per household, or 73 kWh per person per year (assuming five people per household).

Table 20.1: **Number of People in India without Access to Electricity and Relying on Biomass in the Reference and High Growth Scenarios** (million)

	2005*	2015		2030	
		Reference Scenario	High Growth Scenario	Reference Scenario	High Growth Scenario
Without electricity access - rural	380	274	102	59	0
Without electricity access - urban	32	2	0	0	0
Relying on biomass - rural	597	565	529	436	380
Relying on biomass - urban	71	67	51	36	15

* IEA estimate based on 2001 Census of India (www.censusindia.net), TERI for electricity access and NSSO (2007) for reliance on biomass.
Source: IEA analysis.

The number of people relying on biomass is 395 million in 2030 in the High Growth Scenario, 77 million fewer than in the Reference Scenario. But 22% of the population will still be relying on fuelwood and dung for cooking and heating in India in 2030, even with higher growth. This result highlights the urgency of implementing other strategies, such as improving kitchen ventilation and the efficiency of biomass cookstoves in poor households (Box 20.1).

Box 20.1: **Energy Efficiency of Cooking Fuels and Technologies in India**

The most important energy service in rural areas of India is cooking, but policy efforts have been largely unsuccessful at improving the efficiency and cleanliness of this basic service. Biomass is expected to remain the main cooking fuel in rural areas over the *Outlook* period, as it is the cheapest and most widely available fuel. The energy efficiency of biomass cookstoves is, however, very low compared with other fuel options: 8% with dung and agricultural residues and 9% with fuelwood using traditional stoves, compared to 25% with coal and charcoal, and 50% to 60% with natural gas, superior kerosene stoves and LPG.

India's Energy Advisory Board report, *Towards a Perspective on Energy Demand and Supply in India in 2004/05*, for the first time postulated a target for a minimum level of energy consumption; but the report projected no major change in the relative shares of cooking fuels in rural areas. The main

20

focus of government policy is to expand the use of improved cookstoves in rural areas. Improved stoves using traditional biomass can achieve efficiencies of 20% to 30%. In the absence of programmes to encourage switching to cleaner fuels, improved cookstoves are the most practical option for cutting smoke exposure, reducing fuel waste and lowering the burden of gathering fuel for large numbers of poor rural women and children. However, there are difficulties in encouraging households to switch to more efficient stoves, including affordability and lack of public awareness of the health impacts of burning biomass. The success of improved cookstove programmes in India has been impeded in the past by an absence of adequate training and support services, as well as a lack of market research to determine concerns of the women who would be using the stoves and their cooking habits. Another concern is the short life of the stoves, about one year for the most basic. There is wide recognition that a business model for scaling up improved cookstove programmes is required in India.

Measuring Energy Poverty

There are different approaches to measuring energy poverty (Pachauri *et al.*, 2004). We have chosen to devise an energy development index (EDI) to measure progress in the transition to cleaner cooking and heating fuels and the degree of maturity of energy end use. The index, which first appeared in *WEO-2004*, has been updated and modified to compare energy development among Indian states and union territories (UTs) and relative to other developing countries. For this *Outlook*'s EDI, we use three indicators: the share of households using cleaner, more efficient cooking and heating fuels (liquefied petroleum gas, kerosene, electricity and biogas); the share of households with access to electricity; and electricity consumption per capita. The third indicator is used to capture the level of overall energy development.

Figure 20.1 shows the change in India in two of the indicators between 1991 and 2005. The point of intersection of the two straight lines in the figure is the average share in 1991 in India of households relying on biomass, 77%, and the share of households with electricity access, 42%. The shares in each state and union territory are shown relative to this average. In Bihar, Assam, Orissa and Uttar Pradesh, access to both electricity and cleaner cooking fuels in 2005 was still below the average for India in 1991. In 2005, Tripura and Meghalaya were close to the 1991 Indian average for electricity access but still showed heavy reliance on biomass.

Figure 20.1: **Electricity Access and Reliance on Biomass in India**

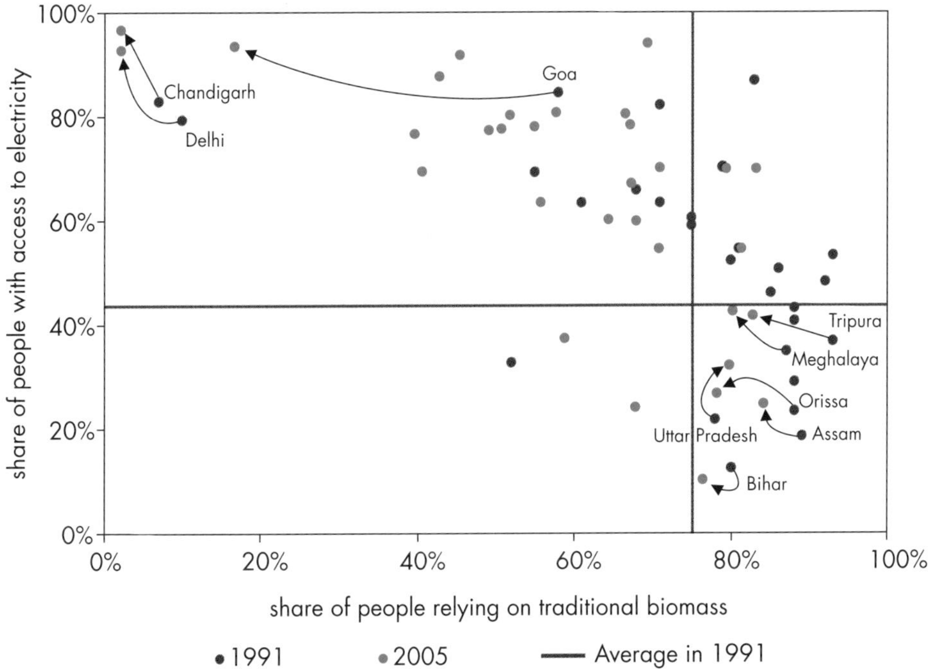

Chart axes:
- y-axis: share of people with access to electricity (0% to 100%)
- x-axis: share of people relying on traditional biomass (0% to 100%)

Labels on chart: Chandigarh, Delhi, Goa, Tripura, Meghalaya, Orissa, Uttar Pradesh, Assam, Bihar

Legend: • 1991   • 2005   —— Average in 1991

The union territories have a higher energy development index on average than the states (Table 20.2). This is because the territories are much more urban and have higher GDP per capita (see Figure 14.1 in Chapter 14).[5] Goa has the highest electricity generation per capita of the states. Himachal Pradesh has the highest level of electricity access of all the states, but because most households rely heavily on biomass for cooking and heating, its energy development index is 0.44. Access to cleaner cooking fuels is much lower in all states compared with most of the richer developing countries and the union territories. An interesting case is Rajasthan where, although electricity access is relatively high, heavy dependence on biomass for cooking puts its energy development index below those of Nicaragua, Indonesia and Nigeria. Assam and Bihar have the lowest energy development indices. Not only have the poorest states experienced the lowest rates of growth in gaining access to electricity and cleaner cooking fuels, they also continue to have the lowest level of household energy use per capita (Pachauri, 2007).

20

5. Although there has been a recent census on slum populations in India, energy access in slums is not included in the Indian census data for access to electricity by state and UT.

The government has focused many programmes and policies on expanding access to electricity. Except for energy subsidies, which have not had the desired effect of benefiting poorer households (see below), much less effort has been made to improve access to clean cooking fuels. The people living in Himachal Pradesh have about the same level of electricity access as Brazil but the clean cooking fuel index is about a third of Brazil's. Similarly, in Gujarat over 80% of households have access to electricity but dependence on biomass is equivalent to that of Senegal where about 30% of people have access to electricity. Bolivia has fewer households with access to electricity than ten Indian states with energy development indices which are lower overall. In most Indian states, household energy consumption patterns differ from other developing countries in that electricity access is relatively high but so is dependence on fuelwood and dung for cooking and heating.[6]

Table 20.2: **Energy Development Index***

Rank	State/UT/country	Clean cooking fuel index	Electricity access index	Electricity generation per capita index	EDI
	Malaysia	1.000	0.979	0.614	0.864
	Chile	0.894	0.978	0.594	0.822
	South Africa	0.778	0.645	1.000	0.808
	Brazil	0.874	0.951	0.383	0.736
1	Delhi (UT)	0.990	0.932	0.279	0.734
2	Goa	0.828	0.940	0.399	0.722
3	Chandigarh (UT)	0.991	0.976	0.153	0.707
	China	0.602	1.000	0.306	0.636
	Thailand	0.577	0.912	0.358	0.616
4	Puducherry (UT)	0.536	0.875	0.372	0.595
5	Punjab	0.507	0.921	0.221	0.550
6	Gujarat	0.436	0.793	0.231	0.487
7	Maharashtra	0.475	0.761	0.152	0.463
8	Sikkim	0.448	0.764	0.172	0.461
9	Andaman and Nicobar (UT)	0.572	0.753	0.051	0.459
	Bolivia	0.658	0.622	0.079	0.453
10	Haryana	0.370	0.821	0.165	0.452
11	Tamil Nadu	0.401	0.768	0.159	0.443
12	Himachal Pradesh	0.239	0.945	0.143	0.442
13	Mizoram	0.561	0.673	0.069	0.434
14	Jammu and Kashmir	0.271	0.795	0.101	0.389

6. Electricity access does not equate with regular supply (see Chapters 16 and 17).

Table 20.2: Energy Development Index* (continued)

Rank	State/UT/country	Clean cooking fuel index	Electricity access index	Electricity generation per capita index	EDI
15	Karnataka	0.264	0.772	0.110	0.382
16	Andhra Pradesh	0.262	0.646	0.130	0.346
17	Nagaland	0.392	0.606	0.020	0.339
18	Uttarakhand	0.295	0.569	0.109	0.324
19	Kerala	0.221	0.679	0.061	0.321
20	Madhya Pradesh	0.128	0.677	0.083	0.296
	INDIA	**0.265**	**0.519**	**0.102**	**0.295**
21	Chhattisgarh	0.085	0.677	0.115	0.292
22	Manipur	0.255	0.566	0.035	0.285
	Nicaragua	0.320	0.416	0.085	0.274
	Indonesia	0.216	0.482	0.090	0.263
23	Arunachal Pradesh	0.223	0.507	0.058	0.263
24	West Bengal	0.358	0.315	0.064	0.246
	Nigeria	0.297	0.398	0.015	0.237
25	Rajasthan	0.106	0.507	0.096	0.236
	Senegal	0.434	0.247	0.025	0.235
26	Meghalaya	0.118	0.373	0.084	0.192
	Cameroon	0.141	0.351	0.034	0.175
27	Jharkhand	0.257	0.168	0.089	0.171
28	Tripura	0.090	0.363	0.052	0.168
	Ghana	0.013	0.438	0.038	0.163
29	Orissa	0.141	0.197	0.125	0.154
30	Uttar Pradesh	0.129	0.253	0.044	0.142
	Bangladesh	0.099	0.254	0.015	0.123
31	Assam	0.076	0.175	0.017	0.089
32	Bihar	0.161	0.012	0.000	0.058
	Tanzania	0.000	0.000	0.004	0.001

* To construct the EDI, a separate index was created for each indicator, using the actual maximum and minimum values for the countries covered. Performance is expressed as a value between 0 and 1, calculated using the following formula:

Dimension index = (actual value – minimum value)/(maximum value – minimum value).

The index is then calculated as the arithmetic average of the three values for each country. The *maximum values* are: per-capita electricity generation 5 375 GWh/capita (South Africa); share of biomass in residential energy demand: 91% (Tanzania); and electrification rate: 99% (China). The *minimum values* are per-capita electricity generation: 75 GWh/capita (Bihar); share of biomass in residential energy demand: 1% (Malaysia); and electrification rate: 9% (Tanzania).

Note: Table excludes the UTs of Dadra and Nagar Haveli, Daman and Diu, and Lakshadweep.

Source: IEA analysis.

20

Expanding Access to Electricity in India

The Electricity Act of 2003 obliges utilities to supply electricity to all areas of India, including villages. The act set out a two-pronged approach, encompassing grid extension and distributed generation. The Integrated Energy Policy 2006 also sets out an objective to provide electricity for all people in India. The greatest challenge is electrification of rural households, especially in remote villages.[7] In 2005, the Ministry of Power introduced the Rajiv Gandhi Grameen Vidhyutikaran Yojana (RGGVY) scheme, which aims to provide electricity to all villages and to all rural households by 2009/10.[8] Before a recent revision, a village was considered electrified in India if electricity was used for any purpose within its boundaries. Under a new definition, a village is considered electrified if 10% of all households have electricity. Other requirements contained in the revision include a provision for distribution transformers and lines to be made available in each village and for power to be available on demand in schools, village council offices, health centres and community centres. The RGGVY scheme is also intended to support electricity for agriculture and for small and medium-sized industries. This would facilitate overall rural development, employment creation and poverty alleviation.

According to the Ministry of Power, using their latest definition (see above), 10% of the households in all of the villages in Andhra Pradesh, Haryana, Maharashtra, Kerala, Punjab, Tamil Nadu, Nagaland and Goa have been electrified. These villages represent 18% of the total population covered under the RGGVY scheme. In contrast, less than 80% of villages in Bihar, Jharkhand, Assam, Orissa, Uttar Pradesh and West Bengal have been electrified. An estimated 380 million people in rural areas did not have access to electricity in 2005. The remoteness of villages and weak infrastructure in poorer states make electrification expensive. As a result, the government offers support for distributed generation systems through a subsidy equivalent to 90% of capital expenditure and soft loans available from the Indian Renewable Energy Development Agency.

Table 20.3 provides an estimate of the cost of providing access to electricity to the 412 million people without access in 2005. The costs are broken down for central grid, mini-grid and off-grid technologies. Off-grid diesel generators have the lowest investment cost per kW, followed by mini-grid hydro-based electrification. However, generators entail an additional annual expense for diesel of about $20 per person. Central-grid electrification involves an additional annual expense of $4 per person.

---

7. There is no specific plan in place for slum electrification.
8. Information about the scheme is available at http://powermin.nic.in. For a comprehensive review of this and other rural electrification schemes in India, see Modi (2005).

Table 20.3: **Costs of Electrifying Households in India**

	Central grid	Mini-grid hydro	Mini-grid biomass gasifier	Mini-grid wind power	Off-grid diesel	Off-grid photo-voltaic
kWh per household per month	30	30	30	30	30	30
kWh per person per year	73	73	73	73	73	73
Installed kW per person	0.02	0.03	0.05	0.04	0.04	0.04
Investment cost per kW ($)*	2 300	1 150	1 200	3 500	700	10 000
Investment cost per person ($)	41	31	57	133	29	417

*Including investment in transmission and distribution.
Note: Wind power and photovoltaic systems are assumed to include batteries for storage.
Source: IEA analysis.

Providing access to electricity for all 412 million people who do not yet have it would require a mix of central grid, mini-grid and off-grid options. Assuming that the investment cost per person is in the range of $40 to $60, the total investment cost would be between $16 billion and $25 billion. This is very low compared with the total power infrastructure investment needs of almost a trillion dollars in the Reference Scenario over the *Outlook* period.

Mini-grids can serve as a node from which schools, health clinics and other public facilities can draw power. One of the main challenges facing these technologies is maintenance. Because mini-grid hydro schemes are often located in remote terrain, providing them with a steady stream of parts and skilled labour is often difficult. Solar PV technology has been successfully employed for lighting at the rural household level in India. The most common type of solar technology promoted at the rural level, however, does not have the capacity to meet the high-load mechanical applications required for agricultural processing, thus its benefits for rural development are limited. Modern biomass systems, in particular small-scale biomass gasifiers, have been successfully used for remote electricity generation in India (Aßmann *et al.*, 2006). The choice of distributed generation technology will depend largely on local conditions.

Health and Energy Poverty

There is a strong correlation between disease, such as chronic bronchitis, tuberculosis, cataracts and acute respiratory infection (ARI), and exposure to indoor air pollution (IAP) from burning biomass fuels on unventilated, inefficient

20

stoves. According to the World Health Organization, the use of biomass for cooking and heating causes over 400 000 premature deaths per year in India (Figure 20.2). The number of such deaths in India each year is equivalent to the population of Luxembourg. Most of the premature deaths are women and children. Women, who are traditionally responsible for cooking, and children suffer most from indoor air pollution because they spend many hours by the cooking fire.

The distribution of particulate matter (PM) in Indian households using biomass is over 2 000 microgrammes per cubic metre ($\mu g/m^3$) (Smith, 2000).[9] This compares with the 150 $\mu g/m^3$ standard set by the US Environmental Protection Agency for good health. During the cooking period, levels in India are much higher and in densely populated communities, high emissions from biomass burning can result in elevated local pollution. Acute respiratory infections make up about one-ninth of the national disease burden in India and are one of the main causes of death in children under five years of age. Such infections in India are the largest single disease category in the world, accounting for 2.5% of the global burden of ill health (WHO, 2007).

Figure 20.2: **Annual Average Premature Deaths from Indoor Air Pollution**

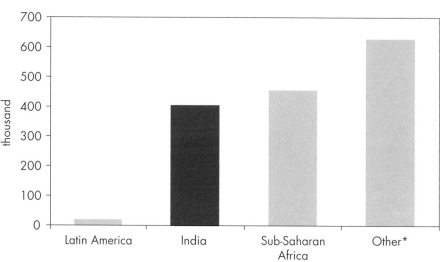

* Includes all other countries except OECD countries, for which WHO does not assess the burden on health from ARI because less than 5% of the population uses solid fuels.
Note: Includes premature deaths among people using coal and biomass for cooking and heating.
Source: World Health Organization (2007).

9. Smith measured particulate matter (PM_{10}) 24-hour concentrations.

Blindness is also more prevalent among people living in households that use biomass than among those living in households using cleaner fuels. The effects are large and statistically significant for both men and women and for urban and rural areas. Some 18% of partial and complete blindness among persons aged 30 and older can be attributed to biomass fuel use in India (Mishra *et al.*, 1999).

Box 20.2: **Energy Poverty and Gender**

Gender issues are attaining increasing prominence in the debate in international forums on sustainable energy development. Women are much more likely to be affected by energy poverty than men. Women suffer most from the negative health impacts of cooking with biomass and from having to walk long distances to collect wood or dung for fuel. In Orissa, for example, women walk up to nine kilometres a day, carrying a load of 35 kilogrammes of fuelwood, to earn Rs 15 ($0.37) from the local sale of fuelwood.

These issues are often discussed as a separate topic rather than integrated with strategies and solutions to energy poverty. In India, the traditional approach to energy in development policy took no account of gender. The situation is changing now. The Planning Commission's Integrated Energy Policy 2006 gives some consideration to gender issues, emphasising that energy subsidies, if properly implemented, could relieve drudgery, reduce health impacts and increase productivity (Planning Commission, 2006). As the major users of biomass resources, women have practical knowledge of how different fuels burn, efficient fire management and fuel-saving techniques. Households in which women are the head are much more likely to use LPG or kerosene for cooking. But it is not just in the use of biomass resources that women have expertise. Women can become managers of fuelwood or oil-seed plantations, retailers of kerosene or LPG, manufacturers of cookstoves or managers of electricity distribution and billing.

The Indian Renewable Energy Development Agency in the Ministry of New and Renewable Energy has been tasked with increasing the participation of poor rural women in integrated approaches to cooking and health. Equal access to credit and training is essential to ensure that clean energy and electricity supplies are available for women's domestic tasks and micro-enterprise activities. Micro-credit for women is still less widely available than in Bangladesh and Nepal, for example. The Indian Working Women's Forum is trying to change this and disbursed Rs 138 million ($3.4 million) to women entrepreneurs in its 14 branches in Tamil Nadu, Karnataka and Andhra Pradesh between April 2005 and March 2006.

20

Subsidies on Kerosene and LPG, and the Poor

Kerosene and LPG are heavily subsidised in India, with the intended aim of shifting fuel consumption patterns away from biomass to cleaner, more efficient fuels. Since it is mostly poor households that rely on biomass and live in rural areas, the subsidies were designed to support energy access for the poor.[10] However, in practice, this objective has not been met. The current subsidy scheme gives greater benefit to the urban sector and richer households and has for the most part failed to shift fuel consumption patterns away from biomass in rural areas. The Energy and Resources Institute in India estimates that 40% of the subsidies for LPG and kerosene benefits the richest 7% of the population (Misra *et al.*, 2005). In 2006, the Committee on Pricing and Taxation of Petroleum Products reported that restricting the kerosene subsidy to households below the poverty line would reduce the quantity of subsidised kerosene by 40% (Rangarajan, 2006).

In per-capita terms, urban areas consume 20% more subsidised kerosene than rural areas (Gangopadhyay *et al.*, 2005). As the per-unit subsidy is largely the same across sectors, this means that urban areas receive more subsidy than rural areas in per-capita terms. Per-capita purchases of public distribution system (PDS) kerosene rise in line with expenditure in general in rural areas, making the rural subsidy regressive. In the urban sector, per-capita purchases of PDS kerosene peak in the middle-income groups and then slowly decline until they fall off sharply among the wealthiest group. This is largely because the higher expenditure groups in urban areas have shifted out of kerosene to LPG.

LPG and kerosene subsidies have been very ineffective in improving the welfare of the poor, particularly in rural areas.[11] There are, however, new schemes under consideration in India, including a system of energy debit cards which could be issued to the targeted households with a monthly expense limit (Planning Commission, 2006). The debit cards could be used to procure LPG cylinders without paying cash. The system would target subsidies directly to the poor. Stronger and better-targeted policies are needed for cleaner cooking fuels to reach the poorest households. Subsidies on the technology instead of the fuel have been applied in Andhra Pradesh (Box 20.3).

---

10. See Box 16.5 in Chapter 16 for an explanation of the kerosene subsidy scheme.
11. Kerosene subsidies have benefited the urban poor, although part of the reason for this is that richer households often give their allotment to their poor servants (Barnes *et al.*, 2005).

Box 20.3: **Deepam LPG Scheme in Andhra Pradesh**

The Deepam Scheme, launched in 1998 by the state government in Andhra Pradesh, provides LPG connections free of charge to poor households. About 1.4 million households have benefited. The oil companies who supply the refill canisters had expected demand for an average of eight to nine refills a year, but most households have been able to afford only two or three. The state government provides a subsidy of Rs 1 000 ($25) towards the connection, but does not subsidise the cost of a refill, which is around Rs 250 ($6) per cylinder. There are strong arguments for subsidising the connection to LPG rather than the fuel itself, but many households have reverted to using traditional biomass for cooking. The state government has introduced smaller, 5-kg LPG cylinders, which require a deposit of only Rs 500 ($12) and a refill cost of Rs 100 to Rs 150 ($2.50 to $3.70). It is hoped that the smaller cylinders will lead to more regular consumption of LPG by the poor, especially in rural areas, at a lower cost to the government for subsidies.

Other policy measures can promote switching to LPG, such as campaigns to enhance public awareness about the health benefits of reducing exposure to indoor air pollution from burning fuelwood, dung and agricultural residues in unventilated stoves. Cleaner cooking fuel schemes should focus first on those areas where the availability of free or cheap biomass is diminishing. This will concentrate limited financial resources on those households already motivated to seek alternatives.

The economic, social and environmental benefits of expanding access to clean cooking fuels are so large (UNDP, 2006) as to justify an integrated approach that cuts across all sectors. The challenge of scaling up successful pilot projects in India is huge, first involving systematic evaluation of advantages to identify the most successful and then widespread communication of the results. More efforts are necessary to delegate to local governments, local communities and women the responsibility for delivering energy services.

Energy Demand in Slums

Slum areas in India's major cities are growing. Despite this, there is no specific federal programme in place for extending energy access to the urban poor. The first-ever census of slums in India was undertaken in 2001. Slum data were reported by 26 of the 35 states and UTs. Some 43 million people lived in slums in 2001, about equivalent to the entire population of Spain. Those in slums constituted 15% of the total urban population and 22.6% of the urban population of the reporting states and UTs. Some 11 million of the total slum population lived in Maharashtra, followed by Andhra Pradesh (5.2 million),

20

Uttar Pradesh (4.4 million) and West Bengal (4.1 million). The Greater Mumbai Municipal Corporation, with 6.5 million slum dwellers, had the highest slum population, followed by Delhi, Kolkata and Chennai. In Mumbai, more than 50% of the population lives in slums (Table 20.4). A survey by the National Sample Survey Organisation in 2002 recorded 52 000 slums in Mumbai. These contain only half the poor – the others live on the streets.

Very little information is available about energy use in slums. The 2001 Census indicates that almost all major slum areas have some electricity, but it is mostly used only for street lighting. Households in officially recognised slums are much more likely to have electricity connections, though most do not. A study by the United Kingdom's Department for International Development (DFID) reveals that infrastructure facilities in Delhi slums are sorely lacking; only 15% of slum households have drinking water, electricity and latrines in their premises (USAID, 2006). Less than 25% of them have sanitation systems. Access to modern energy services is undermined by imprecision over property rights or their total absence. Households without an address often cannot qualify for subsidised kerosene or electricity.

Despite low incomes, reliance on fuelwood for cooking is lower in urban slums than in rural areas because fuelwood, as a scarce resource, is often more expensive than subsidised kerosene.[12] Slum dwellers who are able to obtain a ration card for cooking fuels are likely to use kerosene. The priority uses of electricity are for lighting and for fans for cooling in hot seasons and for controlling mosquitoes.

According to the DFID study, about half of slum households in Delhi had illegal electricity connections. An illegal connection is not necessarily a free connection. Slum dwellers who depend on informal methods for acquiring electricity are vulnerable to exploitation by various middlemen, who charge high prices. It is not uncommon for slum dwellers to pay higher rates for electricity and water than most middle-class residents in their city. In Mumbai, pavement dwellers working with the National Slum Dwellers Federation paid Rs 300 ($7.50) a month for an illegal electricity connection.[13] For a cost of about Rs 900 ($22.40) per household, they could obtain a legal connection and a meter, with the result that their monthly costs would fall to about Rs 100 ($2.50). Willingness to pay for a legal connection is high among those with illegal connections.

Sound policy requires more knowledge of slum energy demand in India. The 2001 Census is a step in the right direction, but more information is needed.

---

12. Barnes *et al.*, (2005) confirmed that fuelwood use in Hyderabad is not an important household fuel, except in the poorest households, and even these use a significant amount of kerosene for cooking.
13. For more information see documentation from the National Slum Dwellers Federation, available at www.sparcindia.org.

Table 20.4: **Cities with More than One Million Inhabitants and Share of Population Living in Slums**

	Population (million)	Share in slums (%)		Population (million)	Share in slums (%)
Greater Mumbai, Maharashtra	12	54	Bhopal, Madhya Pradesh	1.4	9
Delhi	9.9	19	Ludhiana, Punjab	1.4	23
Kolkata, West Bengal	4.6	33	Patna, Bihar	1.4	0.3
Chennai, Tamil Nadu	4.3	19	Vadodara, Gujarat	1.3	14
Bengaluru, Karnataka	4.3	10	Agra, Uttar Pradesh	1.3	10
Hyderabad, Andhra Pradesh	3.6	17	Thane, Maharashtra	1.3	28
Ahmadabad, Gujarat	3.5	14	Kalyan-Dombivli, Maharashtra	1.2	3
Surat, Gujarat	2.4	21	Varanasi, Uttar Pradesh	1.1	13
Kanpur, Uttar Pradesh	2.6	14	Nashik, Maharashtra	1.1	13
Pune, Maharashtra	2.5	20	Meerut, Uttar Pradesh	1.1	44
Jaipur, Rajasthan	2.3	16	Faridabad, Haryana	1.1	47
Lucknow, Uttar Pradesh	2.2	8	Pimpri Chinchwad, Maharashtra	1	12
Nagpur, Maharashtra	2.1	36	Haora, West Bengal	1	12
Indore, Madhya Pradesh	1.5	18			

Source: Census of India (www.censusindia.net).

20

ANNEXES

TABLES FOR REFERENCE AND ALTERNATIVE POLICY SCENARIO PROJECTIONS

General Note to the Tables

The tables show projections of energy demand, electricity generation and energy-related CO_2 emissions for the following regions/countries: World, China, India, OECD, OECD North America, United States, OECD Pacific, Japan, OECD Europe, European Union, Transition economies, Russia, Developing countries, Developing Asia, Latin America, Middle East and Africa. The tables for China and India include power generation capacity data and the World tables include more detail for CO_2 emissions.

For OECD countries and non-OECD countries, the energy demand and CO_2 emissions data up to 2005 are based on IEA statistics, published in *Energy Balances of OECD Countries; Energy Balances of Non-OECD Countries;* and *CO_2 Emissions from Fuel Combustion.*

The definitions for regions, fuels and sectors can be found in Annex B.

Both in the text of this book and in the tables, rounding may cause some differences between the total and the sum of the individual components. Growth rates and changes compared with the Reference Scenario are marked "n.a." when the base year is zero or the value exceeds 200%.

Definitional Note to the Tables

Total primary energy demand is equivalent to *power generation* plus *other energy sector* (formerly called other transformation, own use and losses) excluding electricity and heat, plus *total final consumption* excluding electricity and heat. *Total primary energy demand* does not include ambient heat from heat pumps nor electricity trade. *Power generation* includes electricity and heat production by main activity producers and autoproducers. *Non-energy use* includes some non-specified energy use. The row *of which bunkers* refers to international marine bunkers. Total CO_2 emissions include emissions from *other energy sector*, as well as from *power generation*, and *total final consumption* (as shown in the tables). CO_2 emissions from international marine bunkers are included at the global level, while CO_2 emissions from international aviation are excluded from the tables. CO_2 emissions do not include emissions from industrial waste and non-renewable municipal waste.

Reference Scenario: World

	Energy demand (Mtoe)				Shares (%)			Growth (% p.a.)	
	1990	2005	2015	2030	2005	2015	2030	2005–2015	2005–2030
Total primary energy demand	**8 755**	**11 429**	**14 361**	**17 721**	**100**	**100**	**100**	**2.3**	**1.8**
Coal	2 216	2 892	3 988	4 994	25	28	28	3.3	2.2
Oil	3 216	4 000	4 720	5 585	35	33	32	1.7	1.3
of which bunkers	*114*	*171*	*188*	*217*	*1*	*1*	*1*	*1.0*	*1.0*
Gas	1 676	2 354	3 044	3 948	21	21	22	2.6	2.1
Nuclear	525	721	804	854	6	6	5	1.1	0.7
Hydro	184	251	327	416	2	2	2	2.7	2.0
Biomass and waste	903	1 149	1 334	1 615	10	9	9	1.5	1.4
Other renewables	35	61	145	308	1	1	2	9.0	6.7
Power generation	**2 983**	**4 261**	**5 615**	**7 225**	**100**	**100**	**100**	**2.8**	**2.1**
Coal	1 227	1 954	2 688	3 456	46	48	48	3.2	2.3
Oil	376	298	293	238	7	5	3	-0.2	-0.9
Gas	581	909	1 251	1 737	21	22	24	3.2	2.6
Nuclear	525	721	804	854	17	14	12	1.1	0.7
Hydro	184	251	327	416	6	6	6	2.7	2.0
Biomass and waste	58	77	130	275	2	2	4	5.4	5.2
Other renewables	32	51	123	249	1	2	3	9.1	6.5
Other energy sector	**851**	**1 144**	**1 462**	**1 859**	**100**	**100**	**100**	**2.5**	**2.0**
of which electricity	*181*	*275*	*370*	*488*	*24*	*25*	*26*	*3.0*	*2.3*
Total final consumption	**6 185**	**7 736**	**9 657**	**11 860**	**100**	**100**	**100**	**2.2**	**1.7**
Coal	762	660	893	982	9	9	8	3.1	1.6
Oil	2 503	3 259	3 930	4 767	42	41	40	1.9	1.5
Gas	950	1 226	1 527	1 876	16	16	16	2.2	1.7
Electricity	833	1 291	1 830	2 557	17	19	22	3.5	2.8
Heat	343	270	309	344	3	3	3	1.4	1.0
Biomass and waste	791	1 020	1 148	1 277	13	12	11	1.2	0.9
Other renewables	3	10	22	58	0	0	0	8.3	7.3
Industry and non–energy use	**2 197**	**2 834**	**3 765**	**4 576**	**100**	**100**	**100**	**2.9**	**1.9**
Coal	513	553	784	886	20	21	19	3.6	1.9
Oil	674	873	1 063	1 204	31	28	26	2.0	1.3
Gas	452	563	738	927	20	20	20	2.7	2.0
Electricity	384	549	830	1 140	19	22	25	4.2	3.0
Heat	53	111	125	130	4	3	3	1.2	0.6
Biomass and waste	120	184	224	287	7	6	6	2.0	1.8
Other renewables	0	0	1	3	0	0	0	8.0	7.8
Transport	**1 471**	**2 011**	**2 469**	**3 163**	**100**	**100**	**100**	**2.1**	**1.8**
Oil	1 378	1 895	2 296	2 919	94	93	92	1.9	1.7
Biofuels	6	19	57	102	1	2	3	11.6	7.0
Other fuels	87	96	117	142	5	5	4	1.9	1.6
Residential, services and agriculture	**2 516**	**2 892**	**3 423**	**4 122**	**100**	**100**	**100**	**1.7**	**1.4**
Coal	235	102	105	95	4	3	2	0.3	-0.3
Oil	452	491	571	645	17	17	16	1.5	1.1
Gas	444	592	702	842	20	21	20	1.7	1.4
Electricity	428	721	973	1 384	25	28	34	3.0	2.6
Heat	289	159	183	213	5	5	5	1.4	1.2
Biomass and waste	665	817	867	887	28	25	22	0.6	0.3
Other renewables	3	10	21	55	0	1	1	8.4	7.2

Reference Scenario: World

	Electricity generation (TWh)				Shares (%)			Growth (% p.a.)	
	1990	2005	2015	2030	2005	2015	2030	2005-2015	2005-2030
Total generation	**11 802**	**18 197**	**25 556**	**35 384**	**100**	**100**	**100**	**3.5**	**2.7**
Coal	4 429	7 334	11 081	15 796	40	43	45	4.2	3.1
Oil	1 324	1 186	1 166	929	7	5	3	-0.2	-1.0
Gas	1 725	3 585	5 334	8 068	20	21	23	4.1	3.3
Nuclear	2 013	2 771	3 083	3 275	15	12	9	1.1	0.7
Hydro	2 145	2 922	3 799	4 842	16	15	14	2.7	2.0
Biomass and waste	124	231	407	840	1	2	2	5.8	5.3
Wind	4	111	549	1 287	1	2	4	17.3	10.3
Geothermal	36	52	99	173	0	0	0	6.7	4.9
Solar	1	3	37	161	0	0	0	27.4	16.9
Tide and wave	1	1	1	12	0	0	0	9.9	12.8

	CO_2 emissions (Mt)				Shares (%)			Growth (% p.a.)	
	1990	2005	2015	2030	2005	2015	2030	2005-2015	2005-2030
Total CO_2 emissions	**20 688**	**26 620**	**34 071**	**41 905**	**100**	**100**	**100**	**2.5**	**1.8**
Coal	8 286	10 980	15 091	18 700	41	44	45	3.2	2.2
Oil	8 594	10 304	12 125	14 334	39	36	34	1.6	1.3
of which bunkers	*358*	*543*	*598*	*689*	*2*	*2*	*2*	*1.0*	*1.0*
Gas	3 808	5 336	6 855	8 872	20	20	21	2.5	2.1
Power generation	**7 476**	**10 942**	**14 690**	**18 708**	**100**	**100**	**100**	**3.0**	**2.2**
Coal	4 923	7 856	10 827	13 884	72	74	74	3.3	2.3
Oil	1 196	954	934	757	9	6	4	-0.2	-0.9
Gas	1 358	2 132	2 929	4 067	19	20	22	3.2	2.6
Total final consumption	**11 855**	**13 834**	**17 146**	**20 510**	**100**	**100**	**100**	**2.2**	**1.6**
Coal	3 236	2 925	3 888	4 300	21	23	21	2.9	1.6
Oil	6 482	8 155	9 861	12 068	59	58	59	1.9	1.6
Gas	2 137	2 754	3 397	4 143	20	20	20	2.1	1.6
Industry and non–energy use	**4 516**	**5 180**	**6 845**	**7 956**	**100**	**100**	**100**	**2.8**	**1.7**
Coal	2 190	2 465	3 406	3 861	48	50	49	3.3	1.8
Oil	1 345	1 505	1 860	2 124	29	27	27	2.1	1.4
Gas	981	1 210	1 579	1 971	23	23	25	2.7	2.0
Transport	**3 950**	**5 370**	**6 524**	**8 293**	**100**	**100**	**100**	**2.0**	**1.8**
Coal	50	17	14	8	0	0	0	-1.5	-2.7
Oil	3 775	5 188	6 307	8 034	97	97	97	2.0	1.8
Gas	125	165	202	250	3	3	3	2.0	1.7
Residential, services and agriculture	**3 389**	**3 284**	**3 777**	**4 261**	**100**	**100**	**100**	**1.4**	**1.0**
Coal	995	444	468	430	14	12	10	0.5	-0.1
Oil	1 362	1 462	1 693	1 909	45	45	45	1.5	1.1
Gas	1 031	1 378	1 616	1 922	42	43	45	1.6	1.3

A

Alternative Policy Scenario: World

	Energy demand (Mtoe)		Shares (%)		Growth (% p.a.)		Change vs. RS (%)	
	2015	2030	2015	2030	2005-2015	2005-2030	2015	2030
Total primary energy demand	13 818	15 783	100	100	1.9	1.3	**-3.8**	**-10.9**
Coal	3 643	3 700	26	23	2.3	1.0	-8.7	-25.9
Oil	4 512	4 911	33	31	1.2	0.8	-4.4	-12.1
of which bunkers	*184*	*204*	*1*	*1*	*0.7*	*0.7*	*-2.2*	*-5.8*
Gas	2 938	3 447	21	22	2.2	1.5	-3.5	-12.7
Nuclear	850	1 080	6	7	1.7	1.6	5.8	26.5
Hydro	352	465	3	3	3.4	2.5	7.6	11.6
Biomass and waste	1 359	1 738	10	11	1.7	1.7	1.9	7.6
Other renewables	165	444	1	3	10.4	8.2	13.6	44.1
Power generation	5 354	6 298	100	100	2.3	1.6	**-4.6**	**-12.8**
Coal	2 401	2 438	45	39	2.1	0.9	-10.7	-29.5
Oil	273	203	5	3	-0.8	-1.5	-6.6	-14.5
Gas	1 183	1 388	22	22	2.7	1.7	-5.4	-20.1
Nuclear	850	1 080	16	17	1.7	1.6	5.8	26.5
Hydro	352	465	7	7	3.4	2.5	7.6	11.6
Biomass and waste	164	400	3	6	7.9	6.8	26.0	45.4
Other renewables	130	325	2	5	9.8	7.7	6.3	30.2
Other energy sector	1 418	1 669	100	100	2.2	1.5	**-3.0**	**-10.3**
of which electricity	*351*	*412*	*25*	*25*	*2.5*	*1.6*	*-5.3*	*-15.6*
Total final consumption	9 324	10 623	100	100	1.9	1.3	**-3.4**	**-10.4**
Coal	840	770	9	7	2.4	0.6	-5.9	-21.6
Oil	3 751	4 160	40	39	1.4	1.0	-4.5	-12.7
Gas	1 494	1 742	16	16	2.0	1.4	-2.1	-7.1
Electricity	1 765	2 247	19	21	3.2	2.2	-3.5	-12.1
Heat	300	311	3	3	1.1	0.6	-2.7	-9.4
Biomass and waste	1 139	1 275	12	12	1.1	0.9	-0.7	-0.2
Other renewables	34	119	0	1	13.1	10.4	53.5	103.9
Industry and non–energy use	3 649	4 114	100	100	2.6	1.5	**-3.1**	**-10.1**
Coal	737	694	20	17	2.9	0.9	-5.9	-21.6
Oil	1 025	1 097	28	27	1.6	0.9	-3.5	-8.8
Gas	717	854	20	21	2.4	1.7	-2.9	-7.8
Electricity	809	1 022	22	25	4.0	2.5	-2.5	-10.3
Heat	123	117	3	3	1.0	0.2	-1.9	-10.2
Biomass and waste	236	323	6	8	2.5	2.3	5.5	12.4
Other renewables	1	7	0	0	12.0	11.3	44.7	126.2
Transport	2 369	2 797	100	100	1.7	1.3	**-4.1**	**-11.6**
Oil	2 171	2 481	92	89	1.4	1.1	-5.4	-15.0
Biofuels	78	164	3	6	15.3	9.0	38.0	60.8
Other fuels	120	152	5	5	2.2	1.8	2.5	7.3
Residential, services and agriculture	3 306	3 711	100	100	1.3	1.0	**-3.4**	**-10.0**
Coal	99	74	3	2	-0.3	-1.3	-5.8	-22.3
Oil	554	582	17	16	1.2	0.7	-2.9	-9.8
Gas	689	773	21	21	1.5	1.1	-2.0	-8.2
Electricity	929	1 189	28	32	2.6	2.0	-4.5	-14.1
Heat	177	194	5	5	1.1	0.8	-3.3	-9.0
Biomass and waste	825	788	25	21	0.1	-0.1	-4.9	-11.3
Other renewables	33	112	1	3	13.1	10.3	53.9	102.8

Alternative Policy Scenario: World

	Electricity generation (TWh)		Shares (%)		Growth (% p.a.)		Change vs. RS (%)	
	2015	2030	2015	2030	2005-2015	2005-2030	2015	2030
Total generation	24 654	31 240	100	100	3.1	2.2	–3.5	–11.7
Coal	9 755	10 716	40	34	2.9	1.5	–12.0	–32.2
Oil	1 118	844	5	3	–0.6	–1.4	–4.1	–9.2
Gas	5 150	6 602	21	21	3.7	2.5	–3.4	–18.2
Nuclear	3 262	4 144	13	13	1.6	1.6	5.8	26.5
Hydro	4 089	5 403	17	17	3.4	2.5	7.6	11.6
Biomass and waste	511	1 166	2	4	8.2	6.7	25.6	38.7
Wind	623	1 800	3	6	18.8	11.8	13.5	39.9
Geothermal	100	190	0	1	6.8	5.3	1.2	9.9
Solar	44	352	0	1	29.7	20.6	19.3	118.5
Tide and wave	2	24	0	0	12.3	16.2	24.0	108.1

	CO_2 emissions (Mt)		Shares (%)		Growth (% p.a.)		Change vs. RS (%)	
	2015	2030	2015	2030	2005-2015	2005-2030	2015	2030
Total CO_2 emissions	31 893	33 890	100	100	1.8	1.0	–6.4	–19.1
Coal	13 702	13 630	43	40	2.2	0.9	–9.2	–27.1
Oil	11 581	12 555	36	37	1.2	0.8	–4.5	–12.4
of which bunkers	*585*	*649*	*2*	*2*	*0.7*	*0.7*	*–2.2*	*–5.8*
Gas	6 611	7 705	21	23	2.2	1.5	–3.6	–13.2
Power generation	13 314	13 686	100	100	2.0	0.9	–9.4	–26.8
Coal	9 668	9 784	73	71	2.1	0.9	–10.7	–29.5
Oil	873	649	7	5	–0.9	–1.5	–6.5	–14.3
Gas	2 772	3 253	21	24	2.7	1.7	–5.4	–20.0
Total final consumption	16 386	17 683	100	100	1.7	1.0	–4.4	–13.8
Coal	3 659	3 371	22	19	2.3	0.6	–5.9	–21.6
Oil	9 404	10 479	57	59	1.4	1.0	–4.6	–13.2
Gas	3 323	3 833	20	22	1.9	1.3	–2.2	–7.5
Industry and non–energy use	6 530	6 768	100	100	2.3	1.1	–4.6	–14.9
Coal	3 204	3 025	49	45	2.7	0.8	–5.9	–21.6
Oil	1 793	1 930	27	29	1.8	1.0	–3.6	–9.1
Gas	1 533	1 813	23	27	2.4	1.6	–2.9	–8.0
Transport	6 188	7 102	100	100	1.4	1.1	–5.1	–14.4
Coal	14	8	0	0	–1.6	–3.0	–1.7	–6.0
Oil	5 967	6 826	96	96	1.4	1.1	–5.4	–15.0
Gas	207	268	3	4	2.3	1.9	2.4	7.1
Residential, services and agriculture	3 667	3 812	100	100	1.1	0.6	–2.9	–10.5
Coal	441	337	12	9	–0.1	–1.1	–5.8	–21.7
Oil	1 644	1 722	45	45	1.2	0.7	–2.9	–9.8
Gas	1 583	1 753	43	46	1.4	1.0	–2.1	–8.8

A

Reference Scenario: China

	Energy demand (Mtoe)				Shares (%)			Growth (% p.a.)	
	1990	2005	2015	2030	2005	2015	2030	2005-2015	2005-2030
Total primary energy demand	874	1 742	2 851	3 819	100	100	100	5.1	3.2
Coal	534	1 094	1 869	2 399	63	66	63	5.5	3.2
Oil	116	327	543	808	19	19	21	5.2	3.7
Gas	13	42	109	199	2	4	5	10.0	6.4
Nuclear	0	14	32	67	1	1	2	8.8	6.5
Hydro	11	34	62	86	2	2	2	6.1	3.8
Biomass and waste	200	227	225	227	13	8	6	−0.1	0.0
Other renewables	0	3	12	33	0	0	1	14.4	9.9
Power generation	181	682	1 222	1 774	100	100	100	6.0	3.9
Coal	153	605	1 073	1 487	89	88	84	5.9	3.7
Oil	16	18	18	15	3	1	1	−0.0	−0.8
Gas	1	7	25	64	1	2	4	13.8	9.4
Nuclear	0	14	32	67	2	3	4	8.8	6.5
Hydro	11	34	62	86	5	5	5	6.1	3.8
Biomass and waste	0	3	6	38	0	1	2	6.7	10.2
Other renewables	0	0	6	17	0	0	1	36.5	18.4
Other energy sector	94	204	375	513	100	100	100	6.3	3.8
of which electricity	*12*	*43*	*85*	*118*	*21*	*23*	*23*	*6.9*	*4.1*
Total final consumption	670	1 129	1 808	2 375	100	100	100	4.8	3.0
Coal	315	373	573	605	33	32	25	4.4	1.9
Oil	88	278	480	734	25	27	31	5.6	4.0
Gas	10	32	79	127	3	4	5	9.3	5.6
Electricity	43	175	379	610	15	21	26	8.0	5.1
Heat	13	43	73	93	4	4	4	5.3	3.1
Biomass and waste	200	224	218	189	20	12	8	−0.2	−0.7
Other renewables	0	3	6	16	0	0	1	8.1	7.1
Industry	242	478	833	1 046	100	100	100	5.7	3.2
Coal	177	280	443	473	59	53	45	4.7	2.1
Oil	21	39	52	57	8	6	5	3.1	1.5
Gas	3	12	30	48	3	4	5	9.3	5.5
Electricity	30	117	262	395	24	32	38	8.5	5.0
Heat	11	29	44	51	6	5	5	4.2	2.2
Biomass and waste	0	1	2	22	0	0	2	5.5	13.6
Other renewables	0	0	0	0	0	0	0	–	–
Transport	41	121	240	460	100	100	100	7.0	5.5
Oil	30	115	231	442	95	96	96	7.2	5.5
Biofuels	0	1	2	8	0	1	2	11.6	11.7
Other fuels	10	6	7	9	5	3	2	2.3	1.9
Residential, services and agriculture	333	421	540	642	100	100	100	2.5	1.7
Coal	99	64	70	63	15	13	10	1.0	−0.0
Oil	18	62	95	123	15	18	19	4.4	2.8
Gas	2	13	32	59	3	6	9	9.7	6.3
Electricity	11	45	94	182	11	17	28	7.6	5.7
Heat	2	13	27	41	3	5	6	7.5	4.6
Biomass and waste	200	222	215	159	53	40	25	−0.3	−1.3
Other renewables	0	3	6	16	1	1	2	8.2	7.1
Non–energy use	55	109	196	227	100	100	100	6.0	3.0

Reference Scenario: China

	Electricity generation (TWh)				Shares (%)			Growth (% p.a.)	
	1990	2005	2015	2030	2005	2015	2030	2005-2015	2005-2030
Total generation	650	2 544	5 391	8 472	100	100	100	7.8	4.9
Coal	471	1 996	4 326	6 586	78	80	78	8.0	4.9
Oil	49	61	58	49	2	1	1	−0.5	−0.9
Gas	3	26	98	313	1	2	4	14.2	10.5
Nuclear	0	53	123	256	2	2	3	8.8	6.5
Hydro	127	397	717	1 005	16	13	12	6.1	3.8
Biomass and waste	0	8	17	110	0	0	1	7.7	10.9
Wind	0	2	49	133	0	1	2	37.4	18.2
Geothermal	0	0	2	5	0	0	0	36.1	18.2
Solar	0	0	0	15	0	0	0	–	22.6
Tide and wave	0	0	0	0	0	0	0	–	4.7

	Capacity (GW)			Shares (%)			Growth (% p.a.)	
	2005	2015	2030	2005	2015	2030	2005-2015	2005-2030
Total capacity	517	1 110	1 775	100	100	100	7.9	5.1
Coal	368	814	1 259	71	73	71	8.3	5.0
Oil	12	14	11	2	1	1	1.9	−0.3
Gas	10	31	98	2	3	6	12.0	9.6
Nuclear	7	15	31	1	1	2	8.4	6.3
Hydro	117	215	300	23	19	17	6.3	3.8
Biomass and waste	2	4	18	0	0	1	4.8	8.5
Wind	1	17	49	0	2	3	29.8	15.7
Geothermal	0	0	1	0	0	0	26.8	14.8
Solar	0	0	9	0	0	0	–	21.4
Tide and wave	0	0	0	0	0	0	–	3.9

	CO$_2$ emissions (Mt)				Shares (%)			Growth (% p.a.)	
	1990	2005	2015	2030	2005	2015	2030	2005-2015	2005-2030
Total CO$_2$ emissions	2 244	5 101	8 632	11 448	100	100	100	5.4	3.3
Coal	1 914	4 199	7 067	8 977	82	82	78	5.3	3.1
Oil	304	811	1 342	2 059	16	16	18	5.2	3.8
Gas	26	91	223	413	2	3	4	9.4	6.2
Power generation	652	2 500	4 450	6 202	100	100	100	5.9	3.7
Coal	598	2 424	4 328	5 997	97	97	97	6.0	3.7
Oil	52	59	62	51	2	1	1	0.3	−0.6
Gas	2	16	60	155	1	1	2	13.8	9.4
Total final consumption	1 507	2 400	3 777	4 693	100	100	100	4.6	2.7
Coal	1 265	1 652	2 442	2 572	69	65	55	4.0	1.8
Oil	225	688	1 196	1 897	29	32	40	5.7	4.1
of which transport	*83*	*321*	*649*	*1 243*	*13*	*17*	*26*	*7.3*	*5.6*
Gas	17	60	138	224	3	4	5	8.7	5.4

A

Alternative Policy Scenario: China

	Energy demand (Mtoe)		Shares (%)		Growth (% p.a.)		Change vs. RS (%)	
	2015	2030	2015	2030	2005-2015	2005-2030	2015	2030
Total primary energy demand	**2 743**	**3 256**	**100**	**100**	**4.6**	**2.5**	**-3.8**	**-14.7**
Coal	1 743	1 842	64	57	4.8	2.1	-6.7	-23.2
Oil	518	653	19	20	4.7	2.8	-4.5	-19.2
Gas	126	225	5	7	11.5	6.9	15.0	12.6
Nuclear	44	120	2	4	12.2	9.0	36.0	79.4
Hydro	75	109	3	3	8.2	4.8	21.4	26.4
Biomass and waste	223	255	8	8	-0.2	0.5	-0.7	12.4
Other renewables	14	52	1	2	16.6	11.9	21.2	57.4
Power generation	**1 181**	**1 543**	**100**	**100**	**5.7**	**3.3**	**-3.3**	**-13.0**
Coal	985	1 107	83	72	5.0	2.4	-8.2	-25.5
Oil	18	15	2	1	-0.1	-0.8	-0.6	-0.9
Gas	40	87	3	6	19.5	10.8	62.1	36.0
Nuclear	44	120	4	8	12.2	9.0	36.0	79.4
Hydro	75	109	6	7	8.2	4.8	21.4	26.4
Biomass and waste	13	77	1	5	14.4	13.4	101.7	101.8
Other renewables	7	28	1	2	39.0	20.8	20.0	65.1
Other energy sector	**371**	**463**	**100**	**100**	**6.2**	**3.3**	**-1.2**	**-9.6**
of which electricity	*82*	*101*	*22*	*22*	*6.6*	*3.4*	*-2.9*	*-14.6*
Total final consumption	**1 730**	**1 966**	**100**	**100**	**4.4**	**2.2**	**-4.3**	**-17.2**
Coal	535	450	31	23	3.7	0.8	-6.7	-25.5
Oil	456	584	26	30	5.1	3.0	-4.9	-20.6
Gas	81	132	5	7	9.5	5.8	2.3	3.8
Electricity	371	522	21	27	7.8	4.5	-2.0	-14.5
Heat	69	76	4	4	4.7	2.3	-5.0	-18.4
Biomass and waste	210	179	12	9	-0.6	-0.9	-3.6	-5.5
Other renewables	8	23	0	1	10.3	8.8	22.3	48.8
Industry	**807**	**859**	**100**	**100**	**5.4**	**2.4**	**-3.2**	**-17.9**
Coal	414	352	51	41	4.0	0.9	-6.5	-25.6
Oil	50	48	6	6	2.6	0.8	-4.3	-16.4
Gas	29	39	4	5	8.8	4.7	-4.5	-18.5
Electricity	261	339	32	39	8.4	4.4	-0.7	-14.2
Heat	43	43	5	5	4.0	1.6	-1.5	-15.4
Biomass and waste	10	38	1	4	27.0	16.0	n.a.	69.8
Other renewables	0	1	0	0	13.6	13.3	n.a	n.a
Transport	**232**	**367**	**100**	**100**	**6.7**	**4.5**	**-3.1**	**-20.2**
Oil	220	337	95	92	6.7	4.4	-4.8	-23.7
Biofuels	5	19	2	5	25.4	15.7	n.a	144.5
Other fuels	8	10	3	3	2.7	2.3	3.7	9.6
Residential, services and agriculture	**504**	**545**	**100**	**100**	**1.8**	**1.0**	**-6.5**	**-15.0**
Coal	65	45	13	8	0.2	-1.4	-7.5	-29.0
Oil	91	108	18	20	4.0	2.3	-4.4	-12.1
Gas	33	68	7	12	10.2	6.9	4.9	15.5
Electricity	88	149	17	27	6.9	4.9	-6.3	-18.1
Heat	24	31	5	6	6.3	3.5	-10.9	-22.8
Biomass and waste	196	122	39	22	-1.3	-2.4	-9.2	-23.4
Other renewables	8	23	1	4	10.3	8.7	21.1	44.5
Non-energy use	**187**	**195**	**100**	**100**	**5.6**	**2.4**	**-4.5**	**-14.5**

Alternative Policy Scenario: China

	Electricity generation (TWh)		Shares (%)		Growth (% p.a.)		Change vs. RS (%)	
	2015	2030	2015	2030	2005-2015	2005-2030	2015	2030
Total generation	5 297	7 435	100	100	7.6	4.4	–1.7	–12.2
Coal	3 932	4 736	74	64	7.0	3.5	–9.1	–28.1
Oil	58	48	1	1	–0.5	–0.9	–0.5	–0.8
Gas	170	427	3	6	20.6	11.8	72.7	36.6
Nuclear	168	459	3	6	12.2	9.0	36.0	79.4
Hydro	871	1 270	16	17	8.2	4.8	21.4	26.4
Biomass and waste	36	222	1	3	15.9	14.1	109.0	102.6
Wind	59	207	1	3	40.0	20.3	20.9	55.4
Geothermal	2	7	0	0	36.1	19.3	–	25.8
Solar	3	59	0	1	41.9	29.3	n.a	n.a
Tide and wave	0	0	0	0	–	10.4	–	n.a

	Capacity (GW)		Shares (%)		Growth (% p.a.)		Change vs. RS (%)	
	2015	2030	2015	2030	2005-2015	2005-2030	2015	2030
Total capacity	1 110	1 627	100	100	7.9	4.7	0.0	–8.4
Coal	743	910	67	56	7.3	3.7	–8.8	–27.7
Oil	13	11	1	1	1.6	–0.1	–2.8	4.5
Gas	44	120	4	7	16.0	10.5	42.6	21.5
Nuclear	20	55	2	3	11.8	8.9	36.0	79.4
Hydro	261	380	23	23	8.3	4.8	21.1	26.8
Biomass and waste	7	39	1	2	10.9	11.9	76.1	116.2
Wind	21	79	2	5	32.4	18.0	21.8	62.7
Geothermal	0	1	0	0	26.8	15.8	–	25.5
Solar	2	31	0	2	37.4	27.6	n.a.	n.a.
Tide and wave	0	0	0	0	–	9.4	–	n.a

	CO$_2$ emissions (Mt)		Shares (%)		Growth (% p.a.)		Change vs. RS (%)	
	2015	2030	2015	2030	2005-2015	2005-2030	2015	2030
Total CO$_2$ emissions	8 092	8 877	100	100	4.7	2.2	–6.3	–22.5
Coal	6 550	6 757	81	76	4.5	1.9	–7.3	–24.7
Oil	1 283	1 660	16	19	4.7	2.9	–4.4	–19.3
Gas	259	460	3	5	11.1	6.7	16.2	11.4
Power generation	4 131	4 726	100	100	5.2	2.6	–7.2	–23.8
Coal	3 973	4 465	96	94	5.1	2.5	–8.2	–25.5
Oil	61	50	1	1	0.3	–0.7	–0.6	–0.9
Gas	97	210	2	4	19.4	10.7	62.1	36.0
Total final consumption	3 560	3 647	100	100	4.0	1.7	–5.7	–22.3
Coal	2 280	1 913	64	52	3.3	0.6	–6.6	–25.6
Oil	1 139	1 507	32	41	5.2	3.2	–4.7	–20.6
of which transport	*618*	*948*	*17*	*26*	*6.8*	*4.4*	*–4.8*	*–23.7*
Gas	141	227	4	6	8.9	5.5	1.6	1.4

A

Reference Scenario: India

	Energy demand (Mtoe)				Shares (%)			Growth (% p.a.)	
	1990	2005	2015	2030	2005	2015	2030	2005-2015	2005-2030
Total primary energy demand	**320**	**537**	**770**	**1 299**	**100**	**100**	**100**	**3.7**	**3.6**
Coal	106	208	330	620	39	43	48	4.7	4.5
Oil	63	129	188	328	24	24	25	3.9	3.8
Gas	10	29	48	93	5	6	7	5.2	4.8
Nuclear	2	5	16	33	1	2	3	13.2	8.3
Hydro	6	9	13	22	2	2	2	4.4	3.9
Biomass and waste	133	158	171	194	29	22	15	0.8	0.8
Other renewables	0	1	4	9	0	1	1	23.8	11.7
Power generation	**73**	**191**	**312**	**583**	**100**	**100**	**100**	**5.0**	**4.6**
Coal	58	156	240	444	81	77	76	4.4	4.3
Oil	4	8	10	8	4	3	1	1.5	0.0
Gas	3	13	27	52	7	8	9	7.5	5.8
Nuclear	2	5	16	33	2	5	6	13.2	8.3
Hydro	6	9	13	22	4	4	4	4.4	3.9
Biomass and waste	0	1	3	17	1	1	3	11.9	11.5
Other renewables	0	1	4	7	0	1	1	22.5	11.1
Other energy sector	**19**	**50**	**85**	**150**	**100**	**100**	**100**	**5.5**	**4.5**
of which electricity	*7*	*19*	*32*	*58*	*38*	*38*	*38*	*5.3*	*4.5*
Total final consumption	**252**	**356**	**487**	**804**	**100**	**100**	**100**	**3.2**	**3.3**
Coal	41	38	63	119	11	13	15	5.3	4.7
Oil	54	106	155	287	30	32	36	3.8	4.1
Gas	6	14	19	38	4	4	5	3.1	4.1
Electricity	18	41	82	181	12	17	23	7.1	6.1
Heat	0	0	0	0	0	0	0	n.a.	n.a.
Biomass and waste	133	157	168	178	44	34	22	0.7	0.5
Other renewables	0	0	0	1	0	0	0	n.a.	n.a.
Industry	**69**	**99**	**157**	**271**	**100**	**100**	**100**	**4.7**	**4.1**
Coal	28	29	55	111	30	35	41	6.4	5.4
Oil	10	19	27	38	19	17	14	3.6	2.7
Gas	0	5	7	10	5	4	4	3.0	2.7
Electricity	9	18	39	83	18	25	31	7.9	6.3
Heat	0	0	0	0	0	0	0	n.a.	n.a.
Biomass and waste	22	27	29	30	28	19	11	0.7	0.4
Other renewables	0	0	0	0	0	0	0	n.a.	n.a.
Transport	**28**	**37**	**66**	**162**	**100**	**100**	**100**	**6.0**	**6.1**
Oil	26	35	63	154	96	94	95	5.9	6.1
Biofuels	0	0	1	2	0	1	1	55.7	22.9
Other fuels	3	2	3	7	4	4	4	6.4	5.9
Residential, services and agriculture	**138**	**183**	**219**	**295**	**100**	**100**	**100**	**1.8**	**1.9**
Coal	7	6	6	6	3	3	2	−0.2	−0.0
Oil	12	27	35	45	15	16	15	2.6	2.1
Gas	0	1	2	6	0	1	2	7.5	8.5
Electricity	8	20	39	92	11	18	31	6.8	6.3
Heat	0	0	0	0	0	0	0	n.a.	n.a.
Biomass and waste	111	130	138	146	71	63	49	0.6	0.5
Other renewables	0	0	0	1	0	0	0	n.a.	n.a.
Non–energy use	**16**	**37**	**45**	**75**	**100**	**100**	**100**	**1.9**	**2.9**

Reference Scenario: India

	Electricity generation (TWh)				Shares (%)			Growth (% p.a.)	
	1990	2005	2015	2030	2005	2015	2030	2005-2015	2005-2030
Total generation	**289**	**699**	**1 322**	**2 774**	**100**	**100**	**100**	**6.6**	**5.7**
Coal	192	480	889	1 958	69	67	71	6.4	5.8
Oil	10	31	35	31	4	3	1	1.2	–0.0
Gas	10	62	133	292	9	10	11	7.9	6.4
Nuclear	6	17	60	128	2	5	5	13.2	8.3
Hydro	72	100	154	258	14	12	9	4.4	3.9
Biomass and waste	0	2	6	29	0	0	1	11.9	11.5
Wind	0	6	43	69	1	3	3	21.5	10.2
Geothermal	0	0	0	1	0	0	0	n.a.	n.a.
Solar	0	0	0	8	0	0	0	26.0	35.2
Tide and wave	0	0	0	0	0	0	0	n.a.	n.a.

	Capacity (GW)			Shares (%)			Growth (% p.a.)	
	2005	2015	2030	2005	2015	2030	2005-2015	2005-2030
Total capacity	**146**	**255**	**522**	**100**	**100**	**100**	**5.7**	**5.2**
Coal	77	136	295	53	53	56	5.8	5.5
Oil	10	12	11	7	5	2	1.4	0.4
Gas	16	31	79	11	12	15	6.5	6.5
Nuclear	3	8	17	2	3	3	10.2	7.2
Hydro	34	51	85	23	20	16	4.1	3.7
Biomass and waste	0	1	4	0	0	1	11.0	11.1
Wind	4	17	27	3	6	5	14.4	7.6
Geothermal	0	0	0	0	0	0	27.1	13.9
Solar	0	0	4	0	0	1	–	22.8
Tide and wave	0	0	0	0	0	0	n.a.	n.a.

	CO_2 emissions (Mt)				Shares (%)			Growth (% p.a.)	
	1990	2005	2015	2030	2005	2015	2030	2005-2015	2005-2030
Total CO_2 emissions	**587**	**1 147**	**1 804**	**3 314**	**100**	**100**	**100**	**4.6**	**4.3**
Coal	404	774	1 226	2 284	67	68	69	4.7	4.4
Oil	164	312	475	829	27	26	25	4.3	4.0
Gas	19	62	104	201	5	6	6	5.3	4.8
Power generation	**245**	**659**	**1 022**	**1 870**	**100**	**100**	**100**	**4.5**	**4.3**
Coal	225	604	931	1 723	92	91	92	4.4	4.3
Oil	11	25	30	25	4	3	1	1.5	0.0
Gas	8	30	62	122	5	6	6	7.5	5.8
Total final consumption	**327**	**453**	**719**	**1 356**	**100**	**100**	**100**	**4.7**	**4.5**
Coal	173	168	290	552	37	40	41	5.6	4.9
Oil	144	258	392	730	57	55	54	4.3	4.2
of which transport	*72*	*96*	*174*	*427*	*21*	*24*	*32*	*6.1*	*6.2*
Gas	9	27	36	73	6	5	5	3.1	4.1

A

Alternative Policy Scenario: India

	Energy demand (Mtoe)		Shares (%)		Growth (% p.a.)		Change vs. RS (%)	
	2015	2030	2015	2030	2005-2015	2005-2030	2015	2030
Total primary energy demand	**719**	**1 082**	**100**	**100**	**3.0**	**2.8**	**-6.7**	**-16.7**
Coal	289	411	40	38	3.3	2.8	-12.6	-33.7
Oil	173	272	24	25	3.0	3.0	-7.6	-17.1
Gas	47	89	7	8	5.0	4.6	-1.9	-4.3
Nuclear	19	47	3	4	15.2	9.9	19.1	41.9
Hydro	17	32	2	3	7.2	5.3	29.6	42.3
Biomass and waste	168	211	23	19	0.6	1.2	-1.9	8.5
Other renewables	6	21	1	2	26.7	15.8	26.2	145.5
Power generation	**285**	**483**	**100**	**100**	**4.1**	**3.8**	**-8.7**	**-17.2**
Coal	206	293	72	61	2.9	2.6	-14.0	-34.0
Oil	9	8	3	2	0.6	-0.3	-8.7	-6.5
Gas	25	45	9	9	7.0	5.2	-5.1	-12.6
Nuclear	19	47	7	10	15.2	9.9	19.1	41.9
Hydro	17	32	6	7	7.2	5.3	29.6	42.3
Biomass and waste	4	45	2	9	15.0	16.0	31.3	169.8
Other renewables	5	13	2	3	24.4	13.6	16.7	73.0
Other energy sector	**76**	**99**	**100**	**100**	**4.2**	**2.7**	**-11.1**	**-34.5**
of which electricity	*26*	*36*	*34*	*36*	*3.0*	*2.5*	*-19.5*	*-38.0*
Total final consumption	**463**	**699**	**100**	**100**	**2.7**	**2.7**	**-4.9**	**-13.0**
Coal	58	86	12	12	4.3	3.4	-9.1	-27.5
Oil	142	236	31	34	3.0	3.2	-8.1	-17.8
Gas	19	40	4	6	3.4	4.4	2.5	6.9
Electricity	79	163	17	23	6.8	5.7	-2.9	-10.2
Heat	0	0	0	0	n.a.	n.a.	n.a.	n.a.
Biomass and waste	164	166	35	24	0.4	0.2	-2.5	-6.6
Other renewables	1	8	0	1	n.a.	n.a.	109.9	n.a.
Industry	**149**	**234**	**100**	**100**	**4.2**	**3.5**	**-4.8**	**-13.9**
Coal	50	79	33	34	5.3	4.0	-9.6	-28.3
Oil	25	34	17	14	2.9	2.3	-6.5	-10.2
Gas	7	9	4	4	2.6	2.4	-3.8	-7.8
Electricity	38	78	25	33	7.6	6.0	-2.7	-6.2
Heat	0	0	0	0	n.a.	n.a.	n.a.	n.a.
Biomass and waste	30	33	20	14	0.9	0.8	2.5	9.7
Other renewables	0	0	0	0	n.a.	n.a.	n.a.	n.a.
Transport	**61**	**136**	**100**	**100**	**5.2**	**5.4**	**-7.6**	**-16.3**
Oil	54	115	89	85	4.4	4.8	-13.1	-25.2
Biofuels	3	8	5	6	74.0	30.1	n.a	n.a.
Other fuels	4	13	7	9	10.4	8.8	44.2	95.8
Residential, services and agriculture	**210**	**260**	**100**	**100**	**1.4**	**1.4**	**-4.4**	**-12.0**
Coal	5	5	3	2	-0.8	-0.7	-5.2	-14.9
Oil	33	39	16	15	2.2	1.5	-4.3	-13.0
Gas	1	5	1	2	7.0	7.8	-4.7	-14.3
Electricity	37	78	18	30	6.4	5.6	-3.7	-14.9
Heat	0	0	0	0	n.a.	n.a.	n.a.	n.a.
Biomass and waste	131	125	63	48	0.1	-0.2	-5.0	-14.3
Other renewables	1	8	0	3	n.a.	n.a.	102.5	n.a
Non-energy use	**43**	**70**	**100**	**100**	**1.6**	**2.6**	**-3.6**	**-6.6**

Alternative Policy Scenario: India

	Electricity generation (TWh)		Shares (%)		Growth (% p.a.)		Change vs. RS (%)	
	2015	2030	2015	2030	2005-2015	2005-2030	2015	2030
Total generation	1 221	2 305	100	100	5.7	4.9	–7.6	–16.9
Coal	735	1 261	60	55	4.4	3.9	–17.4	–35.6
Oil	33	29	3	1	0.4	–0.2	–7.7	–5.6
Gas	123	246	10	11	7.0	5.6	–7.7	–15.6
Nuclear	71	182	6	8	15.2	9.9	19.1	41.9
Hydro	200	368	16	16	7.2	5.3	29.6	42.3
Biomass and waste	8	79	1	3	15.0	16.0	31.3	169.8
Wind	50	124	4	5	23.3	12.8	15.7	79.0
Geothermal	0	1	0	0	n.a.	n.a.	–	–
Solar	1	15	0	1	75.8	39.0	n.a.	100.7
Tide and wave	0	0	0	0	n.a.	n.a.	n.a.	n.a.

	Capacity (GW)		Shares (%)		Growth (% p.a.)		Change vs. RS (%)	
	2015	2030	2015	2030	2005-2015	2005-2030	2015	2030
Total capacity	249	477	100	100	5.5	4.9	–2.3	–8.6
Coal	110	187	44	39	3.5	3.6	–19.2	–36.5
Oil	12	11	5	2	1.4	0.4	–	–
Gas	31	65	12	14	6.5	5.7	0.4	–17.4
Nuclear	10	24	4	5	12.1	8.7	19.1	41.9
Hydro	66	120	26	25	6.8	5.2	29.2	41.2
Biomass and waste	1	12	0	3	14.2	15.7	32.8	173.1
Wind	19	48	8	10	16.2	10.2	16.8	81.6
Geothermal	0	0	0	0	27.1	13.9	–	–
Solar	1	9	0	2	38.1	26.3	n.a.	102.4
Tide and wave	0	0	0	0	n.a.	n.a.	n.a.	n.a

	CO_2 emissions (Mt)		Shares (%)		Growth (% p.a.)		Change vs. RS (%)	
	2015	2030	2015	2030	2005-2015	2005-2030	2015	2030
Total CO_2 emissions	1 607	2 415	100	100	3.4	3.0	–10.9	–27.1
Coal	1 069	1 544	67	64	3.3	2.8	–12.8	–32.4
Oil	436	678	27	28	3.4	3.2	–8.1	–18.2
Gas	102	193	6	8	5.1	4.7	–1.8	–4.0
Power generation	886	1 268	100	100	3.0	2.6	–13.3	–32.2
Coal	800	1 137	90	90	2.9	2.6	–14.0	–34.0
Oil	27	24	3	2	0.6	–0.3	–8.7	–6.5
Gas	59	106	7	8	7.0	5.2	–5.1	–12.6
Total final consumption	660	1 072	100	100	3.8	3.5	–8.2	–21.0
Coal	264	400	40	37	4.6	3.5	–9.1	–27.5
Oil	359	591	54	55	3.3	3.4	–8.6	–19.1
of which transport	*151*	*320*	*23*	*30*	*4.6*	*4.9*	*–13.1*	*–25.2*
Gas	38	81	6	8	3.5	4.5	3.9	10.0

A

Reference Scenario: OECD

	Energy demand (Mtoe)				Shares (%)			Growth (% p.a.)	
	1990	2005	2015	2030	2005	2015	2030	2005-2015	2005-2030
Total primary energy demand	**4 518**	**5 542**	**6 180**	**6 800**	**100**	**100**	**100**	**1.1**	**0.8**
Coal	1 063	1 130	1 225	1 318	20	20	19	0.8	0.6
Oil	1 893	2 247	2 385	2 478	41	39	36	0.6	0.4
Gas	844	1 211	1 425	1 654	22	23	24	1.6	1.3
Nuclear	449	611	642	616	11	10	9	0.5	0.0
Hydro	101	109	121	130	2	2	2	1.1	0.7
Biomass and waste	141	194	291	415	4	5	6	4.1	3.1
Other renewables	28	39	91	187	1	1	3	8.8	6.5
Power generation	**1 702**	**2 231**	**2 538**	**2 814**	**100**	**100**	**100**	**1.3**	**0.9**
Coal	751	917	1 021	1 097	41	40	39	1.1	0.7
Oil	149	118	101	64	5	4	2	-1.5	-2.4
Gas	175	382	482	599	17	19	21	2.3	1.8
Nuclear	449	611	642	616	27	25	22	0.5	0.0
Hydro	101	109	121	130	5	5	5	1.1	0.7
Biomass and waste	52	60	93	154	3	4	5	4.5	3.8
Other renewables	25	33	78	154	1	3	5	9.0	6.4
Other energy sector	**379**	**442**	**455**	**526**	**100**	**100**	**100**	**0.3**	**0.7**
Total final consumption	**3 136**	**3 850**	**4 324**	**4 810**	**100**	**100**	**100**	**1.2**	**0.9**
Coal	229	126	118	106	3	3	2	-0.7	-0.7
Oil	1 639	1 999	2 167	2 302	52	50	48	0.8	0.6
of which transport	*961*	*1 255*	*1 370*	*1 493*	*33*	*32*	*31*	*0.9*	*0.7*
Gas	590	739	839	923	19	19	19	1.3	0.9
Electricity	547	769	917	1 103	20	21	23	1.8	1.5
Heat	40	76	73	82	2	2	2	-0.5	0.3
Biomass and waste	89	134	197	261	3	5	5	3.9	2.7
of which biofuels	*0*	*11*	*40*	*62*	*0*	*1*	*1*	*13.5*	*7.0*
Other renewables	3	6	13	33	0	0	1	7.7	7.0

	Electricity (TWh)				Shares (%)			Growth (% p.a.)	
Total generation	**7 569**	**10 391**	**12 228**	**14 597**	**100**	**100**	**100**	**1.6**	**1.4**
Coal	3 060	3 947	4 552	5 398	38	37	37	1.4	1.3
Oil	694	538	467	295	5	4	2	-1.4	-2.4
Gas	770	1 958	2 517	3 363	19	21	23	2.5	2.2
Nuclear	1 725	2 348	2 462	2 364	23	20	16	0.5	0.0
Hydro	1 170	1 270	1 412	1 510	12	12	10	1.1	0.7
Biomass and waste	117	192	304	492	2	2	3	4.7	3.8
Wind	4	102	420	959	1	3	7	15.2	9.4
Geothermal	29	33	58	93	0	0	1	5.9	4.2
Solar	1	3	34	112	0	0	1	29.3	16.3
Tide and wave	1	1	1	11	0	0	0	10.1	12.9

	CO_2 emissions (Mt)				Shares (%)			Growth (% p.a.)	
Total CO_2 emissions	**11 053**	**12 838**	**14 054**	**15 067**	**100**	**100**	**100**	**0.9**	**0.6**
Coal	4 102	4 335	4 733	5 007	34	34	33	0.9	0.6
Oil	5 033	5 701	6 044	6 267	44	43	42	0.6	0.4
Gas	1 918	2 802	3 276	3 794	22	23	25	1.6	1.2
Power generation	**3 911**	**4 987**	**5 588**	**6 044**	**100**	**100**	**100**	**1.1**	**0.8**
Coal	3 029	3 710	4 135	4 438	74	74	73	1.1	0.7
Oil	473	382	326	207	8	6	3	-1.6	-2.4
Gas	409	895	1 126	1 399	18	20	23	2.3	1.8
Total final consumption	**6 550**	**7 219**	**7 820**	**8 308**	**100**	**100**	**100**	**0.8**	**0.6**
Coal	1 011	562	533	481	8	7	6	-0.5	-0.6
Oil	4 196	4 955	5 357	5 705	69	69	69	0.8	0.6
Gas	1 342	1 702	1 930	2 122	24	25	26	1.3	0.9

Alternative Policy Scenario: OECD

	Energy demand (Mtoe)		Shares (%)		Growth (% p.a.)		Change vs. RS (%)	
	2015	2030	2015	2030	2005-2015	2005-2030	2015	2030
Total primary energy demand	6 015	6 289	100	100	0.8	0.5	−2.7	−7.5
Coal	1 097	944	18	15	−0.3	−0.7	−10.5	−28.4
Oil	2 302	2 242	38	36	0.2	−0.0	−3.5	−9.5
Gas	1 390	1 481	23	24	1.4	0.8	−2.5	−10.5
Nuclear	672	751	11	12	1.0	0.8	4.8	21.8
Hydro	123	135	2	2	1.2	0.8	1.7	3.9
Biomass and waste	326	482	5	8	5.3	3.7	12.2	16.1
Other renewables	104	255	2	4	10.3	7.8	14.8	36.0
Power generation	2 459	2 568	100	100	1.0	0.6	−3.1	−8.7
Coal	898	741	37	29	−0.2	−0.9	−12.0	−32.4
Oil	98	58	4	2	−1.8	−2.8	−3.5	−10.6
Gas	470	510	19	20	2.1	1.2	−2.5	−14.8
Nuclear	672	751	27	29	1.0	0.8	4.8	21.8
Hydro	123	135	5	5	1.2	0.8	1.7	3.9
Biomass and waste	115	181	5	7	6.7	4.5	22.8	17.0
Other renewables	83	193	3	8	9.8	7.4	7.3	25.5
Other energy sector	445	491	100	100	0.1	0.4	−2.2	−6.8
Total final consumption	4 213	4 447	100	100	0.9	0.6	−2.6	−7.5
Coal	114	98	3	2	−1.0	−1.0	−3.0	−7.9
Oil	2 089	2 073	50	47	0.4	0.1	−3.6	−9.9
of which transport	*1 310*	*1 317*	*31*	*30*	*0.4*	*0.2*	*−4.4*	*−11.8*
Gas	818	845	19	19	1.0	0.5	−2.5	−8.4
Electricity	888	987	21	22	1.4	1.0	−3.2	−10.5
Heat	72	81	2	2	−0.6	0.2	−1.0	−2.0
Biomass and waste	211	301	5	7	4.6	3.3	7.2	15.7
of which biofuels	*52*	*89*	*1*	*2*	*16.3*	*8.6*	*27.6*	*43.8*
Other renewables	21	62	0	1	12.8	9.6	59.8	84.1

	Electricity (TWh)		Shares (%)		Growth (% p.a.)		Change vs. RS (%)	
Total generation	11 834	13 117	100	100	1.3	0.9	−3.2	−10.1
Coal	3 976	3 561	34	27	0.1	−0.4	−12.7	−34.0
Oil	449	261	4	2	−1.8	−2.9	−4.0	−11.5
Gas	2 451	2 729	21	21	2.3	1.3	−2.6	−18.9
Nuclear	2 579	2 881	22	22	0.9	0.8	4.8	21.8
Hydro	1 436	1 569	12	12	1.2	0.8	1.7	3.9
Biomass and waste	371	565	3	4	6.8	4.4	22.2	14.7
Wind	474	1 237	4	9	16.7	10.5	13.0	29.1
Geothermal	60	102	1	1	6.1	4.6	2.0	9.5
Solar	37	190	0	1	30.4	18.8	8.7	69.3
Tide and wave	2	23	0	0	12.5	16.1	24.3	104.7

	CO$_2$ emissions (Mt)		Shares (%)		Growth (% p.a.)		Change vs. RS (%)	
Total CO$_2$ emissions	13 235	12 540	100	100	0.3	−0.1	−5.8	−16.8
Coal	4 213	3 508	32	28	−0.3	−0.8	−11.0	−29.9
Oil	5 828	5 640	44	45	0.2	−0.0	−3.6	−10.0
Gas	3 194	3 392	24	27	1.3	0.8	−2.5	−10.6
Power generation	5 046	4 364	100	100	0.1	−0.5	−9.7	−27.8
Coal	3 633	2 986	72	68	−0.2	−0.9	−12.1	−32.7
Oil	315	185	6	4	−1.9	−2.8	−3.5	−10.6
Gas	1 098	1 193	22	27	2.1	1.2	−2.5	−14.8
Total final consumption	7 552	7 491	100	100	0.5	0.1	−3.4	−9.8
Coal	517	443	7	6	−0.8	−0.9	−3.0	−7.9
Oil	5 154	5 105	68	68	0.4	0.1	−3.8	−10.5
Gas	1 881	1 943	25	26	1.0	0.5	−2.5	−8.4

A

Reference Scenario: OECD North America

	Energy demand (Mtoe)				Shares (%)			Growth (% p.a.)	
	1990	2005	2015	2030	2005	2015	2030	2005-2015	2005-2030
Total primary energy demand	2 257	2 786	3 166	3 573	100	100	100	1.3	1.0
Coal	486	592	668	758	21	21	21	1.2	1.0
Oil	928	1 153	1 274	1 381	41	40	39	1.0	0.7
Gas	517	632	733	822	23	23	23	1.5	1.1
Nuclear	179	238	254	275	9	8	8	0.6	0.6
Hydro	51	57	58	61	2	2	2	0.2	0.3
Biomass and waste	78	95	138	193	3	4	5	3.8	2.9
Other renewables	19	18	41	83	1	1	2	8.5	6.3
Power generation	845	1 107	1 263	1 430	100	100	100	1.3	1.0
Coal	413	537	612	672	48	48	47	1.3	0.9
Oil	47	54	52	39	5	4	3	−0.3	−1.3
Gas	95	181	209	239	16	17	17	1.5	1.1
Nuclear	179	238	254	275	22	20	19	0.6	0.6
Hydro	51	57	58	61	5	5	4	0.2	0.3
Biomass and waste	41	25	41	75	2	3	5	5.1	4.5
Other renewables	19	16	36	69	1	3	5	8.6	6.1
Other energy sector	191	222	237	305	100	100	100	0.6	1.3
Total final consumption	1 552	1 905	2 192	2 476	100	100	100	1.4	1.1
Coal	59	35	34	30	2	2	1	−0.4	−0.6
Oil	822	1 029	1 160	1 283	54	53	52	1.2	0.9
of which transport	*556*	*723*	*815*	*918*	*38*	*37*	*37*	*1.2*	*1.0*
Gas	360	386	442	476	20	20	19	1.4	0.8
Electricity	271	379	449	549	20	21	22	1.7	1.5
Heat	3	4	6	7	0	0	0	3.3	2.1
Biomass and waste	37	70	97	118	4	4	5	3.3	2.1
of which biofuels	*0*	*8*	*22*	*29*	*0*	*1*	*1*	*10.3*	*5.2*
Other renewables	0	2	5	13	0	0	1	7.7	7.5

	Electricity (TWh)				Shares (%)			Growth (% p.a.)	
Total generation	3 809	5 127	6 013	7 300	100	100	100	1.6	1.4
Coal	1 790	2 293	2 726	3 333	45	45	46	1.7	1.5
Oil	217	229	213	164	4	4	2	−0.8	−1.3
Gas	406	904	1 102	1 363	18	18	19	2.0	1.7
Nuclear	687	914	973	1 057	18	16	14	0.6	0.6
Hydro	593	664	679	708	13	11	10	0.2	0.3
Biomass and waste	90	83	128	228	2	2	3	4.4	4.1
Wind	3	19	142	350	0	2	5	22.0	12.3
Geothermal	21	21	39	62	0	1	1	6.6	4.5
Solar	1	1	12	34	0	0	0	33.6	17.0
Tide and wave	0	0	0	2	0	0	0	9.5	17.8

	CO_2 emissions (Mt)				Shares (%)			Growth (% p.a.)	
Total CO_2 emissions	5 554	6 727	7 527	8 250	100	100	100	1.1	0.8
Coal	1 882	2 276	2 577	2 827	34	34	34	1.2	0.9
Oil	2 485	2 980	3 261	3 530	44	43	43	0.9	0.7
Gas	1 187	1 471	1 690	1 894	22	22	23	1.4	1.0
Power generation	1 991	2 713	3 069	3 331	100	100	100	1.2	0.8
Coal	1 618	2 115	2 411	2 646	78	79	79	1.3	0.9
Oil	151	176	168	126	6	5	4	−0.4	−1.3
Gas	222	422	490	558	16	16	17	1.5	1.1
Total final consumption	3 211	3 645	4 069	4 450	100	100	100	1.1	0.8
Coal	262	148	148	132	4	4	3	−0.0	−0.4
Oil	2 122	2 607	2 904	3 222	72	71	72	1.1	0.9
Gas	827	890	1 018	1 095	24	25	25	1.4	0.8

Alternative Policy Scenario: OECD North America

	Energy demand (Mtoe)		Shares (%)		Growth (% p.a.)		Change vs. RS (%)	
	2015	2030	2015	2030	2005-2015	2005-2030	2015	2030
Total primary energy demand	3 094	3 299	100	100	1.1	0.7	−2.3	−7.7
Coal	630	595	20	18	0.6	0.0	−5.7	−21.5
Oil	1 234	1 251	40	38	0.7	0.3	−3.2	−9.4
Gas	718	760	23	23	1.3	0.7	−2.0	−7.5
Nuclear	254	310	8	9	0.6	1.1	–	12.7
Hydro	59	63	2	2	0.4	0.4	1.6	3.6
Biomass and waste	153	212	5	6	4.9	3.3	10.5	9.8
Other renewables	46	107	1	3	9.8	7.4	13.1	29.5
Power generation	1 234	1 304	100	100	1.1	0.7	−2.3	−8.8
Coal	576	516	47	40	0.7	−0.2	−5.9	−23.2
Oil	51	36	4	3	−0.5	−1.6	−1.1	−7.3
Gas	207	216	17	17	1.4	0.7	−1.4	−9.7
Nuclear	254	310	21	24	0.6	1.1	–	12.7
Hydro	59	63	5	5	0.4	0.4	1.6	3.6
Biomass and waste	50	80	4	6	7.2	4.8	21.5	7.3
Other renewables	38	83	3	6	9.0	6.8	3.5	18.7
Other energy sector	232	286	100	100	0.5	1.0	−1.9	−6.4
Total final consumption	2 141	2 292	100	100	1.2	0.7	−2.3	−7.4
Coal	33	28	2	1	−0.7	−0.9	−2.9	−6.4
Oil	1 120	1 156	52	50	0.9	0.5	−3.4	−9.9
of which transport	*782*	*809*	*37*	*35*	*0.8*	*0.4*	*-4.1*	*-11.9*
Gas	433	444	20	19	1.2	0.6	−2.2	−6.8
Electricity	439	502	20	22	1.5	1.1	−2.4	−8.7
Heat	6	6	0	0	3.0	1.9	−2.7	−5.6
Biomass and waste	103	131	5	6	3.9	2.6	5.9	11.3
of which biofuels	*29*	*46*	*1*	*2*	*13.3*	*7.1*	*31.0*	*55.6*
Other renewables	9	25	0	1	14.9	10.2	90.2	86.7

	Electricity (TWh)		Shares (%)		Growth (% p.a.)		Change vs. RS (%)	
Total generation	5 873	6 680	100	100	1.4	1.1	−2.3	−8.5
Coal	2 549	2 530	43	38	1.1	0.4	−6.5	−24.1
Oil	210	152	4	2	−0.9	−1.6	−1.1	−7.5
Gas	1 088	1 249	19	19	1.9	1.3	−1.2	−8.3
Nuclear	973	1 191	17	18	0.6	1.1	–	12.7
Hydro	690	733	12	11	0.4	0.4	1.6	3.6
Biomass and waste	155	242	3	4	6.4	4.4	21.2	6.0
Wind	155	448	3	7	23.1	13.4	9.4	27.8
Geothermal	39	64	1	1	6.6	4.6	0.0	2.7
Solar	13	69	0	1	34.9	20.4	10.1	105.7
Tide and wave	0	3	0	0	9.5	20.4	–	75.2

	CO$_2$ emissions (Mt)		Shares (%)		Growth (% p.a.)		Change vs. RS (%)	
Total CO$_2$ emissions	7 240	7 135	100	100	0.7	0.2	−3.8	−13.5
Coal	2 429	2 201	34	31	0.7	−0.1	−5.7	−22.1
Oil	3 154	3 183	44	45	0.6	0.3	−3.3	−9.8
Gas	1 657	1 751	23	25	1.2	0.7	−2.0	−7.5
Power generation	2 917	2 655	100	100	0.7	−0.1	−4.9	−20.3
Coal	2 268	2 034	78	77	0.7	−0.2	−5.9	−23.2
Oil	166	117	6	4	−0.6	−1.6	−1.1	−7.3
Gas	483	504	17	19	1.4	0.7	−1.4	−9.7
Total final consumption	3 939	4 031	100	100	0.8	0.4	−3.2	−9.4
Coal	143	124	4	3	−0.3	−0.7	−2.9	−6.5
Oil	2 800	2 886	71	72	0.7	0.4	−3.6	−10.4
Gas	996	1 021	25	25	1.1	0.6	−2.2	−6.8

A

Reference Scenario: United States

	Energy demand (Mtoe)				Shares (%)			Growth (% p.a.)	
	1990	2005	2015	2030	2005	2015	2030	2005-2015	2005-2030
Total primary energy demand	1 924	2 336	2 629	2 925	100	100	100	1.2	0.9
Coal	458	556	624	715	24	24	24	1.2	1.0
Oil	767	952	1 042	1 118	41	40	38	0.9	0.6
Gas	439	508	571	595	22	22	20	1.2	0.6
Nuclear	159	211	221	243	9	8	8	0.5	0.6
Hydro	23	23	25	26	1	1	1	0.7	0.3
Biomass and waste	62	74	115	165	3	4	6	4.5	3.3
Other renewables	14	12	31	64	0	1	2	10.3	7.1
Power generation	745	960	1 080	1 207	100	100	100	1.2	0.9
Coal	391	506	574	634	53	53	53	1.3	0.9
Oil	27	34	32	24	4	3	2	-0.6	-1.4
Gas	90	155	164	160	16	15	13	0.6	0.1
Nuclear	159	211	221	243	22	20	20	0.5	0.6
Hydro	23	23	25	26	2	2	2	0.7	0.3
Biomass and waste	40	22	38	69	2	3	6	5.6	4.7
Other renewables	14	9	26	52	1	2	4	10.8	7.0
Other energy sector	154	155	159	204	100	100	100	0.3	1.1
Total final consumption	1 304	1 596	1 825	2 035	100	100	100	1.4	1.0
Coal	54	30	30	27	2	2	1	-0.2	-0.5
Oil	695	867	967	1 055	54	53	52	1.1	0.8
of which transport	*484*	*625*	*696*	*769*	*39*	*38*	*38*	*1.1*	*0.8*
Gas	303	321	367	387	20	20	19	1.4	0.7
Electricity	226	320	375	454	20	21	22	1.6	1.4
Heat	2	3	4	5	0	0	0	3.2	2.0
Biomass and waste	23	52	77	96	3	4	5	4.0	2.5
of which biofuels	*0*	*8*	*21*	*28*	*1*	*1*	*1*	*10.2*	*5.1*
Other renewables	0	2	4	12	0	0	1	7.6	7.1

	Electricity (TWh)				Shares (%)			Growth (% p.a.)	
Total generation	3 203	4 266	4 959	5 947	100	100	100	1.5	1.3
Coal	1 700	2 154	2 552	3 148	50	51	53	1.7	1.5
Oil	131	141	133	102	3	3	2	-0.6	-1.3
Gas	382	783	858	896	18	17	15	0.9	0.5
Nuclear	611	811	849	933	19	17	16	0.5	0.6
Hydro	273	272	291	297	6	6	5	0.7	0.3
Biomass and waste	86	71	115	209	2	2	4	4.9	4.4
Wind	3	18	119	282	0	2	5	20.9	11.7
Geothermal	16	15	30	49	0	1	1	7.3	4.9
Solar	1	1	12	31	0	0	1	34.3	17.0
Tide and wave	0	0	0	1	0	0	0	n.a.	n.a.

	CO$_2$ emissions (Mt)				Shares (%)			Growth (% p.a.)	
Total CO$_2$ emissions	4 832	5 789	6 392	6 891	100	100	100	1.0	0.7
Coal	1 774	2 131	2 403	2 657	37	38	39	1.2	0.9
Oil	2 047	2 457	2 658	2 847	42	42	41	0.8	0.6
Gas	1 011	1 202	1 330	1 387	21	21	20	1.0	0.6
Power generation	1 829	2 465	2 744	2 945	100	100	100	1.1	0.7
Coal	1 532	1 992	2 257	2 494	81	82	85	1.3	0.9
Oil	88	112	104	78	5	4	3	-0.7	-1.4
Gas	210	361	384	373	15	14	13	0.6	0.1
Total final consumption	2 731	3 066	3 393	3 652	100	100	100	1.0	0.7
Coal	239	126	128	115	4	4	3	0.2	-0.4
Oil	1 795	2 197	2 417	2 645	72	71	72	1.0	0.7
Gas	697	743	848	892	24	25	24	1.3	0.7

Alternative Policy Scenario: United States

	Energy demand (Mtoe)		Shares (%)		Growth (% p.a.)		Change vs. RS (%)	
	2015	2030	2015	2030	2005-2015	2005-2030	2015	2030
Total primary energy demand	**2 574**	**2 712**	**100**	**100**	**1.0**	**0.6**	**−2.1**	**−7.3**
Coal	592	571	23	21	0.6	0.1	−5.2	−20.1
Oil	1 009	1 015	39	37	0.6	0.3	−3.2	−9.2
Gas	562	562	22	21	1.0	0.4	−1.5	−5.6
Nuclear	221	278	9	10	0.5	1.1	–	14.4
Hydro	25	26	1	1	0.7	0.3	0.2	–
Biomass and waste	129	179	5	7	5.7	3.6	12.2	8.2
Other renewables	36	82	1	3	11.9	8.1	16.0	29.0
Power generation	**1 059**	**1 109**	**100**	**100**	**1.0**	**0.6**	**−1.9**	**−8.1**
Coal	543	497	51	45	0.7	−0.1	−5.4	−21.6
Oil	31	22	3	2	−0.7	−1.7	−1.1	−7.7
Gas	165	156	16	14	0.6	0.0	0.3	−2.1
Nuclear	221	278	21	25	0.5	1.1	–	14.4
Hydro	25	26	2	2	0.7	0.3	0.2	–
Biomass and waste	46	70	4	6	7.9	4.8	23.2	1.9
Other renewables	27	60	3	5	11.2	7.7	3.6	15.6
Other energy sector	**156**	**191**	**100**	**100**	**0.1**	**0.8**	**−1.8**	**−6.5**
Total final consumption	**1 784**	**1 889**	**100**	**100**	**1.1**	**0.7**	**−2.2**	**−7.2**
Coal	29	25	2	1	−0.5	−0.8	−2.8	−5.7
Oil	934	953	52	50	0.8	0.4	−3.3	−9.6
of which transport	*668*	*682*	*37*	*36*	*0.7*	*0.4*	*−3.9*	*−11.4*
Gas	359	360	20	19	1.1	0.5	−2.3	−7.0
Electricity	366	415	21	22	1.4	1.0	−2.3	−8.6
Heat	4	5	0	0	2.9	1.7	−2.9	−6.0
Biomass and waste	83	108	5	6	4.7	3.0	6.8	12.8
of which biofuels	*28*	*43*	*2*	*2*	*13.0*	*6.9*	*29.2*	*53.1*
Other renewables	8	22	0	1	14.8	9.9	91.7	89.5

	Electricity (TWh)		Shares (%)		Growth (% p.a.)		Change vs. RS (%)	
Total generation	**4 850**	**5 450**	**100**	**100**	**1.3**	**1.0**	**−2.2**	**−8.4**
Coal	2 402	2 437	50	45	1.1	0.5	−5.9	−22.6
Oil	132	94	3	2	−0.7	−1.6	−1.1	−7.5
Gas	862	886	18	16	1.0	0.5	0.4	−1.1
Nuclear	849	1 067	18	20	0.5	1.1	–	14.4
Hydro	292	297	6	5	0.7	0.3	0.2	–
Biomass and waste	141	213	3	4	7.1	4.5	23.2	1.9
Wind	130	344	3	6	21.9	12.6	8.5	21.9
Geothermal	30	49	1	1	7.3	4.9	–	–
Solar	13	62	0	1	35.3	20.3	7.9	102.8
Tide and wave	0	1	0	0	n.a.	n.a.	–	125.1

	CO$_2$ emissions (Mt)		Shares (%)		Growth (% p.a.)		Change vs. RS (%)	
Total CO$_2$ emissions	**6 160**	**5 986**	**100**	**100**	**0.6**	**0.1**	**−3.6**	**−13.1**
Coal	2 278	2 106	37	35	0.7	−0.0	−5.2	−20.7
Oil	2 572	2 571	42	43	0.5	0.2	−3.3	−9.7
Gas	1 310	1 310	21	22	0.9	0.3	−1.5	−5.6
Power generation	**2 623**	**2 391**	**100**	**100**	**0.6**	**−0.1**	**−4.4**	**−18.8**
Coal	2 136	1 954	81	82	0.7	−0.1	−5.4	−21.6
Oil	102	72	4	3	−0.8	−1.8	−1.1	−7.7
Gas	385	365	15	15	0.6	0.0	0.3	−2.1
Total final consumption	**3 286**	**3 314**	**100**	**100**	**0.7**	**0.3**	**−3.2**	**−9.3**
Coal	125	109	4	3	−0.1	−0.6	−2.8	−5.7
Oil	2 332	2 375	71	72	0.6	0.3	−3.5	−10.2
Gas	829	830	25	25	1.1	0.4	−2.2	−7.0

A

Reference Scenario: OECD Pacific

	Energy demand (Mtoe)				Shares (%)			Growth (% p.a.)	
	1990	2005	2015	2030	2005	2015	2030	2005-2015	2005-2030
Total primary energy demand	639	882	1 016	1 099	100	100	100	1.4	0.9
Coal	138	218	248	246	25	24	22	1.3	0.5
Oil	338	390	400	389	44	39	35	0.2	−0.0
Gas	69	126	168	199	14	17	18	2.9	1.8
Nuclear	66	118	151	191	13	15	17	2.6	2.0
Hydro	11	10	13	13	1	1	1	1.9	1.0
Biomass and waste	10	15	25	38	2	2	3	5.6	4.0
Other renewables	5	6	11	23	1	1	2	6.8	5.6
Power generation	239	382	466	524	100	100	100	2.0	1.3
Coal	61	147	170	168	38	36	32	1.5	0.6
Oil	54	32	25	14	8	5	3	−2.4	−3.3
Gas	40	65	90	108	17	19	21	3.3	2.1
Nuclear	66	118	151	191	31	33	37	2.6	2.0
Hydro	11	10	13	13	3	3	3	1.9	1.0
Biomass and waste	3	6	9	14	1	2	3	4.4	3.6
Other renewables	3	5	9	15	1	2	3	6.4	4.9
Other energy sector	60	72	80	83	100	100	100	1.0	0.6
Total final consumption	437	586	665	714	100	100	100	1.3	0.8
Coal	48	38	38	38	6	6	5	0.0	−0.0
Oil	268	339	357	358	58	54	50	0.5	0.2
of which transport	*116*	*158*	*172*	*174*	*27*	*26*	*24*	*0.9*	*0.4*
Gas	25	57	74	85	10	11	12	2.6	1.6
Electricity	86	136	170	194	23	26	27	2.2	1.4
Heat	0	5	6	7	1	1	1	2.0	1.4
Biomass and waste	7	9	16	25	2	2	3	6.3	4.2
of which biofuels	*0*	*0*	*0*	*1*	*0*	*0*	*0*	*44.7*	*19.2*
Other renewables	2	1	3	7	0	0	1	8.8	7.8

	Electricity (TWh)				Shares (%)			Growth (% p.a.)	
Total generation	1 129	1 779	2 186	2 482	100	100	100	2.1	1.3
Coal	254	665	801	828	37	37	33	1.9	0.9
Oil	273	174	138	73	10	6	3	−2.3	−3.4
Gas	198	332	453	559	19	21	23	3.1	2.1
Nuclear	255	454	581	734	26	27	30	2.5	1.9
Hydro	133	121	146	154	7	7	6	1.9	1.0
Biomass and waste	12	23	33	47	1	1	2	3.8	3.0
Wind	0	3	19	52	0	1	2	18.9	11.6
Geothermal	4	6	9	13	0	0	1	4.0	3.1
Solar	0	0	6	20	0	0	1	46.8	22.2
Tide and wave	0	0	0	1	0	0	0	n.a.	n.a.

	CO_2 emissions (Mt)				Shares (%)			Growth (% p.a.)	
Total CO_2 emissions	1 563	2 064	2 311	2 323	100	100	100	1.1	0.5
Coal	519	823	944	933	40	41	40	1.4	0.5
Oil	887	946	983	945	46	43	41	0.4	−0.0
Gas	157	294	384	445	14	17	19	2.7	1.7
Power generation	544	884	1 030	1 029	100	100	100	1.5	0.6
Coal	281	627	739	730	71	72	71	1.7	0.6
Oil	169	103	80	44	12	8	4	−2.5	−3.3
Gas	94	154	211	254	17	20	25	3.2	2.0
Total final consumption	952	1 106	1 205	1 227	100	100	100	0.9	0.4
Coal	217	179	183	181	16	15	15	0.2	0.1
Oil	678	795	853	852	72	71	69	0.7	0.3
Gas	57	132	170	194	12	14	16	2.5	1.6

Alternative Policy Scenario: OECD Pacific

	Energy demand (Mtoe)		Shares (%)		Growth (% p.a.)		Change vs. RS (%)	
	2015	2030	2015	2030	2005-2015	2005-2030	2015	2030
Total primary energy demand	**996**	**1 044**	**100**	**100**	**1.2**	**0.7**	**–2.0**	**–5.0**
Coal	235	185	24	18	0.8	–0.6	–5.3	–24.7
Oil	386	360	39	34	–0.1	–0.3	–3.4	–7.5
Gas	165	172	17	16	2.8	1.3	–1.8	–13.3
Nuclear	154	222	15	21	2.7	2.6	1.4	16.0
Hydro	13	14	1	1	2.0	1.1	1.3	3.7
Biomass and waste	29	57	3	5	7.3	5.6	17.6	48.1
Other renewables	14	34	1	3	8.7	7.3	19.7	51.0
Power generation	**456**	**494**	**100**	**100**	**1.8**	**1.0**	**–2.2**	**–5.7**
Coal	159	113	35	23	0.8	–1.0	–6.6	–32.7
Oil	22	10	5	2	–3.6	–4.5	–12.0	–27.4
Gas	88	87	19	18	3.1	1.2	–1.7	–19.7
Nuclear	154	222	34	45	2.7	2.6	1.4	16.0
Hydro	13	14	3	3	2.0	1.1	1.3	3.7
Biomass and waste	11	25	2	5	6.9	6.1	27.2	80.6
Other renewables	10	24	2	5	7.6	6.7	12.2	53.8
Other energy sector	**78**	**78**	**100**	**100**	**0.8**	**0.3**	**–2.1**	**–5.9**
Total final consumption	**650**	**677**	**100**	**100**	**1.0**	**0.6**	**–2.2**	**–5.2**
Coal	37	35	6	5	–0.2	–0.4	–2.7	–7.9
Oil	347	333	53	49	0.2	–0.1	–2.9	–6.9
of which transport	*166*	*159*	*25*	*23*	*0.5*	*0.0*	*–3.9*	*–8.6*
Gas	73	80	11	12	2.4	1.3	–1.8	–5.9
Electricity	165	179	25	26	1.9	1.1	–2.9	–7.6
Heat	6	7	1	1	1.7	1.2	–2.8	–5.3
Biomass and waste	19	32	3	5	7.6	5.3	12.6	30.2
of which biofuels	*1*	*3*	*0*	*0*	*62.0*	*24.5*	*n.a.*	*196.4*
Other renewables	4	11	1	2	13.0	9.4	45.4	45.9

	Electricity (TWh)		Shares (%)		Growth (% p.a.)		Change vs. RS (%)	
Total generation	**2 127**	**2 300**	**100**	**100**	**1.8**	**1.0**	**–2.7**	**–7.3**
Coal	743	538	35	23	1.1	–0.8	–7.3	–35.0
Oil	121	52	6	2	–3.6	–4.7	–12.2	–28.6
Gas	445	465	21	20	3.0	1.4	–1.7	–16.9
Nuclear	589	851	28	37	2.6	2.5	1.4	16.0
Hydro	148	160	7	7	2.0	1.1	1.3	3.7
Biomass and waste	40	73	2	3	5.8	4.8	21.9	56.5
Wind	23	102	1	4	21.4	14.6	23.0	95.5
Geothermal	10	16	0	1	4.8	3.8	7.2	17.9
Solar	7	41	0	2	49.1	25.7	17.2	103.8
Tide and wave	0	2	0	0	n.a.	n.a.	n.a.	140.6

	CO_2 emissions (Mt)		Shares (%)		Growth (% p.a.)		Change vs. RS (%)	
Total CO_2 emissions	**2 213**	**1 929**	**100**	**100**	**0.7**	**–0.3**	**–4.2**	**–17.0**
Coal	890	677	40	35	0.8	–0.8	–5.8	–27.4
Oil	947	868	43	45	0.0	–0.3	–3.7	–8.1
Gas	377	384	17	20	2.5	1.1	–1.8	–13.9
Power generation	**968**	**727**	**100**	**100**	**0.9**	**–0.8**	**–6.0**	**–29.4**
Coal	691	491	71	68	1.0	–1.0	–6.6	–32.8
Oil	71	32	7	4	–3.7	–4.6	–12.0	–27.4
Gas	207	204	21	28	3.0	1.1	–1.7	–19.7
Total final consumption	**1 171**	**1 138**	**100**	**100**	**0.6**	**0.1**	**–2.8**	**–7.3**
Coal	178	167	15	15	–0.1	–0.3	–2.7	–7.9
Oil	826	788	71	69	0.4	–0.0	–3.1	–7.5
Gas	167	183	14	16	2.4	1.3	–1.8	–5.8

A

Reference Scenario: Japan

	Energy demand (Mtoe)				Shares (%)			Growth (% p.a.)	
	1990	2005	2015	2030	2005	2015	2030	2005-2015	2005-2030
Total primary energy demand	444	530	589	601	100	100	100	1.1	0.5
Coal	77	112	128	114	21	22	19	1.3	0.1
Oil	251	249	238	211	47	40	35	–0.5	–0.7
Gas	48	72	98	111	14	17	18	3.1	1.7
Nuclear	53	79	98	128	15	17	21	2.1	1.9
Hydro	8	7	8	8	1	1	1	1.6	0.7
Biomass and waste	5	6	13	18	1	2	3	7.5	4.3
Other renewables	3	4	6	11	1	1	2	5.5	4.3
Power generation	172	226	272	297	100	100	100	1.8	1.1
Coal	26	63	73	62	28	27	21	1.5	–0.1
Oil	49	26	20	11	12	7	4	–2.8	–3.5
Gas	33	44	63	73	20	23	25	3.5	2.0
Nuclear	53	79	98	128	35	36	43	2.1	1.9
Hydro	8	7	8	8	3	3	3	1.6	0.7
Biomass and waste	2	4	5	7	2	2	2	3.5	2.3
Other renewables	1	3	5	8	1	2	3	4.9	3.9
Other energy sector	39	47	54	53	100	100	100	1.4	0.5
Total final consumption	304	351	375	373	100	100	100	0.7	0.2
Coal	32	26	25	25	7	7	7	–0.3	–0.2
Oil	190	208	204	187	59	54	50	–0.2	–0.4
of which transport	*75*	*91*	*90*	*79*	*26*	*24*	*21*	*–0.2*	*–0.6*
Gas	14	28	35	36	8	9	10	2.2	1.0
Electricity	65	85	102	111	24	27	30	1.8	1.1
Heat	0	1	1	1	0	0	0	1.7	1.2
Biomass and waste	3	2	7	11	1	2	3	11.9	6.3
of which biofuels	*0*	*0*	*0*	*0*	*0*	*0*	*0*	*n.a.*	*n.a.*
Other renewables	1	1	2	3	0	0	1	7.5	5.6

	Electricity (TWh)				Shares (%)			Growth (% p.a.)	
Total generation	837	1 104	1 302	1 411	100	100	100	1.7	1.0
Coal	117	309	368	315	28	28	22	1.8	0.1
Oil	250	146	110	59	13	8	4	–2.8	–3.5
Gas	166	231	314	375	21	24	27	3.1	2.0
Nuclear	202	314	376	493	28	29	35	1.8	1.8
Hydro	89	78	92	94	7	7	7	1.6	0.7
Biomass and waste	11	20	27	35	2	2	2	3.4	2.3
Wind	0	2	9	26	0	1	2	17.8	11.3
Geothermal	2	3	4	5	0	0	0	2.4	1.9
Solar	0	0	3	8	0	0	1	109.0	39.6
Tide and wave	0	0	0	0	0	0	0	n.a.	n.a.

	CO_2 emissions (Mt)				Shares (%)			Growth (% p.a.)	
Total CO_2 emissions	1 057	1 210	1 291	1 182	100	100	100	0.6	–0.1
Coal	292	419	472	417	35	37	35	1.2	–0.0
Oil	653	620	595	522	51	46	44	–0.4	–0.7
Gas	113	171	224	243	14	17	21	2.7	1.4
Power generation	360	469	543	487	100	100	100	1.5	0.2
Coal	129	278	332	281	59	61	58	1.8	0.0
Oil	153	85	63	34	18	12	7	–2.9	–3.6
Gas	78	106	148	172	23	27	35	3.4	2.0
Total final consumption	653	701	711	667	100	100	100	0.1	–0.2
Coal	147	125	122	119	18	17	18	–0.2	–0.2
Oil	472	511	508	465	73	71	70	–0.1	–0.4
Gas	33	65	81	83	9	11	12	2.2	1.0

Alternative Policy Scenario: Japan

	Energy demand (Mtoe)		Shares (%)		Growth (% p.a.)		Change vs. RS (%)	
	2015	2030	2015	2030	2005-2015	2005-2030	2015	2030
Total primary energy demand	**576**	**567**	**100**	**100**	**0.8**	**0.3**	**−2.1**	**−5.6**
Coal	121	90	21	16	0.8	−0.9	−5.2	−21.5
Oil	229	195	40	34	−0.8	−1.0	−3.5	−7.4
Gas	97	95	17	17	3.0	1.1	−1.1	−14.3
Nuclear	98	139	17	24	2.1	2.3	–	7.9
Hydro	8	8	1	1	1.6	0.8	0.4	1.2
Biomass and waste	15	23	3	4	8.8	5.2	13.5	25.5
Other renewables	8	18	1	3	8.3	6.5	30.1	68.8
Power generation	**265**	**275**	**100**	**100**	**1.6**	**0.8**	**−2.6**	**−7.2**
Coal	68	42	26	15	0.8	−1.6	−6.8	−31.9
Oil	17	7	6	3	−4.3	−5.0	−15.0	−32.3
Gas	62	58	23	21	3.4	1.1	−1.2	−20.5
Nuclear	98	139	37	50	2.1	2.3	–	7.9
Hydro	8	8	3	3	1.6	0.8	0.4	1.2
Biomass and waste	6	9	2	3	5.1	3.3	16.3	29.0
Other renewables	6	12	2	4	6.6	5.9	17.8	62.1
Other energy sector	**53**	**49**	**100**	**100**	**1.2**	**0.2**	**−2.3**	**−7.0**
Total final consumption	**368**	**356**	**100**	**100**	**0.5**	**0.1**	**−2.0**	**−4.7**
Coal	25	23	7	6	−0.6	−0.6	−2.7	−8.1
Oil	199	175	54	49	−0.5	−0.7	−2.6	−6.4
of which transport	*87*	*73*	*24*	*21*	*−0.5*	*−0.9*	*−3.7*	*−7.5*
Gas	35	35	9	10	2.1	0.9	−0.9	−2.5
Electricity	99	103	27	29	1.5	0.8	−2.9	−7.3
Heat	1	1	0	0	1.7	1.2	–	–
Biomass and waste	8	14	2	4	13.1	7.2	11.4	23.5
of which biofuels	*1*	*2*	*0*	*0*	*n.a.*	*n.a.*	*n.a.*	*n.a.*
Other renewables	3	6	1	2	13.0	8.2	65.7	85.3

	Electricity (TWh)		Shares (%)		Growth (% p.a.)		Change vs. RS (%)	
Total generation	**1 264**	**1 307**	**100**	**100**	**1.4**	**0.7**	**−2.9**	**−7.3**
Coal	341	211	27	16	1.0	−1.5	−7.2	−32.8
Oil	93	40	7	3	−4.4	−5.0	−15.0	−32.4
Gas	310	311	25	24	3.0	1.2	−1.1	−17.0
Nuclear	376	532	30	41	1.8	2.1	–	7.9
Hydro	92	95	7	7	1.6	0.8	0.4	1.2
Biomass and waste	32	45	3	3	5.0	3.4	16.8	29.7
Wind	11	46	1	4	20.4	14.0	23.9	81.4
Geothermal	5	8	0	1	3.9	3.4	15.7	44.7
Solar	4	18	0	1	114.4	43.9	29.1	113.3
Tide and wave	0	1	0	0	n.a.	n.a.	n.a.	108.6

	CO$_2$ emissions (Mt)		Shares (%)		Growth (% p.a.)		Change vs. RS (%)	
Total CO$_2$ emissions	**1 238**	**1 002**	**100**	**100**	**0.2**	**−0.8**	**−4.1**	**−15.3**
Coal	446	316	36	32	0.6	−1.1	−5.6	−24.2
Oil	571	480	46	48	−0.8	−1.0	−4.0	−8.2
Gas	221	206	18	21	2.6	0.7	−1.1	−15.3
Power generation	**509**	**351**	**100**	**100**	**0.8**	**−1.1**	**−6.3**	**−27.9**
Coal	309	191	61	54	1.1	−1.5	−6.8	−31.9
Oil	53	23	11	7	−4.5	−5.1	−15.0	−32.3
Gas	146	137	29	39	3.3	1.0	−1.2	−20.5
Total final consumption	**692**	**624**	**100**	**100**	**−0.1**	**−0.5**	**−2.6**	**−6.5**
Coal	119	109	17	17	−0.5	−0.5	−2.7	−8.1
Oil	493	434	71	70	−0.3	−0.6	−2.8	−6.7
Gas	80	81	12	13	2.1	0.9	−0.9	−2.4

A

Reference Scenario: OECD Europe

	Energy demand (Mtoe)				Shares (%)			Growth (% p.a.)	
	1990	2005	2015	2030	2005	2015	2030	2005-2015	2005-2030
Total primary energy demand	1 623	1 874	1 998	2 127	100	100	100	**0.6**	**0.5**
Coal	438	320	310	314	17	15	15	-0.3	-0.1
Oil	627	703	711	709	38	36	33	0.1	0.0
Gas	258	453	525	633	24	26	30	1.5	1.3
Nuclear	204	256	236	150	14	12	7	-0.8	-2.1
Hydro	38	42	50	56	2	3	3	1.9	1.2
Biomass and waste	53	85	127	184	5	6	9	4.2	3.1
Other renewables	4	15	38	82	1	2	4	9.8	7.0
Power generation	618	742	809	860	100	100	100	**0.9**	**0.6**
Coal	277	234	239	257	32	30	30	0.2	0.4
Oil	48	32	24	12	4	3	1	-2.7	-4.0
Gas	40	137	183	252	18	23	29	2.9	2.5
Nuclear	204	256	236	150	34	29	17	-0.8	-2.1
Hydro	38	42	50	56	6	6	6	1.9	1.2
Biomass and waste	8	30	44	66	4	5	8	4.0	3.3
Other renewables	3	12	33	69	2	4	8	10.4	7.2
Other energy sector	128	148	139	139	100	100	100	**-0.6**	**-0.3**
Total final consumption	1 147	1 359	1 467	1 619	100	100	100	**0.8**	**0.7**
Coal	122	53	46	38	4	3	2	-1.5	-1.3
Oil	549	631	650	661	46	44	41	0.3	0.2
of which transport	*289*	*374*	*383*	*400*	*28*	*26*	*25*	*0.2*	*0.3*
Gas	204	296	323	362	22	22	22	0.9	0.8
Electricity	190	254	298	360	19	20	22	1.6	1.4
Heat	37	67	61	68	5	4	4	-1.0	0.1
Biomass and waste	44	55	83	118	4	6	7	4.3	3.1
of which biofuels	*0*	*3*	*18*	*32*	*0*	*1*	*2*	*19.1*	*9.7*
Other renewables	1	3	6	13	0	0	1	7.1	6.2

	Electricity (TWh)				Shares (%)			Growth (% p.a.)	
Total generation	2 632	3 485	4 029	4 815	100	100	100	**1.5**	**1.3**
Coal	1 016	989	1 025	1 237	28	25	26	0.4	0.9
Oil	203	135	117	57	4	3	1	-1.4	-3.4
Gas	167	721	963	1 442	21	24	30	2.9	2.8
Nuclear	782	981	907	574	28	23	12	-0.8	-2.1
Hydro	443	486	587	648	14	15	13	1.9	1.2
Biomass and waste	16	87	144	217	2	4	5	5.2	3.7
Wind	1	79	259	556	2	6	12	12.6	8.1
Geothermal	4	6	10	18	0	0	0	5.4	4.3
Solar	0	2	15	58	0	0	1	24.1	15.0
Tide and wave	1	1	1	9	0	0	0	10.1	12.0

	CO_2 emissions (Mt)				Shares (%)			Growth (% p.a.)	
Total CO_2 emissions	3 936	4 047	4 216	4 493	100	100	100	**0.4**	**0.4**
Coal	1 701	1 235	1 212	1 247	31	29	28	-0.2	0.0
Oil	1 661	1 775	1 801	1 791	44	43	40	0.1	0.0
Gas	574	1 037	1 202	1 454	26	29	32	1.5	1.4
Power generation	1 375	1 390	1 489	1 685	100	100	100	**0.7**	**0.8**
Coal	1 130	968	985	1 061	70	66	63	0.2	0.4
Oil	152	103	78	37	7	5	2	-2.7	-4.0
Gas	93	319	426	587	23	29	35	2.9	2.5
Total final consumption	2 387	2 467	2 545	2 632	100	100	100	**0.3**	**0.3**
Coal	532	235	203	168	10	8	6	-1.5	-1.3
Oil	1 396	1 552	1 600	1 631	63	63	62	0.3	0.2
Gas	458	680	742	833	28	29	32	0.9	0.8

Alternative Policy Scenario: OECD Europe

	Energy demand (Mtoe)		Shares (%)		Growth (% p.a.)		Change vs. RS (%)	
	2015	2030	2015	2030	2005-2015	2005-2030	2015	2030
Total primary energy demand	**1 925**	**1 946**	**100**	**100**	**0.3**	**0.2**	**−3.6**	**−8.5**
Coal	232	163	12	8	−3.2	−2.7	−25.1	−48.0
Oil	682	631	35	32	−0.3	−0.4	−4.0	−10.9
Gas	507	548	26	28	1.1	0.8	−3.4	−13.4
Nuclear	265	219	14	11	0.4	−0.6	12.0	46.2
Hydro	51	58	3	3	2.1	1.3	1.8	4.3
Biomass and waste	144	214	7	11	5.4	3.8	13.0	16.2
Other renewables	44	113	2	6	11.3	8.4	15.0	38.3
Power generation	**769**	**771**	**100**	**100**	**0.4**	**0.1**	**−4.9**	**−10.4**
Coal	163	112	21	15	−3.5	−2.9	−31.5	−56.5
Oil	24	11	3	1	−2.7	−4.1	0.3	−1.5
Gas	175	208	23	27	2.5	1.7	−4.2	−17.4
Nuclear	265	219	34	28	0.4	−0.6	12.0	46.2
Hydro	51	58	7	8	2.1	1.3	1.8	4.3
Biomass and waste	54	76	7	10	6.2	3.8	23.2	14.8
Other renewables	36	87	5	11	11.4	8.2	10.2	26.1
Other energy sector	**135**	**127**	**100**	**100**	**−0.9**	**−0.6**	**−2.6**	**−8.3**
Total final consumption	**1 422**	**1 478**	**100**	**100**	**0.5**	**0.3**	**−3.1**	**−8.7**
Coal	44	34	3	2	−1.8	−1.7	−3.4	−9.0
Oil	622	584	44	40	−0.1	−0.3	−4.4	−11.6
of which transport	*362*	*349*	*25*	*24*	*−0.3*	*−0.3*	*−5.4*	*−12.7*
Gas	313	322	22	22	0.6	0.3	−3.1	−11.1
Electricity	284	306	20	21	1.1	0.7	−4.6	−14.8
Heat	61	67	4	5	−1.0	0.0	−0.6	−1.2
Biomass and waste	90	138	6	9	5.0	3.7	7.7	17.0
of which biofuels	*21*	*41*	*2*	*3*	*21.2*	*10.8*	*19.2*	*28.6*
Other renewables	8	26	1	2	10.9	9.2	42.5	102.8

	Electricity (TWh)		Shares (%)		Growth (% p.a.)		Change vs. RS (%)	
Total generation	**3 834**	**4 137**	**100**	**100**	**1.0**	**0.7**	**−4.8**	**−14.1**
Coal	684	494	18	12	−3.6	−2.7	−33.3	−60.1
Oil	118	56	3	1	−1.3	−3.4	0.4	−1.2
Gas	918	1 015	24	25	2.4	1.4	−4.7	−29.6
Nuclear	1 017	839	27	20	0.4	−0.6	12.0	46.2
Hydro	597	675	16	16	2.1	1.3	1.8	4.3
Biomass and waste	177	250	5	6	7.4	4.3	23.2	14.8
Wind	295	688	8	17	14.1	9.1	14.2	23.6
Geothermal	11	23	0	1	5.9	5.3	4.6	26.7
Solar	16	79	0	2	24.6	16.4	4.2	36.1
Tide and wave	2	18	0	0	12.7	15.3	25.6	107.8

	CO$_2$ emissions (Mt)		Shares (%)		Growth (% p.a.)		Change vs. RS (%)	
Total CO$_2$ emissions	**3 782**	**3 477**	**100**	**100**	**−0.7**	**−0.6**	**−10.3**	**−22.6**
Coal	894	631	24	18	−3.2	−2.7	−26.2	−49.4
Oil	1 727	1 589	46	46	−0.3	−0.4	−4.1	−11.3
Gas	1 161	1 258	31	36	1.1	0.8	−3.4	−13.5
Power generation	**1 161**	**983**	**100**	**100**	**−1.8**	**−1.4**	**−22.0**	**−41.7**
Coal	675	462	58	47	−3.5	−2.9	−31.5	−56.5
Oil	78	36	7	4	−2.7	−4.1	0.3	−1.5
Gas	409	485	35	49	2.5	1.7	−4.2	−17.4
Total final consumption	**2 442**	**2 322**	**100**	**100**	**−0.1**	**−0.2**	**−4.1**	**−11.8**
Coal	196	152	8	7	−1.8	−1.7	−3.4	−9.1
Oil	1 527	1 431	63	62	−0.2	−0.3	−4.6	−12.3
Gas	719	739	29	32	0.6	0.3	−3.1	−11.2

A

Reference Scenario: European Union

	Energy demand (Mtoe)				Shares (%)			Growth (% p.a.)	
	1990	2005	2015	2030	2005	2015	2030	2005-2015	2005-2030
Total primary energy demand	**1 653**	**1 814**	**1 910**	**2 006**	**100**	**100**	**100**	**0.5**	**0.4**
Coal	451	317	291	275	17	15	14	−0.8	−0.6
Oil	626	671	678	670	37	35	33	0.1	−0.0
Gas	295	444	509	610	24	27	30	1.4	1.3
Nuclear	207	260	239	159	14	13	8	−0.8	−2.0
Hydro	25	26	34	37	1	2	2	2.8	1.4
Biomass and waste	46	83	127	182	5	7	9	4.3	3.2
Other renewables	3	13	33	72	1	2	4	10.0	7.2
Power generation	**651**	**733**	**778**	**808**	**100**	**100**	**100**	**0.6**	**0.4**
Coal	293	240	232	231	33	30	29	−0.3	−0.2
Oil	61	34	26	13	5	3	2	−2.8	−3.9
Gas	54	132	172	239	18	22	30	2.7	2.4
Nuclear	207	260	239	159	35	31	20	−0.8	−2.0
Hydro	25	26	34	37	4	4	5	2.8	1.4
Biomass and waste	8	29	44	65	4	6	8	4.1	3.2
Other renewables	3	12	30	65	2	4	8	10.1	7.1
Other energy sector	**129**	**141**	**134**	**132**	**100**	**100**	**100**	**−0.5**	**−0.3**
Total final consumption	**1 157**	**1 302**	**1 395**	**1 528**	**100**	**100**	**100**	**0.7**	**0.6**
Coal	120	44	33	24	3	2	2	−2.8	−2.3
Oil	532	601	618	624	46	44	41	0.3	0.2
of which transport	*281*	*361*	*368*	*383*	*28*	*26*	*25*	*0.2*	*0.2*
Gas	228	295	320	356	23	23	23	0.8	0.8
Electricity	185	237	273	326	18	20	21	1.4	1.3
Heat	54	71	66	74	5	5	5	−0.7	0.2
Biomass and waste	38	54	82	117	4	6	8	4.4	3.2
of which biofuels	*0*	*3*	*18*	*32*	*0*	*1*	*2*	*19.1*	*9.7*
Other renewables	1	1	2	7	0	0	0	8.6	7.7

	Electricity (TWh)				Shares (%)			Growth (% p.a.)	
Total generation	**2 565**	**3 275**	**3 736**	**4 404**	**100**	**100**	**100**	**1.3**	**1.2**
Coal	1 055	1 001	981	1 094	31	26	25	−0.2	0.4
Oil	221	139	117	57	4	3	1	−1.7	−3.5
Gas	191	664	891	1 362	20	24	31	3.0	2.9
Nuclear	795	998	918	610	30	25	14	−0.8	−2.0
Hydro	286	304	399	432	9	11	10	2.8	1.4
Biomass and waste	14	84	146	217	3	4	5	5.7	3.8
Wind	0	78	259	552	2	7	13	12.7	8.1
Geothermal	3	5	8	14	0	0	0	5.0	4.3
Solar	0	2	15	58	0	0	1	24.1	15.0
Tide and wave	0	1	1	9	0	0	0	10.1	11.9

	CO$_2$ emissions (Mt)				Shares (%)			Growth (% p.a.)	
Total CO$_2$ emissions	**4 084**	**3 944**	**4 011**	**4 176**	**100**	**100**	**100**	**0.2**	**0.2**
Coal	1 759	1 223	1 134	1 085	31	28	26	−0.7	−0.5
Oil	1 667	1 706	1 713	1 691	43	43	41	0.0	−0.0
Gas	658	1 015	1 164	1 399	26	29	34	1.4	1.3
Power generation	**1 517**	**1 408**	**1 440**	**1 547**	**100**	**100**	**100**	**0.2**	**0.4**
Coal	1 194	992	957	951	70	66	61	−0.4	−0.2
Oil	196	109	81	40	8	6	3	−2.8	−3.9
Gas	127	307	401	556	22	28	36	2.7	2.4
Total final consumption	**2 398**	**2 353**	**2 405**	**2 470**	**100**	**100**	**100**	**0.2**	**0.2**
Coal	528	198	152	115	8	6	5	−2.7	−2.2
Oil	1 357	1 478	1 518	1 539	63	63	62	0.3	0.2
Gas	513	677	735	816	29	31	33	0.8	0.7

Alternative Policy Scenario: European Union

	Energy demand (Mtoe)		Shares (%)		Growth (% p.a.)		Change vs. RS (%)	
	2015	2030	2015	2030	2005-2015	2005-2030	2015	2030
Total primary energy demand	**1 846**	**1 844**	**100**	**100**	**0.2**	**0.1**	**-3.4**	**-8.1**
Coal	218	142	12	8	-3.7	-3.2	-24.9	-48.5
Oil	650	595	35	32	-0.3	-0.5	-4.0	-11.3
Gas	492	529	27	29	1.0	0.7	-3.2	-13.3
Nuclear	269	230	15	12	0.3	-0.5	12.5	44.9
Hydro	34	39	2	2	2.8	1.6	0.4	4.6
Biomass and waste	143	213	8	12	5.5	3.8	13.0	16.6
Other renewables	38	97	2	5	11.5	8.5	15.5	34.7
Power generation	**743**	**732**	**100**	**100**	**0.1**	**-0.0**	**-4.5**	**-9.4**
Coal	161	102	22	14	-3.9	-3.4	-30.6	-56.0
Oil	25	11	3	2	-2.9	-4.3	-1.0	-9.9
Gas	165	196	22	27	2.3	1.6	-4.0	-17.8
Nuclear	269	230	36	31	0.3	-0.5	12.5	44.9
Hydro	34	39	5	5	2.8	1.6	0.4	4.6
Biomass and waste	54	75	7	10	6.3	3.8	22.3	14.8
Other renewables	33	79	5	11	11.1	7.9	9.9	21.2
Other energy sector	**131**	**121**	**100**	**100**	**-0.8**	**-0.6**	**-2.4**	**-8.3**
Total final consumption	**1 354**	**1 394**	**100**	**100**	**0.4**	**0.3**	**-2.9**	**-8.7**
Coal	32	22	2	2	-3.0	-2.6	-2.2	-8.3
Oil	591	550	44	39	-0.2	-0.4	-4.3	-11.9
of which transport	*347*	*332*	*26*	*24*	*-0.4*	*-0.3*	*-5.7*	*-13.3*
Gas	311	317	23	23	0.5	0.3	-2.9	-10.8
Electricity	261	277	19	20	1.0	0.6	-4.4	-15.0
Heat	65	72	5	5	-0.8	0.1	-1.1	-2.4
Biomass and waste	89	138	7	10	5.2	3.8	8.1	17.7
of which biofuels	*21*	*41*	*2*	*3*	*21.2*	*10.8*	*19.2*	*28.6*
Other renewables	5	18	0	1	15.4	12.0	84.5	162.0

	Electricity (TWh)		Shares (%)		Growth (% p.a.)		Change vs. RS (%)	
Total generation	**3 564**	**3 783**	**100**	**100**	**0.9**	**0.6**	**-4.6**	**-14.1**
Coal	667	445	19	12	-4.0	-3.2	-32.0	-59.3
Oil	117	54	3	1	-1.7	-3.7	-0.3	-5.9
Gas	851	948	24	25	2.5	1.4	-4.5	-30.4
Nuclear	1 033	884	29	23	0.3	-0.5	12.5	44.9
Hydro	401	452	11	12	2.8	1.6	0.4	4.6
Biomass and waste	179	249	5	7	7.8	4.4	22.0	14.8
Wind	292	639	8	17	14.0	8.8	12.8	15.7
Geothermal	8	18	0	0	5.7	5.5	6.2	32.6
Solar	16	77	0	2	24.6	16.3	3.6	33.8
Tide and wave	2	18	0	0	12.7	15.3	25.6	107.5

	CO$_2$ emissions (Mt)		Shares (%)		Growth (% p.a.)		Change vs. RS (%)	
Total CO$_2$ emissions	**3 606**	**3 244**	**100**	**100**	**-0.9**	**-0.8**	**-10.1**	**-22.3**
Coal	837	540	23	17	-3.7	-3.2	-26.2	-50.2
Oil	1 642	1 492	46	46	-0.4	-0.5	-4.1	-11.8
Gas	1 126	1 211	31	37	1.0	0.7	-3.3	-13.5
Power generation	**1 130**	**912**	**100**	**100**	**-2.2**	**-1.7**	**-21.5**	**-41.1**
Coal	664	418	59	46	-3.9	-3.4	-30.6	-56.0
Oil	81	36	7	4	-2.9	-4.3	-1.0	-9.7
Gas	385	457	34	50	2.3	1.6	-4.0	-17.8
Total final consumption	**2 310**	**2 177**	**100**	**100**	**-0.2**	**-0.3**	**-3.9**	**-11.9**
Coal	148	105	6	5	-2.9	-2.5	-2.3	-8.5
Oil	1 448	1 345	63	62	-0.2	-0.4	-4.6	-12.6
Gas	714	727	31	33	0.5	0.3	-2.9	-10.9

A

Reference Scenario: Transition economies

	Energy demand (Mtoe)				Shares (%)			Growth (% p.a.)	
	1990	2005	2015	2030	2005	2015	2030	2005-2015	2005-2030
Total primary energy demand	**1 554**	**1 080**	**1 273**	**1 434**	**100**	**100**	**100**	**1.7**	**1.1**
Coal	362	204	239	229	19	19	16	1.6	0.5
Oil	487	220	256	283	20	20	20	1.5	1.0
Gas	604	539	641	743	50	50	52	1.8	1.3
Nuclear	61	73	81	104	7	6	7	1.0	1.4
Hydro	23	26	30	36	2	2	2	1.4	1.2
Biomass and waste	17	17	20	25	2	2	2	1.4	1.5
Other renewables	0	1	5	13	0	0	1	25.9	13.6
Power generation	**749**	**543**	**622**	**680**	**100**	**100**	**100**	**1.4**	**0.9**
Coal	200	135	158	142	25	25	21	1.6	0.2
Oil	127	26	23	18	5	4	3	-1.4	-1.6
Gas	334	278	319	360	51	51	53	1.4	1.0
Nuclear	61	73	81	104	14	13	15	1.0	1.4
Hydro	23	26	30	36	5	5	5	1.4	1.2
Biomass and waste	4	5	6	8	1	1	1	2.1	2.4
Other renewables	0	0	5	12	0	1	2	30.3	15.2
Other energy sector	**165**	**158**	**181**	**203**	**100**	**100**	**100**	**1.4**	**1.0**
Total final consumption	**1 104**	**689**	**821**	**943**	**100**	**100**	**100**	**1.8**	**1.3**
Coal	116	41	47	51	6	6	5	1.3	0.8
Oil	297	165	200	226	24	24	24	1.9	1.3
of which transport	*139*	*92*	*114*	*128*	*13*	*14*	*14*	*2.2*	*1.3*
Gas	262	227	279	333	33	34	35	2.1	1.5
Electricity	128	94	119	149	14	14	16	2.3	1.8
Heat	288	149	162	167	22	20	18	0.8	0.5
Biomass and waste	13	12	14	16	2	2	2	1.3	1.2
of which biofuels	*0*	*0*	*0*	*0*	*0*	*0*	*0*	*68.6*	*27.1*
Other renewables	0	0	0	1	0	0	0	3.2	4.7

	Electricity (TWh)				Shares (%)			Growth (% p.a.)	
Total generation	**1 910**	**1 554**	**1 931**	**2 365**	**100**	**100**	**100**	**2.2**	**1.7**
Coal	448	339	435	454	22	23	19	2.5	1.2
Oil	270	52	40	29	3	2	1	-2.8	-2.3
Gas	695	574	769	986	37	40	42	3.0	2.2
Nuclear	231	281	312	399	18	16	17	1.0	1.4
Hydro	266	304	347	414	20	18	17	1.4	1.2
Biomass and waste	0	3	11	36	0	1	2	14.5	10.6
Wind	0	0	12	34	0	1	1	53.7	23.9
Geothermal	0	0	5	11	0	0	0	55.6	23.3
Solar	0	0	0	1	0	0	0	69.9	33.2
Tide and wave	0	0	0	0	0	0	0	n.a.	n.a.

	CO_2 emissions (Mt)				Shares (%)			Growth (% p.a.)	
Total CO_2 emissions	**4 017**	**2 538**	**2 988**	**3 230**	**100**	**100**	**100**	**1.6**	**1.0**
Coal	1 347	776	914	872	31	31	27	1.7	0.5
Oil	1 263	553	640	698	22	21	22	1.5	0.9
Gas	1 408	1 210	1 434	1 659	48	48	51	1.7	1.3
Power generation	**1 998**	**1 303**	**1 479**	**1 493**	**100**	**100**	**100**	**1.3**	**0.5**
Coal	812	562	658	593	43	44	40	1.6	0.2
Oil	407	87	73	56	7	5	4	-1.7	-1.7
Gas	779	654	748	844	50	51	56	1.4	1.0
Total final consumption	**1 912**	**1 129**	**1 382**	**1 590**	**100**	**100**	**100**	**2.0**	**1.4**
Coal	529	209	250	273	19	18	17	1.8	1.1
Oil	790	413	505	569	37	37	36	2.0	1.3
Gas	593	507	628	748	45	45	47	2.2	1.6

Alternative Policy Scenario: Transition economies

	Energy demand (Mtoe)		Shares (%)		Growth (% p.a.)		Change vs. RS (%)	
	2015	2030	2015	2030	2005-2015	2005-2030	2015	2030
Total primary energy demand	1 225	1 298	100	100	1.3	0.7	–3.7	–9.5
Coal	227	198	19	15	1.1	–0.1	–5.0	–13.6
Oil	241	250	20	19	0.9	0.5	–5.9	–11.7
Gas	619	640	50	49	1.4	0.7	–3.6	–13.9
Nuclear	83	124	7	10	1.2	2.1	1.8	19.3
Hydro	30	38	2	3	1.4	1.5	0.5	7.6
Biomass and waste	20	29	2	2	1.6	2.2	2.0	16.4
Other renewables	6	17	0	1	26.7	14.8	6.6	31.6
Power generation	598	613	100	100	1.0	0.5	–3.9	–9.8
Coal	149	119	25	19	1.0	–0.5	–5.7	–15.9
Oil	22	16	4	3	–1.5	–1.9	–0.9	–7.7
Gas	302	289	51	47	0.9	0.2	–5.2	–19.7
Nuclear	83	124	14	20	1.2	2.1	1.8	19.3
Hydro	30	38	5	6	1.4	1.5	0.5	7.6
Biomass and waste	6	11	1	2	2.5	3.4	3.9	29.3
Other renewables	5	15	1	2	30.4	16.0	0.4	19.3
Other energy sector	170	179	100	100	0.7	0.5	–6.0	–12.0
Total final consumption	793	856	100	100	1.4	0.9	–3.4	–9.3
Coal	46	45	6	5	1.0	0.4	–3.5	–11.2
Oil	186	198	23	23	1.2	0.7	–6.8	–12.4
of which transport	*103*	*109*	*13*	*13*	*1.2*	*0.7*	*–8.9*	*–14.2*
Gas	273	304	34	36	1.9	1.2	–2.0	–8.7
Electricity	115	134	15	16	2.0	1.4	–3.1	–9.7
Heat	158	154	20	18	0.6	0.1	–2.5	–8.2
Biomass and waste	14	18	2	2	1.4	1.6	1.2	10.2
of which biofuels	*0*	*0*	*0*	*0*	*78.3*	*29.6*	*75.6*	*62.9*
Other renewables	1	2	0	0	12.7	10.8	140.3	n.a.

	Electricity (TWh)		Shares (%)		Growth (% p.a.)		Change vs. RS (%)	
Total generation	1 875	2 180	100	100	1.9	1.4	–2.9	–7.8
Coal	398	328	21	15	1.6	–0.1	–8.5	–27.7
Oil	39	25	2	1	–2.9	–2.9	–1.2	–14.0
Gas	741	784	40	36	2.6	1.3	–3.7	–20.5
Nuclear	317	477	17	22	1.2	2.1	1.8	19.3
Hydro	349	445	19	20	1.4	1.5	0.5	7.6
Biomass and waste	14	46	1	2	17.3	11.7	26.8	27.7
Wind	12	61	1	3	53.9	26.8	1.9	78.6
Geothermal	5	11	0	1	55.6	23.3	–	–
Solar	0	2	0	0	69.9	36.3	–	78.5
Tide and wave	0	0	0	0	n.a.	n.a.	n.a.	n.a.

	CO$_2$ emissions (Mt)		Shares (%)		Growth (% p.a.)		Change vs. RS (%)	
Total CO$_2$ emissions	2 853	2 787	100	100	1.2	0.4	–4.5	–13.7
Coal	867	748	30	27	1.1	–0.1	–5.1	–14.3
Oil	604	616	21	22	0.9	0.4	–5.7	–11.7
Gas	1 382	1 423	48	51	1.3	0.7	–3.6	–14.2
Power generation	1 402	1 229	100	100	0.7	–0.2	–5.2	–17.7
Coal	621	499	44	41	1.0	–0.5	–5.7	–15.8
Oil	72	52	5	4	–1.8	–2.0	–0.8	–7.5
Gas	709	677	51	55	0.8	0.1	–5.2	–19.7
Total final consumption	1 328	1 423	100	100	1.6	0.9	–3.9	–10.5
Coal	241	243	18	17	1.4	0.6	–3.7	–11.1
Oil	472	498	36	35	1.3	0.8	–6.5	–12.4
Gas	616	682	46	48	2.0	1.2	–2.0	–8.8

A

Reference Scenario: Russia

	Energy demand (Mtoe)				Shares (%)			Growth (% p.a.)	
	1990	2005	2015	2030	2005	2015	2030	2005-2015	2005-2030
Total primary energy demand	879	645	766	871	100	100	100	1.7	1.2
Coal	182	103	125	131	16	16	15	1.9	1.0
Oil	272	133	152	166	21	20	19	1.3	0.9
Gas	367	348	416	473	54	54	54	1.8	1.2
Nuclear	31	39	46	68	6	6	8	1.6	2.3
Hydro	14	15	16	17	2	2	2	0.8	0.6
Biomass and waste	12	7	7	6	1	1	1	-0.4	-0.3
Other renewables	0	0	4	8	0	0	1	26.3	13.5
Power generation	445	354	403	450	100	100	100	1.3	1.0
Coal	105	77	93	97	22	23	22	2.0	0.9
Oil	62	16	15	12	5	4	3	-0.7	-1.3
Gas	228	202	226	245	57	56	54	1.1	0.8
Nuclear	31	39	46	68	11	11	15	1.6	2.3
Hydro	14	15	16	17	4	4	4	0.8	0.6
Biomass and waste	4	4	3	3	1	1	1	-1.5	-1.6
Other renewables	0	0	4	8	0	1	2	26.1	13.5
Other energy sector	94	93	109	122	100	100	100	1.6	1.1
Total final consumption	657	420	499	566	100	100	100	1.7	1.2
Coal	55	18	19	20	4	4	3	0.8	0.4
Oil	155	95	112	126	23	22	22	1.6	1.1
of which transport	*84*	*54*	*65*	*74*	*13*	*13*	*13*	*1.9*	*1.3*
Gas	143	128	165	199	30	33	35	2.6	1.8
Electricity	71	56	68	83	13	14	15	2.0	1.6
Heat	224	120	131	135	29	26	24	0.9	0.5
Biomass and waste	8	3	3	3	1	1	1	1.2	1.1
of which biofuels	*0*	*0*	*0*	*0*	*0*	*0*	*0*	*n.a.*	*n.a.*
Other renewables	0	0	0	0	0	0	0	n.a.	n.a.

	Electricity (TWh)				Shares (%)			Growth (% p.a.)	
Total generation	1 077	946	1 126	1 352	100	100	100	1.8	1.4
Coal	157	166	226	293	18	20	22	3.2	2.3
Oil	124	20	18	13	2	2	1	-0.9	-1.7
Gas	511	435	508	550	46	45	41	1.6	0.9
Nuclear	118	149	176	261	16	16	19	1.6	2.3
Hydro	166	173	188	200	18	17	15	0.8	0.6
Biomass and waste	0	3	2	13	0	0	1	-3.7	6.6
Wind	0	0	4	13	0	0	1	88.5	35.0
Geothermal	0	0	4	9	0	0	1	51.9	22.1
Solar	0	0	0	0	0	0	0	n.a.	n.a.
Tide and wave	0	0	0	0	0	0	0	n.a.	n.a.

	CO$_2$ emissions (Mt)				Shares (%)			Growth (% p.a.)	
Total CO$_2$ emissions	2 189	1 528	1 802	1 973	100	100	100	1.7	1.0
Coal	688	429	507	525	28	28	27	1.7	0.8
Oil	634	316	367	395	21	20	20	1.5	0.9
Gas	866	783	928	1 053	51	52	53	1.7	1.2
Power generation	1 163	859	974	1 022	100	100	100	1.3	0.7
Coal	432	327	396	411	38	41	40	1.9	0.9
Oil	199	55	49	38	6	5	4	-1.1	-1.5
Gas	532	476	529	573	55	54	56	1.0	0.7
Total final consumption	970	609	744	854	100	100	100	2.0	1.4
Coal	254	101	108	111	17	15	13	0.8	0.4
Oil	397	225	268	300	37	36	35	1.8	1.2
Gas	318	283	368	443	46	49	52	2.7	1.8

Alternative Policy Scenario: Russia

	Energy demand (Mtoe)		Shares (%)		Growth (% p.a.)		Change vs. RS (%)	
	2015	2030	2015	2030	2005-2015	2005-2030	2015	2030
Total primary energy demand	729	789	100	100	1.2	0.8	−4.8	−9.4
Coal	121	118	17	15	1.6	0.5	−3.2	−10.0
Oil	141	147	19	19	0.6	0.4	−7.4	−11.8
Gas	394	411	54	52	1.3	0.7	−5.3	−13.1
Nuclear	46	77	6	10	1.6	2.8	–	13.5
Hydro	16	19	2	2	0.9	0.9	0.8	7.5
Biomass and waste	7	6	1	1	−0.3	−0.3	0.5	−0.0
Other renewables	4	11	1	1	26.6	14.5	2.6	24.3
Power generation	385	407	100	100	0.9	0.6	−4.6	−9.6
Coal	91	87	24	21	1.7	0.5	−2.5	−10.1
Oil	15	12	4	3	−0.7	−1.4	−0.7	−0.6
Gas	210	200	54	49	0.4	−0.1	−7.2	−18.5
Nuclear	46	77	12	19	1.6	2.8	–	13.5
Hydro	16	19	4	5	0.9	0.9	0.8	7.5
Biomass and waste	3	3	1	1	−1.5	−1.6	–	–
Other renewables	4	10	1	2	26.2	14.2	0.3	15.8
Other energy sector	99	105	100	100	0.7	0.5	−8.8	−14.2
Total final consumption	477	516	100	100	1.3	0.8	−4.3	−8.8
Coal	19	18	4	3	0.4	−0.0	−4.5	−9.6
Oil	102	109	21	21	0.7	0.5	−8.7	−13.2
of which transport	*59*	*63*	*12*	*12*	*0.9*	*0.6*	*−10.0*	*−14.7*
Gas	160	184	33	36	2.3	1.5	−3.2	−7.4
Electricity	65	76	14	15	1.6	1.3	−4.0	−8.4
Heat	128	125	27	24	0.7	0.1	−2.4	−7.5
Biomass and waste	3	3	1	1	1.4	1.1	1.2	−0.1
of which biofuels	*0*	*0*	*0*	*0*	*n.a.*	*n.a.*	*75.6*	*62.9*
Other renewables	0	1	0	0	n.a.	n.a.	176.6	n.a.

	Electricity (TWh)		Shares (%)		Growth (% p.a.)		Change vs. RS (%)	
Total generation	1 088	1 283	100	100	1.4	1.2	−3.4	−5.1
Coal	212	223	19	17	2.5	1.2	−6.4	−23.9
Oil	18	13	2	1	−1.3	−1.9	−4.3	−3.7
Gas	482	484	44	38	1.0	0.4	−5.1	−12.1
Nuclear	176	297	16	23	1.6	2.8	–	13.5
Hydro	189	215	17	17	0.9	0.9	0.8	7.5
Biomass and waste	4	15	0	1	3.9	7.3	113.5	17.0
Wind	4	28	0	2	89.0	39.3	3.0	122.9
Geothermal	4	9	0	1	51.9	22.1	–	–
Solar	0	0	0	0	n.a.	n.a.	n.a.	n.a.
Tide and wave	0	0	0	0	n.a.	n.a.	n.a.	n.a.

	CO$_2$ emissions (Mt)		Shares (%)		Growth (% p.a.)		Change vs. RS (%)	
Total CO$_2$ emissions	1 712	1 735	100	100	1.1	0.5	−5.0	−12.1
Coal	492	473	29	27	1.4	0.4	−3.0	−9.8
Oil	341	349	20	20	0.8	0.4	−6.9	−11.5
Gas	878	912	51	53	1.2	0.6	−5.4	−13.4
Power generation	926	875	100	100	0.8	0.1	−4.9	−14.4
Coal	386	370	42	42	1.7	0.5	−2.5	−10.1
Oil	49	38	5	4	−1.2	−1.5	−0.7	−0.6
Gas	491	467	53	53	0.3	−0.1	−7.2	−18.5
Total final consumption	706	772	100	100	1.5	1.0	−5.1	−9.6
Coal	103	101	15	13	0.3	0.0	−4.6	−9.0
Oil	246	261	35	34	0.9	0.6	−8.2	−13.1
Gas	356	410	50	53	2.3	1.5	−3.1	−7.3

A

Reference Scenario: Developing countries

	Energy demand (Mtoe)				Shares (%)			Growth (% p.a.)	
	1990	2005	2015	2030	2005	2015	2030	2005-2015	2005-2030
Total primary energy demand	**2 569**	**4 635**	**6 720**	**9 270**	**100**	**100**	**100**	**3.8**	**2.8**
Coal	791	1 557	2 523	3 446	34	38	37	4.9	3.2
Oil	721	1 362	1 892	2 606	29	28	28	3.3	2.6
Gas	228	604	977	1 551	13	15	17	4.9	3.8
Nuclear	15	37	81	133	1	1	1	8.2	5.3
Hydro	61	116	175	251	3	3	3	4.2	3.1
Biomass and waste	745	937	1 023	1 175	20	15	13	0.9	0.9
Other renewables	7	22	49	108	0	1	1	8.5	6.6
Power generation	**533**	**1 487**	**2 455**	**3 732**	**100**	**100**	**100**	**5.1**	**3.7**
Coal	276	902	1 509	2 218	61	61	59	5.3	3.7
Oil	100	154	169	156	10	7	4	0.9	0.1
Gas	73	249	450	778	17	18	21	6.1	4.7
Nuclear	15	37	81	133	2	3	4	8.2	5.3
Hydro	61	116	175	251	8	7	7	4.2	3.1
Biomass and waste	2	12	31	112	1	1	3	9.9	9.4
Other renewables	7	18	40	83	1	2	2	8.2	6.3
Other energy sector	**307**	**544**	**825**	**1 130**	**100**	**100**	**100**	**4.3**	**3.0**
Total final consumption	**1 945**	**3 197**	**4 512**	**6 107**	**100**	**100**	**100**	**3.5**	**2.6**
Coal	416	492	728	825	15	16	14	4.0	2.1
Oil	568	1 096	1 563	2 239	34	35	37	3.6	2.9
of which transport	*278*	*549*	*812*	*1 299*	*17*	*18*	*21*	*4.0*	*3.5*
Gas	99	260	409	620	8	9	10	4.6	3.5
Electricity	158	427	794	1 305	13	18	21	6.4	4.6
Heat	14	44	73	94	1	2	2	5.2	3.1
Biomass and waste	689	874	937	1 000	27	21	16	0.7	0.5
of which biofuels	*6*	*7*	*16*	*40*	*0*	*0*	*1*	*8.0*	*7.0*
Other renewables	0	4	9	24	0	0	0	9.6	7.8

	Electricity (TWh)				Shares (%)			Growth (% p.a.)	
Total generation	**2 323**	**6 252**	**11 397**	**18 422**	**100**	**100**	**100**	**6.2**	**4.4**
Coal	921	3 048	6 093	9 944	49	53	54	7.2	4.8
Oil	360	596	659	605	10	6	3	1.0	0.1
Gas	260	1 054	2 048	3 720	17	18	20	6.9	5.2
Nuclear	57	141	310	511	2	3	3	8.2	5.3
Hydro	709	1 348	2 039	2 919	22	18	16	4.2	3.1
Biomass and waste	7	36	92	312	1	1	2	9.7	9.0
Wind	0	9	117	294	0	1	2	28.7	14.8
Geothermal	8	19	36	69	0	0	0	6.7	5.3
Solar	0	1	3	48	0	0	0	15.6	18.6
Tide and wave	0	0	0	0	0	0	0	–	11.5

	CO_2 emissions (Mt)				Shares (%)			Growth (% p.a.)	
Total CO_2 emissions	**5 260**	**10 700**	**16 432**	**22 919**	**100**	**100**	**100**	**4.4**	**3.1**
Coal	2 838	5 869	9 444	12 821	55	57	56	4.9	3.2
Oil	1 940	3 507	4 842	6 680	33	29	29	3.3	2.6
Gas	482	1 325	2 146	3 419	12	13	15	4.9	3.9
Power generation	**1 567**	**4 652**	**7 622**	**11 171**	**100**	**100**	**100**	**5.1**	**3.6**
Coal	1 081	3 584	6 034	8 853	77	79	79	5.3	3.7
Oil	316	485	535	493	10	7	4	1.0	0.1
Gas	170	583	1 054	1 824	13	14	16	6.1	4.7
Total final consumption	**3 393**	**5 487**	**7 944**	**10 611**	**100**	**100**	**100**	**3.8**	**2.7**
Coal	1 696	2 155	3 104	3 546	39	39	33	3.7	2.0
Oil	1 495	2 787	3 999	5 793	51	50	55	3.7	3.0
Gas	201	545	840	1 273	10	11	12	4.4	3.5

Alternative Policy Scenario: Developing countries

	Energy demand (Mtoe)		Shares (%)		Growth (% p.a.)		Change vs. RS (%)	
	2015	2030	2015	2030	2005-2015	2005-2030	2015	2030
Total primary energy demand	**6 395**	**7 991**	**100**	**100**	**3.3**	**2.2**	**−4.8**	**−13.8**
Coal	2 319	2 558	36	32	4.1	2.0	−8.1	−25.8
Oil	1 785	2 214	28	28	2.7	2.0	−5.6	−15.1
Gas	930	1 326	15	17	4.4	3.2	−4.8	−14.5
Nuclear	95	205	1	3	10.0	7.1	18.0	53.7
Hydro	198	291	3	4	5.5	3.8	13.0	16.1
Biomass and waste	1 013	1 226	16	15	0.8	1.1	−1.0	4.4
Other renewables	55	172	1	2	9.7	8.6	12.2	59.9
Power generation	**2 297**	**3 117**	**100**	**100**	**4.4**	**3.0**	**−6.4**	**−16.5**
Coal	1 354	1 577	59	51	4.2	2.3	−10.3	−28.9
Oil	153	130	7	4	−0.0	−0.7	−9.2	−16.8
Gas	411	589	18	19	5.1	3.5	−8.7	−24.4
Nuclear	95	205	4	7	10.0	7.1	18.0	53.7
Hydro	198	291	9	9	5.5	3.8	13.0	16.1
Biomass and waste	43	208	2	7	13.6	12.1	39.8	85.6
Other renewables	42	117	2	4	8.8	7.8	5.1	40.4
Other energy sector	**803**	**999**	**100**	**100**	**4.0**	**2.5**	**−2.7**	**−11.6**
Total final consumption	**4 318**	**5 320**	**100**	**100**	**3.1**	**2.1**	**−4.3**	**−12.9**
Coal	680	627	16	12	3.3	1.0	−6.5	−24.0
Oil	1 476	1 888	34	35	3.0	2.2	−5.6	−15.7
of which transport	*758*	*1 054*	*18*	*20*	*3.3*	*2.6*	*−6.6*	*−18.8*
Gas	403	593	9	11	4.5	3.4	−1.5	−4.4
Electricity	763	1 125	18	21	6.0	3.9	−3.9	−13.8
Heat	70	77	2	1	4.7	2.2	−5.0	−18.2
Biomass and waste	914	955	21	18	0.5	0.4	−2.4	−4.5
of which biofuels	*26*	*75*	*1*	*1*	*13.4*	*9.7*	*63.9*	*86.9*
Other renewables	13	55	0	1	13.5	11.4	42.4	126.6

	Electricity (TWh)		Shares (%)		Growth (% p.a.)		Change vs. RS (%)	
Total generation	**10 945**	**15 943**	**100**	**100**	**5.8**	**3.8**	**−4.0**	**−13.5**
Coal	5 381	6 826	49	43	5.8	3.3	−11.7	−31.4
Oil	630	558	6	3	0.6	−0.3	−4.3	−7.8
Gas	1 959	3 089	18	19	6.4	4.4	−4.3	−17.0
Nuclear	366	786	3	5	10.0	7.1	18.0	53.7
Hydro	2 304	3 389	21	21	5.5	3.8	13.0	16.1
Biomass and waste	125	555	1	3	13.2	11.5	36.7	78.0
Wind	137	501	1	3	30.7	17.2	16.5	70.5
Geothermal	36	77	0	0	6.7	5.8	0.1	11.9
Solar	7	161	0	1	26.4	24.5	145.0	n.a.
Tide and wave	0	1	0	0	–	16.8	–	n.a.

	CO₂ emissions (Mt)		Shares (%)		Growth (% p.a.)		Change vs. RS (%)	
Total CO$_2$ emissions	**15 220**	**17 914**	**100**	**100**	**3.6**	**2.1**	**−7.4**	**−21.8**
Coal	8 622	9 374	57	52	3.9	1.9	−8.7	−26.9
Oil	4 564	5 649	30	32	2.7	1.9	−5.8	−15.4
Gas	2 035	2 890	13	16	4.4	3.2	−5.2	−15.5
Power generation	**6 865**	**8 093**	**100**	**100**	**4.0**	**2.2**	**−9.9**	**−27.6**
Coal	5 414	6 299	79	78	4.2	2.3	−10.3	−28.8
Oil	486	411	7	5	0.0	−0.7	−9.1	−16.7
Gas	964	1 383	14	17	5.2	3.5	−8.5	−24.2
Total final consumption	**7 505**	**8 769**	**100**	**100**	**3.2**	**1.9**	**−5.5**	**−17.4**
Coal	2 901	2 685	39	31	3.0	0.9	−6.6	−24.3
Oil	3 778	4 875	50	56	3.1	2.3	−5.5	−15.8
Gas	826	1 208	11	14	4.2	3.2	−1.7	−5.1

A

Reference Scenario: Developing Asia

	Energy demand (Mtoe)				Shares (%)			Growth (% p.a.)	
	1990	2005	2015	2030	2005	2015	2030	2005-2015	2005-2030
Total primary energy demand	1 600	3 027	4 615	6 427	100	100	100	4.3	3.1
Coal	697	1 423	2 368	3 254	47	51	51	5.2	3.4
Oil	322	728	1 086	1 594	24	24	25	4.1	3.2
Gas	71	216	371	586	7	8	9	5.6	4.1
Nuclear	10	29	66	119	1	1	2	8.3	5.7
Hydro	24	53	91	133	2	2	2	5.5	3.7
Biomass and waste	470	560	593	658	18	13	10	0.6	0.6
Other renewables	6	18	41	83	1	1	1	8.7	6.3
Power generation	331	1 074	1 844	2 799	100	100	100	5.6	3.9
Coal	230	827	1 420	2 098	77	77	75	5.6	3.8
Oil	46	53	61	48	5	3	2	1.3	–0.4
Gas	16	90	156	260	8	8	9	5.7	4.4
Nuclear	10	29	66	119	3	4	4	8.3	5.7
Hydro	24	53	91	133	5	5	5	5.5	3.7
Biomass and waste	0	6	17	76	1	1	3	10.2	10.5
Other renewables	6	15	34	65	1	2	2	8.5	6.1
Other energy sector	171	339	575	805	100	100	100	5.4	3.5
Total final consumption	1 223	2 026	2 996	4 097	100	100	100	4.0	2.9
Coal	390	463	694	788	23	23	19	4.1	2.1
Oil	250	601	920	1 412	30	31	34	4.3	3.5
of which transport	*109*	*269*	*455*	*834*	*13*	*15*	*20*	*5.4*	*4.6*
Gas	33	91	167	264	5	6	6	6.2	4.3
Electricity	85	282	573	957	14	19	23	7.4	5.0
Heat	14	44	73	94	2	2	2	5.2	3.1
Biomass and waste	451	542	562	565	27	19	14	0.4	0.2
of which biofuels	*0*	*1*	*4*	*16*	*0*	*0*	*0*	*23.9*	*14.6*
Other renewables	0	3	7	18	0	0	0	9.3	7.6

	Electricity (TWh)				Shares (%)			Growth (% p.a.)	
Total generation	1 276	4 143	8 233	13 480	100	100	100	7.1	4.8
Coal	731	2 730	5 701	9 364	66	69	69	7.6	5.1
Oil	163	214	243	194	5	3	1	1.3	–0.4
Gas	59	428	804	1 391	10	10	10	6.5	4.8
Nuclear	39	113	252	456	3	3	3	8.3	5.7
Hydro	277	617	1 057	1 547	15	13	11	5.5	3.7
Biomass and waste	0	16	44	204	0	1	2	10.4	10.6
Wind	0	8	101	246	0	1	2	28.5	14.6
Geothermal	7	17	29	48	0	0	0	5.8	4.3
Solar	0	0	2	30	0	0	0	35.8	25.7
Tide and wave	0	0	0	0	0	0	0	–	9.9

	CO_2 emissions (Mt)				Shares (%)			Growth (% p.a.)	
Total CO_2 emissions	3 522	7 690	12 440	17 464	100	100	100	4.9	3.3
Coal	2 534	5 441	8 957	12 194	71	72	70	5.1	3.3
Oil	852	1 791	2 699	4 029	23	22	23	4.2	3.3
Gas	136	458	784	1 241	6	6	7	5.5	4.1
Power generation	1 083	3 669	6 247	9 149	100	100	100	5.5	3.7
Coal	899	3 289	5 684	8 380	90	91	92	5.6	3.8
Oil	146	170	196	156	5	3	2	1.4	–0.4
Gas	38	210	367	613	6	6	7	5.7	4.4
Total final consumption	2 285	3 688	5 591	7 514	100	100	100	4.2	2.9
Coal	1 578	2 026	2 970	3 396	55	53	45	3.9	2.1
Oil	646	1 481	2 303	3 616	40	41	48	4.5	3.6
Gas	60	181	318	502	5	6	7	5.8	4.2

Alternative Policy Scenario: Developing Asia

	Energy demand (Mtoe)		Shares (%)		Growth (% p.a.)		Change vs. RS (%)	
	2015	2030	2015	2030	2005-2015	2005-2030	2015	2030
Total primary energy demand	4 391	5 496	100	100	3.8	2.4	–4.8	–14.5
Coal	2 178	2 411	50	44	4.3	2.1	–8.0	–25.9
Oil	1 023	1 330	23	24	3.5	2.4	–5.7	–16.6
Gas	368	557	8	10	5.5	3.9	–1.0	–4.9
Nuclear	80	186	2	3	10.5	7.6	22.2	56.4
Hydro	111	169	3	3	7.6	4.7	21.8	27.2
Biomass and waste	587	718	13	13	0.5	1.0	–1.0	9.1
Other renewables	45	126	1	2	9.7	8.1	10.0	52.1
Power generation	1 741	2 371	100	100	5.0	3.2	–5.6	–15.3
Coal	1 279	1 497	73	63	4.5	2.4	–10.0	–28.7
Oil	58	46	3	2	0.9	–0.6	–4.2	–5.1
Gas	152	228	9	10	5.4	3.8	–2.7	–12.2
Nuclear	80	186	5	8	10.5	7.6	22.2	56.4
Hydro	111	169	6	7	7.6	4.7	21.8	27.2
Biomass and waste	26	157	1	7	15.2	13.7	56.2	105.8
Other renewables	36	89	2	4	9.1	7.4	5.6	36.4
Other energy sector	557	694	100	100	5.1	2.9	–3.1	–13.8
Total final consumption	2 856	3 499	100	100	3.5	2.2	–4.7	–14.6
Coal	647	593	23	17	3.4	1.0	–6.7	–24.7
Oil	863	1 162	30	33	3.7	2.7	–6.1	–17.7
of which transport	*425*	*664*	*15*	*19*	*4.7*	*3.7*	*–6.5*	*–20.4*
Gas	*169*	*270*	*6*	*8*	*6.3*	*4.4*	*0.9*	*2.4*
Electricity	551	816	19	23	6.9	4.3	–3.9	–14.7
Heat	70	77	2	2	4.7	2.2	–5.0	–18.2
Biomass and waste	547	544	19	16	0.1	0.0	–2.7	–3.7
of which biofuels	*13*	*43*	*0*	*1*	*38.4*	*19.4*	*n.a.*	*173.6*
Other renewables	9	37	0	1	12.3	10.8	31.4	109.9

	Electricity (TWh)		Shares (%)		Growth (% p.a.)		Change vs. RS (%)	
Total generation	7 876	11 510	100	100	6.6	4.2	–4.3	–14.6
Coal	5 049	6 440	64	56	6.3	3.5	–11.4	–31.2
Oil	232	183	3	2	0.8	–0.6	–4.4	–5.5
Gas	776	1 242	10	11	6.1	4.4	–3.6	–10.7
Nuclear	307	712	4	6	10.5	7.6	22.2	56.4
Hydro	1 287	1 968	16	17	7.6	4.7	21.8	27.2
Biomass and waste	70	407	1	4	15.6	13.7	59.5	99.9
Wind	119	405	2	4	30.7	16.9	17.7	64.4
Geothermal	29	53	0	0	5.8	4.7	–	10.0
Solar	6	98	0	1	51.4	31.8	196.4	n.a.
Tide and wave	0	1	0	0	–	15.6	–	n.a.

	CO$_2$ emissions (Mt)		Shares (%)		Growth (% p.a.)		Change vs. RS (%)	
Total CO$_2$ emissions	11 511	13 438	100	100	4.1	2.3	–7.5	–23.1
Coal	8 191	8 921	71	66	4.2	2.0	–8.5	–26.8
Oil	2 546	3 353	22	25	3.6	2.5	–5.7	–16.8
Gas	773	1 164	7	9	5.4	3.8	–1.3	–6.2
Power generation	5 665	6 672	100	100	4.4	2.4	–9.3	–27.1
Coal	5 118	5 983	90	90	4.5	2.4	–10.0	–28.6
Oil	188	148	3	2	1.0	–0.6	–4.1	–5.0
Gas	358	541	6	8	5.5	3.8	–2.4	–11.9
Total final consumption	5 255	6 033	100	100	3.6	2.0	–6.0	–19.7
Coal	2 770	2 552	53	42	3.2	0.9	–6.7	–24.9
Oil	2 165	2 971	41	49	3.9	2.8	–6.0	–17.9
Gas	320	510	6	8	5.8	4.2	0.7	1.6

A

Reference Scenario: Latin America

	Energy demand (Mtoe)				Shares (%)			Growth (% p.a.)	
	1990	2005	2015	2030	2005	2015	2030	2005-2015	2005-2030
Total primary energy demand	340	500	646	873	100	100	100	2.6	2.3
Coal	17	23	28	42	5	4	5	2.1	2.4
Oil	158	227	259	324	45	40	37	1.4	1.4
Gas	54	101	165	259	20	26	30	5.1	3.8
Nuclear	2	4	10	9	1	1	1	8.3	2.9
Hydro	31	53	71	94	11	11	11	2.9	2.3
Biomass and waste	76	90	108	131	18	17	15	1.9	1.5
Other renewables	1	2	4	14	0	1	2	7.3	7.7
Power generation	69	125	189	285	100	100	100	4.3	3.4
Coal	5	8	11	21	6	6	7	3.0	3.9
Oil	14	25	19	10	20	10	4	−2.6	−3.4
Gas	14	27	67	124	22	36	44	9.5	6.3
Nuclear	2	4	10	9	3	5	3	8.3	2.9
Hydro	31	53	71	94	43	37	33	2.9	2.3
Biomass and waste	2	5	8	14	4	4	5	4.2	4.2
Other renewables	1	2	4	12	2	2	4	6.7	7.2
Other energy sector	51	60	74	96	100	100	100	2.2	1.9
Total final consumption	263	393	500	669	100	100	100	2.4	2.1
Coal	7	10	12	15	3	2	2	1.6	1.4
Oil	128	184	219	286	47	44	43	1.8	1.8
of which transport	*71*	*109*	*125*	*169*	*28*	*25*	*25*	*1.4*	*1.8*
Gas	25	58	79	110	15	16	16	3.2	2.6
Electricity	35	63	96	146	16	19	22	4.3	3.4
Heat	0	0	0	0	0	0	0	n.a.	n.a.
Biomass and waste	68	78	93	109	20	19	16	1.8	1.3
of which biofuels	*6*	*7*	*11*	*21*	*2*	*2*	*3*	*4.3*	*4.5*
Other renewables	0	0	0	2	0	0	0	19.4	15.0

	Electricity (TWh)				Shares (%)			Growth (% p.a.)	
Total generation	491	905	1 371	2 058	100	100	100	4.2	3.3
Coal	15	31	44	99	3	3	5	3.5	4.7
Oil	40	84	74	43	9	5	2	−1.2	−2.6
Gas	55	133	348	696	15	25	34	10.1	6.9
Nuclear	10	17	37	34	2	3	2	8.3	2.9
Hydro	364	619	825	1 095	68	60	53	2.9	2.3
Biomass and waste	7	19	31	53	2	2	3	5.0	4.1
Wind	0	0	8	22	0	1	1	35.6	17.6
Geothermal	1	2	4	12	0	0	1	7.9	7.6
Solar	0	0	0	4	0	0	0	n.a.	n.a.
Tide and wave	0	0	0	0	0	0	0	n.a.	n.a.

	CO_2 emissions (Mt)				Shares (%)			Growth (% p.a.)	
Total CO_2 emissions	602	938	1 184	1 627	100	100	100	2.4	2.2
Coal	57	88	104	159	9	9	10	1.8	2.4
Oil	426	619	703	875	66	59	54	1.3	1.4
Gas	119	231	377	592	25	32	36	5.0	3.8
Power generation	98	179	263	411	100	100	100	4.0	3.4
Coal	21	37	46	88	21	17	21	2.2	3.5
Oil	45	78	60	33	44	23	8	−2.5	−3.4
Gas	32	64	157	291	36	60	71	9.5	6.3
Total final consumption	439	675	829	1 097	100	100	100	2.1	2.0
Coal	32	47	55	67	7	7	6	1.6	1.4
Oil	350	501	599	785	74	72	72	1.8	1.8
Gas	56	127	174	245	19	21	22	3.2	2.7

Alternative Policy Scenario: Latin America

	Energy demand (Mtoe)		Shares (%)		Growth (% p.a.)		Change vs. RS (%)	
	2015	2030	2015	2030	2005-2015	2005-2030	2015	2030
Total primary energy demand	623	776	100	100	2.2	1.8	–3.6	–11.0
Coal	25	29	4	4	0.7	0.9	–12.7	–29.9
Oil	246	273	40	35	0.8	0.8	–5.0	–15.7
Gas	158	212	25	27	4.6	3.0	–4.6	–18.2
Nuclear	10	12	2	1	8.3	4.0	–	28.8
Hydro	71	93	11	12	3.0	2.3	0.6	–1.2
Biomass and waste	108	136	17	18	1.9	1.7	0.5	4.1
Other renewables	5	21	1	3	7.7	9.5	3.6	49.8
Power generation	182	245	100	100	3.8	2.7	–3.9	–14.0
Coal	8	11	4	4	–0.2	1.1	–26.8	–48.8
Oil	18	8	10	3	–3.3	–4.3	–7.2	–21.2
Gas	62	90	34	37	8.6	4.9	–7.2	–27.9
Nuclear	10	12	5	5	8.3	4.0	–	28.8
Hydro	71	93	39	38	3.0	2.3	0.6	–1.2
Biomass and waste	9	16	5	6	5.8	4.6	17.1	9.7
Other renewables	4	17	2	7	6.7	8.5	0.1	35.4
Other energy sector	72	87	100	100	1.9	1.5	–2.8	–9.2
Total final consumption	482	598	100	100	2.0	1.7	–3.6	–10.5
Coal	12	13	2	2	1.2	1.0	–3.9	–10.9
Oil	208	240	43	40	1.2	1.1	–5.1	–16.2
of which transport	*117*	*133*	*24*	*22*	*0.7*	*0.8*	*–6.8*	*–21.1*
Gas	77	99	16	17	2.9	2.2	–2.8	–9.7
Electricity	92	127	19	21	3.9	2.9	–4.0	–12.8
Heat	0	0	0	0	n.a.	n.a.	n.a.	n.a.
Biomass and waste	93	113	19	19	1.7	1.5	–0.8	3.7
of which biofuels	*12*	*29*	*2*	*5*	*5.5*	*5.9*	*12.3*	*37.8*
Other renewables	0	5	0	1	24.2	19.2	48.4	147.2

	Electricity (TWh)		Shares (%)		Growth (% p.a.)		Change vs. RS (%)	
Total generation	1 330	1 835	100	100	3.9	2.9	–3.0	–10.8
Coal	32	50	2	3	0.1	1.9	–28.1	–49.5
Oil	68	33	5	2	–2.0	–3.6	–7.6	–22.1
Gas	317	495	24	27	9.1	5.4	–9.0	–28.8
Nuclear	37	44	3	2	8.3	4.0	–	28.8
Hydro	829	1 082	62	59	3.0	2.3	0.6	–1.2
Biomass and waste	35	57	3	3	6.1	4.4	11.6	7.8
Wind	8	43	1	2	35.7	20.9	0.7	98.9
Geothermal	4	14	0	1	7.9	8.2	–	15.2
Solar	0	16	0	1	n.a.	n.a.	n.a.	n.a.
Tide and wave	0	0	0	0	n.a.	n.a.	n.a.	66.7

	CO$_2$ emissions (Mt)		Shares (%)		Growth (% p.a.)		Change vs. RS (%)	
Total CO$_2$ emissions	1 114	1 324	100	100	1.7	1.4	–5.9	–18.6
Coal	88	106	8	8	0.1	0.8	–15.5	–33.0
Oil	666	734	60	55	0.7	0.7	–5.2	–16.1
Gas	360	484	32	37	4.5	3.0	–4.6	–18.4
Power generation	234	279	100	100	2.7	1.8	–11.2	–32.3
Coal	32	43	14	16	–1.4	0.6	–30.2	–50.8
Oil	56	26	24	9	–3.3	–4.3	–7.1	–21.0
Gas	146	209	62	75	8.6	4.9	–7.2	–27.9
Total final consumption	791	937	100	100	1.6	1.3	–4.6	–14.6
Coal	53	60	7	6	1.2	1.0	–3.9	–10.9
Oil	568	657	72	70	1.3	1.1	–5.2	–16.4
Gas	170	221	21	24	2.9	2.2	–2.7	–9.6

A

Reference Scenario: Middle East

	Energy demand (Mtoe)				Shares (%)			Growth (% p.a.)	
	1990	2005	2015	2030	2005	2015	2030	2005-2015	2005-2030
Total primary energy demand	228	503	734	1 027	100	100	100	3.9	2.9
Coal	3	9	14	19	2	2	2	4.4	3.2
Oil	150	274	385	464	55	52	45	3.5	2.1
Gas	72	216	326	529	43	44	51	4.2	3.6
Nuclear	0	0	2	2	0	0	0	n.a.	n.a.
Hydro	1	2	3	4	0	0	0	5.2	3.1
Biomass and waste	1	1	2	6	0	0	1	8.8	7.1
Other renewables	0	1	2	4	0	0	0	7.8	6.4
Power generation	64	166	252	391	100	100	100	4.3	3.5
Coal	2	8	12	17	5	5	4	4.5	3.2
Oil	29	59	76	87	35	30	22	2.6	1.6
Gas	32	98	158	276	59	63	71	4.9	4.2
Nuclear	0	0	2	2	0	1	0	n.a.	n.a.
Hydro	1	2	3	4	1	1	1	5.2	3.1
Biomass and waste	0	0	1	4	0	0	1	n.a.	n.a.
Other renewables	0	0	0	1	0	0	0	76.8	32.4
Other energy sector	20	62	78	108	100	100	100	2.3	2.3
Total final consumption	165	330	488	658	100	100	100	4.0	2.8
Coal	0	1	1	2	0	0	0	6.2	3.5
Oil	116	198	284	341	60	58	52	3.7	2.2
of which transport	59	106	150	167	32	31	25	3.6	1.8
Gas	31	86	133	206	26	27	31	4.5	3.5
Electricity	17	43	67	106	13	14	16	4.5	3.6
Heat	0	0	0	0	0	0	0	n.a.	n.a.
Biomass and waste	1	1	1	2	0	0	0	2.3	2.6
of which biofuels	0	0	0	1	0	0	0	n.a.	n.a.
Other renewables	0	1	1	2	0	0	0	5.3	4.6

	Electricity (TWh)				Shares (%)			Growth (% p.a.)	
Total generation	240	640	972	1 522	100	100	100	4.3	3.5
Coal	10	35	55	84	6	6	6	4.5	3.5
Oil	114	240	295	329	38	30	22	2.1	1.3
Gas	104	343	573	1 033	54	59	68	5.3	4.5
Nuclear	0	0	7	7	0	1	0	n.a.	n.a.
Hydro	12	21	35	45	3	4	3	5.2	3.1
Biomass and waste	0	0	3	9	0	0	1	n.a.	n.a.
Wind	0	0	3	10	0	0	1	73.5	30.0
Geothermal	0	0	0	0	0	0	0	n.a.	n.a.
Solar	0	0	1	6	0	0	0	35.4	22.6
Tide and wave	0	0	0	0	0	0	0	n.a.	n.a.

	CO_2 emissions (Mt)				Shares (%)			Growth (% p.a.)	
Total CO_2 emissions	586	1 238	1 794	2 464	100	100	100	3.8	2.8
Coal	12	34	54	77	3	3	3	4.7	3.3
Oil	413	724	1 010	1 201	58	56	49	3.4	2.0
Gas	161	480	730	1 187	39	41	48	4.3	3.7
Power generation	172	442	653	983	100	100	100	4.0	3.3
Coal	9	30	46	66	7	7	7	4.5	3.2
Oil	89	184	238	273	42	36	28	2.6	1.6
Gas	74	228	369	645	52	56	66	4.9	4.2
Total final consumption	365	694	1 017	1 313	100	100	100	3.9	2.6
Coal	2	4	8	11	1	1	1	6.4	3.6
Oil	297	502	721	858	72	71	65	3.7	2.2
Gas	66	188	288	444	27	28	34	4.4	3.5

Alternative Policy Scenario: Middle East

	Energy demand (Mtoe)		Shares (%)		Growth (% p.a.)		Change vs. RS (%)	
	2015	2030	2015	2030	2005-2015	2005-2030	2015	2030
Total primary energy demand	**681**	**866**	**100**	**100**	**3.1**	**2.2**	**–7.2**	**–15.7**
Coal	11	12	2	1	1.9	1.3	–21.4	–36.7
Oil	365	420	54	49	2.9	1.7	–5.4	–9.4
Gas	295	407	43	47	3.2	2.6	–9.5	–23.0
Nuclear	2	2	0	0	n.a.	n.a.	–	–
Hydro	3	4	0	0	5.7	3.1	5.0	1.1
Biomass and waste	3	9	0	1	10.9	9.0	21.4	57.1
Other renewables	3	12	0	1	13.6	11.4	69.1	n.a.
Power generation	**213**	**270**	**100**	**100**	**2.5**	**2.0**	**–15.5**	**–31.0**
Coal	9	11	4	4	1.7	1.3	–23.9	–38.1
Oil	65	66	30	25	1.0	0.5	–14.8	–23.7
Gas	132	176	62	65	3.1	2.4	–16.3	–36.2
Nuclear	2	2	1	1	n.a.	n.a.	–	–
Hydro	3	4	1	1	5.7	3.1	5.0	1.1
Biomass and waste	2	7	1	3	n.a.	n.a.	46.2	83.8
Other renewables	0	4	0	1	78.0	38.1	6.9	189.2
Other energy sector	**76**	**102**	**100**	**100**	**2.1**	**2.0**	**–2.0**	**–5.5**
Total final consumption	**472**	**611**	**100**	**100**	**3.7**	**2.5**	**–3.1**	**–7.3**
Coal	1	1	0	0	5.5	2.0	–5.6	–30.2
Oil	275	320	58	52	3.3	1.9	–3.2	–6.4
of which transport	*144*	*156*	*30*	*26*	*3.1*	*1.6*	*–4.2*	*–6.5*
Gas	129	186	27	31	4.1	3.1	–3.4	–9.6
Electricity	64	94	14	15	4.1	3.2	–3.8	–10.9
Heat	0	0	0	0	n.a.	n.a.	n.a.	n.a.
Biomass and waste	1	2	0	0	2.2	2.7	–0.5	2.1
of which biofuels	*0*	*1*	*0*	*0*	*n.a.*	*n.a.*	*5.0*	*12.5*
Other renewables	2	8	1	1	12.0	9.7	85.9	n.a

	Electricity (TWh)		Shares (%)		Growth (% p.a.)		Change vs. RS (%)	
Total generation	**947**	**1 357**	**100**	**100**	**4.0**	**3.1**	**–2.6**	**–10.9**
Coal	43	51	5	4	2.0	1.4	–21.8	–39.6
Oil	284	306	30	23	1.7	1.0	–3.8	–7.1
Gas	568	887	60	65	5.2	3.9	–1.0	–14.2
Nuclear	7	7	1	1	n.a.	n.a.	–	–
Hydro	37	45	4	3	5.7	3.1	5.0	1.1
Biomass and waste	4	17	0	1	n.a.	n.a.	46.2	83.8
Wind	4	23	0	2	74.9	34.5	8.3	134.9
Geothermal	0	0	0	0	n.a.	n.a.	–	18.5
Solar	1	22	0	2	35.4	29.4	–	n.a.
Tide and wave	0	0	0	0	n.a.	n.a.	n.a.	n.a.

	CO_2 emissions (Mt)		Shares (%)		Growth (% p.a.)		Change vs. RS (%)	
Total CO_2 emissions	**1 652**	**2 033**	**100**	**100**	**2.9**	**2.0**	**–7.9**	**–17.5**
Coal	43	48	3	2	2.3	1.4	–21.2	–37.0
Oil	951	1 078	58	53	2.8	1.6	–5.9	–10.2
Gas	659	906	40	45	3.2	2.6	–9.8	–23.7
Power generation	**547**	**660**	**100**	**100**	**2.2**	**1.6**	**–16.3**	**–32.9**
Coal	35	41	6	6	1.7	1.3	–23.9	–38.1
Oil	203	208	37	32	1.0	0.5	–14.8	–23.7
Gas	309	411	56	62	3.1	2.4	–16.3	–36.2
Total final consumption	**983**	**1 211**	**100**	**100**	**3.5**	**2.3**	**–3.3**	**–7.7**
Coal	8	7	1	1	5.7	2.1	–5.6	–30.2
Oil	697	802	71	66	3.3	1.9	–3.3	–6.5
Gas	278	401	28	33	4.0	3.1	–3.4	–9.6

A

Reference Scenario: Africa

	Energy demand (Mtoe)				Shares (%)			Growth (% p.a.)	
	1990	2005	2015	2030	2005	2015	2030	2005-2015	2005-2030
Total primary energy demand	400	606	726	943	100	100	100	1.8	1.8
Coal	74	102	113	131	17	16	14	1.1	1.0
Oil	90	133	161	224	22	22	24	1.9	2.1
Gas	31	72	115	177	12	16	19	4.8	3.7
Nuclear	2	3	4	4	0	1	0	2.5	1.0
Hydro	5	8	11	20	1	1	2	3.0	3.8
Biomass and waste	198	287	320	380	47	44	40	1.1	1.1
Other renewables	0	1	2	7	0	0	1	8.5	8.3
Power generation	69	123	169	256	100	100	100	3.2	3.0
Coal	39	59	66	82	48	39	32	1.2	1.3
Oil	11	17	13	10	14	7	4	-2.9	-2.0
Gas	11	35	69	118	28	41	46	7.1	5.0
Nuclear	2	3	4	4	2	2	1	2.5	1.0
Hydro	5	8	11	20	6	6	8	3.0	3.8
Biomass and waste	0	1	5	18	0	3	7	25.1	14.8
Other renewables	0	1	1	4	1	1	2	4.4	6.5
Other energy sector	65	84	99	121	100	100	100	1.6	1.4
Total final consumption	294	447	529	683	100	100	100	1.7	1.7
Coal	19	18	20	21	4	4	3	1.4	0.6
Oil	74	113	140	199	25	27	29	2.2	2.3
of which transport	*39*	*65*	*81*	*129*	*15*	*15*	*19*	*2.2*	*2.7*
Gas	9	25	29	40	6	6	6	1.8	2.0
Electricity	21	39	58	96	9	11	14	3.9	3.7
Heat	0	0	0	0	0	0	0	n.a.	n.a.
Biomass and waste	169	252	280	324	56	53	47	1.1	1.0
of which biofuels	*0*	*0*	*1*	*3*	*0*	*0*	*0*	*n.a.*	*n.a.*
Other renewables	0	0	1	2	0	0	0	132.3	47.4

	Electricity (TWh)				Shares (%)			Growth (% p.a.)	
Total generation	316	563	821	1 362	100	100	100	3.8	3.6
Coal	165	251	293	397	45	36	29	1.6	1.9
Oil	43	58	47	39	10	6	3	-2.0	-1.5
Gas	43	150	322	599	27	39	44	7.9	5.7
Nuclear	8	11	15	15	2	2	1	2.5	1.0
Hydro	56	91	122	232	16	15	17	3.0	3.8
Biomass and waste	0	1	14	46	0	2	3	35.9	18.7
Wind	0	1	5	16	0	1	1	19.8	12.6
Geothermal	0	0	3	9	0	0	1	20.4	12.9
Solar	0	1	0	8	0	0	1	-26.9	11.5
Tide and wave	0	0	0	0	0	0	0	n.a.	n.a.

	CO$_2$ emissions (Mt)				Shares (%)			Growth (% p.a.)	
Total CO$_2$ emissions	550	835	1 013	1 365	100	100	100	2.0	2.0
Coal	235	306	328	391	37	32	29	0.7	1.0
Oil	249	373	430	575	45	42	42	1.4	1.7
Gas	65	156	255	398	19	25	29	5.0	3.8
Power generation	214	363	458	627	100	100	100	2.4	2.2
Coal	152	228	257	319	63	56	51	1.2	1.3
Oil	35	53	40	33	15	9	5	-2.9	-2.0
Gas	26	81	161	275	22	35	44	7.1	5.0
Total final consumption	304	430	506	687	100	100	100	1.7	1.9
Coal	83	78	71	72	18	14	10	-0.9	-0.3
Oil	202	303	376	533	71	74	78	2.2	2.3
Gas	19	49	60	82	11	12	12	2.1	2.1

Alternative Policy Scenario: Africa

	Energy demand (Mtoe)		Shares (%)		Growth (% p.a.)		Change vs. RS (%)	
	2015	2030	2015	2030	2005-2015	2005-2030	2015	2030
Total primary energy demand	**699**	**852**	**100**	**100**	**1.4**	**1.4**	**-3.7**	**-9.6**
Coal	106	106	15	12	0.4	0.1	-6.8	-19.6
Oil	150	190	21	22	1.2	1.4	-6.7	-15.1
Gas	110	150	16	18	4.3	3.0	-4.6	-15.3
Nuclear	4	6	1	1	2.5	2.8	–	55.6
Hydro	13	25	2	3	5.2	4.8	23.5	26.7
Biomass and waste	314	363	45	43	0.9	0.9	-1.9	-4.5
Other renewables	3	13	0	1	11.1	11.0	27.4	87.6
Power generation	**161**	**231**	**100**	**100**	**2.7**	**2.6**	**-4.8**	**-9.9**
Coal	59	60	37	26	0.0	0.1	-11.0	-27.4
Oil	12	9	8	4	-3.1	-2.3	-2.2	-8.8
Gas	65	95	40	41	6.5	4.1	-5.9	-19.6
Nuclear	4	6	2	3	2.5	2.8	–	55.6
Hydro	13	25	8	11	5.2	4.8	23.5	26.7
Biomass and waste	6	28	4	12	27.3	17.0	19.3	60.5
Other renewables	2	7	1	3	5.3	8.7	8.9	67.5
Other energy sector	**98**	**116**	**100**	**100**	**1.5**	**1.3**	**-1.0**	**-3.9**
Total final consumption	**509**	**613**	**100**	**100**	**1.3**	**1.3**	**-3.8**	**-10.4**
Coal	20	19	4	3	1.3	0.3	-1.1	-7.7
Oil	130	167	26	27	1.4	1.6	-7.4	-16.3
of which transport	*72*	*101*	*14*	*16*	*1.0*	*1.7*	*-11.2*	*-21.6*
Gas	29	37	6	6	1.5	1.6	-2.9	-8.0
Electricity	56	88	11	14	3.5	3.3	-3.4	-8.8
Heat	0	0	0	0	n.a.	n.a.	n.a.	n.a.
Biomass and waste	273	296	54	48	0.8	0.6	-2.5	-8.6
of which biofuels	*1*	*3*	*0*	*0*	*n.a.*	*n.a.*	*11.0*	*-2.5*
Other renewables	1	5	0	1	144.6	52.3	67.8	125.9

	Electricity (TWh)		Shares (%)		Growth (% p.a.)		Change vs. RS (%)	
Total generation	**792**	**1 242**	**100**	**100**	**3.5**	**3.2**	**-3.5**	**-8.8**
Coal	257	285	32	23	0.2	0.5	-12.4	-28.3
Oil	46	35	6	3	-2.3	-1.9	-2.3	-9.3
Gas	299	465	38	37	7.1	4.6	-7.2	-22.4
Nuclear	15	23	2	2	2.5	2.8	–	55.6
Hydro	151	293	19	24	5.2	4.8	23.5	26.7
Biomass and waste	16	74	2	6	38.3	21.0	19.2	60.5
Wind	6	30	1	2	22.1	15.4	20.7	87.2
Geothermal	3	11	0	1	20.6	13.7	1.6	17.5
Solar	0	25	0	2	-26.9	16.5	–	n.a.
Tide and wave	0	0	0	0	n.a.	n.a.	n.a.	n.a.

	CO$_2$ emissions (Mt)		Shares (%)		Growth (% p.a.)		Change vs. RS (%)	
Total CO$_2$ emissions	**943**	**1 118**	**100**	**100**	**1.2**	**1.2**	**-6.9**	**-18.1**
Coal	299	298	32	27	-0.2	-0.1	-8.9	-23.8
Oil	400	484	42	43	0.7	1.0	-6.8	-15.8
Gas	243	336	26	30	4.5	3.1	-4.6	-15.7
Power generation	**419**	**483**	**100**	**100**	**1.5**	**1.2**	**-8.5**	**-23.0**
Coal	229	232	55	48	0.0	0.1	-11.0	-27.4
Oil	39	30	9	6	-3.1	-2.3	-2.2	-8.8
Gas	152	221	36	46	6.5	4.1	-5.9	-19.6
Total final consumption	**476**	**587**	**100**	**100**	**1.0**	**1.3**	**-6.1**	**-14.5**
Coal	70	66	15	11	-1.0	-0.6	-1.2	-7.9
Oil	347	445	73	76	1.4	1.5	-7.5	-16.5
Gas	58	76	12	13	1.8	1.8	-2.8	-7.9

A

ABBREVIATIONS, DEFINITIONS AND CONVERSION FACTORS

This annex provides general information on abbreviations, fuel, process and regional definitions, and country groupings used throughout *WEO-2007*. General conversion factors for energy and average conversion factors for oil, gas and coal have also been included. Readers interested in obtaining more detailed information about IEA statistics should consult the annual IEA publications available at www.iea.org/statistics.

Abbreviations

Coal	tce	tonne of coal equivalent
	Mtce	million tonnes of coal equivalent
Energy	toe	tonne of oil equivalent
	Mtoe	million tonnes of oil equivalent
	MBtu	million British thermal units
	GJ	gigajoule (1 joule x 10^9)
	EJ	exajoule (1 joule x 10^{18})
	kWh	kilowatt-hour
	MWh	megawatt-hour
	GWh	gigawatt-hour
	TWh	terawatt-hour
Gas	tcf	thousand cubic feet
	mcm	million cubic metres
	bcm	billion cubic metres
	tcm	trillion cubic metres
Mass	kt	kilotonnes (1 tonne x 10^3)
	Mt	million tonnes (1 tonne x 10^6)
	Gt	gigatonnes (1 tonne x 10^9)
Oil	Mb	million barrels
	b/d	barrels per day
	kb/d	thousand barrels per day
	mb/d	million barrels per day
	mpg	miles per gallon

Oil and Gas	boe	barrels of oil equivalent
Power	W	watt (1 joule per second)
	kW	kilowatt (1 watt x 10^3)
	MW	megawatt (1 watt x 10^6)
	GW	gigawatt (1 watt x 10^9)
	TW	terawatt (1 watt x 10^{12})

Fuel Definitions

Biodiesel

Biodiesel is a diesel-equivalent, processed fuel made from the transesterification (a chemical process which removes the glycerine from the oil) of vegetable oils or animal fats.

Biogas

A mixture of methane and carbon dioxide produced by bacterial degradation of organic matter and used as a fuel.

Biomass and Waste

Solid biomass, gas and liquids derived from biomass, industrial waste and the renewable part of municipal waste.

Brown Coal

Includes lignite and sub-bituminous coal where lignite is defined as non-agglomerating coal with a gross calorific value less than 4 165 kcal/kg and sub-bituminous coal is defined as non-agglomerating coal with a gross calorific value between 4 165 kcal/kg and 5 700 kcal/kg. Oil shale and tar sands are also included in this category.

Clean Coal Technologies

Clean coal technologies (CCTs) are designed to enhance the efficiency and the environmental acceptability of coal extraction, preparation and use.

Coal

Coal includes both primary coal (including hard coal and lignite) and derived fuels (including patent fuel, brown-coal briquettes, coke-oven coke, gas coke, coke-oven gas, blast-furnace gas and oxygen steel furnace gas). Peat is also included in this category.

Coal-bed Methane

Methane found in coal seams. Coal-bed methane is a source of unconventional natural gas.

Coal-to-Liquids

Coal-to-Liquids (CTL) refers to both coal gasification, combined with Fischer-Tropsch synthesis to produce liquid fuels, and the less-developed direct coal liquefaction technologies.

Condensates

Condensates are liquid hydrocarbon mixtures recovered from non-associated gas reservoirs. They are composed of C4 and higher carbon number hydrocarbons and normally have an API between 50° and 85°.

Dimethyl Ether

Clear, odourless gas currently produced by dehydration of methanol from natural gas, but which can also be produced from biomass or coal.

Ethanol

Ethanol is an alcohol made by fermenting any biomass high in carbohydrates. Today, ethanol is made from starches and sugars, but second-generation technologies will allow it to be made from cellulose and hemicellulose, the fibrous material that makes up the bulk of most plant matter.

Gas

Gas includes natural gas (both associated and non-associated with petroleum deposits but excluding natural gas liquids) and gas-works gas.

Gas-to-Liquids

Fischer-Tropsch technology is used to convert natural gas into synthesis gas (syngas) and then, through catalytic reforming or synthesis, into very clean conventional oil products. The main fuel produced in most GTL plants is diesel.

Hard Coal

Coal of gross calorific value greater than 5 700 kcal/kg on an ash-free but moist basis. Hard coal can be further disaggregated into anthracite, coking coal and other bituminous coal.

B

Heavy Petroleum Products

Heavy petroleum products include heavy fuel oil.

Hydropower

Hydropower refers to the energy content of the electricity produced in hydropower plants, assuming 100% efficiency. It excludes output from pumped storage plants.

Light Petroleum Products

Light petroleum products include liquefied petroleum gas (LPG), naphtha and gasoline.

Middle Distillates

Middle distillates include jet fuel, diesel and heating oil.

Modern Renewables

Includes hydropower, biomass (excluding traditional use) and other renewables.

Natural Gas Liquids

Natural gas liquids (NGLs) are the liquid or liquefied hydrocarbons produced in the manufacture, purification and stabilisation of natural gas. These are those portions of natural gas which are recovered as liquids in separators, field facilities, or gas-processing plants. NGLs include but are not limited to ethane, propane, butane, pentane, natural gasoline and condensates.

Nuclear

Nuclear refers to the primary heat-equivalent of the electricity produced by a nuclear plant with an average thermal efficiency of 33%.

Oil

Oil includes crude oil, condensates, natural gas liquids, refinery feedstocks and additives, other hydrocarbons (including emulsified oils, synthetic crude oil, mineral oils extracted from bituminous minerals such as oil shale, bituminous sand and oils from coal liquefaction) and petroleum products (refinery gas, ethane, LPG, aviation gasoline, motor gasoline, jet fuels, kerosene, gas/diesel oil, heavy fuel oil, naphtha, white spirit, lubricants, bitumen, paraffin waxes and petroleum coke).

Other Renewables

Includes geothermal, solar PV, solar thermal, wind, tide and wave energy for electricity and heat generation.

Rest of Renewables

Includes biomass and waste, geothermal, solar PV, solar thermal, wind, tide and wave energy for electricity and heat generation.

Traditional Biomass

Traditional biomass refers to the use of fuelwood, animal dung and agricultural residues in stoves with very low efficiencies.

Process Definitions

Electricity Generation

Electricity generation is the total amount of electricity generated by power plants. It includes own use, and transmission and distribution losses.

Greenfield

The construction of plants or facilities in new areas or where no previous infrastructure exists.

International Marine Bunkers

Covers those quantities delivered to sea-going ships that are engaged in international navigation. The international navigation may take place at sea, on inland lakes and waterways, and in coastal waters. Consumption by ships engaged in domestic navigation is excluded. The domestic/international split is determined on the basis of port of departure and port of arrival, and not by the flag or nationality of the ship. Consumption by fishing vessels and by military forces is also excluded.

Lower Heating Value

Lower heating value is the heat liberated by the complete combustion of a unit of fuel when the water produced is assumed to remain as a vapour and the heat is not recovered.

Natural Decline Rate

The base production decline rate of an oil or gas field without intervention to enhance production.

B

Observed Decline Rate

The production decline rate of an oil or gas field after all measures have been taken to maximise production. It is the aggregation of all the production increases and declines of new and mature oil or gas fields in a particular region.

Other Energy Sector

Other energy sector covers the use of energy by transformation industries and the energy losses in converting primary energy into a form that can be used in the final consuming sectors. It includes losses by gas works, petroleum refineries, coal and gas transformation and liquefaction. It also includes energy used in coal mines, in oil and gas extraction and in electricity and heat production. Transfers and statistical differences are also included in this category.

Residential, Services and Agriculture

This sector also includes energy use in the forestry and fishing sectors. It also theoretically includes military fuel use for all mobile and stationary consumption (*e.g.* ships, aircraft, road and energy used in living quarters) regardless of whether the fuel delivered is for the military of that country or for the military of another country. In practice, many countries find this difficult to report.

Power Generation

Power generation refers to fuel use in electricity plants, heat plants and combined heat and power (CHP) plants. Both main activity producer plants and small plants that produce fuel for their own use (autoproducers) are included.

Total Final Consumption

Total final consumption (TFC) is the sum of consumption by the different end-use sectors. TFC is broken down into energy demand in the following sectors: industry (including manufacturing and mining), transport, other (including residential, commercial and public services, agriculture/forestry and fishing), non-energy use (including petrochemical feedstocks), and non-specified.

Total Primary Energy Demand

Total primary energy demand represents domestic demand only, including power generation, other energy sector, and total final consumption. It excludes international marine bunkers, except for world primary energy demand, where it is included.

Regional Definitions and Country Groupings

Africa

Algeria, Angola, Benin, Botswana, Burkina Faso, Burundi, Cameroon, Cape Verde, Central African Republic, Chad, Comoros, Congo, Democratic Republic of Congo, Côte d'Ivoire, Djibouti, Egypt, Equatorial Guinea, Eritrea, Ethiopia, Gabon, Gambia, Ghana, Guinea, Guinea-Bissau, Kenya, Lesotho, Liberia, Libya, Madagascar, Malawi, Mali, Mauritania, Mauritius, Morocco, Mozambique, Namibia, Niger, Nigeria, Reunion, Rwanda, São Tomé and Principe, Senegal, Seychelles, Sierra Leone, Somalia, South Africa, Sudan, Swaziland, United Republic of Tanzania, Togo, Tunisia, Uganda, Zambia and Zimbabwe.

Annex I Parties to the United Nations Framework Convention on Climate Change

Australia, Austria, Belarus, Belgium, Bulgaria, Canada, Croatia, Czech Republic, Denmark, Estonia, European Community, Finland, France, Germany, Greece, Hungary, Iceland, Ireland, Italy, Japan, Latvia, Liechtenstein, Lithuania, Luxembourg, Monaco, Netherlands, New Zealand, Norway, Poland, Portugal, Romania, Russian Federation, Slovak Republic, Slovenia, Spain, Sweden, Switzerland, Turkey, Ukraine, the United Kingdom and the United States.

China

China refers to the People's Republic of China, including Hong Kong.

Developing Asia

Afghanistan, Bangladesh, Bhutan, Brunei, Cambodia, China, Chinese Taipei, Fiji, French Polynesia, India, Indonesia, Kiribati, the Democratic People's Republic of Korea, Laos, Macau, Malaysia, Maldives, Mongolia, Myanmar, Nepal, New Caledonia, Pakistan, Papua New Guinea, the Philippines, Samoa, Singapore, Solomon Islands, Sri Lanka, Thailand, Tonga, Vietnam and Vanuatu.

Developing Countries

Includes countries in the Africa, Developing Asia, Latin America and Middle East regional groupings.

B

European Union

Austria, Belgium, Bulgaria, Cyprus, the Czech Republic, Denmark, Estonia, Finland, France, Germany, Greece, Hungary, Ireland, Italy, Latvia, Lithuania, Luxembourg, Malta, the Netherlands, Poland, Portugal, Romania, Slovak Republic, Slovenia, Spain, Sweden and the United Kingdom.

Latin America

Antigua and Barbuda, Argentina, Bahamas, Barbados, Belize, Bermuda, Bolivia, Brazil, Chile, Colombia, Costa Rica, Cuba, Dominica, the Dominican Republic, Ecuador, El Salvador, French Guiana, Grenada, Guadeloupe, Guatemala, Guyana, Haiti, Honduras, Jamaica, Martinique, Netherlands Antilles, Nicaragua, Panama, Paraguay, Peru, St. Kitts and Nevis, Saint Lucia, St. Vincent and Grenadines, Suriname, Trinidad and Tobago, Uruguay and Venezuela.

Middle East

Bahrain, Iran, Iraq, Israel, Jordan, Kuwait, Lebanon, Oman, Qatar, Saudi Arabia, Syria, the United Arab Emirates and Yemen. It includes the neutral zone between Saudi Arabia and Iraq.

North Africa

Algeria, Egypt, Libya, Morocco and Tunisia.

OECD Europe

Austria, Belgium, the Czech Republic, Denmark, Finland, France, Germany, Greece, Hungary, Iceland, Ireland, Italy, Luxembourg, the Netherlands, Norway, Poland, Portugal, the Slovak Republic, Spain, Sweden, Switzerland, Turkey and the United Kingdom.

OECD North America

Canada, Mexico and the United States.

OECD Pacific

Australia, Japan, Korea and New Zealand.

Organization of the Petroleum Exporting Countries

Algeria, Angola, Indonesia, Iran, Iraq, Kuwait, Libya, Nigeria, Qatar, Saudi Arabia, the United Arab Emirates and Venezuela.

Rest of Developing Asia
Developing Asia regional grouping excluding China and India.

Sub-Saharan Africa
Africa regional grouping excluding North Africa.

Transition Economies
Albania, Armenia, Azerbaijan, Belarus, Bosnia-Herzegovina, Bulgaria, Croatia, Estonia, Serbia and Montenegro, the former Yugoslav Republic of Macedonia, Georgia, Kazakhstan, Kyrgyzstan, Latvia, Lithuania, Moldova, Romania, Russia, Slovenia, Tajikistan, Turkmenistan, Ukraine and Uzbekistan. For statistical reasons, this region also includes Cyprus and Malta.

General Conversion Factors for Energy

To:	TJ	Gcal	Mtoe	MBtu	GWh
From:	*multiply by:*				
TJ	1	238.8	2.388×10^{-5}	947.8	0.2778
Gcal	4.1868×10^{-3}	1	10^{-7}	3.968	1.163×10^{-3}
Mtoe	4.1868×10^{4}	10^{7}	1	3.968×10^{7}	11 630
MBtu	1.0551×10^{-3}	0.252	2.52×10^{-8}	1	2.931×10^{-4}
GWh	3.6	860	8.6×10^{-5}	3 412	1

Average Conversion Factors

Coal	1 Mtoe = 1.9814 million tonnes
Oil	1 Mtoe = 0.0209 mb/d
Gas	1 Mtoe = 1.2117 bcm

Note: These are world averages for the period 2005 to 2030. Region- and quality-specific factors are used to convert Mtoe data in this publication to other units.

B

ACRONYMS

APS	Alternative Policy Scenario
CAFE	Corporate Average Fuel Economy
CBM	coal-bed methane
CCGT	combined-cycle gas turbine
CCS	CO_2 capture and storage
CDM	clean development mechanism (under the Kyoto Protocol)
CFL	compact fluorescent lamp
CHP	combined heat and power; when referring to industrial CHP, the term co-generation is sometimes used
CIRED	Centre International de Recherche sur l'Environnement et le Développement
CNG	compressed natural gas
CO_2	carbon dioxide
CTL	coal-to-liquids
DME	dimethyl ether
EOR	enhanced oil recovery
ERI	Energy Research Institute of China
EU	European Union
FDI	foreign direct investment
GDP	gross domestic product
GHG	greenhouse gas
GTL	gas-to-liquids
HIV/AIDS	human immunodeficiency virus/acquired immunodeficiency syndrome

IAEA	International Atomic Energy Agency
IEA	International Energy Agency
IGCC	integrated gasification combined cycle
IMF	International Monetary Fund
IOC	international oil company
IPCC	Intergovernmental Panel on Climate Change
IPP	independent power producer
LDV	light duty vehicle
LHV	lower heating value
LNG	liquefied natural gas
LPG	liquefied petroleum gas
MER	market exchange rate
MDG	Millennium Development Goal
NEA	Nuclear Energy Agency
NELP	New Exploration Licensing Policy in India
NIMBY	not-in-my-backyard
NGL	natural gas liquids
NOC	national oil company
OCGT	open-cycle gas turbine
ODI	overseas direct investment
OECD	Organisation for Economic Co-operation and Development
OPEC	Organization of the Petroleum Exporting Countries
PDS	public distribution systems
PPP	purchasing power parity
PSC	production-sharing contract
RS	Reference Scenario
TERI	The Energy and Resources Institute of India
TFC	total final consumption

UNDP	United Nations Development Programme
UNEP	United Nations Environment Programme
UNFCCC	United Nations Framework Convention on Climate Change
USGS	United States Geological Survey
WEM	World Energy Model
WHO	World Health Organization
WTI	West Texas Intermediate
WTO	World Trade Organization

REFERENCES

Introduction

Intergovernmental Panel on Climate Change (IPCC) (2007), *Climate Change 2007: Fourth Assessment Report*, IPCC, Geneva. Available at www.ipcc.ch.

IEA (2007a), *Medium-Term Oil Market Report*, OECD/IEA, Paris.

— (2007b) *Natural Gas Market Review 2007: Security in a Globalising Market to 2015*, OECD/IEA, Paris.

— (2006) *World Energy Outlook 2006*, OECD/IEA, Paris.

International Monetary Fund (IMF) (2007), *World Economic Outlook: Spillovers and Cycles in the Global Economy*, April, IMF, Washington D.C.

United Nations Population Division (UNPD) (2007), *World Population Prospects: The 2006 Revision*, United Nations, New York.

Chapter 1: Global Energy Trends

Cedigaz (2007), *Natural Gas in the World*, Insitut Français du Pétrole, Rueil-Malmaison.

IEA (2007a), *Medium Term Oil Market Review: July 2007*, OECD/IEA, Paris.

— (2007b), *Gas Market Review: Security in a Globalising Market to 2015*, OECD/IEA, Paris.

— (2006), *World Energy Outlook 2006*, OECD/IEA, Paris.

— (2005), *World Energy Outlook 2005: Middle East and North Africa Insights*, OECD/IEA, Paris.

International Energy Agency Clean Coal Centre (IEA-CCC) (2007), *Supply Costs for Internationally Traded Coal*, IEA-CCC, London.

Chapter 2: Energy Trends in China and India

Asian Development Bank (ADB) (2006), *Energy Efficiency and Climate Change Considerations for On-road Transport in Asia*, ADB, Manila.

Chapter 3: International Trade and the World Economy

Australian Bureau of Agricultural and Resource Economics (ABARE) (2007), *Australian Commodities*, June, Canberra.

Bloch, H., A. Dockery, C. Morgan and D. Sapsford (2007), "Growth, Commodity Prices, Inflation and the Distribution of Income", in *Metroeconomica*, Vol. 58, No. 1, pp. 3-44.

Bosworth, B. and S. Collins (2007), *Accounting for Growth: Comparing China and India*, Working Paper 12943, National Bureau of Economic Research, Cambridge.

Broadman, H. (2007), *Africa's Silk Road: China and India's New Economic Frontier*, World Bank, Washington D.C.

Goldstein, A., N. Pinaud, H. Reisen and X. Chen (2006), *The Rise of China and India: What's in it for Africa?*, Development Centre, OECD, Paris.

Hourcade, J.-C., M. Jaccard, C. Bataille and F. Ghersi (2006) eds., "Hybrid Modeling of Energy-Environment Policies: Reconciling Bottom-Up and Top-Down", Special Issue of the *Energy Journal*, November.

International Iron and Steel Institute (IISI) (2007), *International Iron and Steel Institute Steel Statistics*, June, IISI, Brussels.

International Monetary Fund (IMF) (2007), *World Economic Outlook*, April, IMF, Washington D.C.

McKinsey Global Institute (2007), *The US Imbalancing Act: Can the Current Account Deficit Contribute?*, McKinsey, Washington D.C.

MOFCOM (2007), *Report on the Foreign Trade Situation of China*, Spring, Ministry of Commerce, Beijing.

Morgan, T. (2007), *The Quest for Record Barbel and its Impact on the World Economy*, Menecon Publishing, London.

National Bureau of Statistics (NBS, 2007), *Announcement on Preliminary Verified GDP Data in 2006*, July, National Bureau of Statistics, China.

OECD (2007), *Economic Survey of India*, OECD, Paris.

— (2006), *OECD International Investment Perspectives*, OECD, Paris.

Roach, S. (2007), "The China Fix", Statement Before the Senate Finance Committee, Hearing on "Risks and Reforms: The Role of Currency in the US-China Relationship", March.

Sachs, J. and A. Warner (2001), The Curse of Natural Resources, *European Economic Review*, Vol. 45, No. 4-6, pp. 827-38.

Stevens, C. and J. Kennan (2006), How to Identify the Trade Impact of China on Small Countries, in *IDS Bulletin* Vol. 37 No.1, January, Institute of Development Studies, University of Sussex, UK.

Streifel, S. (2006), "Impact of China and India on Global Commodity Markets, Focus on Metals & Minerals and Petroleum", Draft Working Paper, Development Prospects Group, World Bank, Washington D.C.

United Nations Conference on Trade and Development (UNCTAD) (2006), *World Investment Report 2006*, UNCTAD, Geneva.

Winters, L. and S. Yusuf (2007) eds., *Dancing with Giants: China, India and the Global Economy*, World Bank/Institute of Policy Studies, Washington D.C.

World Bank (2007), *Global Economic Prospects 2007: Managing the Next Wave of Globalisation*, International Bank for Reconstruction and Development, Washington D.C.

Chapter 4: The World's Energy Security

Andrews-Speed, P. (2006), "China's Energy Policy and its Contribution to International Stability", in *Facing China's Rise: Guidelines for an EU Strategy*, Chaillot Paper No. 94, December, Institute of Security Studies, Paris.

Douglas, J., B. Matthew and K. Schwartz (2006), *Fuelling the Dragon's Flame: How China's Energy Demand Affects its Relationship in the Middle East*, US-China Economic and Security Review Commission, Washington D.C.

Downs, E. (2006), *Energy Security Series: China*, Brookings Institute, Washington D.C.

IEA (2007a), *Energy Security and Climate Policy: Assessing Interactions*, OECD/IEA, Paris.

— (2007b), *Natural Gas Market Review*, OECD/IEA, Paris.

— (2005), *Saving Oil in a Hurry*, OECD/IEA, Paris.

Madan, T. (2006), *Energy Security Series: India*, Brookings Institute, Washington D.C.

Paik, K-W., V. Marcel, G. Lahn, J. V. Mitchell and E. Adylov (2007), *Trends in Asian NOC Investment Abroad*, Working Background Paper: March, Royal Institute of International Affairs, London.

Rosen, D. and T. Houser (2007), *China Energy: A Guide for the Perplexed*, Center for Strategic and International Studies/Peterson Institute for International Economics, Washington D.C.

US-China Economic and Security Review Commission (USCESRC) (2006), *2006 Report to Congress*, USCESRC, Washington D.C.

D

Chapter 5: Global Environmental Repercussions

Carbon Capture Project-2 (2007), *Communication Strategy: Public Perception of CCS – Prioritized Assessment of Issues and Concerns*, Joint report with the IEA Working Party on Fossil Fuels, OECD/IEA, Paris.

IEA Implementing Agreement on Wind Energy (2006), *Design and Operation of Power Systems with Large Amounts of Wind Power, First Result of IEA Collaboration*, OECD/IEA, Paris.

Intergovernmental Panel on Climate Change (IPCC, 2007), *Climate Change 2007: Fourth Assessment Report*, IPCC, Geneva. Available at www.ipcc.ch.

IEA (2007a), *Tracking Industrial Energy Efficiency and CO_2 Emissions*, OECD/IEA, Paris.

— IEA (2007b), *Legal Aspects of Storing CO_2*, OECD/IEA, Paris.

— IEA (2006a), *World Energy Outlook 2006*, OECD/IEA, Paris.

— IEA (2006b), CO_2 Capture and Storage, in *IEA Energy Technology Essentials*, December, OECD/IEA, Paris.

— IEA (2006c), *Energy Technology Perspectives*, OECD/IEA, Paris.

IEA Greenhouse Gas R&D Programme (IEA-GHG) (2007), *CO_2 Capture Ready Plants*, Technical Study, Report No. 2007/4, London.

Massachusetts Institute of Technology (MIT) (2007), *The Future of Coal – Options for a Carbon Constrained World*, March, MIT, Cambridge. Available at http://web.mit.edu/coal.

UK Department for Business, Enterprise and Regulatory Reform (UK-DBERR) (2007), *Analysis of Carbon Capture and Storage Cost-Supply Curves for the UK*, DBERR, London. Available at www.dti.gov.uk/files/file36782.pdf.

United States Geological Survey (USGS) (2000), *World Petroleum Assessment 2000*, USGS, Washington D.C.

World Energy Council (WEC) (2006), *Carbon Capture and Storage: A WEC Interim Balance*, WEC, London.

Wright, I. (2006), *CO_2 Geological Storage – Lessons Learnt from In Salah (Algeria)*, presentation to the EU-OPEC CCS Conference, September, Riyadh.

Chapter 6: Energy Policy Ramifications

Intergovernmental Panel on Climate Change (IPCC) (2007), *Climate Change 2007: Mitigation of Climate Change – Summary for Policymakers*, Contribution of Working Group III to the Fourth Assessment Report, Cambridge University Press, Cambridge/New York.

International Energy Agency (IEA) (2006), *Energy Policies of IEA Countries*, OECD/IEA, Paris.

— (2004), *World Energy Outlook 2004*, OECD/IEA, Paris.

Stern, N. (2006), *The Economics of Climate Change: The Stern Review*, Report to the Cabinet Office, HM Treasury, Cambridge University Press, Cambridge.

Chapter 7: Political, Economic and Demographic Context

Aziz, J. (2006), *Rebalancing China's economy: what does growth theory tell us?*, IMF Working Paper, Washington D.C.

Bosworth, B. and S. Collins (2007), *Accounting for growth: Comparing China and India*, Working Paper 12943, National Bureau of Economic Research, Cambridge.

CEIC (2007), China Premium Database, CEIC Data Company Ltd., Beijing.

Dunaway, S. and V. Arora (2007), *Pension Reform in China: The Need for a New Approach*, IMF, Washington D.C.

Hu, Albert G.Z., G. Jefferson and J. Qian (2005), "R&D and Technology Transfer: Firm-Level Evidence from Chinese Industry", in *Review of Economics and Statistics,* Vol. 87, Issue 4, MIT press.

International Finance Corporation/Multilateral Investment Guarantee Agency (2006), *Country Partnership Strategy for the People's Republic of China for the Period 2006-2010*, World Bank, Washington D.C.

OECD (2007a), *OECD Employment Outlook*, OECD, Paris.

OECD (2007b), *OECD Environmental Performance Reviews China*, OECD, Paris.

— (2006a), *OECD Information Technology Outlook 2006*, OECD, Paris.

— (2006b), *Challenges for China's Public Spending: Towards Greater Effectiveness and Equity*, OECD, Paris.

— (2005a), *China in the Global Economy: Governance in China*, OECD, Paris.

— (2005b), *Economic Survey of China 2005*, OECD, Paris.

Shu, F. (2006), *Report of the Development of an Overall Well-Off Society*, Social Sciences Academic Press, Beijing.

Sicular, T., X. Yue, B. Gustafsson and L. Shi (2007), "The Urban-Rural Income Gap and Inequality in China", in *Review of Income and Wealth*, Series 53, No. 1, International Association for Research in Income and Wealth, March.

United Nations Population Division (UNPD) (2006), *World Population Prospects 2006*, United Nations, New York.

United Nations Population Fund (UNFPA) (2007), *State of World Population 2007*, UNFPA, New York.

D

United Nations Statistics Division (2007), UN Common Database, New York.

World Bank (2007a), *World Development Indicators 2007*, World Bank, Washington D.C.

— (2007b), *East Asia & Pacific Update*, World Bank, Washington D.C.

— (2007c), *China Quarterly Update (February)*, World Bank, Washington D.C.

— (2006), *China Quarterly Update (August)*, World Bank, Washington D.C.

Chapter 8: Overview of the Energy Sector

China Electricity Council (2006), *National Power Industry Statistical Briefs*, China Electricity Council, Beijing.

IEA (2006). *China's Power Sector Reforms*, OECD/IEA, Paris.

NBS (2007a), *Statistical Communiqué of the People's Republic of China on the 2006 National Economic and Social Development*, NBS, Beijing.

— (2007b), *China Energy Statistical Yearbook 2006*, China Statistical Press, National Bureau of Statistics, Beijing.

NDRC (2007a), *11th Five-Year Plan for Energy Development*, NDRC, Beijing.

— (2007b), *Medium and Long-term Development Plan for Renewable Energy in China*, NDRC, Beijing.

— (2006), *Top 1 000 Enterprises Energy-Efficiency Programme*, NDRC, Beijing.

— (2004), *Medium and Long Term Energy Conservation Plan*, NDRC, Beijing.

Chapter 9: Reference Scenario Demand Projections

An, F. and A. Sauer (2004), *Comparison of Passenger Vehicle Fuel Economy and Greenhouse Gas Emission Standards Around the World*, Pew Center, Washington D.C.

Asian Development Bank (ADB) (2006), *Key Indicators 2006: Measuring Policy Effectiveness in Health and Education*, ADB, Manila.

CEIC (2007), China Premium Database, CEIC Data Company Ltd., Beijing.

Gill, I. and H. Kharas (2007), *An East Asian Renaissance: Ideas for Economic Growth: Overview*, World Bank, Washington D.C.

Global Insight (2006), *Asian Automotive Industry Forecast Report*, Global Insight, Waltham.

The International Council on Clean Transportation (ICCT) (2007), *Passenger Vehicle Greenhouse Gas and Fuel Economy Standards: A Global Update*, ICCT, Washington D.C.

International Energy Agency (IEA, 2007), *Indicators for Industrial Energy Efficiency and CO_2 Emissions*, OECD/IEA, Paris.

International Fertilizer Industry Association (2006), *Energy Efficiency in Ammonia Production: Executive Summary for Policy Makers*, www.fertilizer.org.

Kang, Y. (2005), "Challenge and opportunity for building energy conservation", *Environmental Economy*, Beijing.

Leontief, W. (1936), "Quantitative Input-Output Relations in the Economic System of the United States", *Review of Economics and Statistics*, Vol. 18(3): 105-125.

Merrill Lynch (2007), *China and India: A Dash for Gas, Industry Overview*, Merrill Lynch, Hong Kong.

Miller, R.E. and P.D. Blair (1985), *Input-Output Analysis: Foundations and Extensions*, Prentice-Hall, Englewood Cliffs.

National Bureau of Statistics (NBS), *China Statistical Yearbook*, various years.

National Development and Reform Commission (NDRC) (2007a), *Policy of Natural Gas Utilisation*, NDRC, Beijing.

— (2007b), *Guiding Catalogue for the Adjustment of Industrial Structure*, NDRC, Beijing.

— (2007c), *Medium and Long Term Renewable Energy Development Plan*, NDRC, Beijing.

— (2004), *China Medium and Long Term Energy Conservation Plan*, NDRC, Beijing.

OECD (2007), *OECD Environmental Performance Reviews: China*, OECD, Paris.

REN21 (2006), *Renewables Global Status Report 2006 Update*, REN21 Secretariat, Paris and Worldwatch Institute, Washington D.C.

Shui, B. and R.C. Harriss (2006), "The Role of CO_2 Embodiment in US-China Trade", in *Energy Policy* 34(18):4063-4068.

Sinton, J.E., K.R. Smith, J.W. Peabody, L. Yaping, Z. Xiliang, R. Edwards and G. Quan (2004), "An Assessment of programs to promote improved household stoves in China", *Energy for Sustainable Development*, VIII.

Spautz, L., D. Charron, J. Dunaway, F. Hao and X. Chen (2006), "Spreading Innovative Biomass Stove Technologies Through China and Beyond" in *Boiling Point*, Issue 52, HEDON Household Energy Network, Bromley.

Steel Business Briefing (2005), *China's Steel Industry Development Policy*, Steel Business Briefing Reports, Shanghai.

Tsinghua University (2007), *Annual Report on China Building Energy Efficiency*, China Architecture and Building Press, Beijing.

D

United Nations Environment Programme (UNEP, 2007), *UNEP Risoe CDM/JI Pipeline Analysis and Database*, UNEP, Roskilde.

United Nations Population Division (UNPD, 2006), *World Population Prospects 2006*, United Nations, New York.

United States Geological Survey (USGS, 2007), *Mineral Commodity Summaries*, United States Government Printing Office, Washington D.C.

Wixted, B., N. Yamano *et al.* (2006), *Input-Output Analysis in an Increasingly Globalised World: Applications of OECD's Harmonised International Tables*, OECD Science, Technology and Industry Working Papers, Paris.

World Bank (2007), *East Asia & Pacific Update*, World Bank, Washington D.C.

World Coal Institute (2007) *Coal & Steel Facts 2007 Edition*, available at www.worldcoal.org.

World Health Organisation (WHO) (2007), *China Country Profile of Environmental Burden of Disease*, WHO, Geneva.

— (2006), *Fuel for Life*, WHO, Geneva.

Yeats, A. J. (2001). *Just How Big is Global Production Sharing? Fragmentation: New Production Patterns in the World Economy*, S. W. Arndt and H. Kierzkowski (eds.), Oxford University Press, New York.

Zhou, N. *et al.* (2007), *Energy Use in China: Sectoral Trends and Future Outlook*, working draft, Lawrence Berkeley National Laboratory, Berkeley.

Chapter 10: Reference Scenario Supply Projections

ABARE (2006), *Australian Commodities,* Vol.13, No.3, September quarter, Australian Bureau of Agricultural and Resource Economics, Canberra.

ADB (2004), *Report and recommendations of the President and Board of Directors on a proposed loan to the People's Republic of China for the Coal Mine Development Project*, ADB, Manila.

Barlow Jonker (2007), *China Coal Fourth Edition*, Barlow Jonker, Sydney.

— (2005), *China Coal Third Edition*, Barlow Jonker, Sydney.

Beijing HL Consulting (2006), *China Coal Report, 20 June*, Beijing HL Consulting, Beijing.

Cedigaz (2007), *2006 Natural Gas Year in Review*, Cedigaz, Rueil-Malmaison.

Chadwick, J. (2007), *China Shenhua Coal*, International Mining, Vol.3, No.5, May.

Chew Chong Siang (2006), *Current Status of New and Renewable Energies in China - Introduction of Fuel Ethanol*, IEEJ, Tokyo.

Downs, E. (2007), "The Fact and Fiction of Sino-African Energy Relations", *China Security*.

— (2006), *Energy Security Series: China*, Brookings Institute, Washington D.C.

FACTS Global Energy (2006), *China Energy Series*, FACTS, Singapore.

General Geological Bureau (1999), quoted in Zhu, X. (1999), *China's Mineral Resources, Volume I – Energy Mineral Resources*, Geological Publishing Company, Beijing.

Huang, S. (2004), *China Coal Outlook 2004*, China Coal Information Institute, Beijing.

IEA (2006), *Coal Information 2006*, OECD/IEA, Paris.

IEA Greenhouse Gas R&D Programme (IEA GHG) (2002), *Opportunities for Early Applications of CO_2 Sequestration Technology*, IEA GHG, Cheltenham.

IEA Solar Heating and Cooling Programme (IEA SHC) (2007), *Solar Heat Worldwide*, AEE INTEC, Gleisdorf.

Jianping, Ye, Feng Sanli, Fan Zhiqiang, B. Gunter, S. Wong, D. Law (2005) "CO_2 Sequestration potential in Coal Seams of China", GCEP International Workshop August 2005, Tsinghua University.

Lako, P. (2002), *Options for CO_2 Sequestration and Enhanced Fuel Supply*, monograph in the framework of the VLEEM project, April.

Li X. Y. Liu, B. Bai and Z. Fang (2005), "Ranking and screening of CO_2 saline aquifer storage zones in China", in *Chinese Journal of Rock Mechanics and Engineering*, Vol. 25, No. 5, pp 963 – 968.

McCloskey (2007), *McCloskey Coal Report*, various issues, McCloskey, Petersfield.

Minchener, A. (2007), *Coal Resources for Power Generation in China*, IEA Clean Coal Centre, London.

— (2004), *Coal in China London*, IEA Clean Coal Centre, London.

National Development and Reform Commission (NDRC) (2007a), *11th Five Year Plan on Energy Development*, NDRC, Beijing.

— (2007b), *Notice about end-use and transmission & distribution electricity tariff standards for provincial grid in 2006*, NDRC, Beijing.

— (2007c), *Medium and Long Term Renewable Energy Development Plan*, NDRC, Beijing.

Qian, J. and Y. Fan, (2006), "Summary of Carbon Storage Potential and Activities in China", 13 September, NRDC, Taiyuan, www.chinacleanenergy.org/docs/general/CarbonStoragePotential.pdf

D

Pan, K. (2005), "The Depth Distribution of Chinese Coal Resources", presentation at the School of Social Development and Public Policy, Fudan University, Fudan.

Shisen, Xu (2006), "The Status and Development Trends of IGCC", presentation at Thermal Power Research Institute, 4 July, www.chinaesco.net/PDF_ppt_lt/pdf_dir/_xushisen.pdf

Torrens, I.M. (2007), *National and International Activities Related to CCS and ZETs in China*, Working Paper presented at the IEA Working Party on Fossil Fuels Meeting, 28-29 June, Brasilia.

United Nations Development Programme (UNDP/World Bank) (2004), *Toward a Sustainable Coal Sector in China*, Joint UNDP/World Bank Energy Sector Management Assistance Programme.

United States Geological Survey (USGS) (2000), *World Petroleum Assessment 2000*, USGS, Washington D.C.

World Energy Council (WEC) (2007), *Survey of Energy Resources*, WEC, London.

Chapter 11: Alternative Policy Scenario Projections

Fridley, D. *et al.* (2007), *Impacts of China's Current Appliance Standards and Labelling Program to 2020*, LBNL, Berkeley.

IEA (2007), *Indicators for Industrial Energy Efficiency and CO_2 Emissions*, OECD/IEA, Paris.

IPCC (2007), *Climate Change 2007: Impacts, Adaptation and Vulnerability*, Cambridge University Press, Cambridge.

Jin, M. and Li, A. (2005), *The Implications and Impacts of China Energy Labels*, China National Institute of Standardization, Beijing.

Lawrence Berkely National Laboratory (LBNL) (2007), *Impacts of China's Current Appliance Standards and Labelling Program to 2020*, LBNL, Berkeley.

Lin, J. and D. Fridley, (2007), *Accelerating the Adoption of Second-Tier Reach Standards for Applicable Appliance Products in China*, LBNL, Berkeley.

Liu, M. (2006), *Framework of China's Energy Efficiency Standards Enforcement and Monitoring*, China National Institute of Standardization, Beijing.

NDRC (2007), *China's National Climate Change Programme*, NDRC, Beijing.

OECD (2007), *OECD Environmental Performance Reviews: China*, OECD, Paris.

Tsinghua University (2006), *Greenhouse Gas Mitigation in China*, Center for Clean Air Policy, Beijing.

Williams, R. O., A. McKane *et al.* (2005), *The Chinese Motor System Optimization Experience: Developing a Template for a National Program*, Energy Efficiency in Motor Driven Systems, Heidelberg.

Chapter 12: High Growth Scenario Projections

International Road Federation (2006), *World Road Statistics 2006*, International Road Federation, Geneva.

World Bank (2007), *Cost of Pollution in China*, World Bank, Washington D.C.

Chapter 13: Focus on the Coastal Region

CEIC (2007), China Premium Database, CEIC Data Company Ltd., Beijing.

NBS (2007), *Communiqué on Energy Consumption per Unit of GDP by Regions in 2006*, National Bureau of Statistics, the National Development and Reform Commission, and Office for National Energy Leading Group, Beijing.

NDRC (2007), *Policy of Natural Gas Utilisation,* NDRC, Beijing.

State Council (2006), *Reply to the 11th Five Year Period Provincial Unit Energy Consumption Reduction Target*, State Council, Beijing.

United Nations Population Division (UNPD) (2006), *World Population Prospects 2006*, United Nations, New York.

Chapter 14: Political, Economic and Demographic Context

Bosworth, B. and S. Collins (2007), *Accounting for Growth: Comparing India and China*, Working Paper 12943, National Bureau of Economic Research, Cambridge.

Economist Intelligence Unit (EIU) (2006), *Country Report - India*, December, EIU, London.

Government of India (2006), *Towards Faster and More Inclusive Growth: An Approach to the 11th Five Year Plan*, Planning Commission, Government of India, New Delhi.

International Monetary Fund (IMF) (2007), *Indian Subnational Finances: Recent Performance*, IMF Working Paper No. 07/205, IMF, Washington D.C.

McKinsey Global Institute (2007), *Bird of Gold: India's Rising Consumer Market*, McKinsey, San Francisco.

OECD (2007), *Economic Survey of India*, OECD, Paris.

Poddar, Tushar and Eva Yi (2007), *India's Rising Growth Potential*, Global Economics Paper No. 152, Goldman Sachs Economic Research, New York.

D

World Bank (2007), *Doing Business in South Asia*, World Bank, Washington D.C.

— (2006a), *Financing Firms in India*, World Bank Policy Research Working Paper 3975, World Bank, Washington D.C.

— (2006b), *India, Inclusive Growth and Service Delivery: Building on India's Success*, Development Policy Review, Report No. 34580-IN, World Bank, Washington D.C.

— (2006c), *South Asia: Growth and Regional Integration*, World Bank, Washington D.C.

United Nations Population Division (UNPD) (2007), *World Urbanization Prospects: The 2006 Revision*, United Nations, New York.

Van Ark, B., J. Banister and C. Guillemineau (2006), *Competitive Advantage of 'Low-Wage' Countries Often Exaggerated*, Executive action series, No. 212, Conference Board China Center for Economics and Business, New York.

Chapter 15: Overview of the Energy Sector

Cedigaz (2007), *Natural Gas in the World*, Institut Français de Pétrole, Rueil-Malmaison.

Central Electricity Authority (CEA) (2007), *All India Electricity Statistics: General Review 2007*, Government of India, New Delhi.

— (2006), *All India Electricity Statistics: General Review 2006*, Government of India, New Delhi.

International Monetary Fund (IMF) (2006), *India: Selected Issues*, IMF Staff Country Report No. 06/56, IMF, Washington D.C.

Ministry of Power (2007), *Report of the Working Group on Power for Eleventh Plan (2007-2012)*, Volume II, Main Report, New Delhi.

OECD (2007), *OECD Principles for Private Sector Participation in Infrastructure*, OECD, Paris.

Planning Commission (2006), *Integrated Energy Policy: Report of the Expert Committee*, Planning Commission, Government of India, New Delhi.

Chapter 16: Reference Scenario Demand Projections

Asian Development Bank (ADB) (2006), *Energy Efficiency and Climate Change Considerations for On-road Transport in Asia*, ADB, Manila.

Central Electricity Authority (2006), *All India Electricity Statistics: General Review 2006*, Government of India, New Delhi.

Garg, A. *et al.* (2006), "The sectoral trends of multigas emissions inventory of India", *Atmospheric Environment 40*, Elsevier, London.

IEA (2007), *Tracking Industry Energy Efficiency and CO$_2$ Emissions*, OECD/IEA, Paris.

— (2006), *World Energy Outlook 2006*, OECD/IEA, Paris.

Joint Plant Committee (2005/6), *Survey of Indian Sponge Iron Industry*, Joint Plant Committee, Kolkata.

Karangle, H.S., Rashtriya Chemicals & Fertilizers Ltd. (2007) "Energy Efficiency and CO$_2$ Emissions in the Indian Ammonia Sector". Available at: www.fertilizer.org/ifa/technical_2007_hcmc/2007_tech_hcmc_papers.asp

Lawrence Berkeley National Laboratory (LBNL) (2005), *Potential Benefits from Improved Energy Efficiency of Key Electrical Products: The Case of India*, December, LBNL, Berkeley.

Menon-Choudhari, D. *et al.* (2005), *Assessing Policy Choices for Managing SO$_2$ Emissions from Indian Power Sector*, Rajdhani Art Press, India.

Midrex (2006), *World Direct Reduction Statistics*, Midrex, available at www.midrex.com.

Ministry of Petroleum and Natural Gas (2003), *Auto Fuel Policy*, Ministry of Petroleum and Natural Gas, Government of India, New Delhi.

National Sample Survey Organisation (NSSO, 2007), *Energy Sources of Indian Households for Cooking and Lighting, 2004-05*, Government of India, New Delhi.

Planning Commission (2006), *Integrated Energy Policy: Report of the Expert Committee*, Planning Commission, New Delhi.

Rehman, Ibrahim Hafeezur, Preeti Malhotra, Ram Chandra Pal and Phool Badan Singh (2005), "Availability of Kerosene to Rural Households: a Case Study from India", *Energy Policy* , Vol. 33, Issue 17, November, Elsevier, London.

The Energy and Resources Institute (TERI) (2007), *TERI Energy Data Directory & Yearbook* (TEDDY), TERI, New Delhi.

— (2006), *National Energy Map for India - Technology Vision 2030*, The Energy and Resources Institute, New Delhi.

United Nations Population Division (UNPD) (2007), *World Population Prospects: The 2006 Revision*, United Nations, New York.

World Coal Institute (2007) *Coal & Steel Facts 2007 Edition*, available at www.worldcoal.org.

D

Chapter 17: Reference Scenario Supply Projections

Barlow Jonker (2006), *Thermal Coal Market 2006 - Supply and Demand to 2020*, Barlow Jonker, Sydney.

Beck, R. A., Y. Price, S. Friedmann, J. Wilder, L. Neher, (2007), *Regional Assessment of CO_2 Sources and Sinks for the India Subcontinent*, AAPG Annual Meeting, 1-4 April, Long Beach.

BGR (2007), *Reserves, Resources and Availability of Energy Resources – Annual Report 2005*, Federal Institute for Geosciences and Natural Resources, Hanover, available at: www.bgr.bund.de.

Cedigaz (2007), *2006 Natural Gas Year in Review*, Cedigaz, Rueil-Malmaison.

Central Electricity Authority (2007), *All India Electricity Statistics 2006*, CEA, New Delhi.

— (2006), *Review of Performance of Thermal Power Stations 2005/06*, CEA, New Delhi.

— (2005), *Report on Tapping of Surplus Power from Captive Power Plants*, CEA, New Delhi.

Coal India (2007), *Domestic Price Fixation*, Coal India Ltd, Kolkata.

Coal Industry Advisory Board (CIAB) (2002) *Coal in the energy supply of India*, IEA Coal Industry Advisory Board, OECD/IEA, Paris.

CRISIL/ICRA (2006), *State Power Sector Performance Ratings: Final Report to the Ministry of Power*. Available at: www.powermin.nic.in.

Desai, V. (2004), *Obstacles to Private Power Investments in India*, ADB Institute Discussion Paper No 20. Available at: www.adbi.org

Friedmann, S.J. (2006), *The scientific case for large CO_2 storage projects worlwide: Where they should go, what they should look like, and how much they should cost*, GHGT8 Conference, 19-23 June, Trondheim.

Geological Survey of India (2007), *Coal Resources of India 2007*, Geological Survey of India, Kolkata.

Government of India (2002), *Tenth Five Year Plan 2002-07*, Government of India, Delhi.

IndiaCore (2006), *Overview of Coal Sector in India 2005*, IndiaCore, New Delhi.

IEA (2008, forthcoming), *Global Renewable Energy Markets and Policies – Past Trends and Future Prospects*, OECD/IEA, Paris.

— (2006), *World Energy Outlook 2006*, OECD/IEA, Paris.

— (2003), *World Energy Investment Outlook*, OECD/IEA, Paris.

Malti, G. (2007), *Carbon Capture and Storage technology for sustainable energy future – Meeting Report*, International Workshop on R&D Challenges in CO_2 Capture and Storage Technology for Sustainable Energy Future, 12-13 January, Hyderabad, India.

McCloskey (2007), *McCloskey Coal Report*, various issues, McCloskey, Petersfield.

Ministry of Coal (2007a), *Report of the Expert Committee on Road Map for Coal Sector Reforms*, Part II, April, Ministry of Coal, Government of India, New Delhi.

— (2007b), *Annual Report 2006-2007*, Ministry of Coal, Government of India, New Delhi.

— (2005), *Report of the Expert Committee on Road Map for Coal Sector Reforms*, Part I, December, Ministry of Coal, Government of India, New Delhi.

Ministry of Finance (2007), *Economic Survey 2006-2007*, available at: http://indiabudget.nic.in/

Ministry of Power (2007), *Report of the Working Group on Power for Eleventh Five-Year Plan (2007/12)*, Volume II, Main Report, MoP, New Delhi.

NEA/IAEA (2006), *Uranium 2005: Resources, Production and Demand*, OECD, Paris.

Planning Commission (2006), *Integrated Energy Policy: Report of the Expert Committee*, Planning Commission, New Delhi.

Singh, A, V.A. Mendhe and A. Garg (2006), *CO_2 Sequestration Potential of Geological Formations in India*, GHGT8 Conference, 19-23 June, Trondheim.

Sonde, R.R. (2006), *Prospects for CCS in India*, Technical Workshop on Carbon Capture and Storage, New Delhi.

TERI (2007), *TERI Energy Data Directory and Yearbook 2005/06*, The Energy and Resources Institute, New Delhi.

— (2006), *National Energy Map for India — Technology Vision 2030*, The Energy and Resources Institute, New Delhi.

US Department of Labor, Mine Safety and Health Administration (2007), www.msha.gov/stats/centurystats/coalstats.asp, accessed 26 May.

Chapter 18: Alternative Policy Scenario Projections

IEA (2008, forthcoming), *Assessment of upgrading and replacement of older coal-fired units in major coal using economies*, OECD/IEA, Paris.

Ministry of Heavy Industries and Public Enterprises (2006), *Automotive Mission Plan 2006-2016*, New Delhi.

D

Pucher, J., N. Korattyswaropam, N. Mittal, and N. Ittyerah (2005), "Urban Transport Crisis in India", *Transport Policy*, Vol. 12, Issue 3, June.

Singh, S. (2007), *India Bio-fuels Annual 2007*, Gain Report, New Delhi.

The Energy and Resources Institute (TERI) (2007), *Vulnerability to Climate Variability and Change in India: Assessment of Adaptation Issues and Options*, TERI, New Delhi.

World Energy Council (WEC) (2007), *Survey of Energy Resources*, WEC, London.

Chapter 19: High Growth Scenario Projections

Besley, T., R. Burgess and B. Esteve-Volart (2006), "The Policy Origins of Poverty and Growth in India", in T. Besley and L. Cord (eds), *Delivering on the Promise of Pro-Poor Growth*, Palgrave, New York.

Brookings Institute (2007), "Is India's High Growth Sustainable?", Roundtable in Washington D.C. on 8 March, Anderson Court Reporting, Virginia.

Chapter 20: Focus on Energy Poverty

Aßmann, D., U. Laumanns and D. Uh, eds. (2006), *Renewable Energy: A Global Review of Technologies, Policies and Markets*, Earthscan, London.

Barnes, D., K. Krutilla and W. Hyde (2005), *The Urban Household Energy Transition: Social and Environmental Impacts in the Developing World*, Resources for the Future, Washington D.C.

Gangopadhyay, Shubhashis, Bharat Ramaswami and Wilima Wadhwa (2005), "Reducing Subsidies on Household Fuels in India: How Will it Affect the Poor?", *Energy Policy*, Volume 33, Issue 18, December, Elsevier, London.

IEA (2004), *World Energy Outlook 2004*, OECD/IEA, Paris.

Mishra, V.K., R.D. Retherford, K.R. Smith (1999), "Biomass Cooking Fuels and Prevalence of Blindness in India", *Journal of Environmental Medicine*, pp. 189-199, Hoboken.

Misra, Neha, Chawla Ruchika, Leena Srivastava and R.K. Pauchauri (2005), *Petroleum Pricing in India: Balancing Efficiency and Equity*, The Energy and Resources Institute (TERI) Press, New Delhi.

Modi, Vijay (2005), *Improving Electricity Services in Rural India*, Center on Globalization and Sustainable Development Working Paper No. 30, The Earth Institute at Columbia University, New York.

National Sample Survey Organisation (NSSO, 2007), *Energy Sources of Indian Households for Cooking and Lighting, 2004-05*, Government of India, New Delhi.

Pachauri, S. (2007), *An Energy Analysis of Household Consumption. Changing Patterns of Direct and Indirect Use in India*, Heidelberg, Berlin; Springer, New York.

Pachauri, S., A. Mueller, A. Kemmler and D. Spreng (2004), "On Measuring Energy Poverty in Indian Households", *World Development*, Vol. 32, No. 12, Elsevier, London.

Planning Commission (2006), *Integrated Energy Policy: Report of the Expert Committee*, August, Planning Commission, New Delhi.

Rangarajan, C. (2006), "Report of the Committee on Pricing and Taxation of Petroleum Products", Government of India, New Delhi.

Smith, K. R. (2000), "National Burden of Disease in India from Indoor Air Pollution", *Proceedings of the National Academy of Sciences*, Vol. 97, No. 24, London.

United Nations Development Programme (UNDP) (2006), *Expanding Access to Modern Energy Services: Replicating, Scaling Up and Mainstreaming at the Local Level*, UNDP, New York.

United States Agency for International Development (USAID) (2006), *Slum Electrification and Loss Reduction Program: Country Background Report – India*, January, Nexant, San Francisco.

World Health Organization (WHO, 2007), *Indoor Air Pollution: National Burden of Disease Estimates*, WHO, Geneva.

D

The Online Bookshop